MW00576929

Statistics in the Law

STATISTICS IN THE LAW

Edited by

JOSEPH B. KADANE

UNIVERSITY PRESS
2008

OXFORD
UNIVERSITY PRESS

Oxford University Press, Inc., publishes works that further
Oxford University's objective of excellence
in research, scholarship, and education.

Oxford New York
Auckland Cape Town Dar es Salaam Hong Kong Karachi
Kuala Lumpur Madrid Melbourne Mexico City Nairobi
New Delhi Shanghai Taipei Toronto

With offices in
Argentina Austria Brazil Chile Czech Republic France Greece
Guatemala Hungary Italy Japan Poland Portugal Singapore
South Korea Switzerland Thailand Turkey Ukraine Vietnam

Published by Oxford University Press, Inc.
198 Madison Avenue, New York, New York 10016

www.oup.com

Oxford is a registered trademark of Oxford University Press.

Library of Congress Cataloging-in-Publication Data
Statistics in the law / edited by Joseph B. Kadane.
p. cm.
Includes bibliographical references and index.
ISBN 978-0-19-530923-2
1. Statistics. 2. Law—Statistical methods. I. Kadane, Joseph B.
QA276.S783443 2008
347'.0670245195—dc22   2008002474

1 3 5 7 9 8 6 4 2
Printed in the United States of America
on acid-free paper

For Caroline: wife, best friend, in-house counsel,
champion speller

PREFACE: DECIDING WHETHER
TO BE AN EXPERT WITNESS

Suppose you are a statistician sitting peacefully in your office one day, and you get a telephone call from an attorney asking you to serve as an expert witness in a case. Some statisticians react to this invitation as if they had been invited to their own public hanging, while others see it as an opportunity to apply their craft in a new arena. The purpose of this book is to introduce you to what you need to know to decide how to respond.

The first thing to determine is whether you have a conflict of interest. Who are the parties in the lawsuit? Do you know any of them? Is a relative or close friend one of the parties or one of the attorneys? Is there any other connection you have that would lead a reasonable person to doubt that you are giving your best scientific judgment in court?

Once you and the attorney have mutually determined that you do not have a conflict of interest, the next question is what the case is about. What is the legal theory of the attorney? What factual matter(s) does the attorney need you to address to underpin that theory? Is there data available that addresses the factual issue(s)? If so, it then looks feasible to be involved.

To proceed further, you need to understand what the law is about, and what the role of expert witnesses is in the legal proceeding. Accordingly, the first part of this book reports the experiences of two pioneers in offering statistical expertise in court, a general introduction to the basics of American law, and a detailed treatment of the evolving standards for the offering of expert testimony in US courts. It also gives a brief discussion of the ethical issues involved.

The remainder of the book gives examples of kinds of cases in which statisticians have been called upon to offer expert opinions. The majority of these cases are ones in which I was called upon to offer my opinion.

Part 2 considers perhaps the most frequent, kind of case that statisticians are asked to address: discrimination. The cases reported here address age and race discrimination, but the same principles and kinds of analysis would apply to other kinds of discrimination, such as gender, sexual orientation, or religion. Typically there are one or more plaintiffs alleging that the actions against them by someone else were a product of discriminatory animus because of their membership in a protected group. To show this, they want to bring to court not only the (possibly ambiguous) details of their own personal experience, but also the experiences of others of their group, compared to persons not in their group. Enter the statistician.

Part 3 addresses a special form of discrimination, that involving the choice of jurors to hear cases. Jurors occupy a critical role in the American legal system, since they find the facts in both legal and criminal cases. A jury's factual decision cannot be appealed; however an appeal can be taken on grounds that the judge failed to properly instruct the jury on the law. Accordingly, how jurors are chosen becomes a critical point in maintaining and increasing the fairness of the legal system. The choice of jurors is especially sensitive in cases that might result in the imposition of the death penalty.

Part 4 shows a variety of other cases in which statistical evidence has contributed. Each is its own story, but together they show that as a profession, statistics has ideas to offer that go far beyond cases of discrimination.

The book closes with Part 5, which discusses two interesting cases from the United Kingdom. UK law is slightly different from US law, so I give a short review of the differences that are important for understanding these cases. The first set of cases involve "cot death," otherwise known as SIDS (sudden infant death syndrome). Murder accusations were brought against mothers in families suffering more than one of these unusual deaths. A physician testified how extraordinarily unlikely this would be under an assumption of statistical independence. This evidence led to convictions for murder of several grieving mothers. Later it was discovered that there is likely to be a genetic predisposition to cot death, leading to clusters in genetically vulnerable families. I compare these cases to the famous Collins case in California, in which an expert multiplied frequencies of various circumstantial facts of a robbery in an attempt to show that the defendants must be guilty. The appeals court disallowed the evidence, because there was no showing of independence.

The final case in Part 5 concerns the Adams cases in the UK, in which a distinguished Bayesian statistician tried to show the jury how to use Bayesian calculations to find the weight of evidence. The UK appeals court disallowed this as well.

I thank the American Statistical Association for copyright permission to reprint papers 1A1, 1A2, 1C1, 2A1, 2A2, 2B3, 2C1, 3B1, 3B3, 4B, 4D and 4F, the Royal Statistical Society for permission to reprint papers 2B2, 5B1 and 5C1, John Wiley and Sons for permission to reprint papers 4A1, 2 and 3, California Law Review for permission to reprint paper 3A, Blackwell publishers for permission to reprint paper 4C, and the Lemelson Family Foundation for permission to reprint Figure 1 in paper 4F.

I hope that this book will help the reader choose well (i.e., in a way that maximizes your expected utility) whether to accept an assignment to become an expert witness. By giving examples of the connection between legal theories and statistical analyses, I hope it is also useful to lawyers thinking about whether to make a phone call to approach a statistician about being an expert witness.

TABLE OF CONTENTS

ix

CONTRIBUTORS

Professor Paul Meier
Department of Biostatistics
Mailman School of Public Health
Columbia University

Professor Franklin Fisher
Department of Economics
MIT

Michael Finkelstein, Esq.
New York, NY

Professor Bruce Levin
Department of Biostatistics
Mailman School of Public Health
Columbia University

Norma Terrin, Ph.D.
Tufts-New England Medical Center
Boston, MA

Jason Connor, Ph.D.
Noblesville, IN

Professor David Kairys
Beasley School of Law
Temple University

Professor John Lehoczky
Department of Statistics
Carnegie Mellon University

Marc Ware
Pittsburgh, PA

Joseph C. Bright, Esq.
Wolf, Block, Schorr and
Solis-Cohen, LLP

Professor Daniel Nagin
Heinz School
Carnegie Mellon University

Ilaria DiMatteo, Ph.D.
New York, NY

Professor George G. Woodworth
Department of Statistics
University of Iowa

Professor Ray Hill
Department of Mathematics
University of Salford

Professor Peter Donnelly
Department of Statistics
University of Oxford
Statistics in the Law

Statistics in the Law

Part 1

WHAT'S IT LIKE TO BE AN EXPERT WITNESS?

What is the culture of the legal profession, particularly as it relates to expert witnesses? How are you going to decide if you want to accept the invitation to be an expert witness? To make that decision, it helps to know what the experiences have been of those who have done this before you, and what the rules are that relate to being an expert. In addition, you will want to consider the ethical situation in which you are placing yourself.

Accordingly, section A of Part 1 records the experiences of two of the pioneers as statistical expert witnesses: Paul Meier and Frank Fisher. Their papers give a broad perspective about what they felt about their work as expert witnesses. At the same time, you as a reader will get some idea of their personalities as well. Imagine yourself as a juror hearing testimony from each of them. Which style would you find more appealing and credible? Which style best matches your personality? Which one would you prefer to sound like in court? Which style would you find most disconcerting if used by an opposing expert witness?

There are some fundamental facts about law in the United States that you need to know in order to orient yourself to the case you have been invited to participate in, and to understand the papers in this book. Consequently, section B sets out a framework for understanding the various kinds of cases that arise in the US.

We then address the question of who is permitted to give expert testimony in court, and what the legitimate content of that testimony can be. As it happens, the rules concerning this were changed for federal cases, first by the adoption by Congress of a new set of Federal Rules of Evidence, and then by a series of federal court decisions implementing those new rules. Some states have not adopted the new federal rules concerning expert witnesses, while others have. Accordingly, it is important for you as a budding expert witness to understand what rules are to be used to determine whether you will be allowed to testify, and what you will and will not be allowed to say in court. Paper 1C1, a paper by Finkelstein and Levin, gives a general introduction to this change in rules. Because this matter is of critical importance to preparing you to decide whether to accept the invitation to be an expert witness, I have included the most important court decisions in Section C: the Frye decision (the old standard) and the Daubert and Kumho Tire decisions (establishing the new standard for federal courts).

The criteria the courts will use to decide if you are allowed to serve as an expert witness and what you will be allowed to testify to does not address the matter of how you would be wise to use the privileges thus granted. Section D gives my view of the ethical issues that are likely to arise, and the resolutions to them I find most comfortable. The most delicate of these is how to handle the combined facts that an expert has been hired by an attorney who is trying to prove something and whose job is to present a most compelling case for the client, but the expert's obligation is to the court, to help it understand the issues (for statisticians, the inferential issues) involved in the case. Managing this conflict is essential to being effective as an expert witness.

Finally, there is the issue of what the legal responsibilities of an expert witness are. The Panitz case, which concludes this part, concerns a doctor whose testimony was inconsistent with depositions she had previously given and made available to the lawyers with whom she was working. They refused to pay her, and had sued her for damages because they lost the case. The court held for the doctor, asserting "The primary purpose of expert testimony is not to assist one party or another in winning the case but to assist the trier of the facts in understanding complicated matters." Since she had testified truthfully about her views, she is not liable.

A. PIONEERS

The two papers in this section, "Damned Liars and Expert Witnesses" by Paul Meier, and "Statisticians, Econometricians, and Adversary Proceedings" by Franklin M. Fisher, reflect the experiences of two pioneers in presenting statistical evidence in court. The cases they discuss are different, and yet the circumstances they describe haven't changed much in the intervening years. It's fitting to start this book with the words of two masters of the trade.

1

DAMNED LIARS AND EXPERT WITNESSES

Paul Meier

Until recently the applications of inferential statistics in legal proceedings have been minor and limited. With the advent of civil rights legislation, however, the courts have embraced statistical inference with enthusiasm. The needs of the courts are not well matched with the usual practice of statistics, and this mismatch has serious adverse consequences for both fields. The various sources of difficulty are outlined, and tentative proposals for their amelioration are put forward.

1. INTRODUCTION

Although the field of statistics can find its origins in matters pertaining to society and its governance, statistics as a formal discipline has only recently received special recognition in legal proceedings. To be sure, statistics in the sense of numerical summaries are pervasive—in legal settings as in many others. But statistical inference based on probability models is another matter, and in that respect statistics has had only a minor and restricted role in the law.

The Howland will case of 1867 (Meier and Zabell 1980), in which Benjamin Peirce undertook a statistical analysis of handwriting, is a case in point. The analysis was ingenious, and might even have been persuasive, but the court in that instance found a technical excuse to put it aside. From time to time, most notably in the Collins case a century later (*People v. Collins* 1968), statistical analyses of identification evidence have come before the courts, and generally the courts have rejected them, except in the rather special cases of genetic evidence of paternity and of fingerprint evidence.

Following the passage of the Civil Rights Act of 1964, however, the courts have looked to statistical analysis to decide on the substantiality of evidence of illegal discrimination, and by now the statistical expert witness is definitely in the big time.

It might be thought that this is cause for celebration within our profession. Inference is our field, of course, and what could be more appropriate than a long overdue recognition by the courts of our special expertise? But there is room for second thoughts as well, when we pause to consider the consequences for other professions that have come to occupy a similar role. The situation of psychiatry (to choose a not-at-all random example) is notorious. In the case of the most recent would-be presidential assassin, John Hinckley, neither the prosecution nor the defense had any difficulty in finding distinguished psychiatrists, academics and others, to testify that Hinckley was or was not legally sane at the time he fired the shots. Indeed, it is not stretching matters to say that the courts and the bar, and even the public at large, have come to hold the profession of psychiatry in considerable contempt—as a clan of hired guns, available for a price to whichever side first knocks on the door. That this perception is not altogether fair is beside the point. The statisticians may have cause for congratulation in their newfound status—they also have cause for worry.

Indeed, psychiatry is not alone in its notoriety. The evident ease with which experts in almost any field can be found to testify in support of either side of a case has led to an aphorism in the law that has a familiar ring to statisticians: There are three kinds of liars—liars, damned liars, and expert witnesses. (The origins of this aphorism are uncertain, but it appears in various forms in legal writing during the past century.) Statistics has had a hard time establishing its credibility as a scientific domain, and the credit that it now has may well be threatened by our new-found prominence.

The views expressed here are idiosyncratic, and the interested reader may wish to consult additional sources dealing with the interaction between statistics and law. In particular, the collections edited by Peterson (1983), Monahan and Walker (1985), and DeGroot, Fienberg, and Kadane (1986) present much relevant material and a number of alternative views. Finally, it should be noted that the Committee on National Statistics of the National Academy of Sciences has conducted a three-year study of the applications of statistics in law, and its report is due to appear soon.

The remainder of this article is divided into four sections: the first reviews the different domains in which statistical testimony is sought; the next discusses the environment in which such testimony is given and contrasts that environment with the quite different system prevalent in Europe. The manifold corrupting influences that lead to the unsavory reputation of expert witnesses in American courts are then reviewed, and I close with a very modest proposal for reform.

2. DOMAINS OF APPLICATION OF STATISTICS IN LAW

I start by distinguishing particular domains in which statistical expertise is called upon.

Scientific Sampling

The simplest and, in many ways, the most satisfactory application of statistics in legal proceedings is the use of scientific sampling methods. In this area, W. E. Deming has been the preeminent pioneer (Deming 1954), and he has taken the pains to lay out a clear recipe for satisfactory performance. This consists largely of eschewing any responsibility for choice of population to be sampled or for the evaluation of sampled units. His advice, which I believe to be eminently sensible, is that the sampling expert limit himself to testimony about the inference from the sample to the population, when the same evaluation process is used for both. Deming emphasizes that although the sampling expert may have become familiar with the substantive field and may have given good advice about other aspects of the study, his *professional* expertise is limited, and he should testify only within that area.

Following in Deming's footsteps, on a number of occasions I have assisted in the sampling of railroad traffic, in connection with studies of the effects of a merger between two railroads or of the effects on railroad A of the abandonment of certain lines by railroad B. I have presented such work before administrative law judges of the Interstate Commerce Commission and have often been cross-examined thereon. However, despite strong controversy and aggressive examination of management personnel, the sampling testimony has generally been accepted with minimum fuss, and the cross-examination has ordinarily consisted solely of emphasizing the limits of my responsibility.

Sampling testimony is not always so cut and dried, however. An early case is that of *Sears Roebuck & Co. v. City of Inglewood* (Sprowls 1957), in which the sales to nonresidents had been erroneously subject to tax and Sears was seeking recovery. The expert retained by Sears sampled 33 days from the 826 business days in the period at issue, and all sales slips from each of those 33 days were examined and assessed. The estimated overpayment was $27,000, subject to a standard error of $2,000. No quarrel with the method of sampling was made, but the judge in the case was uneasy about this unfamiliar technique. He ruled that no recovery could be made for individual sales that had not themselves been individually examined, and Sears had to go back and look at each sales slip. The result is a choice teaching example, of course, because it is one of those exceedingly rare cases in which a well-drawn sample is followed by a complete census, illustrating the ultimate validation of the statistician's art. In this case the complete count was surprisingly close to the estimate—a deviation of less than $300.

But even in the clean world of scientific sampling, a witness may find himself in difficulty. He may be asked to comment on the sample drawn by the opposite party—perhaps by a non-statistician—and it may turn out to have been a systematic sample, without randomization of any kind. And here a prudent witness has a problem. The failure to randomize opens the way to possible biases, but as all experienced in sampling are aware, for a great many sampling frames (i.e., those with very little internal structure) the bias in the estimate and even in the calculated standard error is not likely to be large. Should one testify that the job was not competently done and the results should therefore not be given credence? (Counsel for one's own side would believe such testimony is entirely proper and the least that is owed him.) Or should one testify that, although the sample does not adhere to the canons, it is not likely that the result is for that reason wide of the mark? One will be tempted to add that when incompetence is manifest in the visible part of the operation, it is suspect in that which is less visible, and therefore the result should be received with caution. Should one yield to that particular temptation, however, he is most likely to be cut off by an objection from the opposing attorney, protesting testimony that is "mere speculation." Since the court is only interested in the result of your judgment about design, one is likely to either overstate the objections to the systematic sample or so understate them as to make one's client wonder why he was put to the trouble of drawing a random sample in his case. (And what should one say when the sample one is called upon to criticize was drawn by Dr. Deming or Professor Cochran, who—in the exercise of professional judgment—decided that the fuss required to randomize was not worth the trouble in the case at hand? Indeed, Cochran was fond of telling of the occasion when he was called on to carry out a sampling study of, I believe, a class of retail stores, and he instructed that the sample consist of every tenth establishment of that type listed in the Yellow Pages. The judge, he said, welcomed his expert testimony as a learning experience and remarked, after Cochran had been sworn, "I am glad to hear and to learn from Professor Cochran about this scientific sampling business, because I know virtually nothing about it. In fact, about the only thing I *do* know is that you should not just start at the beginning and take every 10th one after that.") I confess that, not being Deming or Cochran, I make it a point when drawing a sample to be sure that the design has as much internal credibility as I can give it, and as little dependence on the quality of my own judgment as I can manage.

Paternity and Fingerprints

In the sampling domain, statistical inference works well because we impose the probability model directly on the situation—through randomization—and our testimony has both the appearance and the substance of relative objectivity. We can feel rather sanguine about our contributions to legal proceedings in this domain.

We have somewhat less security when we turn to certain areas of identification evidence. I refer to blood tests for assessing evidence of paternity and to fingerprint evidence. In the former, at any rate, there is a probability element introduced by Mendelian genetics, and the statistical expert may have a real contribution to make. Unfortunately, the ultimate probability calculation depends on population gene frequencies; even where these are known for the population at large, it is often some subset of the population that is at issue, for which the frequency is not well established, and the expert finds himself on doubly uncertain ground. There are controversies aplenty in this domain, but this is not where most of the action lies.

Observational Data

The broad, almost limitless, domain in which the courts have come more and more to look to statisticians for guidance is in the analysis of observational data.

Consider, for example, the association between cigarette smoking and the subsequent development of lung cancer. First identified as an incidental and highly uncertain association, the accumulated evidence today appears overwhelming, although there are no clinical experiments to support it and only indirect support from cellular biology. The primary evidence is indeed statistical, and it *is* convincing, but not as a result of conventional significance testing. Rather it is the robustness of that association over time, place, and population that is convincing. For the most part, probability-based statistical testing is irrelevant to the strength of our conviction.

The apparent negative association between regular aspirin ingestion and heart attacks is a similar problem with a far more ambiguous outcome. An incidental association reported by a Mississippi clinician appeared to be confirmed by two large hospital-based case-control studies. The association is not of the magnitude of that between smoking and lung cancer, but it appeared to be substantial nonetheless. In this case, a number of clinical experiments have been carried out, and at present we must acknowledge that the degree of protection from heart attacks afforded by aspirin is at best modest, if there is any at all. Again, the probability-based significance level attained by the early case-control studies has no strong bearing on the believability of the aspirin hypothesis.

I do not quite share David Freedman's hard-line position against formal statistical inference for observational data. As explained in a superb elementary textbook (Freedman, Pisani, and Purves 1978), Freedman regards probability-based testing in a situation without a plausible probability model to be at best irrelevant and more likely misleading. I think, in contrast, that such testing serves a useful purpose as a benchmark: If the observed association would not be counted statistically significant had it arisen from a randomized study, it could not be counted as persuasive, when even that foundation is lacking. If the observed association is highly statistically significant, however, the extent of its persuasiveness depends on many uncertain judgments about background factors, and its persuasive value is not at all reflected in the significance level itself.

But the principles of statistical inference relevant to the courts are not the province of the statistics profession alone.

3. THE COURTS, CIVIL RIGHTS, AND STATISTICS

The Supreme Court has canonized formal statistical inference in a series of decisions beginning with a jury discrimination case, *Castaneda v. Partida,* decided in March 1977. Having noted that the population of Hidalgo County was 79% Spanish surnamed, but that the jury panels selected in accordance with the prevailing Texas "key man" system averaged only 39% Spanish surnamed (i.e., 339 of 870 jurors), the Supreme Court itself—more likely one of the Justices' law clerks—calculated the familiar critical ratio according to the binomial distribution—that is, the difference (39% minus 79%, or 40%) divided by the standard error (root pq divided by $n,$ which works out to 1.5%), obtaining a critical ratio of 29. The Court then commented, "as a general rule for such large samples, if the difference between the expected value and the observed number is greater than 2 or 3 standard deviations, then the hypothesis that the jury drawing was random would be suspect to a social scientist."

Formal significance testing next appears in an employment discrimination case, *Hazelwood School District v. United States* (Meier, Sacks, and Zabell 1984), decided three months later. In that case, the proportion of qualified teachers in St. Louis

County (excluding the city of St. Louis) who were black was estimated to be 6%, and during the two-year period at issue, only 15 of 405, or 4%, were black. The Hazelwood court now says, "A precise method of measuring the significance of such statistical disparities was explained in 'Castaneda v. Partida'...," and the opinion goes on to paraphrase the earlier two or three standard deviation rule, but with a slight shift; that is, "...if the difference exceeds 2 or 3 standard deviations, then the hypothesis that teachers were hired *without regard to race* would be suspect" (emphasis added). The reference to randomness is now absent, as is the social scientist. Since it is self-evident that the process of selection is not, nor is it desirable that it be, random, it is far from clear why either the social scientist or the Supreme Court should look upon a standard based on randomness as appropriate to assess the likelihood of purposeful discrimination. To be sure, there was much other evidence in the case, showing explicit discrimination at earlier dates, but the preceding quotation is the only place in the opinion where the relevance of the statistical significance test is in any way explained. Nonetheless, in Hazelwood the court went further, in an obscure remark that pointed clearly to the preeminence of statistics. It said, "Where gross statistical disparities can be shown, they alone may in a proper case constitute prima facie proof of a pattern or practice of discrimination." Thus, in the space of less than half a year, the Supreme Court had moved from the traditional legal disdain for statistical proof to a strong endorsement of it as being capable, on its own, of establishing a prima facie case against a defendant. (It is sometimes argued that the use of doubtful evidence to support a prima facie—that is, preliminary—finding is a matter of small legal consequence. Such a finding merely shifts the burden of proof from the plaintiff to the defendant. In fact, however, there is nothing at all "mere" about this shifting of the burden, since the difficulty of proving oneself innocent of discrimination turns out to be great indeed.)

The accelerating role of statisticians in employment discrimination cases arises from a combination of the statistical significance testing endorsed in Hazelwood with an earlier decision (*Griggs v. Duke Power Company*) in 1971. In Griggs, it was found that the requirements of a high school diploma and a certain score on a standardized IQ test for employment in such jobs as maintenance and laboratory work operated to exclude black applicants far more frequently than they did to exclude white applicants. The court concluded that, in the absence of direct evidence that these criteria related to improved performance on the job, the "adverse impact" of those requirements constituted a violation of Title VII, even though there may have been no intent to discriminate on the grounds of race. (To be sure, there was plenty of evidence of intent in the Griggs case, most especially in the facts that the power company had explicitly excluded blacks prior to passage of the Civil Rights Act and that it had put in the new requirements at the same time that the jobs were first made available to blacks. The principle established in Griggs clearly put the issue of intent aside, however, and the doctrine has been widely applied by lower courts in cases in which there was little or no evidence of invidious intent. Once again, proof of "job relatedness" or, as the Supreme Court says, "business necessity," has proved generally elusive, and a great many employment standards and admissions tests have been found to be in violation of the law for lack of such proof.)

The criterion seems reasonable enough until we are faced, as was the Supreme Court, with a case like *Washington v. Davis* (1976). In this case, Walter Washington, mayor of Washington, D.C., and his police chief had set about to recruit blacks into the D.C. police force, with an aggressive campaign to encourage black applications. The campaign was successful and many blacks were recruited, but among the newly encouraged applicants the written test "operated," in the language of the Griggs decision, "to disqualify Negroes at a substantially higher rate than White applicants." Thus, affirmative action, clearly intended to recruit blacks, fell afoul of the adverse impact principle developed in Griggs. The trial court dismissed the charge, but the appeals court reversed, citing Griggs. The Supreme Court side-stepped the issue. It supported the District program, but it did so on a technicality that did not require it to comment on the general validity of the Griggs principle. In subsequent cases the Griggs principle has continued to guide the lower courts.

It must be acknowledged that there has been some slackening of the tide in Title VII enforcement in the last two or three years, especially in regards to race and sex discrimination. Under the Reagan administration there has been more emphasis on freeing business from government interference and less on righting the wrongs of the oppressed. In response, the courts appear to give somewhat less weight than before to purely statistical evidence.

4. POSITION OF THE STATISTICAL EXPERT

The result of the preceding and related decisions has been to place the statistical expert witness in a most unaccustomed and exalted position. Despite the moderate recent decline in enforcement, the role of the statistical expert remains critical in the cases that are brought. Lawyers gaze with awe as he examines the entrails of complex multiple regression computer output, and they await breathlessly his conclusion that the coefficient of the variable designating sex is indeed more than twice the standard error. Similar attention attends his calculation of continuity-corrected 2×2 chi squares to see whether they are larger or smaller than 3.84. Indeed, largely on these outcomes the case may be won or lost.

That this position is a false one, none can doubt. Certainly the statistical experts know it, and most of them say so to some extent—or at least they assert that the meaningfulness of their numerical results depends on a number of assumptions that they are unable to verify. The courts, however, are not engaged in academic exercises and, having urgent need to come to some conclusion, turn to the Supreme Court instead of to the witness's cautionary phrases for guidance. Those opinions have, intentionally, a somewhat Delphic quality. They tell us that "gross disparity" in pass rates is evidence of illegal discrimination, and they also tell us that the hypothesis of random selection is made to appear doubtful when a difference is larger than two or three standard errors. They do not quite say that statistical significance at the 5% level constitutes gross disparity, but that is how the lower courts read them. The statistical experts cannot help but find this heady stuff, and we should not be surprised to find ourselves speaking with far more assurance about our conclusions than an objective appraisal of the evidence might warrant.

Thus we are led to the unedifying spectacle of two well-qualified statistical witnesses providing analyses that they interpret oppositely, each supporting the interest of the party who introduces him. Other categories of expert witnesses have been there before us, of course—the psychiatrists, medical internists and surgeons, and structural engineers, among others. The courts have urgent need for the assistance of these experts, but they seem uncommonly ill served by them. The point was made clearly in an editorial in the *British Medical Journal:*

> Medical evidence delivered in our courts of law has of late become a public scandal and a professional dishonour. The Bar delights to sneer at and ridicule it; the judge on the bench solemnly rebukes it; and the public stand by in amazement; and honourably minded members of our profession are ashamed of it. (*British Medical Journal* 1863)

This was printed more than a century ago, but little has changed in the intervening years. Statisticians have escaped comparable condemnation because we have been, until recently, too unimportant in the courts to be noticed, not because of any higher ethical standard of our profession.

One cannot help speculating on the possibility of improvement. In fact, I believe that some of the difficulty is structural and that there are ways in which we could function usefully in legal settings without so large a sacrifice of professional integrity. To this end I now discuss the players in the game and the key influences on them. Among the players or participants in the legal ballet, I distinguish three: (a) the courts themselves, most especially the Supreme Court, who together with the Congress set the rules by which the system operates; (b) the lawyers—collectively the bar—who primarily control the direction of play within those rules; and (c) the expert witnesses on whose performance the integrity of the enterprise ultimately depends.

The Courts and the Expert

The obvious objective of the courts in respect to expert testimony is to optimize the search for truth. The courts would like to get the most well-qualified expert, keep him in a situation in which he can devote his best efforts to analyzing the evidence, and have him testify in an atmosphere free of coercion or bias. The courts also want to be sure that the expert is adequately examined to test and verify his qualifications, the adequacy of his preparation, and his objectivity.

To this end, the courts in Germany and France arrange matters very differently from the English and American courts. In cases in which experts are needed, they are in the first instance appointed by and responsible to the court and not to either party. They are first examined by one of the judges and also cross-examined by him. Attorneys for the plaintiff and defendant may also cross-examine, but the proceedings are not generally adversarial as are our own, and the appearance of neutrality, at least, is the rule. Thus the continental system seeks the best witnesses and seeks to put them in a neutral setting, primarily by putting the major responsibilities in the hands of the judges.

The Anglo-American system, in contrast, is based on the proposition that truth is most likely to emerge through the best efforts of adversaries. No point in favor of the defendant will be overlooked or undervalued, it is thought, if responsibility for bringing it out is assigned to the defendant's advocate.

Nonetheless, whatever the merits of the adversary system may be in general, it is well recognized that it wreaks havoc with expert testimony, and proposals for reform appear regularly. Chief among them is to borrow from the continental system and to have the primary expert witnesses appointed by and responsible to the court. This reform was vigorously advocated by last century's revered commentator on legal evidence, John Wigmore (1940), and model codes have been proposed to this end. Indeed, Rule 706 of the Federal Rules of Evidence (Title 28, U.S. Code, annotated) provides explicitly for court-appointed experts. Regardless of the merits, in practice this power is used extremely sparingly. (There may be some cases of court-appointed statisticians in Title VII cases, but I have not heard of any.) One can conceive of many reasons for the ineffectiveness of these "reforms," not least the vulnerability to criticism of a judge who appoints an expert later shown to be inadequate, but it is enough for my purposes to observe that such reforms have not taken hold in this country and that they do not seem likely to become influential in the near future.

The Bar and the Expert

The position of legal counsel, although in principle identical with that of the judge, is in fact quite different. Having committed himself to the adversary system as the best method of reaching a just conclusion, the lawyer for the plaintiff now accepts his position in the system, that of advocate, and leaves to the court the responsibility for discerning the path of justice. To him the expert is simply one of the elements that he must fit into place to make the most effective case. To be sure, any lawyer of competence recognizes that it is usually favorable to his case for the witness to appear to be dispassionate and objective. The best lawyers recognize that a witness will make the best appearance of objectivity if he feels that he is indeed free to go where his research and reflection lead him. This is not to say that these excellent advocates are really in the market for unbiased witnesses who may testify to their side's disadvantage.

John C. Shepherd of St. Louis, a distinguished trial lawyer who was president of the American Bar Association in 1984–1985, spoke to a conference for lawyers on relations with the expert witness (Shepherd 1973), and this is what he said:

> Many people are convinced that the expert who really persuades a jury is the independent, objective, nonarticulate type.... I disagree. I would go into a lawsuit with an objective, uncommitted, independent expert about as willingly as I would occupy a foxhole with a couple of non-combatant soldiers.
>
> If you find the expert you choose is independent and not firmly committed to your theory of the case, be cautious about putting him on the stand. You cannot be sure of his answers on cross-examination. When I put an expert on the stand, he is going to know which side we are on.
>
> The trial lawyer must make of the expert a convincing, persuasive witness. The lawyer deals in words, and he knows how to put the package together to impress the jury favorably. It is his job to instruct the expert, an exercise requiring great tact and firm conviction. (pp. 21–22)

Keep in mind that the lawyer does not need to make bricks without straw. It is perfectly proper for him to consult a great many potential witnesses but to bring to court only that one whose honest convictions fit well with the lawyer's needs.

The phenomenon of "shopping for witnesses" is well recognized by the courts, and it contributes to the wary attitude they have about experts in general. The shopping is done by the lawyers, however, and is thus not subject to exposure in the actual testimony.

Corrupting Influences

As we have just seen, the professional integrity of the expert witness and, through him, of the profession that he represents is not well protected by the courts and hardly at all by counsel. But before we assume too readily that simple morality and personal ethics will be an adequate substitute, we should reflect for a bit on what I call, for lack of a more delicate phrase, *corrupting influences.* Some are inherent in the nature of the situation, and others are special to the adversary situation.

First, there is the fact that the expert witness is playing someone else's game and, inevitably, has to accept the rules as he finds them. His instructor in these matters is, of course, his client's counsel, and the witness is ill-equipped to resist the role of adversary when his lawyer thrusts it upon him. But even supposing the lawyer is less demanding than Shepherd, the expert is beset with temptations.

General

Among the most difficult of the corrupting influences to deal with is what I call *aggrandizement.* In Title VII cases (i.e., those dealing with employment discrimination), the Supreme Court has placed the statistician in the key role. Long ignored and treated with contempt in literature and in the courts, the statistician has been elevated to Olympian levels. Thus the Hazelwood court, quoting its remark in an earlier case, commented:

> We also noted that statistics can be an important source of proof in employment discrimination cases, since, "absent explanation, it is ordinarily to be expected that nondiscriminating hiring practices will in time result in the work force more or less representative of the racial and ethnic composition of the population in the community from which employees are hired." Evidence of long lasting and gross disparity between the composition of the work force and that of the general population thus may be significant even though paragraph 703 (j) makes clear that a work force need not mirror the general population. (*Hazelwood School District v. United States* 1977)

Taken together with the court's embrace of statistical significance testing, the statistician is here given a virtual license for intellectual robbery. Indeed, not only the court but a large contingent of fellow academics (economists numerous among them) give strong endorsement to the particularly magical properties of multiple regression analysis. [Two articles in the *Columbia Law Review*—Fisher (1980) and Finkelstein (1980)—are noteworthy in this regard.] All in all, the statistician is strongly tempted to give the definitive rather than a qualified answer to the key questions. He will be tempted to ignore or to minimize those qualifications that he might emphasize in a more academic setting, he may fail to emphasize the existence of schools of thought other than his own, and he may lay claim to overly broad scope for the inferences he draws.

Adversarial

The adversary system adds a host of additional influences, some quite direct, but others indirect:

1. *Bribery.* The witness is paid by his client and, as often noted, he who pays the piper feels a right to call the tune. To be sure, all the client is entitled to is an honest report of the expert's best effort, but an expert who habitually finds evidence against his client will not be much sought after.

2. *Flattery.* Some, of course, are not bought by money or the prospect of future money—either they already have enough of it or they are sufficiently on guard against that particular type of seduction. Other corruptions await them.

I well recall an occasion on which I was asked to consult in a case at a time that was not especially convenient. I explained that I really could not participate on this occasion. The lawyer, with whom I had worked before and for whom I had a great deal of respect, pled the sorry state of statistical testimony in the courts in general, and in the instant case in particular. He read from the transcript some particularly egregious quotes from the statistical expert for the other side, and he urged the importance for the future of statistics in the domain of public affairs of having corrective testimony. That being a viewpoint I could only share, and tacitly mindful of our shared opinion that I was the ideal candidate to champion the honor of the profession, I reluctantly agreed to testify. Imagine my chagrin when, at a later date, I read some other remarks of the trial lawyer, John Shepherd, whom I quoted earlier. He advises on "Approaching the Expert" as follows:

> Almost every one who considers the subject of experts in court will start with the same thought: The first thing you need to get along with your expert witness is money. But the hiring and successful use of an expert may not be that easy—a lot of good experts are rich. Although you will eventually be talking about money with your expert, it is wiser to begin on another tack. Tell your expert how justice will be served if he will testify on your side of the case. Remind him that the unfortunate situation in our courts today can be improved if we have people of his caliber to help in the administration of justice. That ploy will impress even the rich expert. (Shepherd 1973, p. 19)

3. *Co-option.* To be sure, effective as this ploy may be, it does not in itself lead the expert away from his duty. It establishes an aura of objectivity and mutual respect, however, which may make the expert especially vulnerable to another inevitably corrupting aspect of the adversary system. That is the simple fact that the expert's introduction to the case comes from the client's counsel and will inevitably tend to appear in the light most favorable to the client. He will be introduced to the principals—perhaps a plaintiff, movingly indignant about years of abuse and low pay, perhaps a defendant who truly believes that his cause is just and is worried sick about the distraction of his institutional resources from their proper role into the defense against a baseless charge. This goes along with co-option into advocacy arising when one is asked to review the other side's testimony, point out flaws therein, and assist in the development of effective cross-examination.

F. Downton of the University of Birmingham has written cogently about this latter difficulty in a symposium on statistics and the law (Downton 1977). Downton had been consulting with the police on games of chance, because the law prescribed strict rules for games that, if violated, would allow the police to close the clubs.

Since the clubs were widely regarded as dens of iniquity, this was clearly a public service. Downton wrote:

> As in any other consulting situation, a certain amount of identification with the aims of the client is inevitable; it is fortunate that probability and statistics are basically mathematical in content, since the constraints of mathematics act as a brake on over-enthusiasm. It cannot, however, be denied that a conscious change of attitude was needed to effect the change-over from helpful consultant to objective expert witness.... This ambiguity of roles did create a conflict, which presumably can only be resolved by individual witnesses in their own way. (p. 171)

4. *Gladiatorial Role.* The adversarial environment works against objectivity in yet other ways. The object of cross-examination is not only to expose weaknesses in the expert's analysis but, if possible, to discredit the witness and the weight that should be given to his testimony generally. Thus the cross-examiner may, by adroit framing of questions, force the witness into complex explanations and apparent contradictions. Feeling his credibility slipping away, such a witness may be less likely to give a full and frank answer to a later question that might fairly expose a fact or conclusion operating in favor of the other side. The expert no longer views his interrogator as a fellow searcher for truth, but as an adversary against whom he must defend.

5. *Personal Views.* My final source of corruption is perhaps the most difficult to deal with, and that is the problem of strongly held personal views. Surely there are many cases in which the expert is a priori indifferent between the claims of the contestants, but in other areas, particularly in the great domain opened up by Title VII, there are few of us without strong opinions. In the matter of a contest between a chemical waste disposal company and the residents of a new Love Canal, for example, I would be reluctant to testify on behalf of the company. It might well be, for example, that the evidence of adverse health effects caused by carelessly buried wastes is really nonexistent. Feeling as strongly as I do, however, that such careless behavior is reprehensible and deserving of punishment, I should not like to assist the company's case. I have no problem reconciling my preferences and my professional responsibilities in this case. I am, and should be, free to accept an engagement or not, for whatever personal reason, and reasons of this kind are at least as good as most others.

My problem comes on the other side. Suppose I should be an expert retained by the residents affected by the dump. I find that, in respect of total mortality, there is no evidence of an effect, but in the matter of childhood leukemias there is an excess mortality amounting to 1.8 standard errors greater than the rate in some control group. If I ignore the fact that I am reporting on leukemia because it is the disease category showing the largest difference, and if I adopt the conventional 5% significance level as a standard, and if I urge the relevance here of a one-sided significance test, I may be able to strike a blow for truth and justice, and it would no doubt be tempting to do so. But to paraphrase a major figure in the Watergate investigation, "I could do that, but it would be wrong." I really do not think one-sided 5% level deviations provide convincing evidence one way or the other, and whatever one's views on that, I expect most statisticians would agree with me that it is misleading to the point of dishonesty to quote an unadjusted significance level when I have chosen to present the most extreme of a number of alternative measures.

Perhaps the point can be brought home most forcefully by addressing an even touchier example. There are many of us who view the legacy of slavery as our most appalling and pressing social problem and the effort to explain the low status of the oppressed on the grounds of inherent inferiority as an intolerable offense. Indeed, although the possibility of *some* average difference in intellectual capacity among different groups can never be ruled out, the evidence appears clear that whatever differences there might be in *average* innate ability, they are quite small compared with the variation between individuals. The effects ascribed to race in regression analyses of schoolchild performance, after adjustment for age, years of schooling, mother's socioeconomic status, and the like, are readily explainable as attenuation and other distracting effects that afflict regression analyses generally, and they need not be interpreted as reflecting a real difference due to race.

At the same time, we observe the past and present systematic discrimination against blacks in many areas of employment. Such discrimination has many forms, but its pervasiveness, except where sharply controlled by law, is hardly in doubt. Being confident, then, that a charge of race discrimination likely corresponds to the existence of actual discrimination, what are we to say about a multiple regression in which a salary difference unfavorable to blacks emerges as significant even after adjustment for age, years of schooling, mother's socioeconomic status, and the like? The problems of attenuation apply with equal force, but we may now be reluctant to dismiss the evidence of bias in pay. This time we may believe that there really is discrimination in the system, but it is by no means clear why we should, as statisticians, take different positions in the two situations.

5. WAYS TO DEFEND THE INTEGRITY OF STATISTICAL TESTIMONY

With the variety of assaults on the credibility of expert statistical testimony, I turn again to the question of what possible defensive measures could be implemented. A change to the apparently more neutral continental system is the one answer that has come from the courts, but there seems little likelihood of its adoption. There have been proposals that experts who testify falsely should be punished for perjury, as is an ordinary witness who testifies falsely about an event. This, too, seems far-fetched, since the essential nature of expert testimony is that it is largely a matter of informed opinion.

Professional Codes

There seems to be only one other direction in which to turn, and I have only slender hopes for it. Seeing that neither the bench nor the bar will help us, the only alternative is to help ourselves—that is, to develop limited codes of ethical behavior in the context of legal proceedings that may help to ameliorate the worst excesses.

I come to this conclusion reluctantly, because I have little taste for collective moral instruction and little confidence in its efficacy in general. And yet one cannot deny that codes of ethics for judges, while they do not eliminate venality, are good to have. Violation of these rules can and does lead to discipline, on occasion, as in the case of a distinguished Supreme Court Justice a few years ago, and reminders such as that help to keep others on the right path. Similarly, codes of medical ethics dealing with the proper relationship between physician and patient serve a useful purpose.

A decade ago, Gibbons (1973) reviewed our society's efforts in the direction of developing such codes, and she clearly laid out some of the problems that such codes might help to solve. Evidently there was some movement in that direction in the early 1950s, but momentum was lost, and nothing came of it. The issue of ethical codes continues to elicit debate, a recent instance being the report of the Ad Hoc Committee on Professional Ethics (1983). I see no sign, however, that this or any other code is likely to be adopted as a guide by any of our major professional organizations. Indeed, although discussion of codes of ethics for statisticians continues, I know of only one instance in which such a discussion has had any noticeable practical effect.

The exception is an interesting one, and it may be instructive. As I mentioned earlier, in the context of sample survey design and analysis, W. Edwards Deming established a code that he provided to clients, explaining the reach and the limitation of his methods. The code is notable for its careful restriction of the role of the statistician. In effect, Deming acknowledges that the statistical consultant may come to have a good deal of knowledge about the subject matter under study and that this knowledge may help him to design an effective sample. He makes clear, however, that *responsibility* for the choice of population to be sampled, and for the processing of each sampled unit, belongs to the client and not to the sampling consultant. The consultant undertakes to say only that, had the entire population been processed in the same way that the sampled elements were processed, the sample estimate for the population would be found to be close to the population value, subject to error limits that can be given in the usual probability sense. It might seem that the scope of the sampling expert's testimony is so narrow according to this code that his contribution will have little weight in the proceedings. In fact, of course, the contrary is true. By not reaching beyond well-stated boundaries, the testimony of sampling experts has achieved an enviable level of credibility.

Deming's code, effective as it is in a specific context, gives only a little guidance for the expert testifying in a Title VII case. I submit that a proper code for the latter expert should copy Deming by being specific to the situation and rather restrictive as to the scope of the testimony. I do not think it will pay to start with an ethical code trying to embrace all statistical activities. Let me try to clarify my proposal by being specific. (Here I borrow from Deming where I can.)

I suggest that a statistican asked to testify in court should require that he be given access to all data thought by the client to be relevant and all previous analyses of that data, and he should demand a commitment on the part of the client to a "good faith" effort to supply whatever other data the statistician may judge relevant.

He should advise the client that in his professional role he will remain neutral between the parties (and he should pray for strength when he does this, for he will need it). He undertakes to provide his best effort to analyze the data in ways that seem to him pertinent, and he undertakes further to provide a written report. His report, if it is to be used, must be taken in its entirety.

When testifying, the expert will explain the limitations of his techniques, as seen by a professional statistician, regardless of any statistical principles annointed by the Supreme Court. He will explain the variety of schools of thought within the profession and his place among them.

Doubtless there are a number of other principles to be enunciated, but this is not the time or place for full details. Some will think the principles given are simple and obvious, but they can be assured that they are not obvious to most lawyers. Many lawyers, for example, think it proper to select for attention the principles laid down by the high court as a basis for expert testimony. (I cannot help but wonder if, should a court declare that $\pi = 3$, these lawyers would insist that we accept that too.) Gratuitous testimony about limitations will be especially unwelcome (and in my opinion, especially necessary).

The point is that the expert should be much more his own man and much less the puppet of his client's counsel than is typically the case today.

Consequences

The consequences of such a code, should we adopt and use it, are substantial and not entirely welcome. I do not believe it would help us much in protecting against the influence of our own strongly held social views or against the biases that arise because we are oriented and informed by just one of the adversaries. Nor would it keep us from reflexive defensiveness under hostile cross-examination.

Adherence to such a code is likely to result in a reduction of the pivotal role that statistical analysis has come to play in discrimination law, and we may see the resulting gap filled by others whose competence and good will we question even more than our own. It is conceivable that a more modest posture might lead the courts to seek greater clarity by adopting the reforms, if such they be, of the continental system with court-appointed experts. It is certain, I think, that adherence to such a code would improve the credibility of statistical witnesses.

6. CONCLUSION

The sorry state of expert testimony might lead one to conclude that no honest person should participate in such a scam, but I feel strongly that this conclusion is wrong. Querulous as this discussion has been about the workings of our legal system, it is, after all, the only one we have and, as the courts are fond of proclaiming in their own self-serving way, every citizen owes his best efforts to serve and to improve the system that ultimately is the protector of those freedoms he cherishes. (Shepherd's appeal to the reluctant witness on these grounds is one that *he* views as a ploy, and for him it is exactly that. It nonetheless expresses an important and commendable sentiment.)

In closing I turn to White's (1964) delightful summary of rude comments about statistics and statisticians. After discussing the familiar aphorisms—lies, damned lies, and whatever, many of which canards White then thought to be reasonably apt—he concludes: "Time passes; the new books are written by other kinds of men, and only an occasional author, poet, or politician, or critic fires a blast at the Statisticians. Who knows what the outcome of all this acceptance is going to be?" (p. 17). Since that was written, acceptance in the courts has grown mightily. As matters now stand, prospects for the outcome of all this acceptance are ominous, not to say grim.

Perhaps we can show ourselves to be wiser and more disciplined than the professions that have been before us. We should at least try.

ACKNOWLEDGMENTS

This article is based on the President's Invited Address that was delivered at the Annual Meeting of the American Statistical Association on August 16, 1982. The author is indebted to Sandy Zabell for many helpful discussions on the issues reviewed here. The article was prepared using computer facilities supported in part by National Science Foundation Grant MCS-8404941, awarded to the University of Chicago's Department of Statistics.

REFERENCES

Ad Hoc Committee on Professional Ethics (1983), "Ethical Guidelines for Statistical Practice," *The American Statistician*, 37, 1–20.

British Medical Journal (1863), "Medical Evidence in Courts of Law" (editorial), May 2, 456–457.

Castaneda v. Partida (1977), 430 U.S. 482.

DeGroot, M., Fienberg, S., and Kadane, J. B. (eds.) (1986), *Statistics and the Law*, New York: John Wiley.

Deming, W. Edwards (1954), "On the Presentation of the Results of Sample Surveys as Legal Evidence," *Journal of the American Statistical Association*, 49, 814–825.

Downton, F. (1977), "Experience as an Expert Witness in Gambling Cases," *The Statistician*, 26, 163–172.

Finkelstein, Michael O. (1980), "The Judicial Reception of Multiple Regression Studies in Race and Sex Discrimination Cases," *Columbia Law Review*, 80, 737–754.

Fisher, Franklin (1980), "Multiple Regression in Legal Proceedings," *Columbia Law Review*, 80, 702–736.

Freedman, David, Pisani, Robert, and Purves, Roger (1978), *Statistics*, New York: W. W. Norton.

Gibbons, Jean D. (1973), "A Question of Ethics," *The American Statistician*, 27, 72–76.

Griggs v. Duke Power Company (1971), 401 U.S. 424.

Hazelwood School District v. United States (1977), 433 U.S. 299.

Meier, Paul, Sacks, Jerome, and Zabell, Sandy (1984), "What Happened in Hazelwood: Statistics, Employment Discrimination, and the 80% Rule," *American Bar Foundation Research Journal*, Winter, 139–186.

Meier, Paul, and Zabell, Sandy (1980), "Benjamin Peirce and the Howland Will," *Journal of the American Statistical Association*, 75, 497–506.

Monahan, John, and Walker, Laura (1985), *Social Sciences in Law: Cases and Materials*, Mineola, NY: Foundation Press.

People v. Collins (1968), 68 Cal. 2d 319, 66 Cal. Rptr. 497.

Peterson, Donald W. (1983), "Statistical Inference in Litigation," *Law and Contemporary Problems*, 46, No. 4, 1–303.

Shepherd, John C. (1973), "Relations With the Expert Witness," in *Experts in Litigation*, ed. Grace W. Holmes, Ann Arbor, MI: Institute of Continuing Legal Education.

Sprowls, R. Clay (1957), "The Admissibility of Sample Data Into a Court of Law: A Case History," *UCLA Law Review*, 54, 222–232.

Washington v. Davis (1976), 433 U.S. 229.

Wigmore, John Henry (1940), *Evidence in Trials at Common Law*, Boston: Little, Brown.

White, Colin (1964), "Unkind Cuts at Statisticians," *The American Statistician*, 18, No. 5, 15–17.

2

STATISTICIANS, ECONOMETRICIANS, AND ADVERSARY PROCEEDINGS

Franklin M. Fisher

1. INTRODUCTION

This article concerns the problems and standards of behavior of statisticians acting as consultants and witnesses for lawyers—statisticians in the context of a system that views adversary proceedings as a way of finding the truth. Such an environment differs from that of the professional meeting or the graduate seminar, and the guidelines for successful and appropriate professional behavior (although surely not wholly distinct) are not the same in the two situations. I shall discuss a number of areas in which my own experience suggests problems and modes of behavior. Because I am an economist, the examples used are largely, but not exclusively, drawn from the use of econometrics in adversary proceedings. I believe the discussion to be of relevance for other branches of statistics, however. (For discussions of similar matters, see Royal Statistical Society 1982, including Downton 1982, Fienberg and Straf 1982, and Newell 1982; also see Finkelstein and Levenbach 1983 and Rubinfeld and Steiner 1983.)

My experience in this area has been reasonably extensive, lasting more than 20 years. Twenty years ago, the legal profession viewed econometrics as black magic, and the introduction of econometric evidence was a considerable innovation. Today, although econometrics may still be viewed as black magic, its use is no longer rare, for it is supposed to be capable of rather more than it is.

It is always tempting to seek certainty in quantification rather than to study all of the aspects of a truly complex problem. (The use of market share in antitrust cases is an outstanding example.) The use of statistical methods in general, and of econometric methods in particular, provides that temptation in a form often irresistible to one or the other side of a lawsuit. Such methods provide quantitative, computer-generated answers, which seem invested with a scientific accuracy. When those answers appear to favor their clients, lawyers may be very ready to believe whatever their expert is telling them, whether or not it is truly well founded. Since not all supposed experts are truly skilled, and since not all experts remain truly impartial, this can create serious difficulties. When the opposing side does not itself have expert assistance, the result may be a successful attempt to overawe judge or jury with an apparently scientific answer.

Indeed, a principal reason for capable statisticians to be interested in such activities is the lamentable fact that legal proceedings will contain statistical analyses whether or not competent persons provide them. Unless the area is to be abandoned to practitioners who are poorly trained, poorly skilled, or whose professional standards are not very high, serious and competent statisticians and econometricians must enter it.

The desire to correct misuse of the tools, however, is not the only reason for statisticians—and particularly econometricians—to participate in litigation. Those who have brought statistical methods into legal proceedings, even those who have done so incompetently, have perceived correctly that the techniques of estimation, forecasting, and tests of hypotheses can often be usefully brought to bear on real, practical problems of considerable private and even public interest. Whereas the use of large models or of time series analysis to inform governmental policy is the most exciting, if not always a very successful, application of econometrics to macroeconomics, the use of econometrics in court or regulatory proceedings is, I think, the most exciting application to microeconomics. Econometricians have an opportunity to use their tools on problems that people really care about, to gain access to considerable bodies of data, and to use the expertise of those directly involved. I believe it to be important to accept that opportunity, and much of my own empirical work has come from such participation.

2. EXPLAINING STATISTICAL METHODS

One of the most important problems faced by the statistician who participates in such activity is that of exposition. Direct testimony almost invariably includes an exposition of the statistical methods used as well as a presentation of the results. Hence if all goes well, the statistician will have to explain to a judge or a jury what it is he or she has done and how it should be interpreted. Even if the study is never presented in court, statistical methods will have to be explained to the clients, the lawyers with whom the statistician works. This is not an easy matter, since it is extremely rare that any of these potential audiences has had any technical training (or is even comfortable with high-school algebra). The following are some of the confusions I have encountered.

1. *Parameter Estimation.* At least in general terms, the statistician will have to explain that estimates of the parameters of the model are not chosen arbitrarily. Such explanations can vary from a heuristic, diagram-using discussion of fitting a line to a scatter of points, to the assertion that estimation is done on a computer using a

long-recognized method. At least the statistician's own client is likely to require the more elaborate explanation; whether a jury will do so may be a matter of tactics.

It is important to emphasize this point so that the statistician is not thought to be picking numbers out of the air and the audience will understand that statistical methods provide the best reflection of what the data say. Attorneys and others involved in the case may have their own opinions as to what the values of parameters are; they may not realize that it is not sufficient merely to object that the results do not come out as expected. Furthermore, attorneys and judges tend to view with suspicion any computational method that they do not understand in very elementary terms. That suspicion may be difficult to allay unless examples are well thought out and well presented, for as Edward Levi has remarked, "The basic pattern of legal reasoning is reasoning by example" (Levi 1948, p. 1).

To take a particular example, some years ago I had occasion to study the effect of cable television on local television stations in the United States (Fisher and Ferrall 1966). As part of that study, I estimated the value of a viewing household to a station in terms of advertising revenue. I did this by regressing station revenues on audience size using a cross-section of television stations. The regression passed very close to the origin and had a slope of about $27 per year for each additional household viewing in the average half-hour of prime evening time.

Despite the fact that the regression fit a great many observations nearly perfectly, that estimate was attacked as obviously incorrect by some who failed to understand the units or the fact that an increase in household viewing in prime time in a cross-section of stations was highly correlated with household viewing in other parts of the day, so $27 did not merely represent advertising rates charged in the evening. Such confusions were only resolved by computations made by Martin Seiden, an economist (not an econometrician) hired by the Federal Communications Commission (FCC), the regulatory body in charge of television. (After receiving referee comments on a first draft in which this was not done and after consultation with the editor, I have decided to refer to people by name when they can in any case be readily identified from the public record.) Seiden took all television revenues in the United States and divided them by the number of television households multiplied by 60%, the fraction of prime time that the average household was found in surveys to watch television. He came up with approximately $27 per household.

This was hardly a surprising finding. Consider a two-variable regression that goes through the origin, $y = xb + u$, where b is the least squares estimate of the slope so that u has sample mean zero. Denoting sample means by bars, it is immediate that $b = \bar{y}/\bar{x}$, since $\bar{u} = 0$. Hence Seiden was merely approximately recalculating the same regression coefficient that I had already given. Nevertheless, the fact that he could reach essentially the same result by elementary methods made him, and others, decide that my result was believable after all. Indeed, the number so calculated became perhaps the only quantitative result in my study that was widely accepted.

Certainly, such an experience would be less likely today. The FCC staff of those days had essentially no experience with econometrics or even with computers. (They expressed doubt that the same punchcards they used for accounting purposes could also be used for input for computations.) Nevertheless, even though potential clients are likely to be more sophisticated today, the story is still relevant. Unless care is taken to explain some of the mysteries of the science, the statistician may run into difficulties simply because some of the audience believe that their guess is just as good as the results, and perhaps better.

2. *Ratios.* The same story illustrates a different phenomenon. Most adult Americans, even those exposed to mathematics in college, are extremely uncomfortable with mathematical thinking. Even linear equations involving a single variable and a constant term, let alone equations involving several variables, are likely to be regarded with fear as "higher mathematics."

This requires careful attention by the statistician, especially because a layperson's idea of how to do forecasting may simply be to assume that the ratio of the dependent variable to one of the independent variables is constant. Despite the advantage of simplicity in such a "constant ratio" approach, the statistician generally should firmly resist the temptation to use it. He or she should carefully explain that the constant-ratio assumption is but a special (usually a very special) case and that if the assumption is true it will be found to be so in the context of the more general and satisfactory linear model. Perhaps because of the difficulties of exposition involved in the use of more satisfactory methods, not all expert witnesses resist the temptation to use the simplistic, constant-ratio approach.

3. *Systematic vs. Random Elements: Data Mining.* The tendency for laypersons to believe that they know how the study must come out is related to another problem that is frequently encountered. Statisticians are used to the idea that regression equations do not generally fit the data perfectly. They understand that although the systematic part of a regression equation involves the most important variables that theory or common sense suggests influence the dependent variable, the effects of minor or particular influences is left to the random error term. Although one naturally wants a fit as good as possible, other things equal, it is well recognized that even the best model is but an approximation. There will always be unexplained deviations from the regression plane.

Lawyers may understand this in principle, but they do not like it. Attorneys, presented with a general argument, tend to think in terms of counterexamples, and the fact that not all observations lie in the regression plane may seem to them to provide ammunition for the opponent. They tend to offer ad hoc explanations of at least the larger deviations from the regression plane, even in cases in which the fit is good. Furthermore, attorneys tend to be quite sensitive to the objection that not every plausible argument was taken into account. The result is a tendency to insist that the statistician explicitly model every influence that anyone can think of as possibly important.

This presents both statistician and attorney with a dilemma. On the one hand, unless the statistician explores all of the plausible alternatives, the attorney's fears may be realized: the statistician may be made to look careless on cross-examination by his or her failure to see if any of the suggested influences made a difference. On the other hand, the statistician cannot simply mine the data until a satisfactory model is found to fit the sample. To do so at best vitiates any claim as to the statistical properties of the model and at worst exposes the statistician to the suspicion that models have been rejected until one favoring the client is found. Since not all expert witnesses attempt to maintain full impartiality and since, as I discuss later, subjective honesty is not a simple matter even for those with the best of intentions, care must be taken to guard against such a charge.

The problem is not merely one of appearances. In scientific work, one may want to impose severe limits on exploratory data analysis in order to protect against overfitting. In legal work, on the other hand, the same stringent limits may prevent the examination of potentially relevant evidence. Yet it is dangerous either to ignore such evidence or to depart from usual scientific procedures for the sake of the type of investigation or the client.

I recommend the following mode of operation to deal with such problems. First, the attorney sponsoring the project must be made to understand the problems associated with "data mining." Explanation should carefully be given as to the role of the random error in picking up particular exceptions to a general rule and as to the fact that the estimated equation will describe an average tendency rather than an exact universal law.

Second, particularly in cases in which the data are insufficient to permit segregation of the sample for exploratory model building and for testing, the statistician should set down in advance the versions of the model (few in number) that he or she believes are most likely to be the appropriate ones. (Of course, this should be done after discussion with the attorney and with others knowledgeable about the phenomena being modeled.) If those versions prove to provide a reasonable explanation of the phenomenon under investigation (with no more than the usual reconsideration after the results are seen), the particular factors that the attorney or someone else thinks may also operate can be entered as additional variables. Often such additional factors make little difference to the results.

In this way the statistician can succeed whether or not the particular factors involved are in fact important. If they are, this will be discovered. If they are not important, as I have suggested will be the usual case, the statistician will be able to testify to that lack of importance on direct or cross-examination and will have created a record showing that the model he or she sponsors was not arrived at after massive experimentation on the same set of data.

In some cases, of course, the matter may be easier to resolve, for the statistician may have the luxury of sufficient data with which to explore different possibilities while reserving other data for confirmatory testing. (This is perhaps more likely in discrimination cases, where the data relate to individuals, than in antitrust cases, where a model of the industry is required.) Even in such cases, the procedure just outlined is desirable. Writing the alternatives thought most plausible in advance becomes part of the necessary practice of creating a retrievable record of what decisions were taken and the reasons for them. When data are used for exploration, the reasons for changing or abandoning a model and for following up or not following up some alternative must be set forth. In any case, enough exploration must be done so that the results are shown to be relatively insensitive to plausible alternative specifications and data choices. Only in that way can the statistician protect himself or herself from the temptation to favor the client and from the ensuing cross-examination.

4. *Goodness-of-Fit and the Nature of Models.* Associated with the tendency to believe that a model must fit all cases is the tendency to judge models purely on how well they fit the data, with goodness of fit being measured in the grossest sense of R^2. There is a natural view that models are supposed to do nothing other than predict, and therefore a model that superficially appears to predict well must be

believed. Since it is often possible to fit the data well with models that do not reflect any structural characteristics of the phenomena being investigated, the danger here is that the laypersons involved will be impressed by poor work to the detriment of better models that do not fit or predict quite so well but are in fact informative about the phenomena being investigated.

It is easy to misunderstand what is likely to be involved here. There is a genuine scientific issue regarding the insights into structure that can be gained from structural models that fit poorly. Furthermore, there are occasions on which the structure of the phenomenon being investigated is so poorly understood that attempts to fit a structural model are not helpful. But those are precisely the occasions on which statistical methods are unlikely to provide much guidance for judge or jury. The issue in litigation is unlikely to be whether an autoregressive integrated moving average (ARIMA) model predicts better than a supposedly structural one. Rather, it is likely to be whether to believe the implications of very simple models based on trend-fitting and related techniques that have little to do with the phenomenon being studied. On other occasions with somewhat more sophisticated opponents, the question may be whether to believe a poorly constructed structural model just because it appears to fit the sample well. (See the discussion of the *Corrugated* case in Section 3.)

On the other hand, it is important to remember that, depending on the object in view, it may not be necessary to estimate a full structural model. For example, to test whether a certain firm discriminates in the wages it pays women, one need not estimate a model that attempts to mimic the decision processes of the firm. Abstracting from other difficulties, consider a structural model with two equations describing in detail the assignment of employees to jobs and the assignment of salaries for given jobs, in both cases given gender and qualifications. Such a model, if correct, can be revealing as to how discrimination operates. If the details of the job and salary assignment process are poorly understood, however, one may well do better with a single reduced-form equation that merely seeks to explain salary given qualifications and gender. In a more general context, when interest centers on a particular parameter it is useful to remember and explain that effects that bias the estimates of other parameters are not important even if they prevent a full structural explanation of the process being studied.

5. *Significance Levels and Tests of Hypotheses.* Within the context of a structural regression model, the relevant goodness-of-fit statistics are usually the standard errors of the regression coefficients or the associated t statistics, but these are much less easy for nonstatisticians to understand than are overall measures. Furthermore, unless care is taken to explain how the precision of estimation is measured, a judge or jury can be readily confused by objections raised by the relatively untrained or the unscrupulous.

In my work in the IBM antitrust case, for example, I presented regressions that attempted to relate the price of computer central-processing-unit–memory combinations to characteristics such as speed and memory size. There was a very good fit. Despite the fact that t statistics on the order of 20 were obtained for all of the regression coefficients, Alan K. McAdams, appearing as an expert for the government, testified that colinearity made it impossible reliably to separate the effects of the different independent variables and hence that little reliance could be placed on the result (see Fisher, McGowan, and Greenwood 1983, p. 156).

Whatever else that incident illustrates, it shows that numbers, especially technical magnitudes such as *t* statistics, do not speak for themselves to the lay public. In particular, familiar as they may be to the statistician, the concept of significance levels and the nature of hypothesis testing are not particularly easy for the nonstatistician to grasp, and the statistician must be prepared to explain them carefully.

It is not surprising that the lay public, including the legal profession, has difficulty with hypothesis testing and the interpretation of significance levels. Such matters are not easy to explain to beginning students of statistics, and it is easy to slip into the incorrect usage of saying that significance at the 5% level means that the probability that the null hypothesis is true is less than 5%. It is particularly important to avoid such errors in adversary proceedings because of the importance attached to standards of proof in such proceedings.

Civil cases in the United States generally use what is called a *preponderance of the evidence* standard. This means that if it is more probable than not that a given side is correct, then that side should prevail. [See Fienberg and Kadane (1983) for a discussion of Bayesian interpretations of different standards of proof.] If the significance level is incorrectly interpreted as the probability that the null hypothesis is true, then the preponderance-of-the-evidence standard can lead to the position that the correct significance level to use in testing hypotheses is 50%!

I am not making this up. A recent case involved the question of whether a particular drug has the side effect of increasing the risk of a certain condition found in the population at low levels among both drug users and nonusers. A witness called as an expert in statistics by the plaintiff (who contended that the effect was present) actually took this position, at least in advising the client. Indeed, the witness went further, contending not only that the appropriate standard was a one-tail test but also that any experiment showing a reduction of the condition among drug users compared to the population as a whole must be a mistake and should be disregarded. Needless to say, this produced a predictable anti-drug conclusion.

The problem that this example presents is not that of why the analysis was wrong but of how to convince nonspecialists that it is wrong. This is difficult enough when dealing with high-powered lawyers; it is an order of magnitude harder if one is trying to straighten out a jury.

The approach that I have found most useful in this area is that of using very simple examples. In the drug case just outlined, put aside the question of discarding studies showing a reduction of the condition involved. In that example, it is not hard to see that the use of the 50% significance level with a one-tailed test to test the hypothesis that a particular coin is weighted toward heads leads to an absurd result. The conclusion will be that the coin is weighted if it comes up heads more often than tails no matter how small the number of tosses.

The matter is seldom so easy, however. Without the one-tail feature, it is much harder to find a description of the absurdity of using the 50% significance level that is instantaneously convincing to the untrained person. The use of a two-tailed test at the 50% level to test whether a coin is fair only leads to the result that fair coins will be deemed unfair 50% of the time. To nonstatisticians this sounds like the false proposition that rejection of the null hypothesis of fairness at the 50% level means that the coin is more likely than not to be weighted.

The drug example also illustrates a deeper problem. Classical null hypothesis testing is not really suited to a situation in which retention of the null hypothesis when it is false can have very serious consequences compared to rejection of the null hypothesis when it is true. Most court cases of this type do not concern the initial decision to release the drug or even (save implicitly) the question of whether the drug should continue to be used. Suppose, however, that continued use of the drug were indeed the question at issue. If the possible undesirable side-effects were enough compared to the beneficial effects of the drug, one would not want to adopt a procedure implicitly weighted in favor of retaining the drug. Yet that is what happens if one rejects the null hypothesis of no side-effects only when the observed results would be observed less than 5% of the time if the null hypothesis were true. It is very tempting to suppose that one can cure this problem by using a less demanding significance level. Indeed, if the possible harm is great enough, one can understand the temptation to discard all studies showing a condition-reducing effect of the drug, even though some such results are to be expected if the drug has no side-effects.

Both of these temptations lead to suboptimal procedures, even when the decision is whether to permit continued drug use. Adjusting the significance level used directly adjusts the probability of stopping the use of a drug that does not in fact have harmful side-effects. The effect on the probability of continuing a drug that has such side-effects is only implicit, and it is neither easy to state what the optimum significance level is nor necessarily true that the optimal policy can be parameterized in this way.

Furthermore, the usual litigation context is somewhat different. That context is typically one in which the issue is whether a particular plaintiff should receive damages, not directly whether the drug should continue to be used (although the drug may be withdrawn if damages are heavy enough). The statistician must somehow meet the preponderance-of-the-evidence standard by estimating the probability that a particular plaintiff was injured. Within the framework of classical null hypothesis testing, this may be possible by setting up the null hypothesis that the increased risk caused by the drug is no more than some given amount. More satisfactory answers are likely to be even harder to explain.

Plainly, I have no very good solutions for such expository problems. The concepts involved in hypothesis testing and, perhaps even more so, in the construction of confidence intervals are hard to explain to those with no statistical training.

I can, however, offer a negative prescription. The real need to communicate with attorneys, judges, and juries by speaking language they can readily understand must not be met at the cost of imprecision of speech. It is very easy to slip into locutions such as that which describes a significance level as the probability that the null hypothesis is true. Such slips only lead to trouble when the listeners are called on to make the ultimate decision.

3. PROFESSIONAL AND UNPROFESSIONAL BEHAVIOR

In other professions in which men engage
(Said I to myself, said I),
The Army, the Navy, the Church, and the Stage
(Said I to myself, said I),
Professional license, if carried too far,
Your chance of promotion will certainly mar,
And I fancy the rule might apply to the Bar
(Said I to myself, said I).

W. S. Gilbert, *Iolanthe*

Although exposition occupies a central role, there are other problems that the statistician is likely to encounter. Some such problems concern the statistician's behavior in putting forward a study and some concern the statistician's role in criticizing the studies of others. Indeed, since statisticians (or at least supposed statistical work) may appear on both sides of a case, it is important to distinguish legitimate criticism from obfuscation and to consider how best to make clear to the judge or jury what the difference is between serious statistics and pseudoscience.

1. *Data Management: Protecting Oneself.* The first rule in presenting one's own study ought in any context to be standard procedure but too often is not followed. It is the rule that there must be total familiarity with the data sources and total assurance that the data used are absolutely accurate. Unless one has seen it happen, one cannot imagine the skill and ferocity with which a good cross-examining attorney can make a study appear riddled with error and its author a careless fool if there are many mistakes in the underlying data. Such mistakes are easy for a judge or jury to understand, and even if the mistakes are minor, no amount of professional skill or assumption of expertise on the part of the statistician can overcome the fatal impression of carelessness that such cross-examination can produce.

Fortunately, the very participation in an adversary proceeding that makes detailed data examination likely by the opponents also makes it likely that the statistician will have the resources to make such examination fruitless—resources that unfortunately are not always available in pure research. The necessity of guarding against mistakes will generally lead the attorneys for whom the statistician works to pay for careful data management by a staff of trained assistants. (Since, as discussed later, a primary duty of that staff is to protect the statistician from being swept along by overzealous attorneys, it is best if the staff is not provided by the attorneys themselves or the ultimate client but has an outside relationship with the statistician. Possible sources are consulting firms or academic colleagues and students.)

If properly used, such a staff can provide an invaluable resource to the statistician. One staff member should be given the responsibility for data management. That person must understand that her or his responsibility is to ensure that the backup for every statement made and for all data used is accurate and available. Consistent with the use of staff to protect the statistician's subjective honesty (discussed later), it should be made clear that good performance means finding all possible problems, even if this means the discovery of apparent difficulties that prove not to be real ones.

The data-management job involves responsibility for numerical accuracy. It is advisable to have all data transcription and key punching independently done by more than one person. Such relatively mechanical matters are vitally important; however, the data-management job is not limited to them. The data-management person should also be responsible for keeping a complete record of data sources and of decisions concerning data. (A similar record of all decisions concerning model building should also be kept, as should all computer runs when possible.) Furthermore, a large part of data management can involve the evaluation of whether data really represent what they are supposed to and of whether nonstatistical judgments involved can be supported. It is a mistake to leave data-management responsibility to relatively inexperienced or untrained research assistants.

I cannot stress too heavily the desirability and importance of good, systematically organized data management. It is embarrassing to have to explain why some runs were not kept or to remember imprecisely why some things were done. Even more important, there is nothing that makes it easier to withstand cross-examination about one's work than the correct belief that one knows far more about it than the attorney asking the questions. That belief rests on the invaluable assurance (which only good data management can provide) that the data are absolutely accurate and that the answer to any question about what was done can be ascertained readily.

Even a knowledgeable witness can be trapped by a skillful cross-examining attorney. Especially when tired, it is all too easy to lose one's temper, to ramble, and to misspeak a sentence in a way that will be quoted forever after. This is particularly likely to happen if the cross-examiner can legitimately cast doubt on the accuracy of or the basis for the witness's conclusions. Although there is no such thing as too much preparation for cross-examination, no amount of practice can substitute for a well-founded belief in the performance of the data-management staff.

2. *Data Management: Criticizing Others.* The problems that poor data management can produce may arise in another way. The statistician may be asked to criticize a study put forth by someone on the other side, and data errors are a legitimate focus for such criticism. A story is instructive here, however.

After the study of cable television already mentioned was submitted as testimony to the FCC, the other side (in this case the cable trade association) retained Herbert Arkin, a statistician, to criticize it. He did so in a written submission to the FCC, spending the major part of his paper on the possible (not the actual) presence of data error. After admitting that my qualifications were beyond question, he challenged the capability of the MIT graduate students who assisted me.

This, of course, meant war, but it was a war that Arkin—or at least his attorneys—fought in a strange way. He demanded that a subpoena be issued for the data so that they might be checked for error. This was strange, because the data had already been submitted with my original study, so all that such a demand could succeed in doing was to reveal that the person making it had not done his homework. In the event, the FCC staff, reporting on these issues, observed that there must be some mistake because the data had in fact been made part of the file, whereupon Arkin submitted a further letter demanding that a key to the data be subpoenaed, since data are not always in readily readable form. Needless to say, such a key had also already been provided.

In a recent unpublished letter to the editor of this journal, Arkin stated that his attorney clients misinformed him and that he was told that neither the data nor the key were available.

There are several lessons to be drawn from this tale. The first of these is that it is not enough to suggest that errors may exist in someone else's work. Errors are always possible, but merely to point that out is not to advance at all. Instead, one must establish either that errors do exist in fact (and are of material importance) or at least that the procedures used are likely to result in error. Armchair criticism is at best less effective than actual demonstration.

Arkin was correct to ask his clients for the data and, believing that they were unavailable, to demand that they be produced so that they could be examined for error. That line of attack, however, could only be effective if followed up by actual examination. There was little point to a lengthy and general explanation of the possibility of error and little also to general questioning of the qualifications of the MIT graduate students (who were, in fact, quite highly skilled). Without actual demonstration of the presence of error, such comments could not carry much weight. Especially when the FCC staff's observation (after several weeks) that the data had always been available was not followed by actual examination but by a demand for the also available key, the indelible impression was left that armchair criticism was all that was intended.

The second, related, lesson is more complex. Somebody obviously slipped badly in permitting Arkin to submit the demand for the key after the FCC staff had pointed out that the earlier demand had been unnecessary from the outset. Given that Arkin knew that he had been (deliberately or inadvertently) misled by his attorney clients about the availability of the data, he allowed himself to be used too easily a second time. One of the important things that a testifying expert must bear in mind is that he must be deeply enough involved to be certain that slips like this do not occur. Attorneys are often the easiest source of information and reliance on them is sometimes unavoidable, but the expert witness needs to act independently whenever possible. Even though the attorneys generally have a strong interest in protecting the witness and his or her reputation, that is not their primary goal. Furthermore, when, as here, it becomes evident that the attorneys cannot be relied on, the witness must act, taking that unreliability into account.

Arkin appears to have been a victim of (at least) bungling on the part of his attorneys. As a result, much of his criticism came across as mere obfuscation—a form of attack that is tempting when dealing with an untrained audience but is unlikely to pay off against a really well-prepared opponent.

3. *Illegitimate Criticism Generally.* Naturally, these lessons carry over to areas of possible criticism of others beyond that of data management. Precisely because statistics and econometrics are badly understood by judges and juries, statisticians and econometricians have a special obligation not to "hoodwink a judge who is not overwise" (in the words of Gilbert and Sullivan's Lord Chancellor) by throwing up objections that, even if valid, would not materially affect the results of the study being criticized. There is an obligation to demonstrate that the objection is not merely one that could be made to *any* econometric study and to show that the conclusions to be drawn are likely actually to be affected by the error being discussed.

For example, it is not helpful to object that error distributions may not be normal unless there is some reason to think that they are very far from normal; to do so is to play on the listeners' ignorance of the central limit theorem so as to make the assumption of normality seem very special. Similarly, as in the example from the IBM case referred to before, it is not a good idea to damn a study for colinearity without thinking about whether that colinearity really mattered. To do so is at best to hope that the statistician who did the study is not very good at exposition. Alas, the statistician presenting a study had better be prepared for criticisms like these.

Of course, I do not contend that serious criticism should be withheld; far from it. The statistician testifying in an adversary proceeding should expect to withstand serious cross-examination. The attorney doing that cross-examination is entitled to the best professional help from other statisticians. Only in that way can an adversary system hope to arrive at an informed judgment. What the statistician assisting in cross-examination should avoid is deliberate obfuscation that takes unfair advantage of the fact that the judicial decision maker is untrained. Such obfuscation is unprofessional. Furthermore, if the other side is able, deliberate obfuscation will open an opportunity for a telling rebuttal.

Part of the reason that such tactics can pay off, of course, is the relatively untrained nature of the judge or the jury. This raises the possibility that the court might be aided by a special master or clerk trained in statistics. [A similar procedure was tried long ago by Judge Charles Wyszanski in the *United Shoe Machinery* case (see *United States v. United Shoe Machinery Corporation* 1953) when he employed Carl Kaysen, the economist, as his clerk.] In principle, this seems a good idea; in practice, courts will have a hard time identifying good statisticians (or economists) from bad ones. Furthermore, the better statisticians will be less likely than the poor ones to be willing to put in the long periods of time on relatively short notice that may be required to serve in such a capacity. Since the court is likely to rely heavily on its own expert, these are very real difficulties, although some of the problems might be alleviated if the court's expert could be cross-examined by the parties. In any event, such experiments are relatively rare.

4. *Legitimate Criticism: The Corrugated Case.* The statistician looking for something to criticize in adversary proceedings will probably not have to look far. This is because, as already remarked, the advent of the computer and the success of econometrics have opened the door to poor statistical and econometric testimony masquerading as serious science. I have already related some incidents suggesting that this is true; now I turn to a recent one of considerable importance. (For another discussion, see Finkelstein and Levenbach 1983.)

A recent tripartite series of antitrust cases concerned allegations of price-fixing in the sale of corrugated containers (see *In Re Corrugated Container Antitrust Litigation* 1979, 1980, 1985). The first of these cases was a criminal trial in which all defendants who contested the charge were acquitted. The second, tried in 1980, was a private, civil, class-action suit that only one defendant, the Mead Corporation, contested. The third, tried in 1982 and 1983, was brought by several large corporations who had opted not to participate in the class-action trial and was defended by several manufacturers of corrugated containers.

In the class-action suit, the plaintiffs put forward an econometric model developed by John Beyer, an economist with admittedly no econometric training. That model purported to show that the alleged price-fixing conspiracy had resulted in a general overcharge to purchasers of 7%. The jury found for the plaintiffs, and the result was what appears to be the largest settlement collected on behalf of a class in the history of the Sherman Antitrust Act. Beyer's model and testimony justifiably gained notoriety among members of the antitrust bar.

By the time of the "opt-out" trial, Beyer had "improved" his model, which now showed an overcharge of 26%. This time, however, the defendants, now suitably alerted, decided that they had better retain an econometrician themselves, and they retained me. At the end of the trial, the jury found that although there was evidence of conspiracy, there was no evidence of damages to the plaintiffs—rejecting the use of Beyer's econometric model.

Beyer's model was not well grounded in economic analysis. It was generally agreed that the alleged conspiracy had ended as of sometime in the middle 1970s. (Beyer used three different months in 1975 as an ending date and opted for the one that happened to produce the largest damages.) Beyer did not build a model of the post-conspiracy period and project backwards to see what prices would have been absent the conspiracy (a procedure which, as discussed below, showed no damages). Instead, he purported to model the conspiracy period itself, project forward to the late 1970s to estimate what prices would have been had the conspiracy continued, and then assume that the ratio of conspiratorial to competitive prices would have been the same had there been competition in the earlier period.

Aside from the fact that this procedure provides at best a roundabout way of doing such a study, it suffers from the problem that it requires a model of the conspiratorial period. Models of oligopoly behavior are not notoriously successful, and the reader may wonder what Beyer did to overcome this.

He did nothing at all. Instead, he discussed matters in terms of supply and demand functions, disregarding the fact that the supply function does not exist under oligopoly. He estimated what he referred to as a "reduced form" in which price was regressed on a number of variables that seemed likely to affect either supply or demand, including a cost index, measures of output in container-using industries and of capacity utilization and inventory adjustment in containers, and a dummy variable reflecting price controls. The model was monthly, and lagged price was also included. Finally, despite the overwhelming theoretical basis for working in deflated prices, Beyer used an index of money prices as the dependent variable, taking account of inflation by including a general price index among the regressors. This was, of course, guaranteed to produce a high R^2 over any period with high inflation.

Despite (or perhaps because of) its lack of a sound structural foundation, the model produced results that were superficially convincing. When the model was dynamically simulated, starting in the 1960s, the predicted prices tracked the actual prices quite closely until about 1975, the date that ended both the sample period and the alleged conspiracy. After that, the simulated prices, which supposedly embodied the effects of the alleged conspiracy, rose above the actual prices. This is graphically depicted in Figure 1. To the untrained eye, this picture seems a quite convincing demonstration that something important had happened in early 1975.

Something important had indeed happened. Whether or not the alleged conspiracy ended in January 1975, the sample certainly did. Moreover, the period of the early 1970s was marked by the imposition and then the lifting of price controls, followed by the energy crisis and a recession. As a result, actual container prices rose sharply in the early 1970s and then flattened out. It seemed unlikely that these effects could be handled adequately by the use of an additive dummy variable for the price-control period. If so, then the results were likely to be driven by the events at the end of the sample period that had nothing to do with the alleged conspiracy.

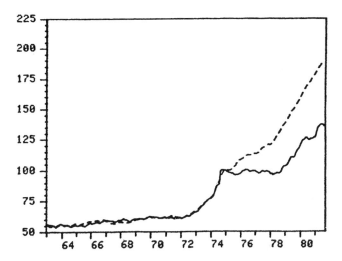

FIGURE I. FBA Corrugated Container Price Index, Actual versus Simulated (December 1974 = 100): Beyer's Model, Estimation Period from 1963:1 to 1975:1 (_ _ _, actual FBA price index; ___, simulated FBA price index).

This suspicion proved to be well founded. Standard Chow (F) tests firmly rejected the null hypothesis that the regression equation was the same for the early 1970s as for the 1960s. More dramatically, reestimation of the same model for different sample periods showed that all of the results crucially depended on the inclusion of the unusual years of the early 1970s in the sample. If those observations are removed and the same model estimated only on data before price controls, a very close fit is again obtained for the sample period. This time, however, the dynamic simulation no longer yields high prices for the late 1970s but produces projections of "conspiratorial" prices *lower* than the prices actually charged after the supposed ending date of the alleged conspiracy. These results are depicted in Figure 2.

Indeed, the model has the characteristic that however one chooses the sample period, the dynamic simulation fits the sample data quite well and then continues along whatever trend the last few sample observations happen to be following. This is true even if one chooses a late sample period and uses dynamic simulation to project backward, the trend in question here being the backward trend of the first few sample observations.

This phenomenon is illustrated in Figure 3, where the model is estimated over the supposed post-conspiracy period and used to backcast what competitive prices would have been up to 1975. Figure 4 illustrates the same phenomenon when a later post-conspiracy period is chosen. (The first few simulated observations in Figures 3 and 4 should be ignored; the simulation is given the actual value of the price index in January 1963 as a starting point.) This is an entirely appropriate test of Beyer's model. As already observed, it is not a model with the structural features of oligopoly and is, if anything, more suited to the straightforward procedure of estimation on a competitive sample period and simulation for the supposed conspiratorial period than to the roundabout procedure for which Beyer used it.

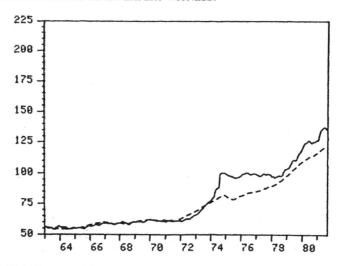

FIGURE 2. FBA Corrugated Container Price Index, Actual versus Simulated (December 1974 = 100): Beyer's Model, Estimation Period from 1963:1 to 1971:7 (___, actual FBA price index; ___, simulated FBA price index).

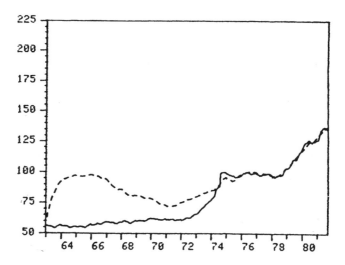

FIGURE 3. FBA Corrugated Container Price Index, Actual versus Simulated (December 1974 = 100): Beyer's Model, Estimation Period from 1975:2 to 1981:8 (___, actual FBA price index; ___, simulated FBA price index).

Note that to make this criticism effective, it was not enough to point out that the supposed "model" was not a structural one so that it was particularly likely that the results merely reflected the particular sample period used; it was necessary to show that this really did matter. A similar statement holds true for a different criticism of the model, although here the materiality of the objection had to be analytically inferred rather than directly empirically demonstrated.

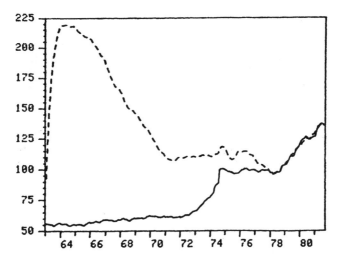

FIGURE 4. FBA Corrugated Container Price Index, Actual versus Simulated (December 1974 = 100): Beyer's Model, Estimation Period from 1979:1 to 1981:8 (_ _ _, actual FBA price index; ____, simulated FBA price index).

As already mentioned, Beyer's "improvements" to his model between the two trials at which he testified had also "improved" the damage estimates that his model produced, raising them from 7% to 26% of the revenues of the defendants. It is never wise in these matters to succumb to the temptation of presenting the result that happens to be best for one's client (a temptation that attorneys are hired to yield to but statisticians should resist); this is particularly true if one has previously testified to a different effect. In any event, the 26% figure was so outrageously large that it was apparent that had prices been lower by the amount of the supposed overcharge, all of the defendants would have been bankrupt long ago.

The plaintiffs' response to the fact that the estimated damage figure was preposterously large took two related forms. First, Beyer pointed out that conspiracy shields firms from the rigors of competition; in particular, it saves them from having to be efficient. He claimed that the costs of the defendants were inflated by the lack of competition, so no inference could be drawn from the fact that prices could not have been 26% lower given their actual, inefficient costs.

Second, a very important component of the costs of producing corrugated containers was the cost of linerboard (a major input). Since many corrugated container firms are integrated backwards into linerboard, the costs of linerboard reflect the internal transfer prices of that input within the integrated firms. According to Beyer, that price was artificially inflated by the alleged conspiracy with profits taken in linerboard rather than in container production.

The problem with advancing arguments such as these, which are possibilities rather than actualities, is that they have implications that can be seen by trained opponents. In the first place, as pointed out by Peter Max, another defense witness, the bankruptcy that 26% lower prices would have brought about would have applied to the integrated firms, taking into account their entire operations. More pertinent for the present discussion, Beyer had used costs as an *exogenous variable* in his own regression. In so doing, he had not only accepted the supposedly fictitious linerboard prices as being real, he had also implicitly assumed that the prices produced by the alleged conspiracy did not themselves affect costs—exactly what he denied in his inefficiency argument. Although general objections as to the possibility of simultaneous equations bias can often be made, here was a case in which that bias was the more serious the more one believed the arguments given by the model builder for accepting his results.

4. CONCLUSION: PROFESSIONAL STANDARDS AND OBJECTIVITY

I have lingered over the story of the *Corrugated* case, partly for its intrinsic interest and partly because it and the other stories already related exemplify a number of the problems that the statistician involved in adversary proceedings is likely to encounter and, perhaps, the principles that he or she ought to follow. (For another discussion of some of the matters in this section, see Fisher et al. 1983, pp. 350–352.)

The first of these has to do with the temptation to pseudo-science. It is tempting to stun the audience with proclaimed expertise together with computer output and professionally drawn visual displays. By pandering to the belief that econometrics is in fact black magic and the statistician the shaman of the computer, it may be possible to overawe the untrained.

Such a temptation should be firmly resisted. On a practical level, such a strategy is only safe to pursue when one is sure that the other side is untrained. An unprofessional job that would not hold up in seminars is unlikely to hold up in court, and it is all too easy to underestimate the intellectual powers of the cross-examining attorney.

Moreover, it is unethical to behave in such a way. As the *Corrugated* story illustrates, the world has quite enough poor statistical and econometric testimony without serious statisticians and econometricians adding to the stock. The court, as well as the statistician's client, is entitled to the statistician's best professional efforts. Anything less is wrong. Statisticians and econometricians have a potentially large contribution to make in litigation; that contribution can only be tainted by the belief that one can get away with less than the best.

Related to this is another temptation, that of telling one's attorney what he or she wants to hear. The attorney is retained to make the best possible argument on behalf of the client. Wise attorneys realize that this can only be done if they also know all of the things that can be said on the other side. Expert witnesses and attorneys who become greedy and listen only to the good news may not find that the news remains good at the end of the case.

More important, statisticians have reputations to protect. For those who are prominent in the profession and are therefore among those whom I am urging to participate in such affairs, their reputation extends beyond the confines of the courtroom or the consulting practice. It is at best shortsighted to overlook this in giving testimony. Furthermore, to do so is to damage the profession itself as well as to mislead the client and the court.

It is not, however, always easy to avoid becoming a "hired gun" (and still harder to avoid being described as one). The danger is sometimes a subtle one, stemming from a growing involvement in the case and friendship with the attorneys. For the serious professional, concerned about preserving his or her standards, the problem is not that one is always being asked to step across a well-defined line by unscrupulous lawyers. Rather, it is that one becomes caught up in the adversary proceeding itself and acquires the desire to win. In so doing, one can lose sight of where the line is and readily acquiesce in language shaded to produce a favorable but not truly defensible result. Particularly because lawyers play by rules that go beyond those of academic fair play, it becomes insidiously easy to see only the apparent unfairness of the other side while overlooking that of one's own. Continuing to regard oneself as objective, one can slip little by little from true objectivity.

I have found it useful to do two things in this regard. First, one should insist at the outset that one must be free to make an independent investigation. If the results of that investigation are unfavorable, then one need not testify, but the results are not to be tailored to the order of the party buying (or bringing) the suit. I have had the experience of having results turn out adversely and it being clear that I could not testify; I do not believe that either the attorneys or I or our relationship suffered for it.

Second, as in data management, it is important to have an independent staff, independent of the lawyers and (to the extent possible) of oneself. Such a staff can be drawn from a consulting firm or from one's academic colleagues and students. Members of the staff (who must themselves have the appropriate technical training) should be charged with the job of protecting the expert from slipping into advocacy. In so doing, they will find it natural to make the best case for the other side; this serves the additional purpose of refining the expert's analysis and preparing him or her for cross-examination. If the staff and the expert work together on different cases (or look forward to doing so), then the staff will particularly understand that they both have a stake in the expert's continuing integrity and reputation.

As the real necessity for the use of such procedures suggests, participation in adversary procedures can be a psychologically wearing experience as well as an exciting one. The question then arises as to why serious statisticians should participate at all. I have already tried to suggest some reasons for participation, and the following story—although only indirectly related to adversary proceedings—may serve to emphasize some of them.

Some years ago, I received a telephone call from a staff member of a prospective panel of the National Academy of Sciences inquiring whether I would be interested in serving on the panel. The panel in question was to review and consider the statistical studies on the question of whether punishment is a significant deterrent to crime (see Blumstein, Cohen, and Nagin 1978). The assistance of an econometrician seemed desirable, and my name had been suggested. I replied that I was very busy and had no particular interest in the subject. My caller asked whether I did not consider it my social duty to participate (a point also urged by my wife) in view of the fact that, among other things, such studies were being used in arguments before the Supreme Court in support of the restoration of capital punishment. I remained unmoved. A day or so later, the staff member called again. "There is something I forgot to tell you," the caller said. "We have reason to believe that the people producing such studies do not understand the identification problem and are misusing least squares." "Why didn't you say so before?" I asked. *"That* is my social duty."

ACKNOWLEDGMENTS

This article is a revised version of a paper presented at the Economic and Social Research Council Econometric Study Group in honor of J. D. Sargan in Oxford, England, March 1984. The author writes, "I have previously attempted to explain to lawyers how to deal with econometricians (Fisher 1980). I am grateful to Ellen P. Fisher for suggesting that it might be useful to approach the same subject from the opposite point of view. The work on the *Corrugated* case owes much to Stephen H. Kalos. I am also indebted to Joseph B. Kadane and two referees for helpful comments. Harry Roberts, in particular, went far beyond the usual call of duty and produced an unusually detailed and helpful report. It is in the nature of a paper such as this, however, that readers and author may disagree, and I am particularly conscious that errors and opinions are my responsibility."

REFERENCES

Blumstein, A., Cohen, J., and Nagin, D. (eds.) (1978), *Deterrence and Incapacitation: Estimating the Effects of Criminal Sanctions on Crime Rates,* Washington, DC: National Academy of Sciences.

Downton, F. (1982), "Legal Probability and Statistics," *Journal of the Royal Statistical Society,* Ser. A, 145, 395–402.

Fienberg, S. E., and Kadane, J. B. (1983), "The Presentation of Bayesian Statistical Analyses in Legal Proceedings," *The Statistician,* 32, 88–98.

Fienberg, S. E., and Straf, M. L. (1982), "Statistical Assessments as Evidence," *Journal of the Royal Statistical Society,* Ser. A, 145, 410–421.

Finkelstein, M. O., and Levenbach, H. (1983), "Regression Estimates of Damages in Price-Fixing Cases," *Law and Contemporary Problems,* 46, 145–169.

Fisher, F. M. (1980), "Multiple Regression in Legal Proceedings," *Columbia Law Review,* 80, 702–736.

Fisher, F. M., and Ferrall, V., Jr. (in association with D. Belsley and B. Mitchell) (1966), "Community Antenna Television Systems and Local Television Station Audience," *Quarterly Journal of Economics,* 80, 227–251.

Fisher, F. M., McGowan, J., and Greenwood, J. (1983), *Folded, Spindled, and Mutilated: Economic Analysis and U.S. v. IBM,* Cambridge, MA: MIT Press.

In Re Corrugated Container Antitrust Litigation (1979), 1980–1 Trade Cas. (CCH), para. 66,163 (S.D. Tex. 1979); aff'd in part and remanded, 643 F.2d 195 (5th Cir. 1981); 84 F.R.D. 40 (S.D. Tex.); aff'd mem., 606 F.2d 319 (5th Cir.); cert. dismissed, 449 U.S. 915 (1980); 620 F.2d 1086 (5th Cir., 1980); 1983 2 Trade Cas. (CCH) para. 65,628 (S.D. Tex.).

—— (1980), 620 F.2d 1086 (5th Cir., 1980); 1983–2 Trade Cas. (CCH) para. 65,628 (S.D. Tex. 1983).

—— (1985), MDL 310; Dockets 83–2281, 83–2486 (5th Circuit, April 4).

Levi, Edward (1948), *An Introduction to Legal Reasoning,* Chicago: Phoenix Books (reprinted by University of Chicago Press in 1962).

Newell, D. (1982), "The Role of the Statistician as an Expert Witness," *Journal of the Royal Statistical Society,* Ser. A, 145, 403–409.

Royal Statistical Society (1982), "Discussion Meeting on 'The Role of the Statistician as an Expert Witness,'" *Journal of the Royal Statistical Society,* Ser. A, 145, 395–438.

Rubinfeld, D. L., and Steiner, P. O. (1983), "Quantitative Methods in Antitrust Litigation," *Law and Contemporary Problems,* 46, 69–141.

United States v. United Shoe Machinery Corporation (1953), 110 F. Supp. 295.

B. A VERY BRIEF INTRODUCTION TO U.S. LAW, AND TO THE ROLE OF EXPERT WITNESSES

This is a short introduction to a very complex subject. It does not substitute for law school, for reading books, or even for internet searches. Instead its idea is to give enough of a background that you can understand the context of the papers to come.

The first thing to understand is that the US system is adversarial. Each side in a legal case is entitled to be represented by an attorney whose job is to make the best case for the side the attorney represents. It is not necessary that the attorney personally endorse the side he or she represents.

Cases are either criminal or civil. A criminal case, originally, was an alleged offense against the king. After the American Revolution, the role of representing the king was transformed into representing the government. Only an authorized government prosecutor can bring a criminal case against someone. Typical criminal matters are murder, rape, burglary, assault, robbery, etc. Only criminal cases can result in imprisonment or execution. A fine, payable to the government, can be imposed on one found guilty in a criminal case.

By contrast a civil case is an issue between two private parties. Examples of this would be breach of contract, landlord-tenant cases, patent and trademark cases, employment cases, divorce and child custody, etc. Almost always the issue in civil cases is money, although other outcomes (*i.e.* custody) can result. Any person, or corporation, can sue any other person or corporation. In addition, people and corporations can sue the government, but only if special laws have been passed permitting such suits. The government can also bring civil cases in certain circumstances. In a civil case, the winning party obtains a judgment against the loser, and the loser is obligated to pay the winner.

The content of the law, and what kinds of civil suits are permitted, is the business of legislatures. What the courts are about is the interpretation of the law to fit the peculiar circumstances of an individual case.

A major complication in US law is the divided nature of sovereignty and hence jurisdiction. After the conclusion of the American Revolution in 1783, the original 13 states were bound together in a loose (and unworkable) confederation. The Constitution of 1787 was a compromise between some who wanted a strong central (federal) government, and those who wanted to limit the powers of the federal government and reserve more power to states and localities. Under the U.S. system, the federal government administers federal laws passed by congressional vote. But each state and city can also pass laws to serve the particular interests of that state's residents (such as, no private store is permitted to sell liquor in Pennsylvania, and no chickens are allowed to be raised or kept within the city limits of Philadelphia). Thus the system of "American law" encompasses various federal, state and local laws, regulations and ordinances.

Nonetheless, the basic idea of the federal constitution is that certain enumerated functions were delegated to the federal government, with all others being held by the states (by the 10th Amendment). Additionally, there are limits on what the federal government can do, notably as governed by the Bill of Rights, in the first ten amendments to the Constitution. With those limitations, Article 6 of the Constitution provides that the Constitution and laws of the federal government are the supreme law of the land, binding on the judges of every state, notwithstanding anything contrary in the Constitution and laws of the states. The principal (early) sources of commentary on what various clauses in the Constitution mean are the Federalist Papers, a series of newspaper articles written by Alexander Hamilton, John Jay and James Madison in the campaign to win approval for the Constitution, and Madison's notes of the debates at the Constitutional Convention.

Thus lawsuits can be divided into criminal and civil cases, and into federal and state cases, leading to four fundamentally different kinds of cases, with different rules that apply to each.

Each case of all four types is heard first by a court of original jurisdiction. For federal cases, this is the Federal District Court. For state cases, the name of the court having original jurisdiction varies (in Pennsylvania, it is the Court of Common Pleas). There are, in addition, even lower state courts having jurisdiction over minor offenses and cases, such as Traffic Court, Small Claims Court, etc.

The Constitution provides for the right to a trial by jury. Thus twelve (usually, although as few as six are sometimes used) citizens are chosen to find the facts of the case, in the context of the judge's statement of the law. In all federal trials and in state court criminal trials, the jury must be unanimous to convict, or to find for the plaintiff. In state civil trials, state law determines the extent of the majority required. A defendant in a criminal case, and both parties in a civil case, may waive the right to a jury trial, in which case the judge decides both the law and the facts. In a criminal case, if the defendant is convicted the judge decides on the penalty, within bounds set by the law. (There is an exception to this for capital cases, *i.e.* those that might result in execution. The jury has to decide on the punishment in such a case.)

Most cases, both criminal and civil, do not come to trial. In civil cases, often the parties agree to a settlement. In criminal cases, often the defendant agrees to a plea bargain, pleading guilty to a lesser crime in return for the prosecution dropping other charges.

The rules of evidence are very different in criminal and civil cases. In a criminal case, the prosecution is required to make the defense aware of any exculpating evidence it may have. The defense must give notice of an alibi defense, witnesses' names and expert reports. Other than that, the evidence to be presented is a surprise to both sides. By contrast, the evidence in a civil case must be made available to the other side if it is to be presented in court. Each side has the right to depose the other side's witnesses, asking them for answers under oath, to whatever questions they want to pose.

Decisions made by Federal District Courts can be appealed to a Circuit Court of Appeals, and their decisions can be appealed to the U.S. Supreme Court. These appeals cannot raise points that the side appealing neglected to bring up in the original trial. Consequently the appeals courts deal with issues of law, not with issues of fact.

Different states have different names for their appeals courts, but they work in a manner similar to the federal system. In Pennsylvania, for example, decisions of the Court of Common Pleas can be appealed to Superior Court, whose decision in turn can be appealed to the Pennsylvania Supreme Court. Parallel to Superior Court, the Pennsylvania Commonwealth Court hears appeals in which the state is a party. Again, their decisions can be appealed to the Pennsylvania Supreme Court.

The courts at all levels are chronically overworked. This leads to the pressure to settle cases, to detailed rules of procedure to speed the work of the courts, to decisions rejecting an appeal without argument and to the general sense that the courts do the best they can, make decisions, and move on.

WITNESSES IN COURT

There are two kinds of witnesses in a court proceeding, fact witnesses and expert witnesses. A fact witness may testify only to what he or she personally saw or heard, like "The blue car ran the red light and hit the green car." An expert witness, who must be specially qualified by education and experience, can give an opinion, such as "Based on the skid marks and the condition of the cars, it is my opinion that the blue car was going 45 miles per hour at the time of the accident." A fact witness can be required to testify, while an expert is hired and paid for the time spent and cannot be forced to be involved in the case.

C. QUALIFICATIONS AND RESPONSIBILITIES OF THE EXPERT WITNESS

The materials in this section are vital for an understanding of what qualifies a person to be designated as an expert by the court, and what the responsibilities of such an expert are.

The introductory article by Finkelstein and Levin is an excellent survey of the qualification issues. The standard for both state and federal courts was set by the Frye decision in 1923. The heart of that standard is general acceptance of the theory of the expert in the relevant community. In 1975, the Congress adopted new Federal Rules of Evidence. Rule 702, on testimony by experts, reads as follows:

> If scientific, technical, or other specialized knowledge will assist the trier of fact to understand the evidence or to determine a fact in issue, a witness qualified as an expert by knowledge, skill, experience, training, or education, may testify thereto in the form of an opinion or otherwise, if (1) the testimony is based upon sufficient facts or data, (2) the testimony is the product of reliable principles and methods, and (3) the witness has applied the principles and methods reliably to the facts of the case.

This led to the Daubert decision in 1993, with a whole new set of criteria, centered around whether the expert's opinion is deemed "reliable" by the court. While the Daubert decision was limited to scientific expertise, the decision in Kumho Tire (1999) extended the Daubert decision to all experts.

1

EPIDEMIOLOGIC EVIDENCE IN THE SILICONE BREAST IMPLANT CASES

Michael O. Finkelstein and Bruce Levin

Last October, an advisory panel of the Food and Drug Administration, after a two-day hearing, approved an application by Inamed Corporation to remarket silicone breast implants. The panel, by a nine-to-six vote, found that Inamed had provided reasonable assurances of safety for such implants. The dissenters had not concluded that silicone implants caused disease, but believed that Inamed's two-year study was not sufficient to demonstrate long-term safety. In January, the FDA, in a surprise move, in effect agreed with the dissenters and deferred a decision on the implants. The FDA found that while information developed over the past ten years had increased its assurance of implant safety, it needed more information on the rates and long-term effects of leaking or rupture for the product to pass the threshold for approval. Inamed responded by vowing to move forward diligently with new studies, in cooperation with the agency. If the FDA ultimately approves the implants, that would be a singular turn of events because, as everyone knows, the last time the implants were widely marketed the result was a legal disaster: the manufacturers were sued by tens of thousands of women who contended that the devices caused connective tissue disease and many other problems. This tsunami of litigation drove one manufacturer, Dow Corning, into bankruptcy. Ultimately, the pressure of litigation forced the manufacturers into global, multibillion dollar settlements—at the time, the largest settlements in American tort law.

What has changed since then is a growing body of professional opinion, based very significantly on meta-analyses of epidemiologic work, that there was no association between breast implants and connective tissue disease. Most of the underlying studies were already available when the breast implant cases were before the courts, but a strong professional consensus had not yet been developed on the subject. In addition, the courts were divided on the relevance of and necessity for epidemiologic evidence. In the federal courts it was negative epidemiology that led to dismissal of claims, while in important state cases it was opinion testimony based on case reports, in vitro experiments, animal studies, or differential diagnoses that led to recoveries for the plaintiffs.

The attitude of the federal courts was a direct outgrowth of a U.S. Supreme Court opinion in a 1993 landmark case, *Daubert v. Merrill Dow Pharmaceuticals, Inc.* In *Daubert,* the Court held that under Rule 702 of the Federal Rules of Evidence the trial judge was required to be a "gatekeeper" for scientific evidence. Before allowing such evidence to be heard by a jury, the judge had to determine that the proposed testimony was not only relevant but also "reliable." The personal opinion of a qualified expert was no longer enough; the expert's pronouncements had to be based on scientific knowledge. The Court coupled this with a non-exclusive set of factors that characterized such knowledge: (1) the theory had been tested: (2) the theory had been peer-reviewed and published; (3) applications of the theory had a known rate of error; (4) the theory had been generally accepted in the relevant scientific community; and (5) the theory was based on facts or data of a type reasonably relied on by experts in the field. *Daubert* became one of the most frequently cited Supreme Court opinions, with over 4,700 citations as of this writing. Federal trial judges responded to this new requirement by holding what became known as Daubert hearings, in which experts from both sides testified and the court then decided whether the proposed testimony was sufficiently relevant and reliable to be admitted.

One of the first Daubert hearings in the breast implant cases was held in 1996 by federal district court judge Robert Jones in Oregon. There were about 70 cases before Judge Jones, with the lead case entitled *Hall v. Baxter Healthcare Corp.* Defendants challenged plaintiffs' proposed expert testimony as unreliable and Judge Jones appointed four experts—an epidemiologist, an immunologist, a rheumatologist, and a biochemist—to assist him in resolving the challenges.

Plaintiffs asserted that silicone had migrated out of the implants and degraded in their bodies causing "atypical connective tissue disease" (ACTD) or "atypical autoimmune disease." The condition was called "atypical" because persons with atypical disease did not have classical connective tissue diseases, such as lupus or rheumatoid arthritis, but a shifting constellation of signs and symptoms, a number of which overlapped with those of chronic fatigue syndrome and fibromyalgia.

At the Daubert hearing before Judge Jones, the parties referenced 16 epidemiological studies of the association between breast implants and connective tissue diseases. In 15 of those studies, the authors found no statistically significant association between breast implants and disease. The remaining study, by Hennekens et al., was by far the largest: it involved almost 400,000 female health professionals and included almost 11,000 women with breast implants. The relative risk was 1.24 and was statistically significant. However, in this study, unlike the others, the women self reported their symptoms and the study was carried out in the midst of intense media attention to the issue. For these reasons, the authors of the study believed that the association was due to bias, and concluded that the results did not show a real elevated risk.

The *Hall* plaintiffs challenged the probative value of the epidemiologic studies because those were focused on classical connective tissue diseases and plaintiffs were claiming an atypical disease. However, although the studies focused on classical diseases, most also included a catch-all category called by various names, e.g., "other autoimmune or rheumatic conditions." These included signs and symptoms of autoimmune or rheumatic disease, such as joint pain, swelling, etc., that did not fulfill the diagnostic criteria for classic disease. The category for "other autoimmune or rheumatic conditions" seemed likely to have covered the constellation of symptoms that plaintiffs claimed for atypical disease. But this was not completely clear because the constellation of symptoms constituting atypical disease was shifting and subjective.

In resolving defendants' motions to exclude plaintiffs' expert's opinions on causation, the court's basic ruling was that because ACTD was undefined, it was "at best an untested hypothesis," and therefore there was "no scientific basis for any testimony as to its causes and presence in the plaintiffs." As for the epidemiology, the court found persuasive that 15 out of 16 studies reported no association between breast implants and classical diseases or disease signs or symptoms, and in the Hennekens study, the relative risk was only 1.24, far below the 2.0 that other courts had held was the minimum relative risk required for a finding of causation. The court excluded any testimony of a general causal link between silicone breast implants and ACTD or other systemic illness or condition, but stayed its decision pending the issuance of a report by the National Science Panel, another group of neutral experts, appointed by another federal district court (which is described at the end of this column).

The rulings by the district court in *Hall* were generally followed in subsequent federal cases. The most important of these was probably *In re Breast Implant Litigation,* an action involving numerous cases brought against the manufacturers and consolidated in the federal district court in Colorado. After a *Daubert* briefing, the court granted defendants' motions to exclude the expert testimony of plaintiffs' witnesses on causation. The court gave critical weight to the negative epidemiology in a series of rulings disposing of the claims.

Specifically, the court held: "epidemiological studies are necessary to determine the cause and effect between breast implants and allegedly associated diseases": the more-probable-than-not burden means that plaintiffs must establish at least a doubling of the risk; the epidemiological evidence demonstrates that there was an absence of doubling of risk; theories of toxic causation "unconfirmed by epidemiologic proof cannot form the basis for causation in a court of law": treating physicians may not use differential diagnosis to conclude that a causal relation exists since they had not established that silicone breast implants were on the list of possible causes from which the others are eliminated; plaintiffs' expert's clinical experience was inadequate for a general causation opinion since "case reports and case studies universally are regarded as an insufficient scientific basis for a conclusion regarding causation because case reports lack controls": a temporal relationship by itself provided no evidence of general or specific causation; evidence by a treating physician of specific causation was inadmissible in the absence of evidence of general causation of supporting epidemiology. Finally, there were statements of various medical societies that there was no connection.

One of plaintiffs' proffered witnesses, a Dr. Kassan, was excluded because, as the court found, he offered no tested or testable theory to explain how, from his limited information, he was able to eliminate all other potential causes of plaintiffs' conditions, nor did he explain how he alone could state as a fact that the plaintiffs' breast implants had caused their problems. Dr. Kassan had not published or peer-reviewed his theories, which had not been objectively tested, had no known rate of error, and were not generally accepted by the medical community. In the absence of proof of general causation, testimony of specific causation no longer "fit" the case; atypical disease had not been defined and so its etiology was untestable; and a suggestion or possibility of a causal relation was insufficient.

Having excluded the evidence of causation, the court dismissed the claims relating to ACTD.

The U.S. Supreme Court's Daubert opinion is not binding on the state courts and the tests of admissibility and sufficiency of scientific evidence of causation may vary widely in those courts, depending on the jurisdiction. Generally the standards have been looser, allowing more speculative testimony to be heard by juries. We note three important cases in which the highest courts in three states approved testimony of causation that would not have passed muster in federal court under *Daubert*. Indeed, some of the same expert witnesses who were excluded from testifying in federal court were allowed to testify in two of those cases.

The first case is *Vassallo v. Baxter Healthcare*, decided by the Massachusetts Supreme Court in 1998. A woman who had received a silicone breast implant sued, claiming that it had caused her ACTD.

One of the witnesses who was precluded from testifying in the Hall case—a Dr. Gershwin—was proffered by the plaintiff in *Vassallo*. An earlier Massachusetts case had established Daubert-like standards for the admissibility of scientific evidence. As in *Hall*, defendants attempted to exclude plaintiff's experts and the record of the Daubert hearing in the Oregon court was put before the court. The arguments that would carry the day for the defendants in *Hall* were repeated. However, the trial judge rejected the challenge and permitted Dr. Gershwin and another plaintiff's expert (who was also excluded from testifying in *Hall*) to give their opinions that there was causation. Plaintiff received a verdict and defendants appealed.

Before the Massachusetts Supreme Judicial Court, defendants objected that Dr. Gershwin's testimony that silicone breast implants caused disease in women, based as it was on case reports and the like, was not admissible without supporting epidemiology. The court unanimously rejected that argument. Referring to the Hall case, it commented that, "although there was conflicting testimony at the Oregon hearing as to the necessity of epidemiological data to establish causation of a disease, the judge appears to have accepted the testimony of an expert epidemiologist that, in the absence of epidemiology, it is 'sound science...to rely on case reports, clinical studies, in vivo tests, [and] animal tests' when epidemiology was lacking or inconclusive." How the court reached that conclusion is mysterious, since Judge Jones had precluded Dr. Gershwin from testifying on that basis. Why was there no epidemiology? In a footnote, the court acknowledged that there had been studies, but none of them had "addressed the disease at issue in this trial, namely, atypical connective tissue disease." Again, the ruling is remarkable in giving ACTD the status of a defined disease when the Hall court had found that at best it was no more than an untested hypothesis, and in ignoring the fact that a number of the epidemiologic studies included signs and symptoms of rheumatic or autoimmune condition as well as classic disease.

The second state case is *Jennings v. Baxter Healthcare,* decided by the Oregon Supreme Court in 2000. In this case, plaintiff's expert, a Dr. Grimm, was a neurologist who had examined the plaintiff. Doctors and lawyers had referred to him 45 to 50 women patients with silicone breast implants and he had noticed that 43 of them had inner ear (balance) problems and tingling and loss of sensation in their fingers. This was a unique pattern that he had not seen before. Based on this clinical experience (and nothing else) he believed that the plaintiff's symptoms were caused by her silicone breast implants. Dr. Grimm admitted that his findings had not been published in any medical literature or subjected to peer review, and further admitted that he didn't understand the mechanism yet. He concluded that plaintiff's condition was caused by her implants by differential diagnosis.

The trial court allowed him to testify as to his observations of the plaintiff and the correlation between the symptoms and breast implants, but refused to permit him to testify as to causation. The jury found for the defendant. On appeal, the intermediate appellate court reversed and held that he should have been permitted to testify fully. A further appeal was taken to the Supreme Court of Oregon.

This case might be a poster child for the use of speculative scientific hypotheses in law because a group of prominent scientists (who had no connection with either side) submitted an amicus brief to the court. The group included two Nobel Laureates. The scientists pointed out that: (1) Dr. Grimm's "non-hypothesis" could be tested or falsified by a controlled study but had not been; (2) it had not been submitted for peer review; (3) the risk of error was considerable; (4) the non-hypothesis had not been presented to the scientific community so it could not be said that it was accepted; and (5) the observations of signs and symptoms were accepted by experts in the field but there was no evidence that the causal interpretation was accepted. In short, Dr. Grimm's personal opinion was not based on reliable scientific knowledge.

The court did not accept the scientists' view of reliable scientific knowledge.

It held that case reports were a sufficient basis for a reliable scientific finding. Although publication and peer review were important, they were not indispensable. As for tests, Dr. Grimm's hypothesis could be tested by another doctor; testing had not occurred because his theories were novel. If it were otherwise scientifically valid, a novel conclusion was admissible. Differential diagnosis was acceptable because it is a generally accepted form of scientific inquiry. Grimm's lack of understanding of the mechanism was not a barrier to his opinion because to require that would be asking too much: "There are many generally accepted hypotheses in science for which the mechanism of cause and effect is not understood fully." The court affirmed the intermediate appellate court and sent the case back for a new trial.

There was one small, but possibly significant, bone thrown to science in the opinion. The court held that although generally a published decision affirming the admissibility of scientific evidence would mean that the proponent of the evidence need not lay a scientific foundation for it again, in this case the court held that its conclusion regarding the admissibility of Grimm's testimony was not necessarily binding on retrial. It applied to the law the bedrock principle that "the scientific enterprise always must remain open to reassessing the validity of past judgments as new evidence develops." This may have been a reference to the report of the

National Science Panel, whose negative findings had been issued too late to be cited by either side in their briefs.

The third case is *Dow Chemical Company v. Mahlum,* decided by the Supreme Court of Nevada in 1998. A woman claimed that her silicone breast implants had caused her ACTD. At the trial, Dr. Gershwin and two of plaintiff's treating physicians took the stand. Dr. Gershwin testified to general causation and the treating physicians testified that they believed that plaintiff's ACTD was caused by her implants. A jury found for the plaintiff and she recovered $4 million.

On appeal, the court said it recognized that causation was "scientifically controversial" in breast implant litigation, but the plaintiff did not have to wait until science developed a consensus that breast implants caused her disease. If she had, it might have been too late to recover. The court specifically declined to adopt the Daubert standard and held that reliability by scientific standards was not required for expert scientific testimony: "The Mahlums' complaint was not tried in the court of scientific opinion, but before a jury of her peers who considered the evidence and concluded that Dow Corning breast implants caused her injuries. The jury in this case was properly instructed to consider the proof by a preponderance of the evidence. There is no evidence that the jury did otherwise. Science may properly require a higher standard of proof before declaring the truth, but that standard did not guide the jury, nor do we use that standard to evaluate the judgment on appeal. For the foregoing reasons, we therefore conclude that the Mahlums provided substantial evidence on the issue of causation."

Mahlum rejected *Daubert,* but its point has resonance for courts that recognize *Daubert.* There is a whiff of paradox that lies at the core of *Daubert's* reliability doctrine. Civil cases are to be decided by the preponderance of the evidence, which means that each element of the case must be found by the jury to be more likely than not. More certainty than that is not required and *Daubert* did not purport to change that standard. But by requiring scientific evidence to be reliable, it may be argued that the Supreme Court has imposed a higher standard of proof, since reliable scientific evidence will usually import a higher degree of certainty than a mere preponderance of probability. Thus, any court that excludes marginal scientific testimony in the name of scientific reliability may be accused of impermissibly raising the substantive standard of proof. This argument was raised in a case in which the issue was whether the drug Parlodel caused strokes. In an amicus brief filed by a prominent law professor and a doctor, they argued that the district court's preclusion of plaintiff's proposed testimony on causation, even if such a ruling complied with strict scientific standards, imposed too great a burden on the plaintiff, who had only to prove her case by a preponderance of the evidence. Nevertheless, the court of appeals affirmed the district court's exclusion of the testimony.

Over 100 years ago a young lawyer, Learned Hand, proposed in an article in the Harvard Law Review, that in cases involving scientific evidence a board of experts or a single expert, not appointed by either side, "advise the jury of the general propositions applicable to the case that lie within his province." Hand went on to become one of the most revered American judges and his suggestion has been repeated many times by legal scholars. Nevertheless, despite numerous calls for court-appointed, neutral experts and a Federal Rule of Evidence 706 expressly empowering a judge to appoint such experts, the practice was almost unknown until recently. Then came the Daubert case, Daubert hearings, and with them the need for judges to understand scientific evidence more deeply than before. When the breast implant cases arrived with their baggage of technical learning, the judges sought help; hence, the appointment of neutral panels.

There were two such panels appointed in the breast implant cases. One was by Judge Jones in *Hall*, as we have already noted, and the other was by federal district court Judge Sam C. Pointer, Jr., in the implant cases assigned to him for pretrial proceedings in Alabama. Both appointments were made in 1996. The experts appointed by Judge Pointer were called a National Science Panel because the cases in which they were assigned had been transferred to Judge Pointer from all over the country. Dr. Barbara S. Hulka was the epidemiologist on the panel.

There is no question that the work of these panels influenced the outcome of the cases. Although he did not accept certain opinions of his technical advisers, Judge Jones found the work of the panel valuable and adopted important parts of it in his opinion dismissing the ACTD claims. Judge Pointer's panel did not address the testimony of particular experts proffered by a group of plaintiffs, but rather the general scientific question. Working for two years, it sifted through an enormous number of scholarly studies, and in the end found no association between breast implants and any established or atypical connective tissue disorder. When the panel finally issued its voluminous report in November 1998, it was too late for the cases we have discussed, but the report stopped the flow of new lawsuits and influenced women with ongoing cases to accept the manufacturers' pending offers of settlement. In the tangled fight over the technical evidence in the implant cases, epidemiologists, with other neutral experts, played a key role. The importance of their contribution should lead to more calls for them and biostatisticians to serve on panels of court-selected, neutral experts in future cases. The challenge is to appoint them early in the life cycle of such mass tort events.

REFERENCES

Daubert v. Merrill Dow Pharmaceuticals, Inc., 509 U.S. 579 (1993).
Dow Chemical Company v. Mahlum, 114 Nev. 1468, 970 P.2d 98 (Sup. Ct. Nev. 1998).
Hall v. Baxter Healthcare Corp., 947 F. Supp. 1387 (D. Ore. 1996).
In re: Breast Implant Litigation, 11 E. Supp. 2d 1217 (D. Col. 1998).
Janowsky, E.C., Kupper, I.L., Hulka, B.S. 2000. "Meta-analyses of the relation between silicone breast implants and the risk of connective-tissue disease." *New Eng. J. Med.* 342, 781.
Jennings v. Baxter Healthcare Corp., 331 Ore. 285, 14 P.3d 596 (Sup. Ct. Ore. 2000).
Vassallo v. Baxter Healthcare, 428 Mass. 1, 696 N.E.2d 909 (1998).

2

FRYE V. UNITED STATES

Court of Appeals of the District of Columbia

54 App. D.C. 46; 293 F. 1013

November 7, 1923, Submitted
December 3, 1923, Decided

OPINION

Before SMYTH, Chief Justice, VAN ORSDEL, Associate Justice, and MARTIN, Presiding Judge of the United States Court of Customs Appeals.

VAN ORSDEL, Associate Justice. Appellant, defendant below, was convicted of the crime of murder in the second degree, and from the judgment prosecutes this appeal.

A single assignment of error is presented for our consideration. In the course of the trial counsel for defendant offered an expert witness to testify to the result of a deception test made upon defendant. The test is described as the systolic blood pressure deception test. It is asserted that blood pressure is influenced by change in the emotions of the witness, and that the systolic blood pressure rises are brought about by nervous impulses sent to the sympathetic branch of the autonomic nervous system. Scientific experiments, it is claimed, have demonstrated that fear, rage, and pain always produce a rise of systolic blood pressure, and that conscious deception or falsehood, concealment of facts, or guilt of crime, accompanied by fear of detection when the person is under examination, raises the systolic blood pressure in a curve, which corresponds exactly to the struggle going on in the subject's mind, between fear and attempted control of that fear, as the examination touches the vital points in respect of which he is attempting to deceive the examiner.

In other words, the theory seems to be that truth is spontaneous, and comes without conscious effort, while the utterance of a falsehood requires a conscious effort, which is reflected in the blood pressure. The rise thus produced is easily detected and distinguished from the rise produced by mere fear of the examination itself. In the former instance, the pressure rises higher than in the latter, and is more pronounced as the examination proceeds, while in the latter case, if the subject is telling the truth, the pressure registers highest at the beginning of the examination, and gradually diminishes as the examination proceeds.

Prior to the trial defendant was subjected to this deception test, and counsel offered the scientist who conducted the test as an expert to testify to the results obtained. The offer was objected to by counsel for the government, and the court sustained the objection. Counsel for defendant then offered to have the proffered witness conduct a test in the presence of the jury. This also was denied.

Counsel for defendant, in their able presentation of the novel question involved, correctly state in their brief that no cases directly in point have been found. The broad ground, however, upon which they plant their case, is succinctly stated in their brief as follows:

"The rule is that the opinions of experts or skilled witnesses are admissible in evidence in those cases in which the matter of inquiry is such that inexperienced persons are unlikely to prove capable of forming a correct judgment upon it, for the reason that the subject-matter so far partakes of a science, art, or trade as to require a previous habit or experience or study in it, in order to acquire a knowledge of it. When the question involved does not lie within the range of common experience or common knowledge, but requires special experience or special knowledge, then the opinions of witnesses skilled in that particular science, art, or trade to which the question relates are admissible in evidence."

Numerous cases are cited in support of this rule. Just when a scientific principle or discovery crosses the line between the experimental and demonstrable stages is difficult to define. Somewhere in this twilight zone the evidential force of the principle must be recognized, and while courts will go a long way in admitting expert testimony deduced from a well-recognized scientific principle or discovery, the thing from which the deduction is made must be sufficiently established to have gained general acceptance in the particular field in which it belongs.

We think the systolic blood pressure deception test has not yet gained such standing and scientific recognition among physiological and psychological authorities as would justify the courts in admitting expert testimony deduced from the discovery, development, and experiments thus far made.

The judgment is affirmed.

3

DAUBERT V. MERRELL DOW
PHARMACEUTICALS

DAUBERT ET UX., INDIVIDUALLY AND AS GUARDIANS
AD LITEM FOR DAUBERT, ET AL. *v.* MERRELL
DOW PHARMACEUTICALS, INC.

CERTIORARI TO THE UNITED STATES COURT OF
APPEALS FOR THE NINTH CIRCUIT

509 U. S. 579 (1993)

No. 92–102. Argued March 30, 1993—Decided June 28, 1993

Petitioners, two minor children and their parents, alleged in their suit against respondent that the children's serious birth defects had been caused by the mothers' prenatal ingestion of Bendectin, a prescription drug marketed by respondent. The District Court granted respondent summary judgment based on a well-credentialed expert's affidavit concluding, upon reviewing the extensive published scientific literature on the subject, that maternal use of Bendectin has not been shown to be a risk factor for human birth defects. Although petitioners had responded with the testimony of eight other well-credentialed experts, who based their conclusion that Bendectin can cause birth defects on animal studies, chemical structure analyses, and the unpublished "reanalysis" of previously published human statistical studies, the court determined that this evidence did not meet the applicable "general acceptance" standard for the admission of expert testimony. The Court of Appeals agreed and affirmed, citing *Frye* v. *United States,* 54 App. D. C. 46, 47, 293 F. 1013, 1014, for the rule that expert opinion based on a scientific technique is inadmissible unless the technique is "generally accepted" as reliable in the relevant scientific community.

Held: The Federal Rules of Evidence, not *Frye,* provide the standard for admitting expert scientific testimony in a federal trial. Pp. 585–597.

   (a) *Frye*'s "general acceptance" test was superseded by the Rules' adoption. The Rules occupy the field, *United States* v. *Abel,* 469 U. S. 45, 49, and, although the common law of evidence may serve as an aid to their application, *id.,* at

51–52, respondent's assertion that they somehow assimilated *Frye* is unconvincing. Nothing in the Rules as a whole or in the text and drafting history of Rule 702, which specifically governs expert testimony, gives any indication that "general acceptance" is a necessary precondition to the admissibility of scientific evidence. Moreover, such a rigid standard would be at odds with the Rules' liberal thrust and their general approach of relaxing the traditional barriers to "opinion" testimony. Pp. 585–589.

(b) The Rules—especially Rule 702—place appropriate limits on the admissibility of purportedly scientific evidence by assigning to the trial judge the task of ensuring that an expert's testimony both rests on a reliable foundation and is relevant to the task at hand. The reliability standard is established by Rule 702's requirement that an expert's testimony pertain to "scientific...knowledge," since the adjective "scientific" implies a grounding in science's methods and procedures, while the word "knowledge" connotes a body of known facts or of ideas inferred from such facts or accepted as true on good grounds. The Rule's requirement that the testimony "assist the trier of fact to understand the evidence or to determine a fact in issue" goes primarily to relevance by demanding a valid scientific connection to the pertinent inquiry as a precondition to admissibility. Pp. 589–592.

(c) Faced with a proffer of expert scientific testimony under Rule 702, the trial judge, pursuant to Rule 104(a), must make a preliminary assessment of whether the testimony's underlying reasoning or methodology is scientifically valid and properly can be applied to the facts at issue. Many considerations will bear on the inquiry, including whether the theory or technique in question can be (and has been) tested, whether it has been subjected to peer review and publication, its known or potential error rate and the existence and maintenance of standards controlling its operation, and whether it has attracted widespread acceptance within a relevant scientific community. The inquiry is a flexible one, and its focus must be solely on principles and methodology, not on the conclusions that they generate. Throughout, the judge should also be mindful of other applicable Rules. Pp. 592–595.

(d) Cross-examination, presentation of contrary evidence, and careful instruction on the burden of proof, rather than wholesale exclusion under an uncompromising "general acceptance" standard, is the appropriate means by which evidence based on valid principles may be challenged. That even limited screening by the trial judge, on occasion, will prevent the jury from hearing of authentic scientific breakthroughs is simply a consequence of the fact that the Rules are not designed to seek cosmic understanding but, rather, to resolve legal disputes. Pp. 595–597.

951 F. 2d 1128, vacated and remanded.

BLACKMUN, J., delivered the opinion for a unanimous Court with respect to Parts I and II–A, and the opinion of the Court with respect to Parts II–B, II–C, III, and IV, in which WHITE, O'CONNOR, SCALIA, KENNEDY, SOUTER, and THOMAS, JJ., joined. REHNQUIST, C. J., filed an opinion concurring in part and dissenting in part, in which STEVENS, J., joined, *post,* p. 598.

COUNSEL

Michael H. Gottesman argued the cause for petitioners. With him on the briefs were *Kenneth J. Chesebro, Barry J. Nace, David L. Shapiro,* and *Mary G. Gillick. Charles Fried* argued the cause for respondent. With him on the brief were *Charles R. Nesson, Joel I. Klein, Richard G. Taranto, Hall R. Marston, George E. Berry, Edward H. Stratemeier,* and *W. Glenn Forrester.*[*]

OPINION OF THE COURT

JUSTICE BLACKMUN delivered the opinion of the Court.

In this case we are called upon to determine the standard for admitting expert scientific testimony in a federal trial.

I

Petitioners Jason Daubert and Eric Schuller are minor children born with serious birth defects. They and their parents sued respondent in California state court, alleging that the birth defects had been caused by the mothers' ingestion of Bendectin, a prescription antinausea drug marketed by respondent. Respondent removed the suits to federal court on diversity grounds.

After extensive discovery, respondent moved for summary judgment, contending that Bendectin does not cause birth defects in humans and that petitioners would be unable to come forward with any admissible evidence that it does. In support of its motion, respondent submitted an affidavit of Steven H. Lamm, physician and epidemiologist, who is a well-credentialed expert on the risks from exposure to various chemical substances.[1] Doctor Lamm stated that he had reviewed all the literature

---

[*] Briefs of *amici curiae* urging reversal were filed for the State of Texas et al. by *Dan Morales,* Attorney General of Texas, *Mark Barnett,* Attorney General of South Dakota, *Marc Racicot,* Attorney General of Montana, *Larry EchoHawk,* Attorney General of Idaho, and *Brian Stuart Koukoutchos;* for the American Society of Law, Medicine and Ethics et al. by *Joan E. Berlin, Marsha S. Berzon,* and *Albert H. Meyerhoff;* for the Association of Trial Lawyers of America by *Jeffrey Robert White* and *Roxanne Barton Conlin;* for Ronald Bayer et al. by *Brian Stuart Koukoutchos, Priscilla Budeiri, Arthur Bryant,* and *George W. Conk;* and for Daryl E. Chubin et al. by *Ron Simon* and *Nicole Schultheis.*

Briefs of *amici curiae* urging affirmance were filed for the United States by *Acting Solicitor General Wallace, Assistant Attorney General Gerson, Miguel A. Estrada, Michael Jay Singer,* and *John P. Schnitker;* for the American Insurance Association by *William J. Kilberg, Paul Blankenstein, Bradford R. Clark,* and *Craig A. Berrington;* for the American Medical Association et al. by *Carter G. Phillips, Mark D. Hopson,* and *Jack R. Bierig;* for the American Tort Reform Association by *John G. Kester* and *John W. Vardaman, Jr.;* for the Chamber of Commerce of the United States by *Timothy B. Dyk, Stephen A. Bokat,* and *Robin S. Conrad;* for the Pharmaceutical Manufacturers Association by *Louis R. Cohen* and *Daniel Marcus;* for the Product Liability Advisory Council, Inc., et al. by *Victor E. Schwartz, Robert P. Charrow,* and *Paul F. Rothstein;* for the Washington Legal Foundation by *Scott G. Campbell, Daniel J. Popeo,* and *Richard A. Samp;* and for Nicolaas Bloembergen et al. by *Martin S. Kaufman.*

Briefs of *amici curiae* were filed for the American Association for the Advancement of Science et al. by *Richard A. Meserve* and *Bert Black;* for the American College of Legal Medicine by *Miles J. Zaremski;* for the Carnegie Commission on Science, Technology, and Government by *Steven G. Gallagher, Elizabeth H. Esty,* and *Margaret A. Berger,* for the Defense Research Institute, Inc., by *Joseph A. Sherman, E. Wayne Taff,* and *Harvey L. Kaplan;* for the New England Journal of Medicine et al. by *Michael Malina* and *Jeffrey I. D. Lewis;* for A Group of American Law Professors by *Donald N. Bersoff;* for Alvan R. Feinstein by *Don M. Kennedy, Loretta M. Smith,* and *Richard A. Oetheimer;* and for Kenneth Rothman et al. by *Neil B. Cohen.*

1. Doctor Lamm received his master's and doctor of medicine degrees from the University of Southern California. He has served as a consultant in birth-defect epidemiology for the National Center for Health Statistics and has published numerous articles on the magnitude of risk from exposure to various chemical and biological substances. App. 34–44.

on Bendectin and human birth defects—more than 30 published studies involving over 130,000 patients. No study had found Bendectin to be a human teratogen (*i.e.*, a substance capable of causing malformations in fetuses). On the basis of this review, Doctor Lamm concluded that maternal use of Bendectin during the first trimester of pregnancy has not been shown to be a risk factor for human birth defects.

Petitioners did not (and do not) contest this characterization of the published record regarding Bendectin. Instead, they responded to respondent's motion with the testimony of eight experts of their own, each of whom also possessed impressive credentials.[2] These experts had concluded that Bendectin can cause birth defects. Their conclusions were based upon "in vitro" (test tube) and "in vivo" (live) animal studies that found a link between Bendectin and malformations; pharmacological studies of the chemical structure of Bendectin that purported to show similarities between the structure of the drug and that of other substances known to cause birth defects; and the "reanalysis" of previously published epidemiological (human statistical) studies.

The District Court granted respondent's motion for summary judgment. The court stated that scientific evidence is admissible only if the principle upon which it is based is "sufficiently established to have general acceptance in the field to which it belongs." 727 F. Supp. 570, 572 (SD Cal. 1989), quoting *United States* v. *Kilgus*, 571 F. 2d 508, 510 (CA9 1978). The court concluded that petitioners' evidence did not meet this standard. Given the vast body of epidemiological data concerning Bendectin, the court held, expert opinion which is not based on epidemiological evidence is not admissible to establish causation. 727 F. Supp., at 575. Thus, the animal-cell studies, live-animal studies, and chemical-structure analyses on which petitioners had relied could not raise by themselves a reasonably disputable jury issue regarding causation. *Ibid.* Petitioners' epidemiological analyses, based as they were on recalculations of data in previously published studies that had found no causal link between the drug and birth defects, were ruled to be inadmissible because they had not been published or subjected to peer review. *Ibid.*

The United States Court of Appeals for the Ninth Circuit affirmed. 951 F. 2d 1128 (1991). Citing *Frye* v. *United States,* 54 App. D. C. 46, 47, 293 F. 1013, 1014 (1923), the court stated that expert opinion based on a scientific technique is inadmissible unless the technique is "generally accepted" as reliable in the relevant scientific community. 951 F. 2d, at 1129–1130. The court declared that expert opinion based on a methodology that diverges "significantly from the procedures accepted by recognized authorities in the field...cannot be shown to be 'generally accepted as

---

2. For example, Shanna Helen Swan, who received a master's degree in biostatistics from Columbia University and a doctorate in statistics from the University of California at Berkeley, is chief of the section of the California Department of Health and Services that determines causes of birth defects and has served as a consultant to the World Health Organization, the Food and Drug Administration, and the National Institutes of Health. *Id.*, at 113–114, 131–132. Stuart A. Newman, who received his bachelor's degree in chemistry from Columbia University and his master's and doctorate in chemistry from the University of Chicago, is a professor at New York Medical College and has spent over a decade studying the effect of chemicals on limb development. *Id.*, at 54–56. The credentials of the others are similarly impressive. See *id.*, at 61–66, 73–80, 148–153, 187–192, and Attachments 12, 20, 21, 26, 31, and 32 to Petitioners' Opposition to Summary Judgment in No. 84–2013–G(I) (SD Cal.).

a reliable technique.'" *Id.,* at 1130, quoting *United States* v. *Solomon,* 753 F. 2d 1522, 1526 (CA9 1985).

The court emphasized that other Courts of Appeals considering the risks of Bendectin had refused to admit reanalyses of epidemiological studies that had been neither published nor subjected to peer review. 951 F. 2d, at 1130–1131. Those courts had found unpublished reanalyses "particularly problematic in light of the massive weight of the original published studies supporting [respondent's] position, all of which had undergone full scrutiny from the scientific community." *Id.,* at 1130. Contending that reanalysis is generally accepted by the scientific community only when it is subjected to verification and scrutiny by others in the field, the Court of Appeals rejected petitioners' reanalyses as "unpublished, not subjected to the normal peer review process and generated solely for use in litigation." *Id.,* at 1131. The court concluded that petitioners' evidence provided an insufficient foundation to allow admission of expert testimony that Bendectin caused their injuries and, accordingly, that petitioners could not satisfy their burden of proving causation at trial.

We granted certiorari, 506 U.S. 914 (1992), in light of sharp divisions among the courts regarding the proper standard for the admission of expert testimony. Compare, *e. g., United States* v. *Shorter,* 257 U. S. App. D. C. 358, 363–364, 809 F. 2d 54, 59–60 (applying the "general acceptance" standard), cert, denied, 484 U. S. 817 (1987), with *DeLuca* v. *Merrell Dow Pharmaceuticals, Inc.,* 911 F. 2d 941, 955 (CA3 1990) (rejecting the "general acceptance" standard).

II

A

In the 70 years since its formulation in the *Frye* case, the "general acceptance" test has been the dominant standard for determining the admissibility of novel scientific evidence at trial. See E. Green & C. Nesson, Problems, Cases, and Materials on Evidence 649 (1983). Although under increasing attack of late, the rule continues to be followed by a majority of courts, including the Ninth Circuit.[3]

The *Frye* test has its origin in a short and citation-free 1923 decision concerning the admissibility of evidence derived from a systolic blood pressure deception test, a crude precursor to the polygraph machine. In what has become a famous (perhaps infamous) passage, the then Court of Appeals for the District of Columbia described the device and its operation and declared:

> "Just when a scientific principle or discovery crosses the line between the experimental and demonstrable stages is difficult to define. Somewhere in this twilight zone the evidential force of the principle must be recognized, and while courts will go a long way in admitting expert testimony deduced from a well-recognized scientific principle or discovery, *the thing from which the deduction is made must be sufficiently established to have gained general acceptance in the particular field in which it belongs."*
> 54 App. D. C., at 47, 293 F., at 1014 (emphasis added).

---

3. For a catalog of the many cases on either side of this controversy, see P. Giannelli & E. Imwinkelried, Scientific Evidence § 1–5, pp. 10–14 (1986 and Supp. 1991).

Because the deception test had "not yet gained such standing and scientific recognition among physiological and psychological authorities as would justify the courts in admitting expert testimony deduced from the discovery, development, and experiments thus far made," evidence of its results was ruled inadmissible. *Ibid.*

The merits of the *Frye* test have been much debated, and scholarship on its proper scope and application is legion.[4] Petitioners' primary attack, however, is not on the content but on the continuing authority of the rule. They contend that the *Frye* test was superseded by the adoption of the Federal Rules of Evidence.[5] We agree.

We interpret the legislatively enacted Federal Rules of Evidence as we would any statute. *Beech Aircraft Corp.* v. *Rainey,* 488 U. S. 153, 163 (1988). Rule 402 provides the baseline:

> "All relevant evidence is admissible, except as otherwise provided by the Constitution of the United States, by Act of Congress, by these rules, or by other rules prescribed by the Supreme Court pursuant to statutory authority. Evidence which is not relevant is not admissible."

"Relevant evidence" is defined as that which has "any tendency to make the existence of any fact that is of consequence to the determination of the action more probable or less probable than it would be without the evidence." Rule 401. The Rules' basic standard of relevance thus is a liberal one.

Frye, of course, predated the Rules by half a century. In *United States* v. *Abel,* 469 U. S. 45 (1984), we considered the pertinence of background common law in interpreting the Rules of Evidence. We noted that the Rules occupy the field, *id.,* at 49, but, quoting Professor Cleary, the Reporter, explained that the common law nevertheless could serve as an aid to their application:

> " 'In principle, under the Federal Rules no common law of evidence remains. "All relevant evidence is admissible, except as otherwise provided...." In reality, of course, the body of common law knowledge continues to exist, though in the

---

4. See, *e.g.,* Green, Expert Witnesses and Sufficiency of Evidence in Toxic Substances Litigation: The Legacy of *Agent Orange* and Bendectin Litigation, 86 Nw. U. L. Rev. 643 (1992) (hereinafter Green); Becker & Orenstein, The Federal Rules of Evidence After Sixteen Years—The Effect of "Plain Meaning" Jurisprudence, the Need for an Advisory Committee on the Rules of Evidence, and Suggestions for Selective Revision of the Rules, 60 Geo. Wash. L. Rev. 857, 876–885 (1992); Hanson, James Alphonzo Frye is Sixty-Five Years Old; Should He Retire?, 16 West St. U. L. Rev. 357 (1989); Black, A Unified Theory of Scientific Evidence, 56 Ford. L. Rev. 595 (1988); Imwinkelried, The "Bases" of Expert Testimony: The Syllogistic Structure of Scientific Testimony, 67 N. C. L. Rev. 1 (1988); Proposals for a Model Rule on the Admissibility of Scientific Evidence, 26 Jurimetrics J. 235 (1986); Giannelli, The Admissibility of Novel Scientific Evidence: *Frye* v. *United States,* a Half-Century Later, 80 Colum. L. Rev. 1197 (1980); The Supreme Court, 1986 Term, 101 Harv. L. Rev. 7, 119, 125–127 (1987).

Indeed, the debates over *Frye* are such a well-established part of the academic landscape that a distinct term—"*Frye*-ologist"—has been advanced to describe those who take part. See Behringer, Introduction, Proposals for a Model Rule on the Admissibility of Scientific Evidence, 26 Jurimetrics J. 237, 239 (1986), quoting Lacey, Scientific Evidence, 24 Jurimetrics J. 254, 264 (1984).

5. Like the question of *Frye*'s merit, the dispute over its survival has divided courts and commentators. Compare, *e.g.,* *United States* v. *Williams,* 583 F. 2d 1194 (CA2 1978) (*Frye* is superseded by the Rules of Evidence), cert. denied, 439 U. S. 1117 (1979), with *Christophersen* v. *Allied-Signal Corp.,* 939 F. 2d 1106, 1111, 1115–1116 (CA5 1991) (en banc) (*Frye* and the Rules coexist), cert. denied, 503 U.S. 912 (1992), 3 J. Weinstein & M. Berger, Weinstein's Evidence ¶ 702[03], pp. 702–36 to 702–37 (1988) (hereinafter Weinstein & Berger) (*Frye* is dead), and M. Graham, Handbook of Federal Evidence § 703.2 (3d ed. 1991) (*Frye* lives). See generally P. Giannelli & E. Imwinkelried, Scientific Evidence § 1–5, at 28–29 (citing authorities).

somewhat altered form of a source of guidance in the exercise of delegated powers.'" *Id.*, at 51–52.

We found the common-law precept at issue in the *Abel* case entirely consistent with Rule 402's general requirement of admissibility, and considered it unlikely that the drafters had intended to change the rule. *Id.*, at 50–51. In *Bourjaily* v. *United States,* 483 U. S. 171 (1987), on the other hand, the Court was unable to find a particular common-law doctrine in the Rules, and so held it superseded.

Here there is a specific Rule that speaks to the contested issue. Rule 702, governing expert testimony, provides:

> "If scientific, technical, or other specialized knowledge will assist the trier of fact to understand the evidence or to determine a fact in issue, a witness qualified as an expert by knowledge, skill, experience, training, or education, may testify thereto in the form of an opinion or otherwise."

Nothing in the text of this Rule establishes "general acceptance" as an absolute prerequisite to admissibility. Nor does respondent present any clear indication that Rule 702 or the Rules as a whole were intended to incorporate a "general acceptance" standard. The drafting history makes no mention of *Frye,* and a rigid "general acceptance" requirement would be at odds with the "liberal thrust" of the Federal Rules and their "general approach of relaxing the traditional barriers to 'opinion' testimony." *Beech Aircraft Corp.* v. *Rainey,* 488 U. S., at 169 (citing Rules 701 to 705). See also Weinstein, Rule 702 of the Federal Rules of Evidence is Sound; It Should Not Be Amended, 138 F. R. D. 631 (1991) ("The Rules were designed to depend primarily upon lawyer-adversaries and sensible triers of fact to evaluate conflicts"). Given the Rules' permissive backdrop and their inclusion of a specific rule on expert testimony that does not mention "general acceptance," the assertion that the Rules somehow assimilated *Frye* is unconvincing. *Frye* made "general acceptance" the exclusive test for admitting expert scientific testimony. That austere standard, absent from, and incompatible with, the Federal Rules of Evidence, should not be applied in federal trials.[6]

B

That the *Frye* test was displaced by the Rules of Evidence does not mean, however, that the Rules themselves place no limits on the admissibility of purportedly scientific evidence.[7] Nor is the trial judge disabled from screening such evidence. To the contrary, under the Rules the trial judge must ensure that any and all scientific testimony or evidence admitted is not only relevant, but reliable.

The primary locus of this obligation is Rule 702, which clearly contemplates some degree of regulation of the subjects and theories about which an expert may testify.

---

6. Because we hold that *Frye* has been superseded and base the discussion that follows on the content of the congressionally enacted Federal Rules of Evidence, we do not address petitioners' argument that application of the *Frye* rule in this diversity case, as the application of a judge-made rule affecting substantive rights, would violate the doctrine of *Erie R. Co.* v. *Tompkins,* 304 U. S. 64 (1938).

7. THE CHIEF JUSTICE "do[es] not doubt that Rule 702 confides to the judge some gatekeeping responsibility," *post,* at 600, but would neither say how it does so nor explain what that role entails. We believe the better course is to note the nature and source of the duty.

"If scientific, technical, or other specialized *knowledge will assist the trier of fact* to understand the evidence or to determine a fact in issue" an expert "may testify *thereto."* (Emphasis added.) The subject of an expert's testimony must be "scientific...knowledge."[8] The adjective "scientific" implies a grounding in the methods and procedures of science. Similarly, the word "knowledge" connotes more than subjective belief or unsupported speculation. The term "applies to any body of known facts or to any body of ideas inferred from such facts or accepted as truths on good grounds." Webster's Third New International Dictionary 1252 (1986). Of course, it would be unreasonable to conclude that the subject of scientific testimony must be "known" to a certainty; arguably, there are no certainties in science. See, *e.g.,* Brief for Nicolaas Bloembergen et al. as *Amici Curiae* 9 ("Indeed, scientists do not assert that they know what is immutably 'true'—they are committed to searching for new, temporary, theories to explain, as best they can, phenomena"); Brief for American Association for the Advancement of Science et al. as *Amici Curiae* 7–8 ("Science is not an encyclopedic body of knowledge about the universe. Instead, it represents a *process* for proposing and refining theoretical explanations about the world that are subject to further testing and refinement" (emphasis in original)). But, in order to qualify as "scientific knowledge," an inference or assertion must be derived by the scientific method. Proposed testimony must be supported by appropriate validation— *i.e.,* "good grounds," based on what is known. In short, the requirement that an expert's testimony pertain to "scientific knowledge" establishes a standard of evidentiary reliability.[9]

Rule 702 further requires that the evidence or testimony "assist the trier of fact to understand the evidence or to determine a fact in issue." This condition goes primarily to relevance. "Expert testimony which does not relate to any issue in the case is not relevant and, ergo, non-helpful." 3 Weinstein & Berger ¶ 702[02], p. 702–18. See also *United States* v. *Downing,* 753 F. 2d 1224, 1242 (CA3 1985) ("An additional consideration under Rule 702—and another aspect of relevancy—is whether expert testimony proffered in the case is sufficiently tied to the facts of the case that it will aid the jury in resolving a factual dispute"). The consideration has been aptly described by Judge Becker as one of "fit." *Ibid.* "Fit" is not always obvious, and scientific validity for one purpose is not necessarily scientific validity for other, unrelated purposes. See Starrs, *Frye v. United States* Restructured and Revitalized: A Proposal to Amend Federal Evidence Rule 702, 26 Jurimetrics J. 249,

---

8. Rule 702 also applies to "technical, or other specialized knowledge." Our discussion is limited to the scientific context because that is the nature of the expertise offered here.

9. We note that scientists typically distinguish between "validity" (does the principle support what it purports to show?) and "reliability" (does application of the principle produce consistent results?). See Black, 56 Ford. L. Rev., at 599. Although "the difference between accuracy, validity, and reliability may be such that each is distinct from the other by no more than a hen's kick," Starrs, *Frye v. United States* Restructured and Revitalized: A Proposal to Amend Federal Evidence Rule 702, 26 Jurimetrics J. 249, 256 (1986), our reference here is to *evidentiary* reliability—that is, trustworthiness. Cf., *e.g.,* Advisory Committee's Notes on Fed. Rule Evid. 602, 28 U. S. C. App., p. 755 (" '[T]he rule requiring that a witness who testifies to a fact which can be perceived by the senses must have had an opportunity to observe, and must have actually observed the fact' is a 'most pervasive manifestation' of the common law insistence upon 'the most reliable sources of information' " (citation omitted)); Advisory Committee's Notes on Art. VIII of Rules of Evidence, 28 U. S. C. App., p. 770 (hearsay exceptions will be recognized only "under circumstances supposed to furnish guarantees of trustworthiness"). In a case involving scientific evidence, *evidentiary reliability* will be based upon *scientific validity.*

258 (1986). The study of the phases of the moon, for example, may provide valid scientific "knowledge" about whether a certain night was dark, and if darkness is a fact in issue, the knowledge will assist the trier of fact. However (absent creditable grounds supporting such a link), evidence that the moon was full on a certain night will not assist the trier of fact in determining whether an individual was unusually likely to have behaved irrationally on that night. Rule 702's "helpfulness" standard requires a valid scientific connection to the pertinent inquiry as a precondition to admissibility.

That these requirements are embodied in Rule 702 is not surprising. Unlike an ordinary witness, see Rule 701, an expert is permitted wide latitude to offer opinions, including those that are not based on firsthand knowledge or observation. See Rules 702 and 703. Presumably, this relaxation of the usual requirement of firsthand knowledge—a rule which represents "a 'most pervasive manifestation' of the common law insistence upon 'the most reliable sources of information,'" Advisory Committee's Notes on Fed. Rule Evid. 602, 28 U. S. C. App., p. 755 (citation omitted)—is premised on an assumption that the expert's opinion will have a reliable basis in the knowledge and experience of his discipline.

C

Faced with a proffer of expert scientific testimony, then, the trial judge must determine at the outset, pursuant to Rule 104(a),[10] whether the expert is proposing to testify to (1) scientific knowledge that (2) will assist the trier of fact to understand or determine a fact in issue.[11] This entails a preliminary assessment of whether the reasoning or methodology underlying the testimony is scientifically valid and of whether that reasoning or methodology properly can be applied to the facts in issue. We are confident that federal judges possess the capacity to undertake this review. Many factors will bear on the inquiry, and we do not presume to set out a definitive checklist or test. But some general observations are appropriate.

Ordinarily, a key question to be answered in determining whether a theory or technique is scientific knowledge that will assist the trier of fact will be whether it can be (and has been) tested. "Scientific methodology today is based on generating hypotheses and testing them to see if they can be falsified; indeed, this methodology is what distinguishes science from other fields of human inquiry." Green 645. See also C. Hempel, Philosophy of Natural Science 49 (1966) ("[T]he statements constituting a scientific explanation must be capable of empirical test"); K. Popper,

---

10. Rule 104(a) provides:

"Preliminary questions concerning the qualification of a person to be a witness, the existence of a privilege, or the admissibility of evidence shall be determined by the court, subject to the provisions of subdivision (b) [pertaining to conditional admissions]. In making its determination it is not bound by the rules of evidence except those with respect to privileges." These matters should be established by a preponderance of proof. See *Bourjaily* v. *United States,* 483 U. S. 171, 175–176 (1987).

11. Although the *Frye* decision itself focused exclusively on "novel" scientific techniques, we do not read the requirements of Rule 702 to apply specially or exclusively to unconventional evidence. Of course, well established propositions are less likely to be challenged than those that are novel, and they are more handily defended. Indeed, theories that are so firmly established as to have attained the status of scientific law, such as the laws of thermodynamics, properly are subject to judicial notice under Federal Rule of Evidence 201.

Conjectures and Refutations: The Growth of Scientific Knowledge 37 (5th ed. 1989) ("[T]he criterion of the scientific status of a theory is its falsifiability, or refutability, or testability") (emphasis deleted).

Another pertinent consideration is whether the theory or technique has been subjected to peer review and publication. Publication (which is but one element of peer review) is not a *sine qua non* of admissibility; it does not necessarily correlate with reliability, see S. Jasanoff, The Fifth Branch: Science Advisors as Policymakers 61–76 (1990), and in some instances well-grounded but innovative theories will not have been published, see Horrobin, The Philosophical Basis of Peer Review and the Suppression of Innovation, 263 JAMA 1438 (1990). Some propositions, moreover, are too particular, too new, or of too limited interest to be published. But submission to the scrutiny of the scientific community is a component of "good science," in part because it increases the likelihood that substantive flaws in methodology will be detected. See J. Ziman, Reliable Knowledge: An Exploration of the Grounds for Belief in Science 130–133 (1978); Relman & Angell, How Good Is Peer Review?, 321 New Eng. J. Med. 827 (1989). The fact of publication (or lack thereof) in a peer reviewed journal thus will be a relevant, though not dispositive, consideration in assessing the scientific validity of a particular technique or methodology on which an opinion is premised.

Additionally, in the case of a particular scientific technique, the court ordinarily should consider the known or potential rate of error, see, *e.g., United States* v. *Smith,* 869 F. 2d 348, 353–354 (CA7 1989) (surveying studies of the error rate of spectrographic voice identification technique), and the existence and maintenance of standards controlling the technique's operation, see *United States* v. *Williams,* 583 F. 2d 1194, 1198 (CA2 1978) (noting professional organization's standard governing spectrographic analysis), cert. denied, 439 U. S. 1117 (1979).

Finally, "general acceptance" can yet have a bearing on the inquiry. A "reliability assessment does not require, although it does permit, explicit identification of a relevant scientific community and an express determination of a particular degree of acceptance within that community." *United States* v. *Downing,* 753 F. 2d, at 1238. See also 3 Weinstein & Berger ¶ 702[03], pp. 702–41 to 702–42. Widespread acceptance can be an important factor in ruling particular evidence admissible, and "a known technique which has been able to attract only minimal support within the community," *Downing,* 753 F. 2d, at 1238, may properly be viewed with skepticism.

The inquiry envisioned by Rule 702 is, we emphasize, a flexible one.[12] Its overarching subject is the scientific validity—and thus the evidentiary relevance and

---

12. A number of authorities have presented variations on the reliability approach, each with its own slightly different set of factors. See, *e.g., Downing,* 753 F. 2d, at 1238–1239 (on which our discussion draws in part); 3 Weinstein & Berger ¶ 702[03], pp. 702–41 to 702–42 (on which the *Downing* court in turn partially relied); McCormick, Scientific Evidence: Defining a New Approach to Admissibility, 67 Iowa L. Rev. 879, 911–912 (1982); and Symposium on Science and the Rules of Evidence, 99 F. R. D. 187, 231 (1983) (statement by Margaret Berger). To the extent that they focus on the reliability of evidence as ensured by the scientific validity of its underlying principles, all these versions may well have merit, although we express no opinion regarding any of their particular details.

reliability—of the principles that underlie a proposed submission. The focus, of course, must be solely on principles and methodology, not on the conclusions that they generate.

Throughout, a judge assessing a proffer of expert scientific testimony under Rule 702 should also be mindful of other applicable rules. Rule 703 provides that expert opinions based on otherwise inadmissible hearsay are to be admitted only if the facts or data are "of a type reasonably relied upon by experts in the particular field in forming opinions or inferences upon the subject." Rule 706 allows the court at its discretion to procure the assistance of an expert of its own choosing. Finally, Rule 403 permits the exclusion of relevant evidence "if its probative value is substantially outweighed by the danger of unfair prejudice, confusion of the issues, or misleading the jury...." Judge Weinstein has explained: "Expert evidence can be both powerful and quite misleading because of the difficulty in evaluating it. Because of this risk, the judge in weighing possible prejudice against probative force under Rule 403 of the present rules exercises more control over experts than over lay witnesses." Weinstein, 138 F. R. D., at 632.

III

We conclude by briefly addressing what appear to be two underlying concerns of the parties and *amici* in this case. Respondent expresses apprehension that abandonment of "general acceptance" as the exclusive requirement for admission will result in a "free-for-all" in which befuddled juries are confounded by absurd and irrational pseudoscientific assertions. In this regard respondent seems to us to be overly pessimistic about the capabilities of the jury and of the adversary system generally. Vigorous cross-examination, presentation of contrary evidence, and careful instruction on the burden of proof are the traditional and appropriate means of attacking shaky but admissible evidence. See *Rock* v. *Arkansas,* 483 U. S. 44, 61 (1987). Additionally, in the event the trial court concludes that the scintilla of evidence presented supporting a position is insufficient to allow a reasonable juror to conclude that the position more likely than not is true, the court remains free to direct a judgment, Fed. Rule Civ. Proc. 50(a), and likewise to grant summary judgment, Fed. Rule Civ. Proc. 56. Cf., *e.g., Turpin* v. *Merrell Dow Pharmaceuticals, Inc.,* 959 F. 2d 1349 (CA6) (holding that scientific evidence that provided foundation for expert testimony, viewed in the light most favorable to plaintiffs, was not sufficient to allow a jury to find it more probable than not that defendant caused plaintiff's injury), cert. denied, 506 U.S. 826 (1992); *Brock* v. *Merrell Dow Pharmaceuticals, Inc.,* 874 F. 2d 307 (CA5 1989) (reversing judgment entered on jury verdict for plaintiffs because evidence regarding causation was insufficient), modified, 884 F. 2d 166 (CA5 1989), cert. denied, 494 U. S. 1046 (1990); Green 680–681. These conventional devices, rather than wholesale exclusion under an uncompromising "general acceptance" test, are the appropriate safeguards where the basis of scientific testimony meets the standards of Rule 702.

Petitioners and, to a greater extent, their *amici* exhibit a different concern. They suggest that recognition of a screening role for the judge that allows for the exclusion of "invalid" evidence will sanction a stifling and repressive scientific orthodoxy and will be inimical to the search for truth. See, *e.g.,* Brief for Ronald

Bayer et al. as *Amici Curiae*. It is true that open debate is an essential part of both legal and scientific analyses. Yet there are important differences between the quest for truth in the courtroom and the quest for truth in the laboratory. Scientific conclusions are subject to perpetual revision. Law, on the other hand, must resolve disputes finally and quickly. The scientific project is advanced by broad and wide-ranging consideration of a multitude of hypotheses, for those that are incorrect will eventually be shown to be so, and that in itself is an advance. Conjectures that are probably wrong are of little use, however, in the project of reaching a quick, final, and binding legal judgment—often of great consequence—about a particular set of events in the past. We recognize that, in practice, a gatekeeping role for the judge, no matter how flexible, inevitably on occasion will prevent the jury from learning of authentic insights and innovations. That, nevertheless, is the balance that is struck by Rules of Evidence designed not for the exhaustive search for cosmic understanding but for the particularized resolution of legal disputes.[13]

IV

To summarize: "General acceptance" is not a necessary precondition to the admissibility of scientific evidence under the Federal Rules of Evidence, but the Rules of Evidence—especially Rule 702—do assign to the trial judge the task of ensuring that an expert's testimony both rests on a reliable foundation and is relevant to the task at hand. Pertinent evidence based on scientifically valid principles will satisfy those demands.

The inquiries of the District Court and the Court of Appeals focused almost exclusively on "general acceptance," as gauged by publication and the decisions of other courts. Accordingly, the judgment of the Court of Appeals is vacated, and the case is remanded for further proceedings consistent with this opinion.

It is so ordered.

CHIEF JUSTICE REHNQUIST, with whom JUSTICE STEVENS joins, concurring in part and dissenting in part.

The petition for certiorari in this case presents two questions: first, whether the rule of *Frye* v. *United States,* 54 App. D. C. 46, 293 F. 1013 (1923), remains good law after the enactment of the Federal Rules of Evidence; and second, if *Frye* remains valid, whether it requires expert scientific testimony to have been subjected to a peer review process in order to be admissible. The Court concludes, correctly in my view, that the *Frye* rule did not survive the enactment of the Federal Rules of Evidence, and I therefore join Parts I and II–A of its opinion. The second question presented in the petition for certiorari necessarily is mooted by this holding, but the Court nonetheless proceeds to construe Rules 702 and 703 very much in the abstract, and then offers some "general observations." *Ante,* at 593.

---

13. This is not to say that judicial interpretation, as opposed to adjudicative factfinding, does not share basic characteristics of the scientific endeavor: "The work of a judge is in one sense enduring and in another ephemeral.... In the endless process of testing and retesting, there is a constant rejection of the dross and a constant retention of whatever is pure and sound and fine." B. Cardozo, The Nature of the Judicial Process, 178–179 (1921).

"General observations" by this Court customarily carry great weight with lower federal courts, but the ones offered here suffer from the flaw common to most such observations—they are not applied to deciding whether particular testimony was or was not admissible, and therefore they tend to be not only general, but vague and abstract. This is particularly unfortunate in a case such as this, where the ultimate legal question depends on an appreciation of one or more bodies of knowledge not judicially noticeable, and subject to different interpretations in the briefs of the parties and their *amici.* Twenty-two *amicus* briefs have been filed in the case, and indeed the Court's opinion contains no fewer than 37 citations to *amicus* briefs and other secondary sources.

The various briefs filed in this case are markedly different from typical briefs, in that large parts of them do not deal with decided cases or statutory language—the sort of material we customarily interpret. Instead, they deal with definitions of scientific knowledge, scientific method, scientific validity, and peer review—in short, matters far afield from the expertise of judges. This is not to say that such materials are not useful or even necessary in deciding how Rule 702 should be applied; but it is to say that the unusual subject matter should cause us to proceed with great caution in deciding more than we have to, because our reach can so easily exceed our grasp.

But even if it were desirable to make "general observations" not necessary to decide the questions presented, I cannot subscribe to some of the observations made by the Court. In Part II–B, the Court concludes that reliability and relevancy are the touchstones of the admissibility of expert testimony. *Ante,* at 590–592. Federal Rule of Evidence 402 provides, as the Court points out, that "[e]vidence which is not relevant is not admissible." But there is no similar reference in the Rule to "reliability." The Court constructs its argument by parsing the language "[i]f scientific, technical, or other specialized knowledge will assist the trier of fact to understand the evidence or to determine a fact in issue,...an expert...may testify thereto...." Fed. Rule Evid. 702. It stresses that the subject of the expert's testimony must be "scientific...knowledge," and points out that "scientific" "implies a grounding in the methods and procedures of science" and that the word "knowledge" "connotes more than subjective belief or unsupported speculation." *Ante,* at 590. From this it concludes that "scientific knowledge" must be "derived by the scientific method." *Ibid.* Proposed testimony, we are told, must be supported by "appropriate validation." *Ibid.* Indeed, in footnote 9, the Court decides that "[i]n a case involving scientific evidence, *evidentiary reliability* will be based upon *scientific validity.*" *Ante,* at 591, n. 9 (emphasis in original).

Questions arise simply from reading this part of the Court's opinion, and countless more questions will surely arise when hundreds of district judges try to apply its teaching to particular offers of expert testimony. Does all of this *dicta* apply to an expert seeking to testify on the basis of "technical or other specialized knowledge"— the other types of expert knowledge to which Rule 702 applies—or are the "general observations" limited only to "scientific knowledge"? What is the difference between scientific knowledge and technical knowledge; does Rule 702 actually contemplate that the phrase "scientific, technical, or other specialized knowledge" be broken down into numerous subspecies of expertise, or did its authors simply pick general descriptive language covering the sort of expert testimony which courts have customarily received? The Court speaks of its confidence that federal judges can make a "preliminary assessment of whether the reasoning or methodology underlying the testimony is scientifically valid and of whether that reasoning or methodology properly can be applied to the facts in issue." *Ante,* at 592–593. The Court then states that a "key question" to be answered in deciding whether something is "scientific knowledge" "will be whether it can be (and has been) tested." *Ante,* at 593. Following this sentence are three quotations from treatises, which not only speak of empirical testing, but one of which states that the " 'criterion of the scientific status of a theory is its falsifiability, or refutability, or testability.' " *Ibid.*

I defer to no one in my confidence in federal judges; but I am at a loss to know what is meant when it is said that the scientific status of a theory depends on its "falsifiability," and I suspect some of them will be, too.

I do not doubt that Rule 702 confides to the judge some gatekeeping responsibility in deciding questions of the admissibility of proffered expert testimony. But I do not think it imposes on them either the obligation or the authority to become amateur scientists in order to perform that role. I think the Court would be far better advised in this case to decide only the questions presented, and to leave the further development of this important area of the law to future cases.

4

KUMHO TIRE CO. V. CARMICHAEL

KUMHO TIRE CO., LTD., ET AL. *v.* CARMICHAEL ET AL.

CERTIORARI TO THE UNITED STATES COURT OF
APPEALS FOR THE ELEVENTH CIRCUIT

526 U. S. 137 (1999)

No. 97–1709. Argued December 7, 1998—Decided March 23, 1999

When a tire on the vehicle driven by Patrick Carmichael blew out and the vehicle overturned, one passenger died and the others were injured. The survivors and the decedent's representative, respondents here, brought this diversity suit against the tire's maker and its distributor (collectively Kumho Tire), claiming that the tire that failed was defective. They rested their case in significant part upon the depositions of a tire failure analyst, Dennis Carlson, Jr., who intended to testify that, in his expert opinion, a defect in the tire's manufacture or design caused the blowout. That opinion was based upon a visual and tactile inspection of the tire and upon the theory that in the absence of at least two of four specific, physical symptoms indicating tire abuse, the tire failure of the sort that occurred here was caused by a defect. Kumho Tire moved to exclude Carlson's testimony on the ground that his methodology failed to satisfy Federal Rule of Evidence 702, which says: "If scientific, technical, or other specialized knowledge will assist the trier of fact..., a witness qualified as an expert...may testify thereto in the form of an opinion." Granting the motion (and entering summary judgment for the defendants), the District Court acknowledged that it should act as a reliability "gatekeeper" under *Daubert* v. *Merrell Dow Pharmaceuticals, Inc.,* 509 U. S. 579, 589, in which this Court held that Rule 702 imposes a special obligation upon a trial judge to ensure that scientific testimony is not only relevant, but reliable. The court noted that *Daubert* discussed four factors—testing, peer review, error rates, and "acceptability" in the relevant scientific community—which might prove helpful in determining the reliability of a particular scientific theory or technique, *id.,* at 593–594, and found that those factors argued against the reliability of Carlson's methodology. On the plaintiffs' motion for reconsideration, the court agreed that *Daubert* should be applied flexibly, that its four factors were simply illustrative, and that other factors could argue in favor of admissibility. However, the court affirmed its earlier order

66

because it found insufficient indications of the reliability of Carlson's methodology. In reversing, the Eleventh Circuit held that the District Court had erred as a matter of law in applying *Daubert*. Believing that *Daubert* was limited to the scientific context, the court held that the *Daubert* factors did not apply to Carlson's testimony, which it characterized as skill or experience based.

Held:

1. The *Daubert* factors may apply to the testimony of engineers and other experts who are not scientists. Pp. 147–153.

   (a) The *Daubert* "gatekeeping" obligation applies not only to "scientific" testimony, but to all expert testimony. Rule 702 does not distinguish between "scientific" knowledge and "technical" or "other specialized" knowledge, but makes clear that any such knowledge might become the subject of expert testimony. It is the Rule's word "knowledge," not the words (like "scientific") that modify that word, that establishes a standard of evidentiary reliability. 509 U. S., at 589–590. *Daubert* referred only to "scientific" knowledge because that was the nature of the expertise there at issue. *Id.,* at 590, n. 8. Neither is the evidentiary rationale underlying *Daubert*'s "gatekeeping" determination limited to "scientific" knowledge. Rules 702 and 703 grant all expert witnesses, not just "scientific" ones, testimonial latitude unavailable to other witnesses on the assumption that the expert's opinion will have a reliable basis in the knowledge and experience of his discipline. *Id.,* at 592. Finally, it would prove difficult, if not impossible, for judges to administer evidentiary rules under which a "gatekeeping" obligation depended upon a distinction between "scientific" knowledge and "technical" or "other specialized" knowledge, since there is no clear line dividing the one from the others and no convincing need to make such distinctions. Pp. 147–149.

   (b) A trial judge determining the admissibility of an engineering expert's testimony *may* consider one or more of the specific *Daubert* factors. The emphasis on the word "may" reflects *Daubert*'s description of the Rule 702 inquiry as "a flexible one." 509 U. S., at 594. The *Daubert* factors do *not* constitute a definitive checklist or test, *id.,* at 593, and the gatekeeping inquiry must be tied to the particular facts, *id.,* at 591. Those factors may or may not be pertinent in assessing reliability, depending on the nature of the issue, the expert's particular expertise, and the subject of his testimony. Some of those factors may be helpful in evaluating the reliability even of experience-based expert testimony, and the Court of Appeals erred insofar as it ruled those factors out in such cases. In determining whether particular expert testimony is reliable, the trial court should consider the specific *Daubert* factors where they are reasonable measures of reliability. Pp. 149–152.

   (c) A court of appeals must apply an abuse-of-discretion standard when it reviews a trial court's decision to admit or exclude expert testimony. *General Electric Co.* v. *Joiner,* 522 U. S. 136, 138–139. That standard applies as much to the trial court's decisions about how to determine reliability as to its ultimate conclusion. Thus, whether *Daubert*'s specific factors are, or are not, reasonable measures of reliability in a particular case is a matter that the law grants the trial judge broad latitude to determine. See *id.,* at 143. The Eleventh Circuit erred insofar as it held to the contrary. Pp. 152–153.

2. Application of the foregoing standards demonstrates that the District Court's decision not to admit Carlson's expert testimony was lawful. The District Court did not question Carlson's qualifications, but excluded his testimony because it initially doubted his methodology and then found it unreliable after examining the transcript in some detail and considering respondents' defense of it. The doubts that triggered the court's initial inquiry were reasonable, as was the court's ultimate conclusion that Carlson could not reliably determine the cause of the failure of the tire in question. The question was not the reliability of Carlson's methodology in general, but rather whether he could reliably determine the cause of failure of *the particular tire at issue.* That tire, Carlson conceded, had traveled far enough so that some of the tread had been worn bald, it should have been taken out of service, it had been repaired (inadequately) for punctures, and it bore some of the very marks that he said indicated, not a defect, but abuse. Moreover, Carlson's own testimony cast considerable doubt upon the reliability of both his theory about the need for at least two signs of abuse and his proposition about the significance of visual inspection in this case. Respondents stress that other tire failure experts, like Carlson, rely on visual and tactile examinations of tires. But there is no indication in the record that other experts in the industry use Carlson's *particular* approach or that tire experts normally make the very fine distinctions necessary to support his conclusions, nor are there references to articles or papers that validate his approach. Respondents' argument that the District Court too rigidly applied *Daubert* might have had some validity with respect to the court's initial opinion, but fails because the court, on reconsideration, recognized that the relevant reliability inquiry should be "flexible," and ultimately based its decision upon Carlson's failure to satisfy either *Daubert's* factors *or any other* set of reasonable reliability criteria. Pp. 153–158.

131 F. 3d 1433, reversed.

BREYER, J., delivered the opinion of the Court, Parts I and II of which were unanimous, and Part III of which was joined by REHNQUIST, C. J., and O'CONNOR, SCALIA, KENNEDY, SOUTER, THOMAS, and GINSBURG, JJ. SCALIA, J., filed a concurring opinion, in which O'CONNOR and THOMAS, JJ., joined, *post,* p. 158. STEVENS, J., filed an opinion concurring in part and dissenting in part, *post,* p. 159.

Joseph P.H. Babington argued the cause for petitioners. With him on the briefs were *Warren C. Herlong, Jr., John T. Dukes, Kenneth S. Geller,* and *Alan E. Untereiner.*

Jeffrey P. Minear argued the cause for the United States as *amicus curiae* urging reversal. With him on the brief were *Solicitor General Waxman, Assistant Attorney General Hunger, Deputy Solicitor General Wallace, Anthony J. Steinmeyer,* and *John P. Schnitker.*

Sidney W. Jackson III argued the cause for respondents. With him on the brief were *Robert J. Hedge, Michael D. Hausfeld, Richard S. Lewis, Joseph M. Sellers,* and *Anthony Z. Roisman.**

OPINION OF THE COURT

JUSTICE BREYER delivered the opinion of the Court.

In *Daubert* v. *Merrell Dow Pharmaceuticals, Inc.,* 509 U. S. 579 (1993), this Court focused upon the admissibility of scientific expert testimony. It pointed out that such testimony is admissible only if it is both relevant and reliable. And it held that the Federal Rules of Evidence "assign to the trial judge the task of ensuring that an expert's testimony both rests on a reliable foundation and is relevant to the task at hand." *Id.,* at 597. The Court also discussed certain more specific factors, such as testing, peer review, error rates, and "acceptability" in the relevant scientific community, some or all of which might prove helpful in determining the reliability of a particular scientific "theory or technique." *Id.,* at 593–594.

This case requires us to decide how *Daubert* applies to the testimony of engineers and other experts who are not scientists. We conclude that *Daubert*'s general holding—setting forth the trial judge's general "gatekeeping" obligation—applies not only to testimony based on "scientific" knowledge, but also to testimony based on "technical" and "other specialized" knowledge. See Fed. Rule Evid. 702. We also conclude that a trial court *may* consider one or more of the more specific factors that *Daubert* mentioned when doing so will help determine that testimony's reliability. But, as the Court stated in *Daubert,* the test of reliability is "flexible," and *Daubert*'s list of specific factors neither necessarily nor exclusively applies to all experts or in every case. Rather, the law grants a district court the same broad latitude when it decides *how* to determine reliability as it enjoys in respect to its ultimate reliability determination. See *General Electric Co.* v. *Joiner,* 522 U.S.

* Briefs of *amici curiae* urging reversal were filed for the American Automobile Manufacturers Association et al. by *Michael Hoenig, Phillip D. Brady,* and *Charles H. Lockwood II;* for the American Insurance Association et al. by *Mark F. Horning* and *Craig A. Berrington;* for the American Tort Reform Association et al. by *Victor E. Schwartz, Patrick W. Lee, Robert P. Charrow, Mark A. Behrens, Jan S. Amundson,* and *Quentin Riegel;* for the Product Liability Advisory Council, Inc., et al. by *Mary A. Wells, Robin S. Conrad,* and *Donald D. Evans;* for the Rubber Manufacturers Association by *Bert Black, Michael S. Truesdale,* and *Michael L. McAllister,* for the Washington Legal Foundation et al. by *Arvin Maskin, Theodore E. Tsekerides, Daniel J. Popeo,* and *Paul D. Kamenar;* for John Allen et al. by *Carter G. Phillips* and *David M. Levy;* and for Stephen N. Bobo et al. by *Martin S. Kaufman.*

Briefs of *amici curiae* urging affirmance were filed for the Association of Trial Lawyers of America by *Jeffrey Robert White* and *Mark S. Mandell;* for the Attorneys Information Exchange Group, Inc., by *Bruce J. McKee* and *Francis H. Hare, Jr.;* for Bona Shipping (U. S.), Inc., et al. by *Robert L. Klawetter* and *Michael F. Sturley;* for the International Association of Arson Investigators by *Kenneth M. Suggs;* for the National Academy of Forensic Engineers by *Alvin S. Weinstein, Larry E. Cohen,* and *David V. Scott;* for Trial Lawyers for Public Justice, P. C., et al. by *Gerson H. Smoger, Arthur H. Bryant, Sarah Posner, William A. Rossbach,* and *Brian Wolfman;* and for Margaret A. Berger et al. by *Kenneth J. Chesebro, Edward J. Imwinkelried, Ms. Berger, pro se, Stephen A. Saltzburg, David G. Wirtes, Jr., Don Howarth, Suzelle M. Smith, Edward M. Ricci, C. Tab Turner, James L. Gilbert,* and *David L. Perry.*

Briefs of *amici curiae* were filed for the Defense Research Institute by *Lloyd H. Milliken, Jr., Julia Blackwell Gelinas, Nelson D. Alexander,* and *Sandra Boyd Williams;* for the National Academy of Engineering by *Richard A. Meserve, Elliott Schulder,* and *Thomas L. Cubbage III;* and for Neil Vidmar et al. by *Ronald Simon, Turner W. Branch, Ronald Motley, Robert Habush,* and *M. Clay Alspaugh.*

136, 143 (1997) (courts of appeals are to apply "abuse of discretion" standard when reviewing district court's reliability determination). Applying these standards, we determine that the District Court's decision in this case—not to admit certain expert testimony—was within its discretion and therefore lawful.

I

On July 6, 1993, the right rear tire of a minivan driven by Patrick Carmichael blew out. In the accident that followed, one of the passengers died, and others were severely injured. In October 1993, the Carmichaels brought this diversity suit against the tire's maker and its distributor, whom we refer to collectively as Kumho Tire, claiming that the tire was defective. The plaintiffs rested their case in significant part upon deposition testimony provided by an expert in tire failure analysis, Dennis Carlson, Jr., who intended to testify in support of their conclusion.

Carlson's depositions relied upon certain features of tire technology that are not in dispute. A steel-belted radial tire like the Carmichaels' is made up of a "carcass" containing many layers of flexible cords, called "plies," along which (between the cords and the outer tread) are laid steel strips called "belts." Steel wire loops, called "beads," hold the cords together at the plies' bottom edges. An outer layer, called the "tread," encases the carcass, and the entire tire is bound together in rubber, through the application of heat and various chemicals. See generally, *e.g.,* J. Dixon, Tires, Suspension and Handling 68–72 (2d ed. 1996). The bead of the tire sits upon a "bead seat," which is part of the wheel assembly. That assembly contains a "rim flange," which extends over the bead and rests against the side of the tire. See M. Mavrigian, Performance Wheels & Tires 81, 83 (1998) (illustrations).

Carlson's testimony also accepted certain background facts about the tire in question. He assumed that before the blowout the tire had traveled far. (The tire was

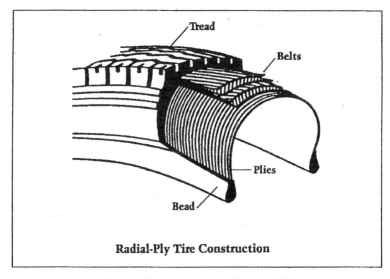

Radial-Ply Tire Construction

FIGURE I. A. Markovich, How to Buy and Care for Tires 4 (1994).

made in 1988 and had been installed some time before the Carmichaels bought the used minivan in March 1993; the Carmichaels had driven the van approximately 7,000 additional miles in the two months they had owned it.) Carlson noted that the tire's tread depth, which was 11/32 of an inch when new, App. 242, had been worn down to depths that ranged from 3/32 of an inch along some parts of the tire, to nothing at all along others. *Id.,* at 287. He conceded that the tire tread had at least two punctures which had been inadequately repaired. *Id.,* at 258–261, 322.

Despite the tire's age and history, Carlson concluded that a defect in its manufacture or design caused the blowout. He rested this conclusion in part upon three premises which, for present purposes, we must assume are not in dispute: First, a tire's carcass should stay bound to the inner side of the tread for a significant period of time after its tread depth has worn away. *Id.,* at 208–209. Second, the tread of the tire at issue had separated from its inner steel-belted carcass prior to the accident. *Id.,* at 336. Third, this "separation" caused the blowout. *Ibid.*

Carlson's conclusion that a defect caused the separation, however, rested upon certain other propositions, several of which the defendants strongly dispute. First, Carlson said that if a separation is *not* caused by a certain kind of tire misuse called "overdeflection" (which consists of underinflating the tire or causing it to carry too much weight, thereby generating heat that can undo the chemical tread/carcass bond), then, ordinarily, its cause is a tire defect. *Id.,* at 193–195, 277–278. Second, he said that if a tire has been subject to sufficient overdeflection to cause a separation, it should reveal certain physical symptoms. These symptoms include (a) tread wear on the tire's shoulder that is greater than the tread wear along the tire's center, *id.,* at 211; (b) signs of a "bead groove," where the beads have been pushed too hard against the bead seat on the inside of the tire's rim, *id.,* at 196–197; (c) sidewalls of the tire with physical signs of deterioration, such as discoloration, *id.,* at 212; and/or (d) marks on the tire's rim flange, *id.,* at 219–220. Third, Carlson said that where he does not find *at least two* of the four physical signs just mentioned (and presumably where there is no reason to suspect a less common cause of separation), he concludes that a manufacturing or design defect caused the separation. *Id.,* at 223–224.

Carlson added that he had inspected the tire in question. He conceded that the tire to a limited degree showed greater wear on the shoulder than in the center, some signs of "bead groove," some discoloration, a few marks on the rim flange, and inadequately filled puncture holes (which can also cause heat that might lead to separation). *Id.,* at 256–257, 258–261, 277, 303–304, 308. But, in each instance, he testified that the symptoms were not significant, and he explained why he believed that they did not reveal overdeflection. For example, the extra shoulder wear, he said, appeared primarily on one shoulder, whereas an overdeflected tire would reveal equally abnormal wear on both shoulders. *Id.,* at 277. Carlson concluded that the tire did not bear at least two of the four overdeflection symptoms, nor was there any less obvious cause of separation; and since neither overdeflection nor the punctures caused the blowout, a defect must have done so.

Kumho Tire moved the District Court to exclude Carlson's testimony on the ground that his methodology failed Rule 702's reliability requirement. The court agreed with

Kumho that it should act as a *Daubert*-type reliability "gatekeeper," even though one might consider Carlson's testimony as "technical," rather than "scientific." See *Carmichael* v. *Samyang Tires, Inc.,* 923 F. Supp. 1514, 1521–1522 (SD Ala. 1996). The court then examined Carlson's methodology in light of the reliability-related factors that *Daubert* mentioned, such as a theory's testability, whether it "has been a subject of peer review or publication," the "known or potential rate of error," and the "degree of acceptance...within the relevant scientific community." 923 F. Supp., at 1520 (citing *Daubert,* 509 U. S., at 589–595). The District Court found that all those factors argued against the reliability of Carlson's methods, and it granted the motion to exclude the testimony (as well as the defendants' accompanying motion for summary judgment).

The plaintiffs, arguing that the court's application of the *Daubert* factors was too "inflexible," asked for reconsideration. And the court granted that motion. *Carmichael* v. *Samyang Tires, Inc.,* Civ. Action No. 93–0860–CB–S (SD Ala., June 5, 1996), App. to Pet. for Cert. 1c. After reconsidering the matter, the court agreed with the plaintiffs that *Daubert* should be applied flexibly, that its four factors were simply illustrative, and that other factors could argue in favor of admissibility. It conceded that there may be widespread acceptance of a "visual-inspection method" for some relevant purposes. But the court found insufficient indications of the reliability of

> "the component of Carlson's tire failure analysis which most concerned the Court, namely, the methodology employed by the expert in analyzing the data obtained in the visual inspection, and the scientific basis, if any, for such an analysis." *Id.,* at 6c.

It consequently affirmed its earlier order declaring Carlson's testimony inadmissible and granting the defendants' motion for summary judgment.

The Eleventh Circuit reversed. See *Carmichael* v. *Samyang Tire, Inc.,* 131 F. 3d 1433 (1997). It "review[ed]...*de novo*" the "district court's legal decision to apply *Daubert.*" *Id.,* at 1435. It noted that "the Supreme Court in *Daubert* explicitly limited its holding to cover only the 'scientific context,'" adding that "a *Daubert* analysis" applies only where an expert relies "on the application of scientific principles," rather than "on skill or experience-based observation." *Id.,* at 1435–1436. It concluded that Carlson's testimony, which it viewed as relying on experience, "falls outside the scope of *Daubert,*" that "the district court erred as a matter of law by applying *Daubert* in this case," and that the case must be remanded for further (non-*Daubert*-type) consideration under Rule 702. 131 F. 3d, at 1436.

Kumho Tire petitioned for certiorari, asking us to determine whether a trial court "may" consider *Daubert*'s specific "factors" when determining the "admissibility of an engineering expert's testimony." Pet. for Cert. i. We granted certiorari in light of uncertainty among the lower courts about whether, or how, *Daubert* applies to expert testimony that might be characterized as based not upon "scientific" knowledge, but rather upon "technical" or "other specialized" knowledge. Fed. Rule Evid. 702; compare, *e.g., Watkins* v. *Telsmith, Inc.,* 121 F. 3d 984, 990–991 (CA5 1997), with, *e.g., Compton* v. *Subaru of America, Inc.,* 82 F. 3d 1513, 1518–1519 (CA10), cert. denied, 519 U. S. 1042 (1996).

II

A

In *Daubert,* this Court held that Federal Rule of Evidence 702 imposes a special obligation upon a trial judge to "ensure that any and all scientific testimony...is not only relevant, but reliable." 509 U. S., at 589. The initial question before us is whether this basic gatekeeping obligation applies only to "scientific" testimony or to all expert testimony. We, like the parties, believe that it applies to all expert testimony. See Brief for Petitioners 19; Brief for Respondents 17.

For one thing, Rule 702 itself says:

> "If scientific, technical, or other specialized knowledge will assist the trier of fact to understand the evidence or to determine a fact in issue, a witness qualified as an expert by knowledge, skill, experience, training, or education, may testify thereto in the form of an opinion or otherwise."

This language makes no relevant distinction between "scientific" knowledge and "technical" or "other specialized" knowledge. It makes clear that any such knowledge might become the subject of expert testimony. In *Daubert,* the Court specified that it is the Rule's word "knowledge," not the words (like "scientific") that modify that word, that "establishes a standard of evidentiary reliability." 509 U. S., at 589–590. Hence, as a matter of language, the Rule applies its reliability standard to all "scientific," "technical," or "other specialized" matters within its scope. We concede that the Court in *Daubert* referred only to "scientific" knowledge. But as the Court there said, it referred to "scientific" testimony "because that [wa]s the nature of the expertise" at issue. *Id.,* at 590, n. 8.

Neither is the evidentiary rationale that underlay the Court's basic *Daubert* "gatekeeping" determination limited to "scientific" knowledge. *Daubert* pointed out that Federal Rules 702 and 703 grant expert witnesses testimonial latitude unavailable to other witnesses on the "assumption that the expert's opinion will have a reliable basis in the knowledge and experience of his discipline." *Id.,* at 592 (pointing out that experts may testify to opinions, including those that are not based on firsthand knowledge or observation). The Rules grant that latitude to all experts, not just to "scientific" ones.

Finally, it would prove difficult, if not impossible, for judges to administer evidentiary rules under which a gatekeeping obligation depended upon a distinction between "scientific" knowledge and "technical" or "other specialized" knowledge. There is no clear line that divides the one from the others. Disciplines such as engineering rest upon scientific knowledge. Pure scientific theory itself may depend for its development upon observation and properly engineered machinery. And conceptual efforts to distinguish the two are unlikely to produce clear legal lines capable of application in particular cases. Cf. Brief for National Academy of Engineering as *Amicus Curiae* 9 (scientist seeks to understand nature while the engineer seeks nature's modification); Brief for Rubber Manufacturers Association as *Amicus Curiae* 14–16 (engineering, as an " 'applied science,' " relies on "scientific reasoning and methodology"); Brief for John Allen et al. as *Amici Curiae* 6 (engineering relies upon "scientific knowledge and methods").

Neither is there a convincing need to make such distinctions. Experts of all kinds tie observations to conclusions through the use of what Judge Learned Hand called "general truths derived from…specialized experience." Hand, Historical and Practical Considerations Regarding Expert Testimony, 15 Harv. L. Rev. 40, 54 (1901). And whether the specific expert testimony focuses upon specialized observations, the specialized translation of those observations into theory, a specialized theory itself, or the application of such a theory in a particular case, the expert's testimony often will rest "upon an experience confessedly foreign in kind to [the jury's] own." *Ibid.* The trial judge's effort to assure that the specialized testimony is reliable and relevant can help the jury evaluate that foreign experience, whether the testimony reflects scientific, technical, or other specialized knowledge.

We conclude that *Daubert*'s general principles apply to the expert matters described in Rule 702. The Rule, in respect to all such matters, "establishes a standard of evidentiary reliability." 509 U. S., at 590. It "requires a valid…connection to the pertinent inquiry as a precondition to admissibility." *Id.,* at 592. And where such testimony's factual basis, data, principles, methods, or their application are called sufficiently into question, see Part III, *infra,* the trial judge must determine whether the testimony has "a reliable basis in the knowledge and experience of [the relevant] discipline." 509 U. S., at 592.

B

Petitioners ask more specifically whether a trial judge determining the "admissibility of an engineering expert's testimony" *may* consider several more specific factors that *Daubert* said might "bear on" a judge's gatekeeping determination. Brief for Petitioners i. These factors include:

- Whether a "theory or technique…can be (and has been) tested";
- Whether it "has been subjected to peer review and publication";
- Whether, in respect to a particular technique, there is a high "known or potential rate of error" and whether there are "standards controlling the technique's operation"; and
- Whether the theory or technique enjoys "general acceptance" within a "relevant scientific community." 509 U. S., at 592–594.

Emphasizing the word "may" in the question, we answer that question yes.

Engineering testimony rests upon scientific foundations, the reliability of which will be at issue in some cases. See, *e.g.,* Brief for Stephen N. Bobo et al. as *Amici Curiae* 23 (stressing the scientific bases of engineering disciplines). In other cases, the relevant reliability concerns may focus upon personal knowledge or experience. As the Solicitor General points out, there are many different kinds of experts, and many different kinds of expertise. See Brief for United States as *Amicus Curiae* 18–19, and n. 5 (citing cases involving experts in drug terms, handwriting analysis, criminal *modus operandi,* land valuation, agricultural practices, railroad procedures, attorney's fee valuation, and others). Our emphasis on the word "may" thus reflects *Daubert*'s description of the Rule 702 inquiry as "a flexible one." 509 U. S., at 594. *Daubert* makes clear that the factors it mentions do *not* constitute a "definitive checklist or test." *Id.,* at 593. And *Daubert* adds that the gatekeeping inquiry must be "tied to the facts" of a particular "case." *Id.,* at 591 (quoting *United States* v. *Downing,* 753 F. 2d 1224, 1242 (CA3 1985)). We agree with the Solicitor General that "[t]he factors identified in *Daubert* may or may not be pertinent in assessing reliability, depending on the nature of the issue, the expert's particular expertise, and the subject of his testimony." Brief for United States as *Amicus Curiae* 19. The conclusion, in our view, is that we can neither rule out, nor rule in, for all cases and for all time the applicability of the factors mentioned in *Daubert,* nor can we now do so for subsets of cases categorized by category of expert or by kind of evidence. Too much depends upon the particular circumstances of the particular case at issue.

Daubert itself is not to the contrary. It made clear that its list of factors was meant to be helpful, not definitive. Indeed, those factors do not all necessarily apply even in every instance in which the reliability of scientific testimony is challenged. It might not be surprising in a particular case, for example, that a claim made by a scientific witness has never been the subject of peer review, for the particular application at issue may never previously have interested any scientist. Nor, on the other hand, does the presence of *Daubert*'s general acceptance factor help show that an expert's testimony is reliable where the discipline itself lacks reliability, as, for example, do theories grounded in any so-called generally accepted principles of astrology or necromancy.

At the same time, and contrary to the Court of Appeals' view, some of *Daubert*'s questions can help to evaluate the reliability even of experience-based testimony. In certain cases, it will be appropriate for the trial judge to ask, for example, how often an engineering expert's experience-based methodology has produced errone-ous results, or whether such a method is generally accepted in the relevant engineer-ing community. Likewise, it will at times be useful to ask even of a witness whose expertise is based purely on experience, say, a perfume tester able to distinguish among 140 odors at a sniff, whether his preparation is of a kind that others in the field would recognize as acceptable.

We must therefore disagree with the Eleventh Circuit's holding that a trial judge may ask questions of the sort *Daubert* mentioned only where an expert "relies on the application of scientific principles," but not where an expert relies "on skill- or experience-based observation." 131 F. 3d, at 1435. We do not believe that Rule 702 creates a schematism that segregates expertise by type while mapping certain kinds of questions to certain kinds of experts. Life and the legal cases that it generates are too complex to warrant so definitive a match.

To say this is not to deny the importance of *Daubert*'s gatekeeping requirement. The objective of that requirement is to ensure the reliability and relevancy of expert testimony. It is to make certain that an expert, whether basing testimony upon pro-fessional studies or personal experience, employs in the courtroom the same level of intellectual rigor that characterizes the practice of an expert in the relevant field. Nor do we deny that, as stated in *Daubert,* the particular questions that it mentioned will often be appropriate for use in determining the reliability of challenged expert testimony. Rather, we conclude that the trial judge must have considerable leeway in deciding in a particular case how to go about determining whether particular expert testimony is reliable. That is to say, a trial court should consider the specific factors identified in *Daubert* where they are reasonable measures of the reliability of expert testimony.

C

The trial court must have the same kind of latitude in deciding *how* to test an expert's reliability, and to decide whether or when special briefing or other proceedings are needed to investigate reliability, as it enjoys when it decides *whether* that expert's relevant testimony is reliable. Our opinion in *Joiner* makes clear that a court of appeals is to apply an abuse-of-discretion standard when it "review[s] a trial court's decision to admit or exclude expert testimony." 522 U. S., at 138–139. That standard applies as much to the trial court's decisions about how to determine reliability as to its ultimate conclusion. Otherwise, the trial judge would lack the discretionary authority needed both to avoid unnecessary "reliability" proceedings in ordinary cases where the reliability of an expert's methods is properly taken for granted, and to require appropriate proceedings in the less usual or more complex cases where cause for questioning the expert's reliability arises. Indeed, the Rules seek to avoid "unjustifiable expense and delay" as part of their search for "truth" and the "jus[t] determin[ation]" of proceedings. Fed. Rule Evid. 102. Thus, whether *Daubert*'s specific factors are, or are not, reasonable measures of reliability in a particular case is a matter that the law grants the trial judge broad latitude to determine. See *Joiner, supra,* at 143. And the Eleventh Circuit erred insofar as it held to the contrary.

III

We further explain the way in which a trial judge "may" consider *Daubert*'s factors by applying these considerations to the case at hand, a matter that has been briefed exhaustively by the parties and their 19 *amici*. The District Court did not doubt Carlson's qualifications, which included a master's degree in mechanical engineering, 10 years' work at Michelin America, Inc., and testimony as a tire failure consultant in other tort cases. Rather, it excluded the testimony because, despite those qualifications, it initially doubted, and then found unreliable, "the methodology employed by the expert in analyzing the data obtained in the visual inspection, and the scientific basis, if any, for such an analysis." Civ. Action No. 93–0860–CB–S (SD Ala., June 5, 1996), App. to Pet. for Cert. 6c. After examining the transcript in "some detail," 923 F. Supp., at 1518–1519, n. 4, and after considering respondents' defense of Carlson's methodology, the District Court determined that Carlson's testimony was not reliable. It fell outside the range where experts might reasonably differ, and where the jury must decide among the conflicting views of different experts, even though the evidence is "shaky." *Daubert,* 509 U. S., at 596. In our view, the doubts that triggered the District Court's initial inquiry here were reasonable, as was the court's ultimate conclusion.

For one thing, and contrary to respondents' suggestion, the specific issue before the court was not the reasonableness *in general* of a tire expert's use of a visual and tactile inspection to determine whether overdeflection had caused the tire's tread to separate from its steel-belted carcass. Rather, it was the reasonableness of using such an approach, along with Carlson's particular method of analyzing the data thereby obtained, to draw a conclusion regarding *the particular matter to which the expert testimony was directly relevant.* That matter concerned the likelihood that a defect in the tire at issue caused its tread to separate from its carcass. The tire in question, the expert conceded, had traveled far enough so that some of the tread had been worn bald; it should have been taken out of service; it had been repaired (inadequately) for punctures; and it bore some of the very marks that the expert said indicated, not a defect, but abuse through overdeflection. See *supra,* at 143–144; App. 293–294. The relevant issue was whether the expert could reliably determine the cause of *this* tire's separation.

Nor was the basis for Carlson's conclusion simply the general theory that, in the absence of evidence of abuse, a defect will normally have caused a tire's separation. Rather, the expert employed a more specific theory to establish the existence (or absence) of such abuse. Carlson testified precisely that in the absence of *at least two* of four signs of abuse (proportionately greater tread wear on the shoulder; signs of grooves caused by the beads; discolored sidewalls; marks on the rim flange), he concludes that a defect caused the separation. And his analysis depended upon acceptance of a further implicit proposition, namely, that his visual and tactile inspection could determine that the tire before him had not been abused despite some evidence of the presence of the very signs for which he looked (and two punctures).

For another thing, the transcripts of Carlson's depositions support both the trial court's initial uncertainty and its final conclusion. Those transcripts cast considerable doubt upon the reliability of both the explicit theory (about the need for two signs of abuse) and the implicit proposition (about the significance of visual inspection in this case). Among other things, the expert could not say whether the tire had traveled more than 10, or 20, or 30, or 40, or 50 thousand miles, adding that 6,000 miles was "about how far" he could "say with any certainty." *Id.,* at 265. The court could reasonably have wondered about the reliability of a method of visual and tactile inspection sufficiently precise to ascertain with some certainty the abuse-related significance of minute shoulder/center relative tread wear differences, but insufficiently precise to tell "with any certainty" from the tread wear whether a tire had traveled less than 10,000 or more than 50,000 miles. And these concerns might have been augmented by Carlson's repeated reliance on the "subjective[ness]" of his mode of analysis in response to questions seeking specific information regarding how he could differentiate between a tire that actually had been overdeflected and a tire that merely looked as though it had been. *Id.,* at 222, 224–225, 285–286. They would have been further augmented by the fact that Carlson said he had inspected the tire itself for the first time the morning of his first deposition, and then only for a few hours. (His initial conclusions were based on photographs.) *Id.,* at 180.

Moreover, prior to his first deposition, Carlson had issued a signed report in which he concluded that the tire had "not been . . . overloaded or underinflated," not because of the absence of "two of four" signs of abuse, but simply because "the rim flange impressions . . . were normal." *Id.,* at 335–336. That report also said that the "tread depth remaining was 3/32 inch," *id.,* at 336, though the opposing expert's (apparently undisputed) measurements indicate that the tread depth taken at various positions around the tire actually ranged from .5/32 of an inch to 4/32 of an inch, with the tire apparently showing greater wear along *both* shoulders than along the center, *id.,* at 432–433.

Further, in respect to one sign of abuse, bead grooving, the expert seemed to deny the sufficiency of his own simple visual-inspection methodology. He testified that most tires have some bead groove pattern, that where there is reason to suspect an abnormal bead groove he would ideally "look at a lot of [similar] tires" to know the grooving's significance, and that he had not looked at many tires similar to the one at issue. *Id.,* at 212–213, 214, 217.

Finally, the court, after looking for a defense of Carlson's methodology as applied in these circumstances, found no convincing defense. Rather, it found (1) that "none" of the *Daubert* factors, including that of "general acceptance" in the relevant expert community, indicated that Carlson's testimony was reliable, 923 F. Supp., at 1521; (2) that its own analysis "revealed no countervailing factors operating in favor of admissibility which could outweigh those identified in *Daubert,*" App. to Pet. for Cert. 4c; and (3) that the "parties identified no such factors in their briefs," *ibid.* For these three reasons *taken together,* it concluded that Carlson's testimony was unreliable.

Respondents now argue to us, as they did to the District Court, that a method of tire failure analysis that employs a visual/tactile inspection is a reliable method, and they point both to its use by other experts and to Carlson's long experience working for Michelin as sufficient indication that that is so. But no one denies that an expert might draw a conclusion from a set of observations based on extensive and specialized experience. Nor does anyone deny that, as a general matter, tire abuse may often be identified by qualified experts through visual or tactile inspection of the tire. See Affidavit of H. R. Baumgardner 1–2, cited in Brief for National Academy of Forensic Engineers as *Amicus Curiae* 16 (tire engineers rely on visual examination and process of elimination to analyze experimental test tires). As we said before, *supra,* at 153–154, the question before the trial court was specific, not general. The trial court had to decide whether this particular expert had sufficient specialized knowledge to assist the jurors "in deciding the particular issues in the case." 4 J. McLaughlin, Weinstein's Federal Evidence ¶ 702.05[1], p. 702–33 (2d ed. 1998); see also Advisory Committee's Note on Proposed Fed. Rule Evid. 702, Preliminary Draft of Proposed Amendments to the Federal Rules of Civil Procedure and Evidence: Request for Comment 126 (1998) (stressing that district courts must "scrutinize" whether the "principles and methods" employed by an expert "have been properly applied to the facts of the case").

The particular issue in this case concerned the use of Carlson's two-factor test and his related use of visual/tactile inspection to draw conclusions on the basis of what seemed small observational differences. We have found no indication in the record that other experts in the industry use Carlson's two-factor test or that tire experts such as Carlson normally make the very fine distinctions about, say, the symmetry of comparatively greater shoulder tread wear that were necessary, on Carlson's own theory, to support his conclusions. Nor, despite the prevalence of tire testing, does anyone refer to any articles or papers that validate Carlson's approach. Cf. Bobo, Tire Flaws and Separations, in Mechanics of Pneumatic Tires 636–637 (S. Clark ed.

1981); C. Schnuth, R. Fuller, G. Follen, G. Gold, & J. Smith, Compression Grooving and Rim Flange Abrasion as Indicators of Over-Deflected Operating Conditions in Tires, presented to Rubber Division of the American Chemical Society, Oct. 21–24, 1997; J. Walter & R. Kiminecz, Bead Contact Pressure Measurements at the Tire-Rim Interface, presented to the Society of Automotive Engineers, Inc., Feb. 24–28, 1975. Indeed, no one has argued that Carlson himself, were he still working for Michelin, would have concluded in a report to his employer that a similar tire was similarly defective on grounds identical to those upon which he rested his conclusion here. Of course, Carlson himself claimed that his method was accurate, but, as we pointed out in *Joiner*, "nothing in either *Daubert* or the Federal Rules of Evidence requires a district court to admit opinion evidence that is connected to existing data only by the *ipse dixit* of the expert." 522 U. S., at 146.

Respondents additionally argue that the District Court too rigidly applied *Daubert*'s criteria. They read its opinion to hold that a failure to satisfy any one of those criteria automatically renders expert testimony inadmissible. The District Court's initial opinion might have been vulnerable to a form of this argument. There, the court, after rejecting respondents' claim that Carlson's testimony was "exempted from *Daubert*-style scrutiny" because it was "technical analysis" rather than "scientific evidence," simply added that "none of the four admissibility criteria outlined by the *Daubert* court are satisfied." 923 F. Supp., at 1521. Subsequently, however, the court granted respondents' motion for reconsideration. It then explicitly recognized that the relevant reliability inquiry "should be 'flexible,'" that its "'overarching subject [should be]...validity' and reliability," and that "*Daubert* was intended neither to be exhaustive nor to apply in every case." App. to Pet. for Cert. 4c (quoting *Daubert*, 509 U. S., at 594–595). And the court ultimately based its decision upon Carlson's failure to satisfy either *Daubert*'s factors *or any other* set of reasonable reliability criteria. In light of the record as developed by the parties, that conclusion was within the District Court's lawful discretion.

In sum, Rule 702 grants the district judge the discretionary authority, reviewable for its abuse, to determine reliability in light of the particular facts and circumstances of the particular case. The District Court did not abuse its discretionary authority in this case. Hence, the judgment of the Court of Appeals is

Reversed.

JUSTICE SCALIA, with whom JUSTICE O'CONNOR and JUSTICE THOMAS join, concurring.

I join the opinion of the Court, which makes clear that the discretion it endorses— trial-court discretion in choosing the manner of testing expert reliability—is not discretion to abandon the gatekeeping function. I think it worth adding that it is not discretion to perform the function inadequately. Rather, it is discretion to choose among *reasonable* means of excluding expertise that is *fausse* and science that is junky. Though, as the Court makes clear today, the *Daubert* factors are not wholly writ, in a particular case the failure to apply one or another of them may be unreasonable, and hence an abuse of discretion.

JUSTICE STEVENS, concurring in part and dissenting in part.

The only question that we granted certiorari to decide is whether a trial judge "[m]ay...consider the four factors set out by this Court in *Daubert* v. *Merrell Dow Pharmaceuticals, Inc.,* 509 U. S. 579 (1993), in a Rule 702 analysis of admissibility of an engineering expert's testimony." Pet. for Cert. i. That question is fully and correctly answered in Parts I and II of the Court's opinion, which I join.

Part III answers the quite different question whether the trial judge abused his discretion when he excluded the testimony of Dennis Carlson. Because a proper answer to that question requires a study of the record that can be performed more efficiently by the Court of Appeals than by the nine Members of this Court, I would remand the case to the Eleventh Circuit to perform that task. There are, of course, exceptions to most rules, but I firmly believe that it is neither fair to litigants nor good practice for this Court to reach out to decide questions not raised by the certiorari petition. See *General Electric Co.* v. *Joiner,* 522 U. S. 136, 150–151 (1997) (STEVENS, J., concurring in part and dissenting in part).

Accordingly, while I do not feel qualified to disagree with the well-reasoned factual analysis in Part III of the Court's opinion, I do not join that Part, and I respectfully dissent from the Court's disposition of the case.

D. ETHICAL ISSUES IN BEING AN EXPERT WITNESS
Joseph B. Kadane

An expert witness is different from a fact witness. A fact witness testifies to what that person saw, heard or experienced. 'I saw the green car turn left and hit the man on the bicycle' is the kind of testimony such a witness might give. By contrast, an expert witness is allowed to give opinion testimony to a court. An expert's qualifications are examined and accepted by the judge before he or she may give such opinions. However, the opinions an expert is permitted to express in court are limited to those in the area of one's expertise, in the case of most of you, to statistics. In particular, it is unwise and unwelcome to express opinions about the law, as the court will not have accepted you as an expert in the law, and regards it as something for the lawyers and the judge to discuss.

The principal legal distinction between cases is that some are criminal cases, while some are civil. In a criminal case, someone's liberty (or life) is in peril; in a civil case, only money or property is generally at stake. In a criminal case, neither the prosecution nor the defence is obligated to disclose to the other what its evidence will be, although the prosecution must disclose any exculpatory evidence to the defence. In a civil case, each side has the right to advance discovery of almost all evidence to be produced by the opponent at trial. This includes the right to depose, i.e. to question under oath, the other side's experts and fact witnesses. Both parties have access to anything written by the expert (including notes, etc.).

There are, in addition, various administrative hearings, each with its own set of rules. An expert should ask what the rules are that apply to the particular setting he or she will be involved in.

In almost all cases, the expert is hired by one side or the other, sometimes to give direct testimony, sometimes to rebut the testimony given by others or sometimes both. The attorney who hires you will make it clear to what they want you to testify.

The hiring attorney will also make it clear how you are to be paid. An expert, unlike an attorney, cannot have a financial stake in the outcome of the case, and hence cannot be paid more if the case is successful than if it is not. Hence, most experts are paid by the hour. (A typical trap question is 'Are you paid for your opinion?,' to which the appropriate answer is 'No, I am paid for my time.')

Sometimes (in my experience rarely, but it does happen), an attorney asks for testimony that I feel I cannot give, because what is being asked for is not true. For example, I remember a case in which the attorney, defending a client accused of Medicare fraud, wanted me to testify that sampling of patient records could not possibly be used to ascertain whether excessive testing had been ordered by the physician defendant who owned the laboratory in which the tests were to be done. I explained, to the attorney's dismay, that such sampling could indeed be done, and if done correctly could save enormous amounts of investigator time. The case against the physician was settled by a plea bargain.

In this instance, I was telling the attorney, in private, the unpalatable truth so that she could knowledgeably guide her client's case. I did my job correctly. Usually when such unwelcome advice is given, the expert's involvement with the case soon ends. The attorney is free to try to find another expert with a different point of view. However, the expert is not free to change sides, as the expert is presumably privy to client secrets.

Usually, however, what one is asked to testify to is not obviously contrary to well-established statistical principles. In that case, an expert is likely to be asked for a report, detailing the sources used, the conclusions reached and the support for these conclusions. In a civil case, this report would be made available to the other side, but not in a criminal case. In writing such a report, it is very important to understand the legal theory your side is using and possibly the other side's as well. You should understand the role the facts you are asked to substantiate will play in what the attorney is trying to argue.

In a civil deposition, the other side will ask questions. At the end, the attorney for your side may ask you questions to clarify the record if need be. In a trial, however, the attorney for your side goes first, asking you questions to lead you through your qualifications, your conclusions and the reasoning that gets you to these conclusions. Then the other side gets to ask you questions. This is known as 'cross-examination,' and is often the most intellectually taxing part of the experience. You should be aware that the purpose of cross-examination is often to confuse the record, get you to say things you may not really mean and, perhaps, to destroy your credibility.

In all of this, there is an underlying tension between your responsibilities to your client and your responsibilities to the court. The oath you take is 'to tell the truth, the whole truth, and nothing but the truth,' not 'only those aspects of the truth that help my client.' Thus, you must answer each question carefully and honestly. As described above, if there are weaknesses in your analysis, you should have explained them to the attorney who hired you before getting to court. If, on cross-examination, a great chasm of logic opens before you, that is what cross-examining lawyers hope for. You will generally not have the opportunity to volunteer information unrequested by either side, however.

I favour the Bayesian approach to statistics. As a result, often the analyses I present are Bayesian. The reason I do so is that, if I am asked the question, 'Is this, in your judgment, the best analysis that can be performed on the data?,' I want to be able to answer 'yes'. Indeed, if I cannot answer 'yes,' what am I doing in court? Thus, my obligation to the court to give my best professional opinion leads, because of my philosophical leanings, to give Bayesian analyses.

I have sometimes encountered the argument that a Bayesian analysis would be difficult to explain, and hard for the judge or jury to understand. I think it is my job to explain whatever analysis I do in such a way that the decision-maker, whether judge or jury, has a feel for what was done, and what it means. To report less than the real reasons for my opinions would be to take the attitude 'You're too stupid to understand my full analysis, and I'm so inarticulate that I can't explain it to you.' This is not a professional attitude with which to approach a court (or any other forum). I think a statistician is better advised to go to court reflecting what he or she believes to be best statistical practice, whether Bayesian, likelihood, frequentist or whatever. Part of what makes testimony an interesting and engaging activity is the clash of statistical styles that emerges.

Of course, the decision-makers, whether judge or jury, are not expert statisticians. Instead, they are supposed to approach their decision with general common sense, and sort out whom to believe. They face the same difficulty in understanding statistics as they do with any other form of specialized testimony.

I find it useful to publish the story of many of my cases. Often there is a matter of confidentiality to be negotiated to do so. However, I find it useful to have in my mind that anything I say in court, I will write and defend to all of my statistical colleagues in print. This restrains my competitive instinct to think and act as a lawyer, which in fact would reduce my usefulness as an expert witness.

In writing about my cases, one question I have had to confront is whether to mention the names of my opponent expert witness, particularly when I have criticisms to make. This is partly, of course, a matter of personal style and there is a reasonable view that would criticize the argument without using the name of the person making it. However, there is more to it than that. A foolish argument put forward by someone holding themselves out as a statistical expert reflects on our profession. In America, we do not have formal mechanisms to certify statisticians (which I think saves us untold grief), so publication with names is the only way for statisticians to police themselves. It is fair, in that anyone who feels unwarrantedly criticized can write in response.

Testifying in court is among the most demanding work that I do, both technically and ethically. It requires thinking through not just the details of the case, but also how I stand philosophically about statistics, and how I stand ethically and morally. And that is why I find it a welcome challenge.

Elaine B. **PANITZ,** M.D., Appellee v. Kenneth W. **BEHREND** and Barbara **Behrend** Ernsberger, Individually and T/A **Behrend** and Ernsberger, Appellants

No. 399 Pittsburgh, 1993

Superior Court of Pennsylvania

429 Pa. Super. 273; 632 A.2d 562;

September 1, 1993, Argued

October 13, 1993, Filed

COUNSEL: Kenneth R. Behrend, Pittsburgh, for appellants.

Alan S. Gold, Jenkintown, for appellee.

JUDGES: Rowley, President Judge, and Wieand and Cirillo, JJ.

OPINION BY: WIEAND

OPINION

Elaine B. Panitz, a medical doctor who regularly offers her services as an expert medical witness, was hired by Kenneth W. Behrend, Barbara Behrend Ernsberger and the law firm of Behrend and Ernsberger to give testimony on behalf of clients whom the law firm represented in a personal injury action. When an unfavorable verdict was returned, the lawyers refused to pay the expert witness the balance of the moneys which they allegedly had agreed to pay. Panitz sued to recover these moneys. The law firm thereupon filed an answer to the complaint which contained a counterclaim for damages resulting from the unfavorable verdict. This, it was alleged, had been caused by gross negligence and misrepresentation regarding the substance of Panitz's testimony at trial. To this counterclaim Panitz filed preliminary objections in the nature of a demurrer. When the trial court sustained the preliminary objections and dismissed the counterclaim, the defendant law firm appealed.

When reviewing an appeal from an order sustaining preliminary objections in the nature of a demurrer to a pleading, we accept as true all well–pleaded facts and all reasonable inferences to be drawn therefrom. The decision of the trial court will be affirmed only if there is no legal theory under which a recovery can be sustained on the facts pleaded. *Allegheny County v. Commonwealth,* 507 Pa. 360, 372, 490 A.2d 402, 408 (1985); *Rutherfoord v. Presbyterian–University Hospital,* 417 Pa.Super. 316, 321–322, 612 A.2d 500, 502–503 (1992).

In the underlying action, the Behrend firm had represented the Charney family whose members, allegedly, had been exposed to formaldehyde in building materials and had sustained formaldehyde sensitization reactions. Panitz was employed to support the alleged cause of action. It was expected that she would be cross–examined about the lack of such sensitization in cigarette smokers who regularly are exposed to much greater concentrations of formaldehyde than were the Charneys. In preparation for trial, Panitz provided the Behrend firm with a transcript of depositions in a prior case in which she had postulated on the lack of sensitization in smokers. Panitz testified at trial, as anticipated, that in her opinion the Charneys' injuries had been caused by formaldehyde present in building materials. When cross–examined about the lack of sensitization in cigarette smokers, however, Panitz conceded that she could not explain the apparent inconsistency. After trial, Panitz explained that she had come to realize prior to trial that the reasoning upon which she had relied in earlier depositions was inaccurate.

As a general rule there is no civil liability for statements made in the pleadings or during trial or argument of a case so long as the statements are pertinent. *Post v. Mendel,* 510 Pa. 213, 221, 507 A.2d 351, 355 (1986); *Greenberg v. Aetna Ins. Co.,* 427 Pa. 511, 516, 235 A.2d 576, 577 (1967), *cert. denied, Scarselletti v. Aetna Ins. Co.,* 392 U.S. 907, 88 S.Ct. 2063, 20 L.Ed.2d 1366 (1968); *Moses v. McWilliams,* 379 Pa. Super. 150, 163, 549 A.2d 950, 956 (1988) *allocatur denied,* 521 Pa. 631, 558 A.2d 532 (1989); *Pelagatti v. Cohen,* 370 Pa.Super. 422, 436, 536 A.2d 1337, 1344 (1987), *allocatur denied,* 519 Pa. 667, 548 A.2d 256 (1988). The privilege, which includes judges, lawyers, litigants and witnesses, had its origin in defamation actions premised upon statements made during legal actions, but it has now been extended to include all tort actions based on statements made during judicial proceedings. Thus, in *Clodgo by Clodgo v. Bowman,* 411 Pa.Super. 267, 601 A.2d 342 (1992), the judicial or testimonial privilege was held to insulate a court appointed medical expert witness from liability premised upon malpractice. See also: *Moses v. McWilliams, supra; Brown v. Delaware Valley Transplant Program,* 371 Pa.Super. 583, 538 A.2d 889 (1988). "The form of the cause of action is not relevant to application of the privilege. Regardless of the tort contained in the complaint, if the communication was made in connection with a judicial proceeding and was material and relevant to it, the privilege applies." *Clodgo by Clodgo v. Bowman, supra,* 411 Pa.Super. at 273, 601 A.2d at 345. The privilege is equally applicable where the cause of action is stated in terms of misrepresentation or a contractual requirement to exercise due care.

The purpose for the privilege is to preserve the integrity of the judicial process by encouraging full and frank testimony. This was recognized by the Supreme Court of the United States in *Briscoe v. LaHue,* 460 U.S. 325, 103 S.Ct. 1108, 75 L.Ed.2d 96 (1983), where the Court said:

> "[T]he claims of the individual must yield to the dictates of public policy, which requires that the paths which lead to the ascertainment of truth should be left as free and unob-structed as possible." *Calkins v. Sumner,* 13 Wis. 193, 197 (1860). A witness' apprehen-sion of subsequent damages liability might induce two forms of self–censorship. First, witnesses might be reluctant to come forward to testify. See *Henderson v. Broomhead, supra,* 578–579, 157 Eng. Rep., at 968. And once a witness is on the stand, his testimony might be distorted by the fear of subsequent liability. See *Barnes v. McCrate,* 32 Me. 442, 446–447 (1851)....A witness who knows that he might be forced to defend a sub-sequent lawsuit, and perhaps to pay damages, might be inclined to shade his testimony in favor of the potential plaintiff, to magnify uncertainties, and thus to deprive the finder of fact of candid, objective, and undistorted evidence. See Veeder, Absolute Immunity in Defamation: Judicial Proceedings, 9 Colum.L.Rev. 463, 470 (1909).

Id. 460 U.S. at 332–333, 103 S.Ct. at 1114, 75 L.Ed.2d at 103–104. Similarly, the Supreme Court of Pennsylvania, in *Binder v. Triangle Publications, Inc.,* 442 Pa. 319, 275 A.2d 53 (1971), explained:

> The reasons for the absolute privilege are well recognized. A judge must be free to administer the law without fear of consequences. This independence would be impaired were he to be in daily apprehension of defamation suits. The privilege is also extended to parties to afford freedom of access to the courts, to witnesses to encourage their complete and unintimidated testimony in court, and to counsel to enable him to best represent his client's interests.

Binder v. Triangle Publications, Inc., supra at 323–324, 275 A.2d at 56. See also: *Moses v. McWilliams, supra,* 379 Pa.Super. at 164, 549 A.2d at 957. The privilege, thus, serves the salutary purpose of encouraging witnesses to give frank and truth-ful testimony. Having testified truthfully in the judicial process, a witness should not thereafter be subjected to civil liability for the testimony which he or she has given.

"[W]itness immunity should and does extend to pretrial communications. The pol-icy of providing for reasonably unobstructed access to the relevant facts is no less compelling at the pre-trial stage of judicial proceedings." *Moses v. McWilliams, supra* 379 Pa.Super. at 166, 549 A.2d at 958. Thus, the privilege is not to be avoided by the disingenuous argument that it was not the in–court testimony that caused the loss but the pre-trial representations about what the in–court testimony would be. The privilege includes all communications "issued in the regular course of judicial proceedings and which are pertinent and material to the redress or relief sought." *Post v. Mendel, supra,* 510 Pa. at 221, 507 A.2d at 355 (emphasis omitted). See also: *Kahn v. Burman,* 673 F.Supp. 210, 213 (E.D. Mich. 1987), *aff'd,* 878 F.2d 1436 (6th Cir. 1989); *Greenberg v. Aetna Ins. Co., supra* at 516, 235 A.2d at 577 (applying privilege to statements in pleadings); *Moses v. McWilliams, supra* (applying privilege to statements made by witness to attorney during pre-trial conference); *Pelagatti v. Cohen, supra,* 370 Pa.Super. at 436, 536 A.2d at 1344 (applying privilege to statements in pre-trial affidavits). The "expert's courtroom testimony is the last act in a long, complex process of evaluation and consultation with the litigant. There is no way to distinguish the testimony from the acts and communications on which it is based." *Bruce v. Byrne–Stevens & Associates Engineers, Inc.,* 113 Wash.2d 123, 135, 776 P.2d 666, 672 (1989). See also: *Middlesex Concrete Products v. Carteret Ind. Ass'n,* 68 N.J.Super. 85, 92, 172 A.2d 22, 25 (1961).

There also is no reason for refusing to apply the privilege to friendly experts hired by a party. The policy of encouraging frank and objective testimony, without fear of civil liability therefor, "obtains irrespective of the manner by which the witness comes to court." *Bruce v. Byrne–Stevens & Associates Engineers, Inc., supra,* 113 Wash.2d at 129, 776 P.2d at 669. The primary purpose of expert testimony is not to assist one party or another in winning the case but to assist the trier of the facts in understanding complicated matters.

In *Bruce v. Byrne–Stevens & Associates Engineers, Inc., supra,* a claim was made against an engineering witness on grounds that the expert had negligently miscalculated the cost of restoring lateral support to plaintiff's land. In dismissing the action, the Supreme Court of Washington held that (1) an expert witness was entitled to the privilege even if he or she had been retained and compensated by a party rather than appointed by the court, and (2) immunity extended not only to in–court testimony but also to acts and communications which had occurred in connection with preparing that testimony. The court reasoned:

> While it may be that many expert witnesses are retained with the expectation that they will perform as "hired guns" for their employer, as a matter of law the expert serves the court. The admissibility and scope of the expert's testimony is a matter within the court's discretion. *Orion Corp. v. State,* 103 Wash.2d 441, 462, 693 P.2d 1369 (1985). That admissibility turns primarily on whether the expert's testimony will be of assistance to the finder of fact. ER 702. The court retains the discretion to question expert witnesses. ER 614(b). The mere fact that the expert is retained and compensated by a party does not change the fact that, as a witness, he is participant in a judicial proceeding.

Id. 113 Wash.2d at 129–130, 776 P.2d at 669. A contrary holding, the court concluded, would result in a loss of objectivity in expert testimony generally and would discourage anyone who was not a professional witness from testifying. Although an intermediate appellate court in California has held that the privilege should not apply to a "friendly" expert, see: *Mattco Forge, Inc. v. Arthur Young & Co.,* 5 Cal.App.4th 392, 6 Cal.Rptr.2d 781 (1992), and an intermediate appellate court in Arizona has implied in dictum that it would not be inclined to extend the privilege to "friendly" experts, see: *Lavit v. Superior Court,* 173 Ariz. 96, 839 P.2d 1141 (Ariz.Ct.App.1992), we conclude that the better view is that followed by the Supreme Court of Washington. To allow a party to litigation to contract with an expert witness and there by obligate the witness to testify only in a manner favorable to the party, on threat of civil liability, would be contrary to public policy. *Griffith v. Harris,* 17 Wis.2d 255, 116 N.W.2d 133 (1962), *cert. denied,* 373 U.S. 927, 83 S.Ct. 1530, 10 L.Ed.2d 425 (1963). See also: *Curtis v. Wolfe,* 160 Il.App.3d 588, 513 N.E.2d 1139,

112 Il.Dec. 530 (1987), *appeal denied,* 118 Ill.2d 542, 520 N.E.2d 384, 117 Ill.Dec. 223 (1988). "Fundamentally, no witness can be required to testify, and no witness should be expected to testify, to anything other than the truth as he [or she] sees it and according to what he [or she] believes it to be. The same is expected of expert witnesses." *Schaffer v. Donegan,* 66 Ohio App.3d 528, 538, 585 N.E.2d 854, 860 (1990), *jurisdictional motion overruled,* 55 Ohio St.3d 722, 564 N.E.2d 500 (1990). An expert witness will not be subjected to civil liability because he or she, in the face of conflicting evidence or during rigorous cross–examination, is persuaded that some or all of his or her opinion testimony has been inaccurate.

Because appellee was immune from civil liability for testimony which she gave, the trial court properly dismissed the counterclaim against her.

Affirmed.

Part 2

DISCRIMINATION

This part addresses cases exemplifying three kinds of discrimination: employment discrimination in Section A, discrimination in the enforcement of the law (Section B), and housing discrimination (Section C).

SECTION A: DISCRIMINATION IN EMPLOYMENT

Generally, federal law prohibits decisions by employers disadvantaging employees because of their race, sex, religion, national origin, and age (40 and older). There are two parts to the case law on this subject. A disparate treatment case deals with situations in which a rule on its face treats a protected class disadvantageously. Such cases are now rare, as employers have learned that discriminatory rules are illegal. The second kind of case is disparate impact, in which the protected class is disadvantaged, even perhaps inadvertently, by the employer. Almost inevitably, statistics plays a large role in such a case.

Both of the papers here deal with discrimination on the basis of age. In a typical age discrimination case, there is a continuum of calender time to deal with. Sometimes the employer's acts in question (here involuntarily terminating, *i.e.,* firing workers), occur at a few discrete points in time. In other cases, the employer's acts are gradual over a period of time that can be several years in length.

The first paper in this section, Kadane (1990), reports on a case with four firing waves, so it is an example of acts at discrete time points. The second paper, Kadane and Woodworth (2004), concerns two cases in which the firings occur gradually over a period of time.

Kadane (1990) does an extensive comparison of the two dominant frameworks currently used in statistics: the classical, or sampling, view and the Bayesian view. To help the reader appreciate the distinction, I give here a brief explanation.

There are two fundamental quantities for statisticians: the data, typically denoted with Roman letters, like x, and parameters, typically denoted with Greek letters, like θ. The likelihood, often written $f(x \mid \theta)$, is the probability distribution of the data x given the value of the parameter θ. One way of understanding the distinction between Bayesian and classical statistics is to focus on what is regarded as random and what is fixed. To a Bayesian, the data, once observed, are fixed at the observed values, but the value of the parameter, being uncertain, has a probability distribution. To a classical statistician, the data, even after they are observed, are random, while the parameter is a fixed (but unknown) value. Kadane (1990) compares these approaches in the context of an employment case, and generally finds the Bayesian approach more useful in addressing the legal questions posed. For more on the comparison between approaches, the reader is referred to the very useful book by Barnett (1999).

Bayesian statistics requires the statement of a prior distribution of the parameter θ, reflecting opinion before the data are observed. Obviously this is a sensitive point, since a choice of a prior distribution that favors one side or the other could bias the result. Both papers go to great pains to explain why the choice of prior is reasonable, and does not favor either party.

A second major technical issue in Bayesian statistics is the computation of the distribution of θ after seeing the data x (this distribution is called the posterior distribution). In Kadane (1990) this issue is simple to resolve, because there θ is one-dimensional, and hence straight-forward numerical integration suffices. However, when time is viewed continuously, θ is then a process that evolves over time. This complicates the calculations; the Kadane and Woodworth (2004) paper explains one way to do them. I hope that those for whom the computational details are daunting can nevertheless appreciate the goal of the calculations.

SECTION B: DRIVING WHILE BLACK

The United States has a history of legalized slavery which existed until the end of the Civil War in 1865, and of Jim Crow legalized discrimination until the 1960's. A modern form of racism is differential enforcement of the law, under which minorities, especially African-Americans, are dealt with more harshly than others for identical behavior. Differential enforcement is illegal under the 14th Amendment to the US Constitution, which says in part, "No state shall ... deny to any person within its jurisdiction the equal protection of the laws." The extent to which differential enforcement persists is hotly debated. The case discussed in this section addresses such an allegation, brought by the New Jersey Public Defender's Office against the New Jersey State Police. It alleged that the New Jersey State Police were stopping cars driven by African-Americans at greater rates than those driven by others, with no evidence of greater rates of wrong-doing.

The first item in this section is the court decision in the case, which gives a summary of the evidence and the judge's view of it. The second item (Kadane and Terrin (1997)) is a paper addressing the major inferential issue in the case, namely whether the high level of missing race data in the police records rendered the matter too uncertain to address. The third paper (Terrin and Kadane (1998)) addresses certain arguments put forward by the state in rebuttal to the Public Defender's reports.

It might be useful to report what happened after Judge Francis released his decision. The New Jersey Attorney General's Office took great exception to Judge Francis's decision, and announced their decision to appeal. About a year later, a week before the appeal was to be heard, the New Jersey Attorney General announced that the state was abandoning the appeal. They had sponsored their own study of which the fourth item is an official summary. For the full report, see Interim Report of the State Police Review Team Regarding Allegations of Racial Profiling (www.state. nj.us/lps/itm_419.pdf (April 20, 1999)) that found what we found. A consent decree was negotiated between the State of New Jersey and the Civil Rights Division of the US Department of Justice (*US will monitor New Jersey State Police on Racial Profiling,* New York Times, Dec. 23, 1999, p. A1, David Kocleniewski). The extent to which the New Jersey State Police have in fact changed their behavior remains to be seen.

SECTION C: RACIAL STEERING

Discrimination by refusal to rent or sell housing because of race is illegal under the Federal Fair Housing Act of 1968, and under some state laws. This paper addresses the issue of whether an apartment manager illegally discriminated against minority applicants by steering minorities to less desirable units. The objective of such a practice may include the segregation of minorities in less desirable buildings, or the exclusion of minorities as renters. The statistical issue is whether new renters were steered by management to rent apartments so as to maintain a racially-segregated apartment complex. The way that might work is in showing only a few of the available apartments, a potential African-American renter might systematically be shown apartments in one part of the complex, while a non-African-American might systematically be shown apartments in another part.

The data available to address this issue were weekly records of the available apartments of various sizes, the units rented and the race of the renter. There were no records of which units had been shown to which prospective renters. Thus a natural model is to estimate the probability that a new tenant rents in one of two sections as a function of the number of available units and the race of the renter. We also need to account for the possibility that one section is more popular than another (to everyone). The easiest model for this purpose is a logistic regression, detailed in the article.

Although our estimates lead us strongly to suspect that racial steering occurred, we cannot be sure because of the lack of data on what units were shown to what potential tenants.

REFERENCES

Barnett, V. (1999). Comparative Statistical Inference, 3rd edition, J. Wiley & Sons, Chichester.

A. DISCRIMINATION IN EMPLOYMENT

1

A STATISTICAL ANALYSIS OF ADVERSE IMPACT OF EMPLOYER DECISIONS

Joseph B. Kadane

1. INTRODUCTION

Federal law (U.S. Code, Title 29, Chapter 14 § 626) forbids discrimination against people 40 years of age and older with respect to employment decisions. Under the doctrine of adverse impact, statistics can be used to establish a prima facie case of discrimination. Under this doctrine it is not necessary to prove that the employer had discriminatory intent, but only to show that his actions had the actual effect of disadvantaging a disproportionate number of people of protected age or race. See United States v. Hazlewood School District, 443 U.S. 299 (1977) (race discrimination) and Geller v. Markham, 635 F.2d 1027 (2d Cir. 1980). To overcome such a prima facie showing, an employer can demonstrate a nondiscriminatory business reason for the decisions.

What must be shown statistically to demonstrate adverse impact? This question is analyzed using the data from a recent case, for which I served as an expert witness for the plaintiff, and which was settled just before the start of testimony. In order to protect the privacy of the parties involved, the plaintiff will be referred to as the Employee, and the defendant as the Company.

The Company was adversely affected by the downturn in the basic metals markets in the early 1980s. In that period the Company reduced the size of its work force in a series of moves that were the occasion for the lawsuit. The Employee is a union member who had received several promotions leading to a very responsible management position. In the fourth of four firing waves, his job was "abolished" (divided among his two former subordinates) and he used his union seniority rights to bump

Journal of the American Statistical Association, December 1990, Vol. 85, No. 412, Applications and Case Studies.

someone from a lower-paying union position. Because there was another aspect of his case, involving actions on the Employee's part in the nature of whistle-blowing, his suit was brought individually against the Company, and not as a class-action proceeding on behalf of all fired management employees. In the adverse impact portion of his lawsuit the Employee alleged that the Company had disproportionately fired people 40 and over (himself included). The case was a civil suit; such cases are decided on "the preponderance of the evidence."

In the remainder of this article, Section 2 describes issues concerning the data base, Section 3 treats the analysis, and Section 4 contains my conclusions.

2. DATABASE ISSUES

In civil litigation, each side is entitled to discovery, that is, to whatever records and analyses the other side intends to use in court and, additionally, all other "relevant" records. In this case, the plaintiff's preparation of the case was hampered by the poor state of the defendant's personnel records. The Employee's position after he was dismissed from his management job involved handling personnel records; he often knew of errors or incompleteness in the service record cards made available by the defendant. Ultimately it was possible for the plaintiff to assemble a list, alleged to be complete, of all management employees during the period in question, their birth dates, and the dates they left management, if they had. Where ambiguities or doubts arose, the benefit of the doubt was given to the service record cards, because they were the documentary evidence made available by the Company in discovery. In this way, errors in the data base, if any, would be the responsibility of the Company.

A cursory examination of the list revealed that many employees had left management on four specific dates: 6/30/82, 11/30/82, 5/31/83, and 6/28/84. I was informed by the Employee's attorney that the Company had announced to these employees that their jobs were to be abolished, that unless they had union rights they would no longer be employed by the Company, and that they could choose to retire and receive a pension from the company if they were eligible. In view of the involuntary nature of these departures from management service, I refer to them as "firings" even though some, with union rights, might still have been employed by the Company in other positions.

As is natural in any human population, there were, in addition, various other departures from service during the two-year period in question. One employee died, several retired, and some resigned to take positions with other firms. There were very few, but some, hirings during the period as well. I ignored these departures on other dates, because they were the results of decisions not by the Company, but rather, in the main, by employees. Thus to include them would be either to credit or to blame the Company for decisions it did not make. For more on this point, see Michelson (1986), who distinguishes between "situations" (which are not relevant for discrimination suits) and "events" (which are relevant).

The first step in my analysis was simply to count those retained and fired in each of the four firing waves, divided by whether the employee was older than 40 or not at the time. These counts are given in Table 1.

TABLE 1. Ages of Those Fired and Retained
in Four Firing Waves by the Company

Age	Fired	Retained
	6/30/82 Firings	
40+	18	129
39–	0	102
	11/30/82 Firings	
40+	26	105
39–	10	83
	5/31/83 Firings	
40+	13	92
39–	14	66
	6/28/84 Firings	
40+	13	81
39–	2	52

Shortly before the trial the Company submitted its own version of the data, and an analysis of those data by its own expert. The data were prepared by the Company's Director of Personnel and Public Relations, who alleged that the data base used by the plaintiff was "inaccurate and incomplete." He presented his results in two data bases. In the first data base, he reported his counts of both voluntary and involuntary terminations, divided into four periods: 6/30/82, 7/01/82 through 11/30/82, 12/01/82 through 5/31/83, and 6/01/83 through 6/24/84. The second data base recorded his counts of involuntary terminations for those four periods. Both of these data bases were simply counts; he did not explain why, or on what basis, he disagreed with the Employee's categorization of the individual terminations. The latter data base did not coincide with the Employee's for several reasons:

1. There were several young employees who were permitted to hold management jobs for a short time, and then furloughed. I did not include these as involuntary terminations in my counts, since the circumstances were that these furloughs occurred while the employees were still on probation. The question of whether the Company had discriminated against its older management employees seemed to me to be independent of whether it had also temporarily promoted several younger workers and then furloughed them back to union positions.
2. The Personnel Director alleged that the entire first wave consisted only of voluntary departures from service.
3. The Personnel Director alleged various other errors and omissions in the data base I had been furnished, without specifying what they were, or to which employee they attached.

The struggle over the data base would have been a major feature of the trial had it occurred. The Company's attorney would have tried to portray the Employee, who would have been presenting the data base on his own behalf, as an error-prone and biased witness. The Employee's attorney would have tried to portray the Personnel Director as disingenuous. That the Personnel Director's information had not been made available to the Employee in discovery would have been used in a legal effort to have his testimony quashed. If the Personnel Director's testimony had been allowed, he would have been asked to identify those departures that he claimed were voluntary. This would have permitted the Employee, in surrebuttal, to call those former employees as witnesses, to testify about the circumstances of their departure from the Company. It is difficult to guess what the effect of all this would have been on the jury.

3. ANALYSES

My expert's report presented three kinds of analyses: a Fisher exact test, a Bayesian analysis, and an analysis based on the Mann–Whitney–Wilcoxon statistic. I would have put least weight on the Fisher exact test; in contrast, the Company's statistician would have put virtually all his weight on Fisher's exact tests at the .05 level.

3.1 Fisher's Exact Tests

The significance levels for the one-tailed Fisher exact test for the four firing waves reported in Table 1 are .0001, .0485, .8821, and .0407, respectively. Thus wave I is highly significant, waves II and IV are marginally significant at the .05 level, and wave III is not significant. The Company's statistician told me after the settlement that with respect to wave I he would have relied in court on the Company's Personnel Director's explanation that those departures were voluntary, and that with respect to waves II and IV the shift of a single fired person from the over-40 to the under-40 group would have changed the Fisher exact test significance levels to .0906 and .1193, respectively, thus allowing him to say that the case rested on a single birth date in each group. See Gastwirth (1988, pp. 226–227) for discussion of this argument in another case.

In rebuttal I would have pointed out that the shift of a single fired employee in the other direction, from the under-40 to the over-40 group, would have changed the significance levels to .0234 and .0091, respectively. I also would have said that I think statisticians should analyze the data sets they have, not make up new ones whose conclusions they like better. Nonetheless, his robustness argument probably would have had some appeal or would have made the jury abandon all hope of understanding the statistics.

In general, tests of significance, including the Fisher exact test, are vulnerable in court to the following sort of fantasy cross-examination:

Q: In your analysis, you used 40 years of age as a threshold between young people and old people. Why did you use 40 years, as opposed to 35 or 45, say?
A: It is my understanding that the law protects exactly persons 40 years of age and older. Consequently, to be relevant to the case I think that it is essential to use 40 years of age as the threshold.
Q: So your use of 40 years of age and older is because you think the law requires it?
A: Yes.

Q: You also use the number .05 as a threshold between data you call "significant" and data you do not call "significant." Why do you use the number .05? Would .03 or .07 do as well?

A: I use .05 because it is a traditional number to use in statistics. The Federal Government Equal Employment Opportunity Commission uses it as a threshold level of significance [29 C.F.R. § 1607.4(D)].

Q: To the best of your knowledge, is the number .05 used in the written law?

A: Not to the best of my knowledge. Court decisions that use it include the Federal Supreme Court in Albemarle Paper Co. v. Moody (1975).

Q: What is the origin of the use of .05 as a significance level?

A: I believe it goes back to Sir Ronald Fisher, the same statistician who invented the Fisher exact test.

Q: Why did Fisher use .05 rather than .03 or .07?

A: Fisher doesn't really say. He points out that .05 is 1 in 20, but similar equivalences could be found for .03 and .07.

Q: Do you agree with Raiffa and Schlaifer (1961, p. vi) when they write "the numbers .05 and .01 [are treated] in statistics with the same superstitious awe that is usually reserved for the number 13"?

A: I use the number .05 because it is the traditional number.

Q: But if you used .07 or .03 you would come to rather different results in this case, is that true?

A: Yes, it is.

Q: Now I want to ask you about the meaning of a significance test, perhaps using the Fisher exact test as an example. What does it mean to say that the data are significant at the .05 level?

A: There are two different meanings given to such a statement. According to Fisher (1959), the meaning is that, if the null hypothesis of independence between age and being fired is true, the probability of seeing data as or more discriminatory than the data observed is less than .05. According to Neyman and Pearson (1967), the number .05 is a property of the decision procedure. It says that if I use this procedure many times when the null hypothesis is true, in only .05 of the times will I make an error in rejecting the null hypothesis.

Q: Let's take each of these meanings in turn. Do I understand correctly that, with respect to the Fisher interpretation, the calculation assumes that the null hypothesis is true, that is, it assumes that the Company did not discriminate?

A: Yes, that is correct.

Q: Since your calculation assumes that the Company did not discriminate, how can it be used to shed light on whether the Company discriminated?

A: Fisher would say that with a significance test, one faces a disjunction. If significance is found, either something rather unusual has happened, or the Company discriminated against older people.

Q: Does the Fisher theory allow you to say which of these has occurred, given that one has?

A: No, it does not.

Q: Does it allow you to give a probability that the Company discriminated?

A: No, it does not.

Q: Does it allow you to say anything in the case that the data are not significant?

A: No, it does not.

Q: What would happen if you made the opposite assumption, that the Company does discriminate?

A: One could do that. It is called a "power analysis," and would depend on exactly what you assume about the extent to which the Company discriminates. This is really part of the Neyman–Pearson, as opposed to Fisher, view of significance tests. Neyman and Pearson refer to such tests as hypothesis tests to distinguish their interpretation from Fisher's.

Q: Then let's now turn to the Neyman–Pearson interpretation. When you say that the Neyman–Pearson view is that .05 is a property of a procedure, do I understand you to mean that it is not a property of any particular use of the procedure?

A: Yes, that is correct.

Q: So, under the Neyman–Pearson interpretation, .05 has to do with a long-run sequence of use and not with this particular use?

A: Yes.

Q: So, for example, a procedure that accepted the null hypothesis .95 proportion of the time and rejected it .05 proportion of the time, without looking at the data, would be a valid .05 level test according to the Neyman–Pearson theory?

A: Yes, it would. Other criteria would be introduced to show that it is not very sensible; in particular, it has poor power compared with some other tests.

Q: So, under the Neyman–Pearson theory, the hypothesis test tells us nothing about this particular use of it, but only about what would happen, hypothetically, if we used it in many cases?

A: Yes.

Q: Have you performed any analyses that do not assume the innocence of your client and that *are* relevant to this particular case?

What this line of questioning shows is that while the language of significance testing is wonderful (who in court wants his data sneered at because of alleged insignificance?), its philosophical underpinnings are weak. There has to be some doubt about how long statisticians can go to court to testify on significance tests using the justification of tradition, which comes down to the idea that many others in statistics make the same mistake.

3.2 A Bayesian Analysis

A Bayesian analysis is a formal procedure for modeling an opinion concerning the issue being decided and then showing how that opinion is changed in light of the data. Since Bayesian analyses measure the transformation of opinion brought about by the data, the first critical question is whose opinion is to be modeled. There are several possibilities: the Employee's, the Company's, the judge or jury's, or my own. Since the Employee and the Company are parties to the conflict, they are likely to have convinced themselves of the justice of their cause. Neither judge nor jury is available for probability elicitation. Furthermore, since in at least some sense they know what they think, presenting them with my model of their beliefs seems convoluted and probably insulting. While as an expert witness my opinions are admissible in court, it is not clear why anyone would or should particularly care about my private views, especially since, having had access only to the plaintiff's side of the case, I am no longer impartial. I would prefer to think that the opinions to be modeled are those of a neutral statistical arbitrator working for the court [see Coulom and Fienberg (1986) for a case study of one such referee]. Thus I am modeling impartiality, which may not represent my real opinion.

The data presented earlier in Table 1 are in the form of four 2 × 2 tables. Certainly, just before each of the decision dates, it is reasonable to consider the age structure of the work force as fixed, and thus to condition on the number of management workers over and under 40. Consideration of the other margin raises issues that once were, and sometimes still are, hotly debated in statistics. Fisher (1958, pp. 96, 97) took the position that both marginal totals were ancillary and consequently contained no relevant information. Barnard (1946) showed that each of several sampling models might be appropriate, depending on the experimental setting. See also Seidenfeld (1979). While, in general, I agree with Barnard's argument, I think that in this instance the most useful way to treat the data is by conditioning on both margins. The right of the Company to fire management workers at a time of financial stringency is not being challenged in this lawsuit, only whom they chose to fire. While conceivably the number of employees fired and retained might depend on the extent of discrimination against people over 40, the dependence is likely to be weak and masked by the legitimate right of the Company to reduce its management work force. Consequently a neutral statistician–referee would want, I think, to condition on the other margin as well, the number of management workers fired and retained.

An alternative model would be to think of the Company as wishing to reduce its salary bill by a fixed amount. This would lead to a different linear constraint on the people fired. However, it is obvious that salary and age tend to be highly correlated. The case law [Metz v. Transit Mix, Inc., C.A. 7 (Ind. 1987), 828 F.2d 1202 on remand 692 F. Supp. 987; Leftwich v. Harris-Stowe State College Bd. of Regents, 540F. Supp. 37 (E. D. Mo. 1982); Geller v. Markham, 635 F.2d 1027, 1034 (2d Cir. 1980), cert. denied, 451 U.S. 945, 101 S. Ct. 2028, 68 L.Ed.2d 332 (1981) (Rehnquist, J., dissenting); Marshall v. Arlene Knitwear, Inc., 454 F. Supp. 715, 728 (E.D.N.Y. 1978); Laugeson v. Anaconda Co., 510 F.2d 307, 316 (6th Cir. 1975)] holds that to fire the highest paid employees is not a sound business reason for discriminating against workers over 40. A neutral statistician would wish to avoid a model that prejudices the case against the Company before the data are even considered.

Conditioning on both margins leads to consideration of four doubly constrained 2 × 2 tables. The likelihood for a single table is given (see Plackett 1981, p. 38) by

$$\frac{\binom{n}{n_{11}, n_{1+}-n_{11}, n_{+1}-n_{11}, n-n_{1+}-n_{+1}+n_{11}} \times p_{11}^{n_{11}} p_{12}^{n_{1+}-n_{11}} p_{21}^{n_{+1}-n_{11}} p_{22}^{n-n_{+1}-n_{1+}+n_{11}}}{\sum_{j=\max(0,n_{1+}+n_{+1}-n)}^{\min(n_{1+},n_{+1})} \binom{n}{j, n_{1+}-j, n_{+1}-j, n-n_{1+}-n_{+1}+j} \times p_{11}^{j} p_{12}^{n_{1+}-j} p_{21}^{n_{+1}-j} p_{22}^{n-n_{1+}-n_{+1}+j}}$$

$$\frac{\displaystyle\binom{n}{n_{11}, n_{1+} - n_{11}, n_{+1} - n_{11}, n - n_{1+} - n_{+1} + n_{11}}\lambda^{n_{11}}}{\displaystyle\sum_{j=\max(0, n_{1+} + n_{+1} - n)}^{\min(n_{1+}, n_{+1})}\binom{n}{j, n_{1+} - j, n_{+1} - j, n - n_{1+} - n_{+1} + j}\lambda^{j}}.$$

Here $n_{i,j}$ is the number of people in age group i whose employment fate is j. Age group 1 is over 40, and group 2 is under 40. Employment fate 1 is to be fired, and fate 2 is to be retained. Similarly $p_{i,j}$ is the probability that a person will fall in category (i, j) in a given firing wave, where $\sum_{i=1}^{2} \sum_{i=1}^{2} p_{ij} = 1$. Then $\lambda = p_{11}p_{22}/p_{12}p_{21}$. Finally $n_{+1} = n_{11} + n_{21}$ and $n_{1+} = n_{11} + n_{12}$.

Thus the likelihood, although formally a function of $\mathbf{p} = (P_{11}, P_{12}, P_{21}, P_{22})$, is a function only of $\lambda(\mathbf{p})$. Consequently the entire prior-to-posterior analysis can be conducted on λ, as the conditional distribution of $\mathbf{p}$ given λ will be unaffected by the data. See Kadane (1975) for a discussion of a similar situation occurring in the Bayesian theory of simultaneous equations in econometrics.

The odds ratio λ is, in addition, a natural and convenient measure of the extent of discrimination against older workers. However, there is one inconvenience to the odds ratio. An odds ratio of 2 would be transformed into an odds ratio of ½ by a relabeling of the rows (or of the columns). Thus to think properly about odds ratios requires attention to multiplicative symmetry around 1. It is far less awkward to transform to the log odds ratio L and attend to additive symmetry around 0. Relabeling now merely changes the sign of the log odds ratio.

The point $L = 0$ corresponds to a policy of the Company that discriminates neither for nor against its older workers. Positive values for L correspond to discrimination against older workers, and negative values correspond to discrimination against younger workers. Thus it certainly seems reasonable for the prior of an impartial statistician to be centered at a log odds ratio L of 0. Furthermore, a log odds ratio of log 2, corresponding to the example discussed previously, is as discriminatory as a log odds ratio of log ½ = −log 2, so symmetry around 0 seems to be a natural condition. Additionally, it seems reasonable to me to require that the prior be unimodal, so that points closer to 0 have density at least as high as that for points far away. I use the normal family, not because I think normality has anything to do with impartiality in this problem, but because I think calculations done with normal priors are typical of what I would get with other shapes symmetric around 0.

Within the family of normal distributions centered at 0, the only parameter left is the variance. When the prior variance is zero, all the mass is at $L = 0$. But this would say that the impartial statistician is so sure that the company did not discriminate as to be uninterested in and uninfluenced by the data, which is unreasonable. Because the likelihood is positive and continuous at $L = 0$, as the prior variance approaches 0 the limiting probability that L is positive approaches ½, regardless of the data. Again, this is not reasonable. Consequently I choose a "large" variance, by which I do not mean an infinite variance, but rather one large enough to allow the data to dominate in the calculation of the posterior. In practice, I use a variety of variances and show that the choice among them does not materially affect the conclusions.

One could argue that newer workers, usually junior, are on a kind of probation for some period after their formal six-month probation has expired. Consequently one might expect a higher natural dismissal rate for junior than for senior employees, absent discrimination. If one ignores this phenomenon by using priors symmetric around 0, the Bayesian analyses given here may understate the extent of the Company's discrimination against people over 40.

There are alternative priors that I considered but did not use. One of them puts a lump of probability, say ½, at $L = 0$ and spreads the rest, perhaps as a normal distribution, with a mean of 0. This would have the anomalous effect of placing ¾ prior probability on the innocence of the Company, which does not seem consonant with impartiality. Another possibility puts such a lump at 0, and perhaps a half-normal distribution on positive L. Now there would be a ½ prior probability on the innocence of the Company, which seems appropriate, but the expected amount of discrimination (L) in the prior is positive, which does not.

One could imagine using Bayesian analysis in another way. Suppose, instead of thinking of a prior-to-posterior transformation by describing the posterior, one asks what characteristics of the prior would be implied by $Pr(L < 0) = ½$. If the normal family is accepted, there would be a curve of prior means and variances such that the posterior would indicate a probability of discrimination of exactly .5. A similar contour could be found for .6, .7, etc. I chose not to do this here because I wish to stress the idea of impartiality of the prior, but this is a legitimate alternative use of Bayesian ideas to express the import of the data.

The statistical arbitrator might first want to know whether there is a pattern of age discrimination in the firings, taken as a whole. This suggests using the sequence of priors specified above, and a likelihood consisting of the product of the likelihoods for each of the four periods. In addition to the posterior distribution in general, the amount of probability falling below 0 (corresponding to discrimination against people under 40), and above 0 (corresponding to discrimination against people over 40) are of special interest. The results are given in Table 2. They show overall a definite pattern of discrimination by the Company against people over 40. It is notable that this conclusion is not sensitive to the standard deviation chosen for the prior. The modal log odds ratio of .72 corresponds to an odds ratio of 2.05. Thus the odds of an over-40 employee being fired are roughly twice those of an employee under 40, indicating substantial discrimination.

It is also of interest to examine each firing wave individually. I consider them chronologically, taking the firing wave of 6/30/82 first. As shown in Table 3, the probability that the log odds ratio is positive is virtually 1, indicating a virtual certainty of discrimination against people 40 years old and older. Because all people fired in this wave were over 40, the data are consistent with arbitrarily large extents of discrimination L against people over 40. It is only the prior distribution that constrains L. As the prior variance on L increases, the posterior mode of L increases.

TABLE 2. Combined Likelihood Summary

		Prior standard deviation			
	1	2	4	8	∞(reference)
Mode of L	.68	.71	.71	.72	.72
$Pr(L > 0)$	.999	.999	.999	.999	.999

If the prior variance increased without limit, so would the posterior mode of L. It is for this reason that Table 2 does not report results for a reference prior for the firing wave of 6/30/82. The data from this wave indicate a truly extraordinary degree of discrimination against employees over 40.

The rhetoric of various authors would suggest something canonical about a prior proportional to Lebesgue measure. Jeffreys (1961, p. 49) wants a "simplicity postulate...sufficiently precise to give exact prior probabilities to all laws." Box and Tiao (1973) support many of the same priors as Jeffreys on the ground that they are "data translated." Bernardo (1979) justifies priors on grounds of a connection with information theory. While I do not find appealing the arguments supporting these priors as canonically correct, often, but not always, such priors yield about the same conclusions as would many other, proper priors better grounded in reasonable opinion that is informed about the problem at hand. In the case of the firing wave of 6/30/82, however, because none of the employees fired was under 40, the likelihood function is ill suited to such a prior. Nonetheless, the inference important here, the probability of discrimination against people 40 and older, is not sensitive to the choice of prior variance, as Table 3 shows.

The second firing wave is that of 11/30/82. In Table 4 the probability of discrimination against people older than 40 is again very high (96%–97%). Again the log odds ratio is about .72, indicating an odds of 2.05. Thus, as in the combined case, this indicates that the odds of an over-40 employee's being fired on 11/30/82 were twice those of a person under 40.

The third wave of firings, that of 5/31/83, shows a very different pattern. As shown in Table 5, the probability of discrimination against people older than 40 drops to about 16%, and the estimated log odds drops to −.40, which corresponds to an odds of .67. Thus for this wave of firings, the odds of an over-40 person's being fired are roughly ⅔ of those of a person under 40. This wave therefore does not show evidence of discrimination against people over 40 in the allocation of firings.

Finally, I examine the fourth wave, that of 6/28/84, in Table 6. This firing wave may have special significance since this was the wave in which the Employee was

TABLE 3. Firing Wave of 6/30/82

	Prior standard deviation			
	1	2	4	8
Mode of L	1.88	2.91	4.01	5.16
$Pr(L > 0)$	1.000	1.000	1.000	1.000

TABLE 4. Firing Wave of 11/30/82

	Prior standard deviation				
	1	2	4	8	∞(reference)
Mode of L	.62	.69	.71	.72	.72
$Pr(L > 0)$	.960	.967	.969	.970	.970

TABLE 5. Firing Wave of 5/31/83

	\multicolumn{5}{c}{Prior standard deviation}
	1	2	4	8	∞(reference)
Mode of L	−.34	−.39	−.40	−.40	−.40
Pr(L > 0)	.183	.169	.165	.164	.164

TABLE 6. Firing Wave of 6/28/84

	\multicolumn{5}{c}{Prior standard deviation}
	1	2	4	8	∞(reference)
Mode of L	.94	1.24	1.37	1.41	1.42
Pr(L > 0)	.965	.981	.985	.986	.987

fired, having survived the firing waves of 6/30/82, 11/30/82, and 5/31/83. Here the probability that there was discrimination against those over 40 was .987 (for the reference prior). The modal estimate for the log odds ratio is 1.42, corresponding to an odds ratio of 4.14. Thus in the firing wave that affected the Employee, the odds of his being dismissed were over four times as great as those of his co-workers under 40 years of age.

My conclusions from this Bayesian study are as follows:

1. As to overall pattern, the data show that the Company did engage in a pattern of discrimination in firing management employees aged 40 years and older.
2. The first firing wave of 6/30/82 shows extreme discrimination. The fourth wave, of 6/28/84, in which the Employee was fired, also shows very substantial discrimination. The second wave, of 11/30/82, shows substantial discrimination. Finally, the wave of 5/31/83 does not show much of a pattern in either direction, but there is perhaps a slight hint of discrimination the other way, against people under 40 years of age.

One might argue that if there were a consistent pattern of discrimination on the basis of age, especially if there were some continuing mechanism at work, all four firing waves should demonstrate the same pattern. The exception of firing wave III, then, casts doubt on this interpretation of events. To establish adverse impact, one should not be required to show that every act of the Company, or even its decisions about whom to fire in each wave, were discriminatory. If a general pattern of age discrimination is found, it should be considered sufficient to shift the burden of proof to the Company of explaining the pattern in a nondiscriminatory way. However, if the case were brought as a class-action suit on behalf of all over-40 employees fired in the four waves, the data might be sufficient to support exclusion from the class of those fired in the third wave. Excluding the one wave in which the Company appears not to have discriminated would only make the combination of the remaining waves look more discriminatory, of course.

TABLE 7. Combination of Waves II–IV

	\multicolumn{5}{c}{Prior standard deviation}				
	1	2	4	8	∞(reference)
Mode of L	.360	.380	.390	.390	.390
$Pr(L > 0)$	.930	.935	.936	.937	.937

TABLE 8. Combination of Waves II–IV (Altered Data)

	\multicolumn{5}{c}{Prior standard deviation}				
	1	2	4	8	∞(reference)
Mode of L	.260	.270	.270	.270	.270
$Pr(L > 0)$	.853	.859	.860	.861	.861

How might, the Bayesian analysis be criticized? A likely line of attack is through the data base. What would be the consequence of excluding firing wave I entirely, taking the view that all these departures from management ranks were voluntary retirements rather than involuntary retirements? The answers are given in Table 7. The probability that the Company discriminated against its older workers drops from .999 in Table 2 to .937 here, not nearly enough of a drop to disturb my conclusions under a standard of the preponderance of the evidence. What does change is the modal log odds ratio, which drops from .72 in Table 2 to .39 here, corresponding to a drop in the odds ratio from 2.05 to 1.48.

Finally one could ask what the effect would be of changing one fired worker in each of waves II and IV from over 40 to under 40. The results are given in Table 8, and indicate a modest further shift in the results. Remembering that no evidence supports this alteration of the data, these calculations suggest reasonable robustness of the Bayesian analysis.

Faced with two methods of analyzing the same data set, one may naturally want to compare them, both in principle and numerically. Altham (1969) showed an equivalence between the Fisher exact test and a Bayesian analysis. However, the Bayesian analysis in Altham's equivalence has only one margin fixed (hence two independent binomial populations), and independent beta priors with parameters (0, 1) on each. This is quite an astonishing result, since the Fisher calculation is a summation over the sample space, while the Bayesian calculation is an integration over the parameter space. Because the Bayesian model used by Altham differs from the one used here both in likelihood and in prior, it is useful to compare the results numerically, as is done in Table 9.

The first four rows are the data from the four firing waves. The fifth and sixth rows result from shifting one fired person from the over-40 to the under-40 group in firing waves II and IV. Conversely, the seventh and eighth rows result from shifting one fired person from the under-40 to the over-40 group in those firing waves. Generally the Bayesian and the Fisher results are parallel, but not the same.

TABLE 9. Fisher's Exact Test Significance Levels and
Bayesian Probabilities of Discrimination

Data	Fisher's exact test	Bayes (ref. prior) Pr(L < 0)
(18, 0, 129, 102)	.0001	.000
(26, 10, 105, 83)	.0485	.030
(13, 14, 92, 66)	.8821	.846
(13, 2, 81, 52)	.0407	.013
(25, 11, 105, 83)	.0906	.061
(12, 3, 81, 52)	.1193	.056
(27, 9, 105, 83)	.0234	.013
(14, 1, 81, 52)	.0091	.002

The strength of a Bayesian analysis in this context, it seems to me, is that it answers the relevant legal question, namely, what the probability is that the Company's policy (L) discriminated against people over 40. Unlike the classical analysis, it does not assume that either side is correct, and it is relevant to the particular case.

Paired Observations

The preceding analysis convinces me that my hypothetical neutral statistical arbitrator would find that the Employee has met the requirement of showing disparate impact of the Company's decisions against people 40 and older, thus placing the burden of proof on the Company to show sound, nondiscriminatory business reasons for its actions. However, it leaves open the question of whether the effects observed truly reflect age discrimination or are an artifact of defining the protected group to be people aged 40 and older. One way of looking at that question would be to perform analyses similar to those of Sections 3.1 and 3.2, varying the age cutoff. However, this does not deal with the question in a continuous way.

I think a more natural analysis of this question would proceed as follows: Suppose we form, conceptually, all pairs of employees, one of whom is fired and one of whom is not, in a given firing wave. If we picked one such pair at random, what is the probability that the older would be the one fired? This proportion is the Mann–Whitney–Wilcoxon statistic, throught of as an estimate of the probability that a member of the fired population is older than a member of the retained population (see Hoeffding 1948). In this application, however, the entire population of management employees is available for analysis, so it seems wrong to conceive of the data as a random sample from some larger population.

The results are given in Table 10, for probabilities, log odds, and odds. Exact nondiscrimination corresponds to probability of ½, log odds of 0, and odds of 1.

There are several ways of conceiving of the combined analysis. The method used in Table 10 (and also Table 11) is to constrain the pairs of employees, whose ages are being compared, so that one was fired and one was retained in the same firing wave. This seems most consonant with an urn conception of the probability process (pour the four urns together and draw again). Also, it maintains the legal

TABLE 10. Paired Analysis

Firing wave	Probability older person fired		
	Probability	Log odds	Odds
I	.983	4.053	57.57
II	.625	0.512	1.669
III	.479	−0.085	0.919
IV	.711	0.902	2.465
Combined	.685	0.778	2.178

TABLE 11. Restricted Paired Analysis

Firing wave	Probability older person fired		
	Probability	Log odds	Odds
I	1.00	∞	∞
II	.672	.72	2.055
III	.400	−.41	.667
IV	.806	1.43	4.173
Combined	.689	.794	2.211

interpretation that age comparisons between fired and retained workers must be limited to the same firing waves.

Again, the results of Table 10 confirm the earlier analysis: Wave I was very discriminating against older people, waves II and IV were substantially discriminatory, and wave III was not discriminatory. Overall, a pattern of discrimination against older people in these firings is confirmed.

These probabilities, log odds, and odds do not have uncertainty measures attached because every pair of employees, one of whom was fired and the other not, is considered. Consequently these are exactly the probabilities, the log odds, and the odds, up to limits of numerical rounding.

Perhaps because of the absence of uncertainty, an analysis of this sort has an appeal. Might it be reformulated so as to respect, once again, the age 40 and over restriction of the protected class? One way to do this is to limit the pairs to those in which one employee is over 40 and one under 40. So now the question is, of all pairs of employees, one under 40 and one over, one fired and one not, what is the proportion of pairs in which the person fired is over 40? The results are given in Table 11.

Thus the results are quite similar to those in Table 10. Again there are no uncertainty measures here, for the same reason as in Table 10.

Let C be the proportion of pairs of employees, one under 40 and one over, one fired and one not, in which the person fired is over 40. Then

$$C = \frac{n_{11}n_{22}}{n_{11}n_{22} + n_{12}n_{21}}.$$

The odds of C are

$$O(C) = \frac{C}{1-C} = \frac{n_{11}n_{22}}{n_{12}n_{21}}$$

$$= \frac{n_{11}n_{22}/n}{n_{12}n_{21}/n},$$

which is the Mantel–Haenszel (1959) statistic. However, for several tables combined,

$$C = \frac{\sum_k n_{11k}n_{22k}}{\sum_k n_{11k}n_{22k} + \sum_k n_{12k}n_{21k}},$$

where k indexes the tables. Here the odds of C are

$$O(C) = \frac{C}{1-C} = \frac{\sum_k n_{11k}n_{22k}}{\sum_k n_{12k}n_{21k}},$$

which is not, in general, equal to the Mantel–Haenszel statistic

$$\frac{\sum_k (n_{11k}n_{22k}/n_k)}{\sum_k (n_{12k}n_{21k}/n_k)}.$$

The Mantel–Haenszel statistic is used in articles discussing discrimination in hiring under a model of two or more independent populations (Gastwirth 1984; Gastwirth and Greenhouse 1987; Louv and Littel 1986).

4. CONCLUSIONS

What are the appropriate responsibilities of a statistician in a legal setting with respect to the data base used? It is interesting to note that I and the statistician working for the Company took different approaches to this question. The Company produced the 2 × 2 tables for analysis by the statistician. Thus the conclusions he could have reached would have been limited to the accuracy of those tables. In contrast, I based my choice of analysis on a list of employees, their birth dates, hiring dates, and firing or departure dates. This might have exposed me to cross-examination about conditions in the Company, about which I would not be very knowledgeable. I do not know what is the best policy for a statistician in such an environment.

If I had been acting as a neutral statistical arbitrator, the Court would have been asked whether the evidence suffices to establish a prima facie case. If so, the Company would have been asked to give an accounting of its policy on firing. The Court would then have been asked whether the explanation offered would, or might, suffice as a sound business reason if sustained by the data. If the answer to this question were positive, then, and only then, would it make sense to consider covariates with a view toward examining the extent to which the Company's explanation is supported by the data.

I find it interesting that the legal context impinges on the data analysis in several places. While it is to be expected that the application would have a strong influence in every applied problem, it is somewhat surprising that an analysis done in a legal context might be substantially different from an analysis done with a solely scientific aim.

It is certainly a fortunate feature of this data set that several different analyses lead to very similar substantive conclusions. To the extent that different substantive conclusions are reached by different analyses, this only serves to sharpen the debate over the meaning of the analyses. It should not be a surprise that a legal case would confront statisticians with deep problems about the meaning of the various techniques proposed, because the adversary structure leads to sharper questioning than statisticians generally confront.

ACKNOWLEDGMENTS

The author thanks Caroline Mitchell and the *Employee* for involving him in the case. He also thanks George Duncan, Stephen Feinberg, John Lehoczky, Michael Meyer, Allan Sampson, and Teddy Seidenfeld for helpful conversations, and, most particularly, Thomas Short for computational assistance.

REFERENCES

Altham, P. M. E. (1969), "Exact Bayesian Analysis of a 2 × 2 Contingency Table, and Fisher's 'Exact' Significance Test," *Journal of the Royal Statistical Society,* Ser. B, 31, 261–269.

Barnard, G. (1946), "Significance Tests for 2 × 2 Tables," *Biometrika,* 34, 123–138.

Bernardo, J. M. (1979), "Reference Posterior Distributions for Bayesian Inference" (with discussion), *Journal of the Royal Statistical Society,* Ser. B, 41, 113–148.

Box, G. E. P., and Tiao, G. (1973), *Bayesian Inference in Statistical Analysis,* Reading, MA: Addison-Wesley.

Coulom, R. F., and Fienberg, S. E. (1986), "The Use of Court Appointed Statistical Experts," in *Statistics and the Law,* eds. M. H. DeGroot, S. E. Fienberg, and J. B. Kadane, New York: John Wiley, pp. 305–332.

Fisher, R. A. (1958), *Statistical Methods for Research Workers* (13th ed.), New York: Hafner Press.

———. (1959), *Statistical Methods and Scientific Inference* (2nd ed.), Edinburgh: Oliver & Boyd.

Gastwirth, J. L. (1984), "Statistical Methods for Analyzing Claims of Employment Discrimination," *Industrial and Labor Review,* 38, 75–86.

———. (1988), *Statistical Reasoning in Law and Public Policy,* Boston: Academic Press.

Gastwirth, J. L., and Greenhouse, S. W. (1987), "Estimating a Common Relative Risk: Application in Equal Employment," *Journal of the American Statistical Association,* 82, 38–45.

Hoeffding, W. (1948), "A Class of Statistics With Asymptotically Normal Distribution," *Annals of Mathematical Statistics,* 19, 293–325.

Jeffreys, H. (1961), *Theory of Probability* (3rd. ed.), Oxford, U.K.: Oxford University Press.

Kadane, J. B. (1975), "The Role of Identification in Bayesian Theory," in L. J. Savage Memorial Volume *Studies in Bayesian Statistics and Econometrics,* S. Fienberg and A. Zellner, eds., pp. 175–191.

Louv, W. C., and Littel, R. C. (1986), "Combining One-Sided Binomial Tests," *Journal of the American Statistical Association,* 81, 550–554.

Mantel, N., and Haenszel, W. (1959), "Statistical Aspects of the Analysis of Retrospective Studies of Disease," *Journal of the National Cancer Institute,* 22, 719–748.

Michelson, S. (1986), "Comment," in *Statistics and the Law,* eds. M. H. DeGroot, S. E. Fienberg, and J. B. Kadane, New York: John Wiley, pp. 169–181.

Neyman, J., and Pearson, E. S. (1967), *Joint Statistical Papers,* Cambridge, U.K.: Cambridge University Press.

Plackett, R. L. (1981), *The Analysis of Categorical Data,* New York: Macmillan.

Raiffa, H., and Schlaifer, R. (1961), *Applied Statistical Decision Theory,* Cambridge, MA: MIT Press.

Seidenfeld, T. (1979), *Philosophical Problems of Statistical Inference: Learning from R. A. Fisher,* Dordrecht: D. Reidel.

2

HIERARCHICAL MODELS FOR EMPLOYMENT DECISIONS

Joseph B. Kadane and George G. Woodworth

1. INTRODUCTION

Federal law forbids discrimination against employees or applicants because of an employee's race, sex, religion, national origin, age (40 or older), or handicap. General discrimination law—say discrimination by race or sex—offers two somewhat distinct legal theories. A disparate treatment case involves policies that on their face treat individuals differently depending on their (protected) group membership, such as a rule prohibiting women from being firefighters.

A disparate impact case, however, permits evidence that a facially neutral policy—say a height requirement for firefighters—has the effect of making it relatively more difficult for women than men to obtain such employment. If the data show a pattern of unfavorable actions (firing, failure to hire, failure to promote, low raises, etc.) disproportionately against the protected group, this can establish or help to establish a prima facie case against the defendant. A prima facie case does not establish the defendant's liability. Instead it shifts the burden of producing evidence to the defendant to explain the business necessity of the disproportionately adverse actions taken against the protected group. In such a case the employer would have to justify the requirement in terms of the needs of the job, and the fact finder (judge or jury) would have to determine whether the justification is a pretext for discrimination or not (Gastwirth 1992; Kadane and Mitchell 1998).

Race and sex discrimination cases fall under Title VII of the Civil Rights Act of 1964, whose provisions permit a prima facie case to be made by statistical evidence that members of the protected class are more likely to experience the adverse outcome of an employment decision. However, age discrimination cases are heard under the Age Discrimination in Employment Act of 1967, whose provisions allow differential treatment of employees based on "reasonable factors other than age," which could be interpreted as barring a disparate impact age discrimination case. The Supreme Court in *Hazen Paper v. Biggins*, 123 L.Ed.2d 338, 113 S.Ct. 1701 (1993), explicitly declined to decide this matter. Various courts and judges have discussed it [Judge Greenberg in *DiBiase v. Smith Kline Beecham Corp.*, 48 F.3d 719 (1995), Judge Posner in *Finnegan et al. v. Transworld Airlines*, 967 F.2d 1161 (7th Cir., 1992), and the references cited there].

However this legal debate is resolved, we expect that statistical evidence of how an employer's policies affect older workers will continue to be relevant, in the legal sense, for the following reason. Federal Rule of Evidence 401 defines relevant evidence as evidence that has "any tendency to make the existence of any fact that is of consequence to the determination of the action more probable or less probable that it would be without the evidence." The issue in disparate treatment cases is establishing the intent of the employer. If an analysis shows that the facially neutral policy of the employer did differentially harm older workers, that reasonably makes it more probable that the employer intended the harm. Thus, we expect our analyses to continue to be relevant, regardless of the fate of the doctrine of disparate impact in age discrimination cases.

In this article we advocate the use of Bayesian analysis of employment decisions, which raises a second legal issue concerning the admissibility of such analysis to age discrimination cases. The rules on what constitutes admissible expert testimony in U.S. courts have changed. Under the Frye rule [*Frye v. United States,* 54 App.D.C. 46, 47, 293 F. 1013, 1014 (1923)], expert opinion based on a scientific technique is inadmissible unless the technique is "generally accepted" as reliable in the scientific community. Congress adopted new Federal Rules of Evidence in 1975. Rule 702 provides "[i]f scientific, technical or other specialized knowledge will assist the trier of fact to understand the evidence or to determine a fact in issue, a witness qualified as an expert by knowledge, skill, experience, training or education, may testify thereto in the form of an opinion or otherwise." In the case of *Daubert, et ux. etc. et al. v. Merrell Dow Pharmaceuticals, Inc.*, 509 U.S. 579, 113 S.Ct. 2786 (1993), the Supreme Court unanimously held that Federal Rule 702 superseded the Frye test.

The *Daubert* decision, continuing with dicta (of lesser standing than holdings) of seven Supreme Court justices, goes on to define "scientific knowledge." "The adjective 'scientific' implies a grounding in the methods and procedures of science. Similarly, the word 'knowledge' connotes more than subjective belief or unsupported speculation." Thus, the *Daubert* decision might be read as casting doubt on the admissibility of Bayesian analyses in federal court, because the priors (and likelihoods) are intended to express subjective belief. We think that this reading of *Daubert* is occasioned by a misinterpretation of what Bayesian statisticians mean by subjectivity. The alternative, that is, claims of objectivity in the sense that anyone who disagrees is either a fool or a knave, is without basis, and appears to be an attempt at proof by verbal intimidation. To hold as we do, that every model (including frequentist models) reflects and expresses subjective opinions, however, is not to hold that every such opinion has an equal claim on the attention of a reader or a court. For an analysis to be most useful, it should be persuasive to a fact finder that an analysis done with his or her own model—likelihood and prior—would result in similar conclusions. This can be done with a combination of arguments based on reasons for the chosen likelihood and prior (other data, scientific theory, etc.) and robustness (the conclusions would be similar with other models "not too far" from the one analyzed). Thus, our view is that an analysis ought neither to be admissible nor inadmissible because it uses a subjective Bayesian approach. Instead its admissibility ought to depend on its persuasiveness in explaining, with a combination of specific arguments and robustness, why the conclusions of a trier of facts might be similar to, and hence influenced by, the analysis offered.

An alternative interpretation of the same activity is that the statistician is providing the fact finder with "scientific methodology" for combining information, including subjective information. The unavailability of the fact finder for elicitation means that the statistician has to present results in the form of "If you believe this, then the results of the data analysis would be these." We believe that under either interpretation a properly grounded and explained Bayesian model, of the kind proposed here, is both admissible and relevant in age discrimination cases. We confine our discussion to binary employment decisions such as hiring, job assignment, promotion, layoff, or termination. The outcome of such a decision is either favorable or unfavorable to the employee, who may or may not be in a legally protected class. We use age discrimination in termination decisions to illustrate our ideas because age discrimination cases dominate our experience. Age discrimination over a short period of time, for example, when an employer makes a large reduction in the workforce over a matter of a few days or weeks as the result of a single policy decision, is comparatively easy to analyze (Kadane 1990) mainly because it is reasonable to assume that the odds ratio is constant; however, age discrimination over an extended time period is more difficult to model, both because the same individual can over time move from the unprotected to the protected class and because, unlike gender or race, age is a continuous characteristic and consequently the hazard rate may vary within the protected class.

Following Finkelstein and Levin (1994), we find that proportional hazards (Cox regression) models provide the flexibility to deal with these issues.

2. PROPORTIONAL HAZARDS MODELS

Suppose that we wish to analyze the employment decisions (e.g., involuntary terminations) of a firm over a given period of observation. The kind of analysis we propose requires data sufficient to determine for each day during the period of observation, the status (protected or unprotected) of each employee and the number of involuntary terminations (if that is the decision to be analyzed) of protected and unprotected employees. Perhaps it is most convenient to obtain this information in the form of flow data for each individual who was employed at any time during the period of observation.

Flow data consist of beginning and ending dates of each employee's period of employment, that employee's birth date, and the reason for separation from employment (if it occurred). We have seen no examples in which employees were rehired for nonoverlapping terms, but such cases could easily be handled by entering one data record for each distinct period of employment. Table 1 is a fragment of a dataset gathered in a hypothetical age discrimination case. Data were obtained on all persons employed by the firm any time between 01/01/94 and 01/31/96. Entry Date is the later of 01/03/1994 or the date of hire. The first record is right censored; that is, that employee was still in the workforce as of 1/31/96, and we are consequently unable to determine the time or cause of his or her eventual separation from the firm (involuntary termination, death, retirement, etc.).

The plaintiff obtains such data from the employer in the pretrial discovery phase. It is generally necessary for the plaintiff's attorney to justify the need for obtaining data over a particular time period—for example, it might be the period from the imposition of a particular policy to the end of the plaintiff's employment. Frequently the defendant can convince the court to narrow the scope of the data provided, arguing, for example, that retrieving records more than 5 years old or linking records involving employee transfers between divisions would be burdensome.

In many litigated cases the observation period is short, and corresponds to one large-scale reduction in force (RIF) in which a substantial number of employees were terminated in a comparatively short period. Kadane (1990) discussed such a case involving four massive firing waves. Data of this sort can be treated as an analysis of the odds ratios (odds on termination of protected versus unprotected employees) in a small number of two-by-two contingency tables. Kadane considered two models for the prior distribution of the odds ratios. In the homogeneous, common odds ratio model, he gave the log odds ratio, β, a normal distribution with zero mean and fixed precision. He computed the posterior probability of adverse impact ($\beta > 0$) of the employer's policy on the protected class for various values of the prior precision. For the inhomogeneous odds ratio model, he assumed independent distributions for the log odds ratios for the four waves of terminations and computed the probability of adverse impact separately for each wave.

This article is an attempt to tackle the analysis of terminations occurring at a comparatively low rate, perhaps one or two employees at a time, over a long time period. The problem with this sort of situation is that the disaggregated data consist of numerous two-by-two tables, each involving a small number of terminations but any aggregation of the data (quarterly, semiannually, etc.) into more substantial two-by-two tables is arbitrary and somewhat distorts the numbers at risk because some employees will not have been in the workforce for the entire period represented by a given aggregated table. A second, and more important issue is how to deal with the possibility of inhomogeneous odds ratios.

TABLE 1. Flow Data for the Period January 1, 1994 to December 31, 1996.

Employee ID	Birth date	Entry date	Separation date	Reason
01	05/23/48	07/27/94		
02	12/17/31	01/03/94	11/20/94	Involuntary termination
03	03/14/48	06/29/94	07/27/94	Involuntary termination
04	02/26/40	10/05/94	06/07/95	Resigned
...	...	...	...	...

Finkelstein and Levin (1994) suggested that proportional hazards (Cox regression) models could be used to deal with disaggregated employment decisions; however, they assumed a constant log odds ratio over the observation period. We like their idea and in this article show how to allow for the possibility that the relative risk of termination varies over the observation period.

Cox (1972) considered a group of individuals at risk for a particular type of failure (involuntary termination) for all or part of an observation period. The jth person enters the risk set at time h_j (either the date of hire or the beginning of the observation period) and leaves the risk set at time T_j either by failure (involuntary termination) or for other reasons (death, voluntary resignation, reassignment, retirement, or the end of the observation period). The survival function $S_j(t) = P(T_j > t)$ is the probability that the jth employee is involuntarily terminated sometime after time t. The hazard function, $\lambda_j(t)$, is the conditional probability that person j is terminated at time t given survival to time t, that is,

$$\lambda_j(t) = \frac{-s_j(t)}{S_j(t)} = -\frac{d}{dt}\log(S_j(t)),$$ (1)

where s_j is the derivative of S_j. Integrating (1) produces

$$S_j(t) = \exp\left(-\int_{h_j}^{t} \lambda_j(t)dt\right).$$ (2)

The Cox proportional hazards model is

$$\lambda_j(t) = \lambda(t) \exp\left(\beta(t)\, z_j(t)\right),$$

where $\lambda(t)$ is the (unobserved) base hazard rate; $z_j(t)$ is an observable, time-varying characteristic of the jth person; and $\beta(t)$ is the unobserved, continuous, time-varying log-relative hazard. In our application $z_j(t) = 1(0)$ if person j is (is not) protected, that is, is (is not) aged 40 or older at time t, and $\beta(t)$ is the logarithm of the odds ratio at time t. The parameter $\beta(t)$ is the instantaneous log odds ratio:

$$\beta(t) = \lim_{dt \to 0} \ln\left(\left[\frac{P\left(t \le T_j < t+dt \mid z_j(t)=1\right)}{P(t+dt \le T_j \mid z_j(t)=1)}\right] \times \left[\frac{P\left(t \le T_j < t+dt \mid z_j(t)=0\right)}{P(t+dt \le T_j \mid z_j(t)=0)}\right]^{-1}\right).$$ (3)

The observed data are (h_j, T_j, c_j, z_j), $1 \le j \le M$, where M is the number of individuals who were in the workforce at any time during the observation period, and $c_j = 0$ if the jth employee was terminated at time T_j and $c_j = 1$ if the employee left the workforce for some other reason. The likelihood function is

$$l(\lambda, \beta \mid Data) = \prod_{j=1}^{M} (\lambda_j(T_j))^{(1-c_j)} S_j(T_j) = \prod_{c_j=0} \lambda(T_j) e^{\beta(T_j)z_j(T_j)}$$
$$\times \exp\left(-\sum_{j=1}^{M} \int_{h_j}^{T_j} e^{\beta(t)z_j(t)}\lambda(t)dt\right).$$ (4)

In practice, times are not recorded continuously, so let us rescale the observation period to the interval [0,1] and assume that time is measured on a finite grid,

$0 = t_0 < t_1 < \cdots < t_p = 1$. A sufficiently fine grid is defined by the times at which something happened (someone was hired, or left the workforce, or reached age 40). The data are reduced to N_i and n_i, the numbers of employees and protected employees at time t_{i-1}, and k_i and x_i, the numbers of employees and protected employees involuntarily terminated in the interval $(t_{i-1}, t_i]$. For data recorded at this resolution, the likelihood is

$$l(\lambda, \beta) = \prod_{i=1}^{p} e^{\beta_i \cdot x_i} \Lambda_i^{k_i} \exp\left(\Lambda_i \left(n_i e^{\beta_i} + (N_i - n_i)\right)\right), \tag{5}$$

where $\beta_i = \beta(t_i)$ and $\Lambda_i = \Lambda(t_i) - \Lambda(t_{i-1}) = \int_{t_{i-1}}^{t_i} \lambda(t)dt$. The function $\Lambda(t) = \int_0^t \lambda(t)dt$ is called the cumulative base rate. The likelihood depends on the log odds ratio function and the cumulative base rate function only through a finite number of values, $\beta = (\beta_1, \ldots \beta_p)'$ and $\Lambda = (\Lambda_1, \ldots, \Lambda_p)'$.

2.1 Hierarchical Priors for Time-Varying Coefficients

Sargent (1997) provided an excellent review of penalized likelihood approaches to modeling time-varying coefficients in proportional hazards models. He argued that these are equivalent to Bayes methods with improper prior distributions on $\beta(\cdot)$ and proposed a "flexible" model with independent first differences $\beta(t_{i+1}) = \beta(t_i) + u_{i+1}$, where the innovations u_{i+1} are mutually independent, normal random variables with mean 0 and precision τ/dt_{i+1} and $dt_{i+1} = t_{i+1} - t_i$. Under this model $\beta(t)$ is nowhere smooth—policy, in effect, changes abruptly at every instant. However, in the employment context, we expect changes to be gradual and smooth in the absence of identifiable causes such as a change in top management. For that reason we propose a smoothness prior for the log odds ratio, $\beta(t)$. The smooth model that we describe later is an integrated Wiener process with linear drift. Lin and Zhang (1998) used this prior for their generalized additive mixed models: however, their quasi-likelihood approach based on the Laplace approximation appears to fail for employment decision analyses when, for one or more time bins, all or none of the involuntary terminations are in the protected class.

2.2 Smoothness Priors

Let $\beta(t)$ be a Gaussian process, let $\beta_i = \beta(t_i)$, $0 \le t_i \le 1$, $1 \le i \le M$, and define $\beta = (\beta_1, \ldots, \beta_M)$. Specifying a "smoothness" prior requires that we have an opinion about the second derivative of $\beta(t)$ (see Gersch 1982 and the references cited there). To this end we use the integrated Wiener process representation (Wahba 1978):

$$\beta(t) = \beta_0 + \beta_0' t + \sqrt{1/\tau} \int_0^t W(t)dt, \tag{6}$$

where $W(\cdot)$ is a standard Wiener process on the unit interval, τ (the precision or "smoothness" parameter) has a proper prior distribution, and the initial state (β^0, β_0') has a proper prior distribution independent of $W(\cdot)$ and τ. Integrating by parts, we obtain the equivalent representation:

$$\beta(t) = \beta_0 + \beta_0'(t - t_0) + \frac{1}{\sqrt{\tau}} \int_0^t (t - s)\, dW(s). \tag{7}$$

Because $dW(s)$ is Gaussian white noise, the conditional covariance function of $\beta(\cdot)$ is

$$\text{cov}\big(\beta(t), \beta(t+d)\,|\,\beta_0, \beta_0', \tau\big)$$

$$= \frac{1}{\tau} E\Big(\int_0^1 \int_0^{t+d} (t+d-u)(t-v)\, dw\,(u)\, dW(v) \Big)$$

$$= \frac{1}{\tau} \int_0^t (t+d-u)\,(t-u)\, du \tag{8}$$

$$= \frac{t^3}{3\tau} + \frac{dt^2}{2\tau}.$$

2.3 Forming an Opinion About Smoothness

The remaining task in specifying the prior distribution of the log odds ratio is to specify prior distributions for the initial state (β_0, β_0') and for the smoothness parameter, τ. We have found that the posterior distribution of $\beta(\cdot)$ is not sensitive to the prior distribution of the initial state, so we give the initial state a diffuse but proper bivariate normal distribution. However, the smoothness parameter τ requires more care.

As $\tau \to \infty$, the Wiener process part of (7) disappears, so this would express certainty that $\beta(\cdot)$ is exactly linear in time. As $\tau \to 0$, the variance of the $\beta(\cdot)$ around the linear-in-time mean goes to ∞, so a good point estimate of $\beta(\cdot)$ would go through each of the sample points exactly, which offers no smoothness at all. It should come as no surprise, then, that what is essential about a prior on τ is not to allow too much probability close to 0. In the employment discrimination context, in the examples we have studied the data do not carry much information about smoothness and it is necessary to have an informative opinion about this parameter. In eliciting opinions about how fast an odds ratio might change, we find it easiest to think about what might happen during a business quarter. We will use as a reference the prior distribution of a person who thinks that, absent any change in business conditions or management turnover, there is a small probability that the odds of terminating a protected employee relative to an unprotected employee would change more than 15% in a single quarter. For example, if at the beginning of a quarter a protected employee is 5% more likely to be terminated than an unprotected employee, it would be surprising to see a 20% disparity at the beginning of the next quarter. Our purpose here is to demonstrate a way to develop a reference prior distribution consistent with easily stated assumptions. When such analyses are used in litigation, it will be important for the expert to be able to state that his or her conclusions are robust over a wide range of prior opinions, which is true in the first case (Case K) that we present but is not in the second case (Case W).

To see what the "no more than 15% change per quarter" assumption implies about the prior distribution of the smoothness parameter, consider the central second difference $\Delta'' = \beta(t + d) - 2\beta(t) + \beta(t - d)$, where d represents a half-quarter expressed in resealed time (i.e., as a fraction of the total observation interval). From the covariance function (8), it is easy to compute, variance$(\Delta'') = 2d^3/3\tau$. It is easy to see that if β changes by at most .15 over the interval $(t - d, t + d)$, then $|\Delta''| \le .30$. Because we regard values larger than this to be improbably a priori, we can treat .30 as roughly two standard deviations of Δ''. Thus, the prior distribution of τ should place high probability on the event $2d^3/3\tau < .15^2$, that is, $30d^3 < \tau$. Thus, for example, if the observation period is about 16 quarters, then a rescaled half-quarter is $d = 1/32$. So the prior distribution should place high probability on the event, $\tau > 30/32^3 \approx .0005$. A gamma distribution with mean .005 and shape 1 places about 90% of its mass above .0005.

The important thing about a prior on τ is where it puts most of its weight. We believe that other choices of the underlying density would not change the conclusions much. However, shifts in the mean are important, because such shifts control how much smoothing is done. Hence, we study sensitivity mainly by varying the mean, holding the rest of the distributional specification unchanged.

2.4 Posterior Distribution

Employers have an absolute right to terminate employees; what they do not have is the right to discriminate on the basis of age without a legitimate business reason unrelated to age. Thus, the base rate is irrelevant to litigation, and we believe that a neutral analyst should give it a flexible, diffuse but proper prior distribution. For convenience, we have chosen to use a gamma process prior with shape parameter α and scale parameter $\alpha\gamma$. In other words, disjoint increments $\Lambda_i = \Lambda(t_i) - \Lambda(t_i - 1) = \int_{t_{i-1}}^{t_i} \lambda(t)dt$ are independent and have gamma distributions with shape parameter $\alpha(t_i - t_{i-1}) = \alpha\, dt_i$ and scale parameter $\alpha\gamma$. The hyperparameters α and γ have diffuse but proper log-normal distributions. Consequently, the posterior distribution is proportional to

$$l(\beta, \Lambda)p(\beta|\beta_0, \beta_0', \tau)p(\beta_0, \beta_0')p(\tau)p(\Lambda + \alpha, \gamma). \tag{9}$$

The goal is to compute the posterior marginal distributions of the log odds ratios β_i, $1 \le i < p$, in particular, to compute the probability that the employer's policy discriminated against members of the protected class at time t_i; that is, $P(\beta_i > 0|\text{Data})$,

$$P(\beta_i > 0 \mid \text{Data}) = \frac{\int_{\beta_i > 0} dP(\beta, \Lambda, \tau, \beta_0, \beta_0', \alpha, \gamma \mid \text{Data})}{\int dP(\beta, \Lambda, \tau, \beta_0, \beta_0', \alpha, \gamma \mid \text{Data})}. \tag{10}$$

Closed-form integration in (10) is not feasible. Owing to the high dimensionality of the parameter space, numerical quadrature is out of the question and the Laplace approximation (Kass, Tierney, and Kadane 1988) would require the maximization of a function over hundreds of arguments. For these reasons we chose to approximate moments and tail areas of the posterior distribution by Markov chain Monte Carlo (MCMC) methods (Tierney 1994; Gelman, Carlin, Stern, and Rubin 1995).

MCMC works by generating a vector-valued Markov chain that has the posterior distribution of the parameter vector as its stationary distribution. An algorithm that generates such a Markov chain is colloquially called a *sampler*. Let θ denote the parameter vector, let $f(\theta|\text{data})$ denote the posterior distribution, and let $\theta_1, \ldots, \theta_M$ be successive realizations of θ generated by the sampler. The ergodic theorem (a weak law of large numbers for Markov chains) implies that

$$\frac{1}{M}\sum_{i=1}^{M} h(\theta_i) \xrightarrow{p} \int h(\theta)f(\theta \mid \text{data})$$

for any integrable function $h(\cdot)$. In particular, if we select $h(\cdot)$ to be the indicator function of an event such as $\beta_{15} > 0$, then the ergodic theorem states that the relative frequency of that event in the sequence $\theta_1, \ldots, \theta_M$ is a consistent estimate of the posterior probability of that event. However, if the successive realizations generated by the sampler are highly correlated, then the relative frequency may approach the limit very slowly; in this situation the Markov chain is said to "mix" slowly.

In our initial attempts to apply MCMC, we found that the Markov chain did mix very slowly, probably because of the highly collinear covariance matrix of the β vector. We were able to reduce the collinearity, with a resulting improvement in the rate of convergence of the Markov chain, by reexpressing β as a linear combination of the initial state vector and the dominant principal components of the integrated Wiener process.

2.5 Reexpressing β

The conditional prior distribution of β is

$$p(\beta|\beta_0, \beta_0', \tau) = N_p\left(L\begin{bmatrix}\beta_0\\\beta_0'\end{bmatrix}, \frac{1}{\tau}V\right) = N_p\left(\mu_\beta, \frac{1}{\tau}V\right),\tag{11}$$

where the matrix V depends only on the observation times,

$$v_{i,j} = \frac{t_i^3}{3} - \frac{(t_j - t_i)t_i^2}{2}, \quad i \le j$$

[see (8)]. The spectral decomposition is $V = U \operatorname{diag}(w)U'$, where U is the orthonormal eigenvector matrix and w is the vector of eigenvalues. Suppose that the first r eigenvalues account for, say, $1 - \varepsilon^2$ of the total variance and let $T = U_r \operatorname{diag}\left(\sqrt{w_r}\right)$ and $z = \operatorname{diag}(w_r^{-.5})U_r'(\beta - \mu_\beta)\sqrt{\tau}$, where U_r is the first r columns of U and w_r is the first r components of w. Clearly, the components of z are iid standard normal and

$$E\left[\| \tau(\beta - \mu_\beta) - Tz \|^2\right] = \epsilon^2 E\left[\| \tau(\beta - \mu_\beta) \|^2\right].$$

Consequently,

$$\beta \approx \mu_\beta + \frac{1}{\sqrt{\tau}}Tz \approx L\begin{bmatrix}\beta_0\\\beta_0'\end{bmatrix} + \frac{1}{\sqrt{\tau}}Tz.\tag{12}$$

3. EXAMPLES

The data in two of these examples come from cases we were involved in—we call them Case K and Case W. In each case one or more plaintiffs were suing a former employer for age discrimination in his or her dismissal. Data for these cases are available in StatLib (Kadane and Woodworth 2001). The third example is a reanalysis of a class action against the U.S. Postal Service (USPS) reported in Freidlin and

Gastwirth (2000). All three cases were analyzed via WinBUGS 1.3 (Spiegelhalter, Thomas, and Best 2000). For Cases K and W we used the principal component representation (12).

3.1 Case K

In this case flow data for all individuals employed by the defendant at any time during a 1,557-day period (about 17 quarters) were available to the statistical expert. During that period 96 employees were involuntarily terminated, 79 of whom were age 40 or above at the time of termination. The data were aggregated into 288 time intervals, or bins, bounded by times at which one or more employees entered the workforce, left the workforce, or reached a 40th birthday. The median bin width was 4 days, the mean was about 5 days, and the maximum was 24 days. Based on the discussion in Section 3.2, we prefer that the prior distribution for the smoothness parameter place most of its mass above .0007, so we selected a gamma prior with shape parameter 1 and mean .007; the other parameters were given diffuse but proper priors. Table 2 shows the data, mean, standard deviation, and positive tail area of the log odds ratios for bins with one or more involuntary terminations.

In Figure 1 we show how the smooth model fits the unsmoothed underlying data. To do this, we grouped cases by quarters and computed 95% equal-tail posterior-density credible intervals for the log odds assuming a normal prior with mean 0 and standard deviation 8. We like a standard deviation of 8 for this sort of descriptive display because it is fairly diffuse (there is, for example, 16% prior probability that the odds ratio exceeds 3,000) yet prevents infinite credible intervals when one category or the other has no terminations. Because this prior pulls the posterior distribution toward 0, it would be difficult for the respondent (the firm) to argue that it is biased in favor of the plaintiff.

Sensitivity to the prior mean of the smoothness parameter is explored in Figure 2, and sensitivity to the shape of the prior distribution of the smoothness parameter is explored in Figure 3. Between days 528 and 1,322 the probability of discrimination $P(\beta(t) > 0|\text{data})$ exceeds .99 and is insensitive to the prior distribution of the smoothness parameter. The plaintiff in Case K had been dismissed within that interval at day 766. It is not surprising that greater sensitivity to the smoothness parameter is shown at the start and the end of the period. Thus, if the date of termination of the plaintiff is near the start or the end of the observation period, the conclusions will be more sensitive to how much smoothing is assumed. This suggests the desirability of designing data collection so that it includes a period of time surrounding the event or events in question.

In addition, we analyzed Case K with Sargent's (1997) first-difference prior. The first-difference prior models the log odds ratio as a linear function plus a Wiener process with precision τ. Thus, the log odds ratio is continuous but not smooth. To scale the prior distribution of the precision of the log odds ratio, we again argue from our prior opinion, that the log odds ratio is unlikely to change more than 15% within a quarter. We interpret this as two standard deviations of the first difference, $\beta(t + d) - \beta(t)$. The variance of a one-quarter first difference is d/τ, where $d \approx 1/19$ is one quarter expressed as a fraction of the total observation period. Thus, $2\sqrt{d/\tau} \leq .15$, which implies that the prior should place most of its mass on $\tau \geq 9.4$.

TABLE 2. Data and Posterior Marginal Distributions of the Log Odds Ratio for Case K

Day	N	n	k	x	Posterior distribution		
					Mean	SD	$P(\beta > 0)$
3	190	102	1	1	−1.5	1.3	.111
175	208	110	1	0	−.95	.82	.110
406	273	150	1	1	.25	.51	.702
444	273	151	1	1	.50	.47	.858
507	283	159	1	0	.91	.41	.985
528	283	159	2	1	1.05	.39	.994
535	284	161	1	1	1.10	.39	.997
555	286	164	1	0	1.23	.38	.999
567	289	167	1	0	1.30	.37	1.000
582	287	167	1	1	1.40	.36	1.000
605	293	170	10	9	1.53	.36	1.000
661	290	165	1	1	1.82	.35	1.000
668	290	165	1	1	1.85	.35	1.000
696	287	163	9	8	1.96	.35	1.000
703	278	155	4	3	1.98	.35	1.000
710	276	152	1	1	2.00	.35	1.000
731	272	150	1	1	2.07	.35	1.000
752	270	147	2	2	2.12	.35	1.000
766	269	143	1	1	2.15	.35	1.000
784	269	141	1	1	2.18	.35	1.000
797	266	138	2	2	2.20	.35	1.000
846	264	136	1	1	2.25	.35	1.000
847	263	135	2	2	2.25	.35	1.000
850	261	133	3	3	2.25	.35	1.000
857	258	130	4	4	2.25	.35	1.000
863	245	120	1	1	2.25	.35	1.000
864	244	119	1	0	2.26	.35	1.000
927	247	121	1	0	2.25	.35	1.000
955	245	120	1	1	2.24	.35	1.000
980	242	118	4	4	2.23	.35	1.000
1,008	236	113	3	3	2.20	.35	1.000
1,017	234	111	1	0	2.19	.35	1.000
1,018	233	111	5	5	2.19	.35	1.000
1,025	227	105	1	1	2.18	.35	1.000
1,037	227	105	1	1	2.16	.35	1.000
1,095	230	105	1	1	2.05	.36	1.000
1,102	226	101	2	1	2.04	.36	1.000
1,106	224	101	1	1	2.03	.36	1.000
1,113	223	100	3	2	2.01	.36	1.000
1,116	220	98	1	0	2.00	.36	1.000
1,141	220	97	1	1	1.94	.36	1.000
1,200	217	93	1	1	1.77	.38	1.000
1,214	216	91	1	1	1.72	.38	1.000
1,224	215	90	2	2	1.69	.39	1.000
1,225	213	88	1	1	1.69	.39	1.000
1,253	215	89	1	1	1.58	.40	1.000
1,256	214	88	1	1	1.56	.40	1.000
1,284	213	89	1	1	1.44	.41	1.000
1,319	210	89	1	1	1.27	.44	.998

TABLE 2. (Continued)

Day	N	n	k	x	Posterior distribution		
					Mean	SD	P(β > 0)
1,322	207	88	1	0	1.26	.44	.997
1,361	207	89	1	0	1.06	.48	.982
1,375	204	89	1	1	.99	.50	.972
1,557	205	90	2	1	.09	.93	.568

NOTE: Data and posterior distributions for bins with one or more involuntary terminations. N, workforce; n, protected; k, involuntary terminations; x, involuntary terminations of protected employees. The smoothness parameter, τ, had a Gamma(1) prior distribution with mean .007, as described in the text. Two chains were run with 50,000 replications each, discarding the first 4,000. The Gelman–Rubin statistic indicated that the chains had converged.

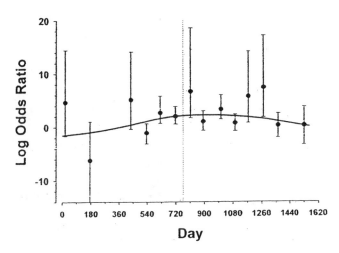

FIGURE 1. Case K: Posterior Mean and Probability of Discrimination for Our Preferred Prior Distribution of the Smoothness Parameter. Vertical bars are 95% posterior highest density regions for quarterly aggregates with iid N(0, 64) priors. Vertical dotted line indicates date of plaintiff's dismissal.

A gamma prior with mean 100 and shape 1 places about 90% of its mass above 9.5. Figure 4 compares the posterior mean log odds ratio for Sargent's smooth model and our continuous model. We found that the posterior distribution for Sargent's model was not sensitive to the prior means of τ between 1 and 100.

We conducted what we believe would be an acceptable frequentist analysis via proportional hazards regression to model the time to involuntary termination. The initial model included design variables for membership in the protected group and for the linear interaction of this variable with time. The linear interaction was insignificant, so we presume that a frequentist would opt for a constant odds model. The maximum likelihood estimate (and asymptotic standard error) of the log odds ratio is 1.51 (.27), which approximates the posterior mean and standard deviation of this parameter for large values of the smoothness parameter (see Table 3).

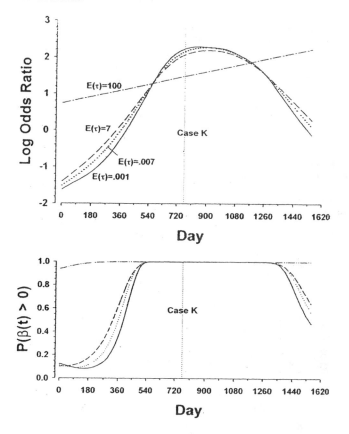

FIGURE 2. Case K: Effect of Varying the Prior Mean of the Smoothness Parameter (all distributions are Gamma with shape parameter 1).

Although the frequentist and Bayesian analyses reach the same conclusion regarding the presence of discrimination in Case K, we believe that the frequentist analysis does not produce probability statements relevant to the particular case in litigation and in no way constitutes a gold standard for our analysis. On the bases of these analyses, a statistical expert would be able to report that there is strong, robust evidence that discrimination against employees aged 40 and above was present in terminations between days 528 and 1,322 and, in particular, on the day of the plaintiff's termination.

3.2 Case W

Two plaintiffs, terminated about a year apart, brought separate age discrimination suits against the employer. The plaintiffs' attorneys requested data on all individuals who were in the defendant's workforce at any time during an approximately 4.5-year observation period containing the termination dates of the plaintiffs. Dates of hire and separation and reason for separation were provided by the employer as well as age in years at entry into the dataset (the first day of the observation period

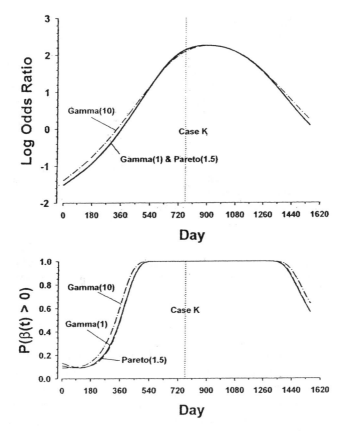

FIGURE 3. Case K: Effect of Varying the Shape of the Prior Distribution of the Smoothness Parameter (all distributions have mean .007 and the indicated shape parameters).

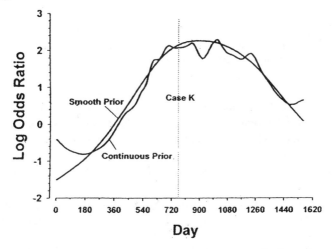

FIGURE 4. Case K: Posterior Mean for Continuous (Wiener process, prior mean smoothness, 100) and Smooth (integrated Wiener process, prior mean smoothness, .007) Distributions of the Log Odds Ratio.

TABLE 3. Sensitivity Analysis[a] for Case K

Model	Prior[b]	Posterior distribution for Case K		
		Mean	SD	$P(\beta > 0/\text{Data})$
Smooth	$E(\tau) = .001$	2.22	.37	1.0000
	$E(\tau) = .007$			
	Gamma(1)	2.15	.35	1.0000
	Gamma(10)	2.10	.33	1.0000
	Pareto(1.5)	2.13	.35	1.0000
	$E(\tau) = .07$	2.06	.37	1.0000
	$E(\tau) = 100$	1.46	.28	1.0000
Continuous	$E(\tau) = 100^c$	2.11	.49	1.0000
Maximum likelihood[d]	Linear log odds ratio	1.44	.28	n/a
	Constant log odds ratio	1.51	.27	n/a

[a] $P(\beta > 0)$ for Case K (day 766) is robust to specification of the prior distribution of the log odds ratio.
[b] Where not indicated otherwise, the smoothness parameter, τ, had a Gamma(1) prior with the indicated mean.
[c] The shape parameter was 1 except for the smooth prior with mean 100, which had shape parameter 5.
[d] Maximum likelihood estimates were computed by proportional hazards regression; asymptotic approximations to the one-sided p values are less than .0005.

or the date of hire) and at separation. The data request was made before the expert statistician was retained and it failed to ask for dates of birth; however, from the ages at entry and exit, it was possible to determine a range of possible birth dates for each employee. Thus, at any given time during the observation period, there is some uncertainty about whether the handful of nonterminated employees near the protected age (40 and older) were or were not in the protected class. We did not attempt to incorporate that uncertainty into this analysis and resolved ambiguities by assuming the birth date was at the center of the interval of dates consistent with the reported ages.

Over an observation period of about 1,600 days, the workforce was reduced by about two-thirds; 103 employees were involuntarily terminated in the process. A new CEO took over at day 862, near the middle of the observation period. The plaintiffs asserted that employees aged 50 (or 60) and above were targeted for termination under the influence of the new CEO. So in our reanalysis we have divided the protected class into two subclasses: ages 40–49 and ages 50–64 and estimated separate log odds ratios for each of the protected subclasses relative to the unprotected class. Here we present a fully Bayesian analysis of two models, one with smoothly time-varying odds ratios for each protected subclass and one with smoothly time-varying odds ratios in two phases—before and after the arrival of the new CEO.

The personnel data were aggregated by status (involuntarily terminated, other) into 171 time bins as described in Case K and three age categories (< 40, 40–49, 50–64). Aggregated data along with posterior means, standard deviations, and probabilities of discrimination for the two protected subclasses for bins containing at least one termination are reported in Table 4. Table 4 also reports posterior means, standard deviations, and probabilities of discrimination for our preferred

TABLE 4. Aggregated Flow Data and Posterior Distributions for Case W

| | Workforce | | | Terminated | | | Posterior distributions of log odds ratios* | | | | | |
| | | | | | | | Age 40–49 vs. unprotected | | | Age 50–64 vs. unprotected | | |
Day	< 40	40–49	50–64	< 40	40–49	50–64	Mean	SD	P(β > 0)	Mean	SD	P(β > 0)
13	77	58	58	0	1	0	.18	.52	.642	.20	.49	.662
35	77	57	58	1	1	1	.15	.48	.632	.19	.46	.667
38	76	56	57	2	0	0	.15	.47	.631	.19	.45	.668
39	74	56	57	1	0	0	.15	.47	.630	.19	.45	.668
42	73	56	57	1	1	0	.14	.47	.628	.19	.45	.669
75	70	55	58	0	1	0	.10	.42	.601	.18	.40	.678
80	70	54	58	3	3	6	.09	.42	.594	.18	.40	.680
81	67	51	52	0	0	1	.09	.41	.593	.18	.40	.680
84	67	51	51	0	0	1	.08	.41	.589	.18	.40	.679
115	68	51	50	1	0	0	.04	.39	.548	.17	.37	.679
161	67	52	50	1	1	0	-.03	.38	.478	.14	.35	.666
164	66	51	50	0	1	0	-.04	.38	.474	.14	.35	.666
175	66	51	50	1	0	0	-.05	.38	.457	.14	.35	.661
209	65	50	49	0	1	0	-.10	.39	.411	.12	.35	.643
252	61	53	49	0	0	1	-.15	.41	.372	.11	.35	.623
259	61	53	48	2	0	0	-.16	.41	.367	.11	.35	.620
263	59	53	48	1	0	0	-.17	.42	.363	.10	.36	.619
266	58	53	48	1	0	0	-.17	.42	.361	.10	.36	.619
353	56	52	49	0	1	4	-.21	.45	.332	.07	.37	.579
357	56	51	45	0	0	1	-.21	.45	.334	.07	.37	.576
490	50	54	44	1	1	1	-.11	.43	.419	-.02	.39	.487
571	47	54	44	0	1	0	.00	.40	.507	-.06	.39	.451
573	47	54	44	0	1	0	.00	.40	.511	-.06	.39	.450
658	51	56	44	0	1	0	.10	.37	.601	-.05	.38	.455
665	51	55	44	1	0	0	.10	.37	.610	-.05	.37	.457
733	47	54	50	6	6	6	.15	.36	.661	-.02	.36	.489

continued

TABLE 4. (Continued)

Day	Workforce			Terminated			Posterior distributions of log odds ratios[*]					
							Age 40–49 vs. unprotected			Age 50–64 vs. unprotected		
	<40	40–49	50–64	<40	40–49	50–64	Mean	SD	$P(\beta > 0)$	Mean	SD	$P(\beta > 0)$
735	41	48	43	0	1	0	.15	.36	.662	-.01	.36	.491
773	40	48	40	1	0	0	.16	.37	.668	.01	.36	.523
838	40	46	39	3	1	0	.16	.39	.653	.08	.36	.590
New CEO arrives												
901	35	43	37	0	1	0	.11	.41	.599	.16	.36	.681
907	35	42	35	0	1	1	.11	.41	.592	.17	.36	.688
924	34	42	34	0	0	1	.08	.42	.570	.20	.36	.711
948	35	41	34	0	1	0	.05	.43	.537	.23	.36	.746
1,017	34	36	33	0	1	0	-.13	.45	.389	.34	.36	.830
1,029	34	36	34	1	1	0	-.17	.45	.358	.36	.36	.842
1,092	30	36	34	3	3	6	-.42	.48	.185	.44	.36	.885
1,113	27	32	27	0	0	0	-.52	.49	.139	.46	.37	.893
1,121	27	31	26	0	0	1	-.56	.50	.124	.47	.37	.896
1,148	27	31	25	1	0	0	-.70	.54	.082	.49	.38	.901
1,162	25	31	25	0	0	1	-.78	.56	.066	.50	.39	.901
1,173	25	31	23	0	0	1	-.84	.58	.058	.50	.39	.904
1,239	23	30	22	0	0	1	-1.23	.74	.024	.51	.41	.895
1,390	24	28	22	0	0	1	-2.21	1.29	.009	.43	.49	.816
1,397	24	28	20	0	0	1	-2.25	1.32	.009	.43	.50	.809
1,438	23	28	19	1	0	0	-2.53	1.50	.008	.39	.54	.773
1,579	20	30	19	1	0	0	-3.48	2.21	.009	.27	.77	.655
1,612	19	30	19	0	0	1	-3.71	2.40	.010	.24	.85	.628

[*] The smoothness parameter had a Gamma(τ) prior with mean .007. Computations were via WinBUGS 1.3. There were 25,000 replications of two chains discarding the first 4,000.

choice of the prior distribution of the smoothness parameters—gamma with shape parameter 1 and mean .007. Figure 5 shows posterior means and probabilities of discrimination for different choices of the prior mean of the smoothness parameter. Figure 6 contrasts two-phase and noninterrupted models for employees aged 50–64.

The figures make it clear that the log odds ratios for either protected subclass were close to 0 before the new CEO arrived. After his arrival it appears that terminations of employees aged 40–49 declined and terminations of employees aged 50–64 increased. This is clearest in the interrupted model, but present to some extent in all models for all smoothness parameter values.

Two plaintiffs, indicated by vertical dotted lines in Figures 5 and 6, brought age discrimination suits against the employer. Plaintiff W1, who was between 50 and 59 years of age, was one of 12 employees involuntarily terminated on day 1,092. His theory of the case was that the new CEO had targeted employees aged 50 and above for termination. Under the two-phase model, which corresponds to the plaintiff's theory of the case, the probability of discrimination at the time of this plaintiff's termination was close to 1.00. However, the posterior probability

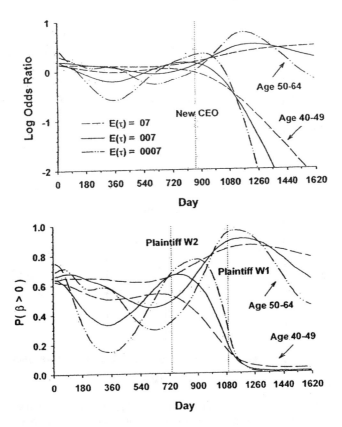

FIGURE 5. Case W: Sensitivity to the Smoothness Parameter.

of discrimination in this case is somewhat sensitive to the choice of model and smoothness parameter (Fig. 6).

In the original case the plaintiff's statistical expert tabulated involuntary termination rates for each calendar quarter and each age decade. He reported that, "[involuntary] separation rates for the [period beginning at day 481] averaged a little above three percent of the workforce per quarter for ages 20–49, but jumped to six and a half percent for ages 50–59. The 50–59 year age group differed significantly from the 20–39 year age group (signed-rank test, $p = .033$, one sided)." Our reanalysis is consistent with that conclusion (Fig. 6, two-phase model). The plaintiff alleged and the defendant denied that the new CEO had vowed to weed out older employees. The case was settled before trial.

The case of plaintiff W2 went to trial. This 60-year-old plaintiff was one of 18 employees involuntarily terminated on day 733. On that day three of eight employees (37.5%) aged 60 and up were terminated compared to 15 of 136 (11.0%) employees terminated out of all other age groups (one-sided hypergeometric $p = .0530$).

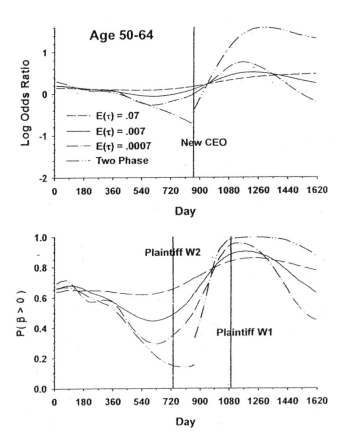

FIGURE 6. Comparison of Noninterrupted and Two-Phase Odds Ratio Models. For the two-phase model the prior distributions of the four smoothness parameters were gamma with shape 1 and mean .007.

Although the plaintiff had been terminated in the quarter prior to the arrival of the new CEO, the plaintiff's theory was that the new CEO had been seen on site before he assumed office and had influenced personnel policy decisions prior to his official arrival date.

The defense statistician presented several analyses of the quarterly aggregated data involving different subsets of the observation period and different subgroups of protected and unprotected employees. Based on two-sided p values, he reported no significant differences between any subgroups; however, one-sided p values are more appropriate in age discrimination cases and several of these are "significant" or nearly so. According to his analysis, for the period after the new CEO was hired, employees aged 50–59 were terminated at a significantly higher rate than employees aged 20–39 ($p = .053$, one sided) and employees aged 40–49 ($p = .050$, one sided); for the period beginning with the new CEO's second quarter in office the one-sided p values were .039 and .038, respectively. The defense expert did not analyze the interval beginning one quarter prior to the arrival of the new CEO—the quarter in which plaintiff W2 was terminated. Thus, the defense expert's analysis generally agrees with the reanalysis presented in this article (Fig. 2.6). The defense expert also reported several discriminant analyses meant to demonstrate that the mean age of involuntarily terminated employees was not different from the mean age of the workforce.

In response to the latter analysis, the plaintiff's statistician argued that this was accounted for by a high rate of termination of employees in their first year of service ("short-term employees") and presented the results of a proportional hazards regression analysis with constant odds ratios over the entire observation period. The model involved design variables for employees in the first year of service and for employees aged 50 and above (thus, the reference category was long-term employees under the age of 50). Employees in their first year of service were terminated at a significantly higher rate relative to the reference category (odds ratio = 2, $p = .01$, two sided) as were employees aged 50 and older (odds ratio = 1.58, $p = .03$, two sided).

The plaintiff's theory of the case, as we understand it, had three components: (1) that the new CEO had been seen on site on several occasions in the quarter before he assumed office and was presumably an active participant in personnel policy decisions prior to his official arrival date, (2) that the CEO had stated his intent to weed out older employees, and (3) that this had an adverse impact on employees aged 60 and above. The preceding proportional hazards analysis and the analysis of the 18 terminations on day 733 (37% of 60-year-olds terminated versus 11% of all other employees) support the disparate impact theory.

The plaintiff had refused the defendant's offer of a different job at lower pay. The judge instructed the jury that the plaintiff had a duty under the law to exercise reasonable diligence to minimize his damages and if they found that he had not done so, then they should reduce his damages by the amount he reasonably could have avoided if he had sought out or taken advantage of such an opportunity. The jury found that the plaintiff had proven that age was a determining factor for his discharge but that he had failed to mitigate his damages. Therefore, the award was the difference between what he would have earned at his original salary prior to discharge minus the amount he would have earned had he accepted the lower salaried job. The defendant appealed the case, but settled prior to trial of the appeal.

Our reanalysis, with time-varying odds ratios, does not support a theory of adverse impact against employees aged 50–64 prior to the arrival of the new CEO; however, the plaintiff's specific claim was discrimination against employees aged 60 and older and there does seem to be evidence of this at the time of the plaintiff's termination (one-sided hypergeometric $p = .053$).

One Sided or Two Sided? Bayesians compute probabilities of relevant events, that is, $P(\beta(t) > 0|\text{data})$, the conditional probability of discrimination at time t given the data. Frequentists favor two-sided p values (roughly speaking, the conditional probability, given the assumption of no discrimination, of getting the data we got plus the probability of getting even more deviant hypothetical data). As Bayesians, we think such probabilities are legally irrelevant. Even within the frequentist paradigm, however, we think that the use of two-sided p values is wrong in age discrimination cases. We say this because frequentist inference claims to control "Type I error," that is, to control the conditional probability that an "innocent" employer will be found to discriminate. In age discrimination cases the Type I errors can occur on one side only because only protected employees have the right to sue under the age discrimination act. Type I errors in the "other tail" would be produced by evidence of discrimination against the unprotected class but the unprotected class has no legal right to relief.

VALENTINO v. UNITED STATES POSTAL SERVICE

Freidlin and Gastwirth (2000) discussed a case in which the plaintiff filed a charge of sex discrimination in promotion at the U.S. Postal Service after she was denied a promotion in mid-1976. The judge certified the women employed at grade 17 and higher as a class. The underlying data, raw log odds ratios, and posterior distributions of the log odds ratios for three different specifications of the prior are shown in Table 5.

Using frequentist methods, Freidlin and Gastwirth (2000) reported that the p value for discrimination against women was .0006 in period 06/74–03/75, .020 in period 03/75–01/76, and greater than .5 in subsequent periods. The authors advocated using CUSUM methods "...to determine the time period when the pattern [of discrimination] remained the same. If the original complaint was filed during a period of statistically significant [discrimination] before the change to fair [employment practices] occurred, then the data are consistent with the plaintiff's claim." They reported that their CUSUM tests showed a significant change in discrimination over time and that this effect was concentrated in grades 17–19 and 23–25. Apparently they argued from the significant CUSUM and the pattern of p values for individual time periods that there was discrimination in 1974–1975 and 1975–1976 but not later. However, no formal test or estimate of the location of the changepoint was offered; instead, "...the graph of [the CUSUM test statistic against time] helps to identify the time of the change if one exists...." Thus, what we appear to be offered is a test of inhomogeneity of the odds ratio over time combined with inspecting a CUSUM graph and a list of p values for individual time periods.

We have reanalyzed the data using three specifications of the joint prior distribution of promotion probabilities for each period, grade, and gender. In the completely

TABLE 5. Aggregated Data and Posterior Distributions for Valentino v. United States Postal Service

		Males		Females			Independent[b]			Exchangeable			AR(1) × AR(1)		
Time	Grade	N[a]	x	N	x	Lor	μ	σ	P(−)[c]	μ	σ	P(−)	μ	σ	P(−)
06/74–03/75	17–19	229	67	73	5	−1.73	−1.80	.50	1.000	−1.55	.44	1.000	−1.51	.43	1.000
	20–22	360	74	48	9	−.11	−.14	.40	.631	−.17	.38	.661	−.16	.38	.655
	23–25	703	132	33	2	−1.28	−1.47	.81	.986	−1.00	.60	.966	−.94	.59	.961
	26–28	236	28	7	1	.21	−.13	1.30	.491	−.21	.87	.580	−.14	.84	.553
	29–31	82	8	1	0	−∞	−7.75	10.17	.800	−.55	1.21.	.675	−.45	1.22	.636
03/75–01/76	17–19	205	40	89	6	−1.21	−1.27	.47	.999	−1.12	.43	.998	−1.10	.42	.998
	20–22	373	39	43	5	.12	.05	.52	.439	−.04	.49	.514	−.03	.48	.513
	23–25	716	41	36	1	−.75	−1.16	1.22	.845	−.58	.75	.776	−.53	.73	.759
	26–28	277	85	9	1	−1.26	−1.64	1.26	.938	−.89	.76	.888	−.83	.74	.877
	29–31	85	7	1	0	−∞	−7.55	10.18	.785	−.50	1.21	.661	−.43	1.23	.632
01/76–01/77	17–19	233	31	101	10	−.33	−.36	.39	.820	−.34	.37	.820	−.33	.36	.818
	20–22	396	32	52	4	−.05	−.13	.58	.571	−.12	.51	.582	−.11	.50	.574
	23–25	721	54	36	5	.69	.63	.52	.116	.54	.48	.133	.55	.48	.126
	26–28	271	28	9	2	.91	.77	.89	.180	.45	.74	.263	.46	.72	.252
	29–31	85	5	2	0	−∞	−8.08	10.15	.823	−.28	1.15	.584	−.21	1.17	.554
01/77–01/78	17–19	200	43	86	18	−.03	−.04	.32	.548	−.03	.31.	.531	−.03	.31	.534
	20–22	377	80	52	9	−.25	−.28	.39	.760	−.24	.37	.734	−.23	.37	.726
	23–25	680	57	35	6	.82	.77	.48	.062	.66	.45	.077	.66	.45	.076
	26–28	262	18	8	1	.66	.32	1.28	.350	.16	.86	.408	.18	.85	.397
	29–31	89	14	3	0	−∞	−9.65	10.06	.933	−.53	1.06	.681	−.48	1.11	.653
01/78–01/79	17–19	196	29	90	8	−.58	−.61	.43	.928	−.48	.40	.889	−.43	.40	.861
	20–22	325	45	50	7	.01	−.03	.45	.511	.04	.42	.444	.09	.41	.396
	23–25	685	3	35	14	5.02	5.15	.71	.000	3.49	.51	.000	3.51	.51	.000
	26–28	252	14	9	1	.75	.41	1.28	.324	.40	.86	.304	.57	.84	.234
	29–31	78	6	3	1	1.79	1.64	1.49	.127	.72	1.00	.227	.91	1.02	.176

Posterior moments and tail areas

[a] N is the number of employees; x is the number promoted.

[b] Prior models are independent log odds ratios, exchangeable main effects of time and grade, exchangeable interactions, and AR(1) main effects of time and grade and exchangeable interactions.

[c] $P(−)$ is the posterior probability that the log odds ratio is negative—that is, women are less likely to be promoted.

131

independent model, each probability has a prior beta (.1, .1) distribution. In the random effects model, the log odds ratio (female versus male) for each year and grade consists of a year effect, a grade effect, and an interaction. In the exchangeable model, each class of effects (time, grade, interaction) has an exchangeable multivariate normal prior distribution. In the AR(1) × AR(1) model, time and grade effects have multivariate normal priors with AR(1) covariance structure and the interactions are exchangeable.

Table 5 shows the posterior mean and standard deviation of the log odds ratio and the probability of discrimination (negative log odds ratio) for each year and grade. We agree with Freidlin and Gastwirth that there is strong evidence of discrimination against women in grades 17–19 in periods 1974–1975 and 1975–1976 and against women in grades 23–25 in year 1974–1975 and not much evidence of discrimination elsewhere. However, we do not see the relevance of a formal changepoint test. If there is inhomogeneity, then it should be incorporated into the model. Our analysis does this and it shows that not all members of the certified class have equal claims for relief. We believe that a neutral statistician analyzing these data would report that the evidence of discrimination is not uniform over grades or time periods, but is concentrated in grades 17–19 and 23–25 in year 1974–1975 and in grades 17–19 in year 1975–1976. We believe that this is precisely the information that the court needs to determine how the award (if any) should be distributed among members of the certified class.

DISCUSSION

A standard criticism of Bayesian analyses is that the prior assumptions are arbitrary. One response is, "Compared to what?" Bayesian analysis can be explained to a jury in less convoluted ways than frequentist analyses and makes explicit the necessity to think about sensitive assumptions, rather than covering them with a mantle of false objectivity. The assumption of constant odds ratios in particular and the functional form of a model in general are examples of unexamined subjectivity. The Bayesian approach to model specification involves specifying a prior distribution over a more general class of time-varying odds ratio models. The subjective component of model specification resides in the prior distribution of the smoothness parameter. To some the need to think carefully about the prior distribution of the smoothness parameter may seem fatally to open the analysis to attack by opposing counsel on the grounds of arbitrariness. To that we respond that the assumption of a constant or (piece-wise) linear odds ratio is not only arbitrary but implausible on its face and that a more realistic analysis has a better chance of prevailing.

ACKNOWLEDGMENTS

The research of Joseph B. Kadane was partially supported by NSF grant DMS-93–03557. We thank Michael Finkelstein, Joseph Gastwirth, Christopher Genovese, Bruce Levin, John Rolph, and Ashish Sanil for their helpful comments on an earlier draft. We also thank Kate Cowles for a key insight in specifying the prior distribution of the smoothness parameter. Finally, we thank Caroline Mitchell for her help with the legal aspects of the problem.

REFERENCES

Cox, D. R. (1972), "Regression Models and Life-Tables," *Journal of the Royal Statistical Society,* Ser. B. 34, 187–220.

Finkelstein, M. O., and Levin, B. (1994), "Proportional Hazard Models for Age Discrimination Cases," *Jurintetrics Journal,* 34, 153–171.

Freidlin, B., and Gastwirth, J. (2000), "Changepoint Tests Designed for the Analysis of Hiring Data Arising in Employment Discrimination Cases," *Journal of Business & Economic Statistics,* 18, 315–322.

Gastwirth, J. (1992), "Employment Discrimination: A Statistician's Look at Analysis of Disparate Impact Claims," *Law and Inequality; A Journal of Theory and Practice,* 11, 151–179.

Gelman, A., Carlin, J. B., Stern, H. S., and Rubin, D. B. (1995), *Bayesian Data Analysis,* London: Chapman & Hall.

Gersch, W. (1982), "Smoothness Priors," in *Encyclopedia of Statistical Sciences,* Vol. 8, eds. S. Kotz, N. L. Johnson, and C. B. Read, New York: Wiley, pp. 518–526.

Kadane, J. B. (1990), "A Statistical Analysis of Adverse Impact of Employer Decisions," *Journal of the American Statistical Association,* 85, 925–933.

Kadane, J., and Mitchell, C. (1998), "Statistics in Proof of Employment Discrimination Cases," in *Controversies in Civil Rights: The Civil Rights Act of 1964 in Perspective,* ed. B. Grofman, Charlottesville: University of Virginia Press.

Kadane, J., and Woodworth, G. (2001), "Employment Discrimination Data," available at http: //lib.stat.cmu.edu/datasets/caseK.txt and /caseW.txt.

Kass, R. E., Tierney, L., and Kadane, J. B. (1988), "Asymptotics in Bayesian Computation," in *Bayesian Statistics* 3, eds. J. M. Bernardo, M. H. DeGroot, D. V. Lindley, and A. F. M. Smith, Oxford: Oxford University Press, pp. 261–278.

Lin, X., and Zhang, D. (1998), "Semiparametric Stochastic Mixed Models for Longitudinal Data," *Journal of the American Statistical Association,* 93, 710–719.

Sargent, D. J. (1997), "A Flexible Approach to Time-Varying Coefficients in the Cox Regression Setting," *Lifetime Data Analysis,* 3, 13–25.

Spiegelhalter, D., Thomas, A., and Best, N. (2000), *WinBUGS Version 1.3 User Manual,* London: MRC Biostatistics Unit, available at http://www.mrcbsu.cam.ac.uk/bugs.

Tierney, L. (1994), "Markov Chains for Exploring Posterior Distributions" (with discussion), *Annals of Statistics,* 22, 1701–1762.

Wahba, G. (1978), "Improper Priors, Spline Smoothing and the Problem of Guarding Against Model Errors in Regression," *Journal of the Royal Statistical Society,* Ser. B, 40, 364–372.

B. DRIVING WHILE BLACK

1

STATE OF NEW JERSEY V. SOTO

STATE OF **NEW JERSEY** v. **PEDRO SOTO,** DELORES
BRASWELL, LARNIE BODDY, CHAUNCEY DAVIDSON,
MILTON LUMPKIN, ALFRED S. POOLE, SAM GANT,
DONALD CREWS, KIM HARRIS, OCIE NORMAN,
ANTOINE PETERS, FLOYD PORTER THEOTIS
WILLIAMS A/K/A WALTER DAY, PAUL DACOSTA,
RONNIE LOCKHART, TERRI MONROE AND KEVIN
JACKSON, DEFENDANTS.

SUPERIOR COURT OF NEW JERSEY, LAW DIVISION,
GLOUCESTER COUNTY

324 N.J. Super. 66; 734 A.2d 350;

March 4, 1996, Decided

Approved for Publication July 15, 1999.

COUNSEL: *P. Jeffrey Wintner,* Deputy Public Defender I, for
defendants Pedro Soto, Delores Braswell, Larnie Boddy, Chauncey
Davidson, Milton Lumpkin and Alfred S. Poole (*Susan L. Reisner,*
Public Defender, attorney).

Wayne E. Natale, Deputy Public Defender II, for defendant Sam
Gant (*Susan L. Reisner,* Public Defender, attorney).

Carrie D. Dingle, Assistant Deputy Public Defender I, for defendants
Donald Crews, Kim Harris, Ocie Norman, Antoine Peters, Floyd
Porter, and Theotis Williams a/k/a Walter Day (*Susan L. Reisner,*
Public Defender, attorney).

William H. Buckman, for defendants Paul DaCosta,
Ronnie Lockhart and Terri Monroe.

Justin Loughry, for defendant Kevin Jackson (*Tomar, Simonoff,
Adourian, O'Brien, Kaplan, Jacoby & Graziano,* attorneys).

John M. Fahy, Senior Deputy Attorney General (*Deborah T. Poritz,* Attorney General of New Jersey, attorney) and Brent Hopkins, Assistant Gloucester County Prosecutor (*Harris Y. Cotton,* Gloucester County Prosecutor, attorney), for the State of New Jersey.

JUDGES: ROBERT E. FRANCIS, J.S.C.

OPINION BY: R.E. FRANCIS

OPINION: ROBERT E. FRANCIS, J.S.C.

These are consolidated motions to suppress under the equal protection and due process clauses of the Fourteenth Amendment.[1] Seventeen defendants of African ancestry claim that their arrests on the New Jersey Turnpike south of exit 3 between 1988 and 1991 result from discriminatory enforcement of the traffic laws by the New Jersey State Police.[2] After a lengthy hearing, I find defendants have established a *prima facie* case of selective enforcement which the State has failed to rebut requiring suppression of all contraband and evidence seized.

Defendants base their claim of institutional racism primarily on statistics. During discovery, each side created a database of all stops and arrests by State Police members patrolling the Turnpike between exits 1 and 7A out of the Moorestown Station for thirty-five randomly selected days between April 1988 and May 1991 from arrest reports, patrol charts, radio logs and traffic tickets. The databases are essentially the same. Both sides counted 3060 stops which the State found to include 1212 race identified stops (39.6%), the defense 1146 (37.4%).

To establish a standard against which to compare the stop data, the defense conducted a traffic survey and a violator survey. Dr. John Lamberth, Chairman of the Psychology Department at Temple University who I found is qualified as an expert in statistics and social psychology, designed both surveys.

The traffic survey was conducted over twenty-one randomly selected two and one-half hour sessions between June 11 and June 24, 1993 and between 8:00 a.m. and 8:00 p.m. at four sites, two northbound and two southbound, between exits 1 and 3 of the Turnpike. Teams supervised by Fred Last, Esq., of the Office of the Public Defender observed and recorded the number of vehicles that passed them except for large trucks, tractortrailers, buses and government vehicles, how many contained a "black" occupant and the state of origin of each vehicle. Of the 42,706 vehicles

---

1. The motions also include claims under the Fourth Amendment, but they were severed before the hearing to await future proceedings if not rendered moot by this decision.

2. Originally, twenty-three defendants joined in the motions. On the first day of the hearing, November 28, 1994, I dismissed the motions of Darrell Stanley, Roderick Fitzgerald, Fred Robinson, Charles W. Grayer, Keith Perry and Alton Williams due to their unexplained nonappearances.

counted, 13.5% had a black occupant. Dr. Lamberth testified that this percentage is consistent with the 1990 Census figures for the eleven states from where almost 90% of the observed vehicles were registered. He said it is also consistent with a study done by the Triangle Group for the U.S. Department of Transportation with which he was familiar.

The violator survey was conducted over ten sessions in four days in July 1993 by Mr. Last traveling between exits 1 and 3 in his vehicle at sixty miles per hour on cruise control after the speedometer had been calibrated and observing and recording the number of vehicles that passed him, the number of vehicles he passed and how many had a black occupant. Mr. Last counted a total of 2096 vehicles other than large trucks, tractortrailers, buses and government vehicles of which 2062 or 98.1% passed him going in excess of sixty miles per hour including 306 with a black occupant equaling about 15% of those vehicles clearly speeding. Multiple violators, that is those violating the speed limit and committing some other moving violation like tailgating, also equaled about 15% black. Dr. Lamberth testified that the difference between the percentage of black violators and the percentage of black travelers from the surveys is statistically insignificant and that there is no evidence traffic patterns changed between the period April 1988 to May 1991 in the databases and June–July 1993 when the surveys were done.

Using 13.5% as the standard or benchmark against which to compare the stop data, Dr. Lamberth found that 127 or 46.2% of the race identified stops between exits 1 and 3 were of blacks constituting an absolute disparity of 32.7%, a comparative disparity of 242% (32.7% divided by 13.5%) and 16.35 standard deviations. By convention, something is considered statistically significant if it would occur by chance fewer than five times in a hundred (over two standard deviations). In case I were to determine that the appropriate stop data for comparison with the standard is the stop data for the entire portion of the Turnpike patrolled by the Moorestown Station in recognition of the fact that the same troopers patrol between exits 3 and 7A as patrol between exits 1 and 3, Dr. Lamberth found that 408 or 35.6% of the race identified stops between exits 1 and 7A were of blacks constituting an absolute disparity of 22.1%, a comparative disparity of 164% and 22.1 standard deviations.[3] He opined it is highly unlikely such statistics could have occurred randomly or by chance.[4]

Defendants also presented the testimony of Dr. Joseph B. Kadane, an eminently qualified statistician. Among his many credentials, Dr. Kadane is a full professor of statistics and social sciences at Carnegie Mellon University, headed the Department

3. Dr. Lamberth erred in using 13.5% as the standard for comparison with the stop data. The violator survey indicates that 14.8%, rounded to 15%, of those observed speeding were black. This percentage is the percentage Dr. Lamberth should have used in making statistical comparisons with the stop data in the databases. Nonetheless, it would appear that whatever the correctly calculated disparities and standard deviations are, they would be nearly equal to those calculated by Dr. Lamberth.

4. In this opinion I am ignoring the arrest data in the databases and Dr. Lamberth's analysis thereof since neither side produced any evidence identifying the Turnpike population between exits 1 and 3 or 1 and 7A eligible to be arrested for drug offenses or otherwise. *See Wards Cove Packing Co. v. Atonio,* 490 U.S. 642, 109 S. Ct. 2115, 104 L. Ed. 2d 733 (1989).

of Statistics there between 1972 and 1981 and is a Fellow of the American Statistical Association, having served on its board of directors and a number of its committees and held various editorships on its Journal. Dr. Kadane testified that in his opinion both the traffic and violator surveys were well designed, carefully performed and statistically reliable for analysis. From the surveys and the defense database, he calculated that a black was 4.85 times as likely as a white to be stopped between exits 1 and 3. This calculation led him to "suspect" a racially non-neutral stopping policy. While he noted that the surveys were done in 1993 and compared to data from 1988 to 1991, he was nevertheless satisfied that the comparisons were useable and accurate within a few percent. He was not concerned that the violator survey failed to count cars going less than sixty miles per hour and travelling behind Mr. Last when he started a session. He was concerned, however, with the fact that only 37.4% of the stops in the defense database were race identified.[5] In order to determine if the comparisons were sensitive to the missing racial data, he did calculations performed on the log odds of being stopped. Whether he assumed the probability of having one's race recorded if black and stopped is the same as if white and stopped or two or three times as likely, the log odds were still greater than .99 that blacks were stopped at higher rates than whites on the Turnpike between exits 1 and 3 during the period April 1988 to May 1991. He therefore concluded that the comparisons were not sensitive to the missing racial data.

Supposing that the disproportionate stopping of blacks was related to police discretion, the defense studied the traffic tickets issued by State Police members between exits 1 and 7A on the thirty-five randomly selected days broken down by State Police unit.[6] There are 533 racially identified tickets in the databases issued by either the now disbanded Radar Unit, the Tactical Patrol Unit or general road troopers ("Patrol Unit"). The testimony indicates that the Radar Unit focused mainly on speeders using a radar van and chase cars and exercised limited discretion regarding which vehicles to stop. The Tac-Pac concentrates on traffic problems at specific locations and exercises somewhat more discretion as regards which vehicles to stop. Responsible to provide general law enforcement, the Patrol Unit exercises by far the most discretion among the three units. From Mr. Last's count, Dr. Lamberth computed that 18% of the tickets issued by the Radar Unit were to blacks, 23.8% of the tickets issued by the Tac-Pac were to blacks while 34.2% of the tickets issued by the Patrol Unit were to blacks. South of exit 3, Dr. Lamberth computed that 19.4% of the tickets issued by the Radar Unit were to blacks, 0.0% of the tickets issued by the Tac-Pac were to blacks while 43.8% of tickets issued by the Patrol Unit were to blacks. In his opinion, the Radar Unit percentages are statistically consistent with the standard established by the violator survey, but the differences between

---

5. That 62.6 percent of the stops in the defense database are not race identified is a consequence of both the destruction of the radio logs for ten of the thirty-five randomly selected days in accordance with the State Police document retention policy and the frequent dereliction of State Police members to comply with S.O.P. F3 effective July 13, 1984 requiring them to communicate by radio to their respective stations the race of all occupants of vehicles stopped prior to any contact.

6. Of the 3060 stops in the databases, 1292 are ticketed stops. Hence, no tickets were issued for nearly 60% of the stops.

the Radar Unit and the Patrol Unit between both exits 1 and 3 and 1 and 7A are statistically significant or well in excess of two standard deviations.

The State presented the testimony of Dr. Leonard Cupingood to challenge or refute the statistical evidence offered by the defense. I found Dr. Cupingood is qualified to give expert testimony in the field of statistics based on his Ph.D in statistics from Temple and his work experience with the Center for Forensic Economic Studies, a for profit corporation headquartered in Philadelphia. Dr. Cupingood collaborated with Dr. Bernard Siskin, his superior at the Center for Forensic Economic Studies and a former chairman of the Department of Statistics at Temple. Dr. Cupingood had no genuine criticism of the defense traffic survey. Rather, he centered his criticism of the defense statistical evidence on the violator survey. Throughout his testimony he maintained that the violator survey failed to capture the relevant data which he opined was the racial mix of those speeders most likely to be stopped or the "tail of the distribution." He even recommended the State authorize him to design a study to collect this data, but the State declined. He was unclear, though, how he would design a study to ascertain in a safe way the vehicle going the fastest above the speed limit at a given time at a given location and the race of its occupants without involving the credibility of State Police members. In any event, his supposition that maybe blacks drive faster than whites above the speed limit was repudiated by all State Police members called by the State who were questioned about it. Colonel Clinton Pagano, Trooper Donald Nemeth, Trooper Stephen Baumann and Detective Timothy Grant each testified that blacks drive indistinguishably from whites. Moreover, Dr. Cupingood acknowledged that he knew of no study indicating that blacks drive worse than whites. Nor could he reconcile the notion with the evidence that 37% of the unticketed stops between exits 1 and 7A in his database were black and 63% of those between exits 1 and 3. Dr. James Fyfe, a criminal justice professor at Temple who the defense called in its rebuttal case and who I found is qualified as an expert in police science and police procedures, also testified that there is nothing in the literature or in his personal experience to support the theory that blacks drive differently from whites.[7]

Convinced in his belief that the defense 15% standard or benchmark was open to question, Dr. Cupingood attempted to find the appropriate benchmark to compare with the databases. He did three studies of presumedly race-blind stops: night stops versus day stops; radar stops versus non-radar stops and drinking driving arrests triggered by calls for service.

In his study of night stops versus day stops, he compared the percentage of stops of blacks at night between exits 1 and 7A in the databases with the percentage of

---

7. During the hearing the State did attempt to introduce some annual speed surveys conducted on the Turnpike by the New Jersey Department of Transportation which the State represented would contradict a conclusion of the violator survey that 98.1% of the vehicles observed travelled in excess of sixty miles per hour. Besides noting that the State knew of these surveys long before the hearing and failed to produce them in discovery, I denied the proffer mainly because the surveys lacked racial data and also because there was a serious issue over their trustworthiness for admission under *N.J.R.E.* 803(c)(8) since the surveys were done to receive federal highway dollars. The critical information here is the racial mix of those eligible to be stopped, not the percentage of vehicles exceeding the speed limit.

stops of blacks during daytime and found that night stops were 37.3% black versus 30.2% for daytime stops. Since he presumed the State Police generally cannot tell race at night, he concluded the higher percentage for night stops of blacks supported a standard well above 15%. His premise that the State Police generally cannot recognize race at night, however, is belied by the evidence. On July 16, 1994 between 9:40 p.m. and 11:00 p.m. Ahmad S. Corbitt, now an assistant deputy public defender, together with Investigator Minor of the Office of the Public Defender drove on the Turnpike at 55 miles per hour for a while and parked perpendicular to the Turnpike at a rest stop for a while to see if they could make out the races of the occupants of the vehicles they observed. Mr. Corbitt testified that the two could identify blacks versus whites about 80% of the time in the moving mode and close to 100% in the stationary mode. Over and above this proof is the fact the databases establish that the State Police only stopped an average of eight black occupied vehicles per night between exits 1 and 7A. Dr. Cupingood conceded a trooper could probably identify one or two black motorists per night.

Next, in his study of radar stops versus non-radar stops, Dr. Cupingood focused on the race identified tickets where radar was used in the databases and found that 28.5% of them were issued to blacks. Since he assumed that radar is race neutral, he suggested 28.5% might be the correct standard. As Dr. Kadane said in rebuttal, this study is fundamentally flawed because it assumes what is in question or that the people stopped are the best measure of who is eligible to be stopped. If racial prejudice were afoot, the standard would be tainted. In addition, although a radar device is race-blind, the operator may not be. Of far more significance is the defense study comparing the traffic tickets issued by the Radar, Tac-Pac and Patrol Units which shows again that where radar is used by a unit concerned primarily with speeders and acting with little or no discretion like the Radar Unit, the percentage of tickets issued to blacks is consistent with their percentage on the highway.

And lastly in his effort to find the correct standard, Dr. Cupingood considered a DUI study done by Lieutenant Fred Madden, Administrative Officer of the Records and Identification Section of the State Police. Lt. Madden tabulated DUI arrests between July 1988 and June 1991 statewide, excluding the State Police, for Troop D of the State Police which patrols the entire length of the Turnpike, for Moorestown Station of Troop D and for Moorestown Station south of exit 3 broken down by race and between patrol related versus calls for service (i.e. accidents, motorist aids and other—the arrested motorist coming to the attention of the State Police by a toll-taker or civilian). Since Dr. Cupingood believed DUI arrests from calls for service were race neutral, he adopted the percentage of DUI arrests of blacks for the Moorestown Station from calls for service of 23% as a possible standard. Like his radar versus non-radar stop study, his use of the DUI arrest study is fundamentally flawed because he assumed what is in question. Further, he erred in assuming that DUI arrests from calls for service involve no discretion. While the encounters involve no discretion, the arrests surely do. He admitted that race/discretion may explain the following widespread statistics in the DUI arrest study:

Statewide (all departments)	12% black
Statewide (excluding State Police)	10.4% black

State Police	16% black
Troop D	23% black
Moorestown Station	34% black
Moorestown Station patrol related	41% black
Moorestown Station patrol related south of exit 3	50% black

After hearing the testimony of Kenneth Ruff and Kenneth Wilson, two former troopers called by the defense who were not reappointed at the end of their terms and who said they were trained and coached to make race based "profile" stops to increase their criminal arrests, the State asked Dr. Cupingood to study the race identified stops in his database and see how many possessed the profile characteristics cited by Ruff and Wilson, particularly how many were young (30 or under), black and male. Dr. Cupingood found that only 11.6% of the race identified stops were of young black males and only 6.6% of all stops were of young black males.

The defense then conducted a profile study of its own. It concentrated on the race identified stops of just blacks issued tickets and found that an adult black male was present in 88% of the cases where the gender of all occupants could be determined and that where gender and age could be determined, a black male 30 or younger was present in 63% of the cases. The defense study is more probative, because it does concentrate on just stops of blacks issued tickets eliminating misleading comparisons with totals including whites or whites and a 62.6% group of race unknowns. Neither side, of course, could consider whether the blacks stopped and not issued tickets possessed profile characteristics since the databases contain no information about them.

Dr. Cupingood's so-called Mantel-Haentzel analysis ended the statistical evidence. He put forward this calculation of "expected black tickets" in an attempt to disprove the defense study showing the Patrol Unit, the unit with the most discretion, ticketed blacks at a rate not only well above the Radar and Tac-Pac Units, but also well above the standard fixed by the violator survey. The calculation insinuates that the Patrol Unit issued merely 5 excess tickets to blacks beyond what would have been expected. The calculation is worthless. First and foremost, Dr. Cupingood deleted the non-radar tickets which presumably involved a greater exercise of discretion. The role police discretion played in the issuance of tickets to blacks was the object of the defense study. Under the guise of comparing only things similarly situated, he thereupon deleted any radar tickets not issued in one of the four time periods he divided each of the thirty-five randomly selected days into for which there was not at least one race identified radar ticket issued by the Patrol Unit and at least one by the combined Radar and Tac-Pac Unit. He provided no justification for either creating the 140 time periods or combining the tickets of the Radar and Tac-Pac Units. To compound his defective analysis, he pooled the data in each time period into a single number and employed the resultant weighted average of the two units to compute the expected and excess, if any, tickets issued to blacks. By using weighted averages, he once again assumed the answer to the question he purported to address. He assumed the Patrol Unit gave the same number of tickets to blacks as did the Radar and Tac-Pac Units, rather than test to see if it did. Even after "winnowing" the data, the comparison between the Patrol Unit and the Radar and Tac-Pac Units is marginally statistically significant. Without winnowing, Dr. Kadane found the comparison of the radar tickets issued by the Patrol Unit to blacks with the radar tickets issued by the Radar and Tac-Pac Units to blacks constituted 3.78 standard deviations which is distinctly above the 5% standard of statistical significance.

The defense did not rest on its statistical evidence alone. Along with the testimony of former troopers Kenneth Ruff and Kenneth Wilson about having been trained and coached to make race based profile stops but whose testimony is weakened by bias related to their not having been reappointed at the end of their terms, the defense elicited evidence through cross-examination of State witnesses and a rebuttal witness, Dr. James Fyfe, that the State Police hierarchy allowed, condoned, cultivated and tolerated discrimination between 1988 and 1991 in its crusade to rid New Jersey of the scourge of drugs.

Conjointly with the passage of the Comprehensive Drug Reform Act of 1987 and to advance the Attorney General's Statewide Action Plan for Narcotics Enforcement issued in January 1988 which "directed that the enforcement of our criminal drug laws shall be the highest priority law enforcement activity," Colonel Pagano formed the Drug Interdiction Training Unit (DITU) in late 1987 consisting of two supervisors and ten other members, two from each Troop selected for their successful seizure statistics, "...to actually patrol with junior road personnel and provide critical on-the-job training in recognizing potential violators." State Police Plan For Action dated July 7, 1987, at p. 14. According to Colonel Pagano, the DITU program was intended to be one step beyond the existing coach program to impart to newer troopers insight into drug enforcement and the "criminal program" (patrol related arrests) in general. DITU was disbanded in or around July 1992.

No training materials remain regarding the training DITU members themselves received, and few training materials remain regarding the training DITU members provided the newer troopers except for a batch of checklists.[8] Just one impact study was ever prepared regarding the effectiveness of the DITU program rather than periodic impact evaluations and studies as required by S.O.P. F4 dated January 12, 1989, but this one undated report marked D-62 in evidence only provided statistics about the number of investigations conducted, the number of persons involved and the quantity and value of drugs seized without indicating the race of those involved or the number of fruitless investigations broken down by race. In the opinion of Dr. Fyfe, retention of training materials is important for review of the propriety of the training and to discern agency policy, and preparation of periodic impact evaluations and studies is important not only to determine the effectiveness of the program from a numbers standpoint, but more than that to enable administration to monitor and control the quality of the program and its impact on the public, especially a crackdown program like DITU which placed so much emphasis on stopping drug transportation by the use of "consents" to search following traffic stops in order to prevent constitutional excesses.

8. Although DITU kept copies of all arrest, operations and investigation reports and consent to search forms growing out of encounters with the public during training for a time at its office, the copies were destroyed sometime in 1989 or 1990 and before they were sought in discovery. The originals of these reports and forms were filed at the trainee's station and incorporated into the State Police "traditional reporting system" making them impossible to ferret out now.

Despite the paucity of training materials and lack of periodic and complete impact evaluations and studies, a glimpse of the work of DITU emerges from the preserved checklists and the testimony of Sergeants Brian Caffrey and David Cobb. Sergeant Caffrey was the original assistant supervisor of DITU and became the supervisor in 1989. Sergeant Cobb was an original member of DITU and became the assistant supervisor in 1989. Sergeant Caffrey left DITU sometime in 1992, Sergeant Cobb sometime in 1991. Both testified that a major purpose of DITU was to teach trainees tip-offs and techniques about what to look for and do to talk or "dig" their way into a vehicle after, not before, a motor vehicle stop to effectuate patrol related arrests. Both denied teaching or using race as a tip-off either before or after a stop. Nevertheless, Sergeant Caffrey condoned a comment by a DITU trainer during the time he was the supervisor of DITU stating:

> "Trooper Fash previously had DITU training, and it showed in the way he worked. He has become a little reluctant to stop cars in lieu [sic] of the Channel 9 News Report. He was told as long as he uses Title 39 he can stop any car he wants. He enjoys DITU and would like to ride again."

As the defense observes in its closing brief, "Why would a trooper who is acting in a racially neutral fashion become reluctant to stop cars as a result of a news story charging that racial minorities were being targeted [by the New Jersey State Police]?" Even A. A. G. Ronald Susswein, Deputy Director of the Division of Criminal Justice, acknowledged that this comment is incomplete because it fails to add the caveat, "as long as he doesn't also use race or ethnicity." Further, Sergeant Caffrey testified that "ethnicity is something to keep in mind" albeit not a tip-off and that he taught attendees at both the annual State Police in-service training session in March 1987 and the special State Police in-service training sessions in July and August 1987 that Hispanics are mainly involved in drug trafficking and showed them the film Operation Pipeline wherein the ethnicity of those arrested, mostly Hispanics, is prominently depicted. Dr. Fyfe criticized Sergeant Caffrey's teaching Hispanics are mainly involved and his showing Operation Pipeline as well as the showing of the Jamaican Posse film wherein only blacks are depicted as drug traffickers at the 1989 annual State Police in-service training session saying trainers should not teach what they do not intend their trainees to act upon. At a minimum, teaching Hispanics are mainly involved in drug trafficking and showing films depicting mostly Hispanics and blacks trafficking in drugs at training sessions worked at cross-purposes with concomitant instruction pointing out that neither race nor ethnicity may be considered in making traffic stops.

Key corroboration for finding the State Police hierarchy allowed and tolerated discrimination came from Colonel Pagano. Colonel Pagano was Superintendent of the State Police from 1975 to February 1990. He testified there was a noisy demand in the 1980s to get drugs off the streets. In accord, Attorney General Cary Edwards and he made drug interdiction the number one priority of law enforcement. He helped formulate the Attorney General's Statewide Action Plan for Narcotics Enforcement and established DITU within the State Police. He kept an eye on DITU through conversations with staff officers and Sergeants Mastella and Caffrey and review of reports generated under the traditional reporting system and D-62 in evidence. He had no thought DITU would engage in constitutional violations. He knew all State

Police members were taught that they were guardians of the Constitution and that targeting any race was unconstitutional and poor police practice to boot. He recognized it was his responsibility to see that race was not a factor in who was stopped, searched and arrested. When he became Superintendent, he formed the Internal Affairs Bureau to investigate citizen complaints against State Police members to maintain the integrity of the Division. Substantiated deviations from regulations resulted in sanctions, additional training or counseling.

More telling, however, is what Colonel Pagano said and did, or did not do, in response to the Channel 9 expose entitled "Without Just Cause" which aired in 1989 and which troubled Trooper Fash and what he did not do in response to complaints of profiling from the NAACP and ACLU and these consolidated motions to suppress and similar motions in Warren and Middlesex Counties. He said to Joe Collum of Channel 9 that "[violating rights of motorists was] of serious concern [to him], *but no where near* the concern that I think we have got to look to in trying to correct some of the problems we find with the criminal element in this State" and "the bottom line is that those stops were not made on the basis of race *alone*" (emphasis added). Since perhaps these isolated comments were said inadvertently or edited out of context, a truer reflection of his attitude about claims of racism would appear to be his videotaped remarks shown all members of the State Police at roll call in conjunction with the WOR series. Thereon he clearly said that he did not want targeting or discriminatory enforcement and that "[w]hen you put on this uniform, you leave your biases and your prejudices behind." But he also said as regarded the charge of a Trenton school principal named Jones that he had been stopped on the Turnpike and threatened, intimidated and assaulted by a trooper, "We know that the teacher assaulted the trooper. He didn't have a driver's license or a registration *for his fancy new Mercedes*" (emphasis added). And he called Paul McLemore, the first African-American trooper in New Jersey and now a practicing attorney and who spoke of discrimination within the ranks of the State Police, "an ingrate." And he told the members to "keep the heat on" and then assured them:

> "...[H]ere at Division Headquarters we'll make sure that when the wheels start to squeak, we'll do whatever we can to make sure that you're supported out in the field....Anything that goes toward implementing the Drug Reform Act is important. And, we'll handle the squeaky wheels here."

He admitted the Internal Affairs Bureau was not designed to investigate general complaints, so he could not refer the general complaints of discrimination to it for scrutiny. Yet he never requested the Analytical Unit to investigate stop data from radio logs, patrol charts and tickets or search and seizure data from arrest reports, operations reports, investigation reports and consent to search forms, not even after the Analytical Unit informed him in a report on arrests by region, race and crime that he had requested from it for his use in the WOR series that "...arrests are not a valid reflection of stops (data relative to stops with respect to race is not compiled)." The databases compiled for these motions attest, of course, to the fact that race identified stop data could have been compiled. He testified he could not launch an investigation into every general complaint because of limited resources and that there was insufficient evidence of discrimination in the Channel 9 series, the NAACP and ACLU complaints and the various motions to suppress for him to spend his "precious" resources. In short, he left the issue of discrimination up to the courts and months of testimony in this and other counties at State expense.

The right to be free from discrimination is firmly supported by the Fourteenth Amendment to the United States Constitution and the protections of Article I, paragraphs 1 and 5 of the New Jersey Constitution of 1947. To be sure, "[t]he eradication of the 'cancer of discrimination' has long been one of our State's highest priorities." *Dixon v. Rutgers, The State University of N.J.,* 110 N.J. 432, 451, 541 A.2d 1046 (1988). It is indisputable, therefore, that the police may not stop a motorist based on race or any other invidious classification. *See State v. Kuhn,* 213 N.J. Super. 275, 517 A.2d 162 (1986).

Generally, however, the inquiry for determining the constitutionality of a stop or a search and seizure is limited to "whether the conduct of the law enforcement officer who undertook the [stop or] search was objectively reasonable, without regard to his or her underlying motives of intent." *State v. Bruzzese,* 94 N.J. 210, 463 A.2d 320 (1983). Thus, it has been said that the courts will not inquire into the motivation of a police officer whose stop of a vehicle was based upon a traffic violation committed in his presence. *See United States v. Smith,* 799 F.2d 704, 708–709 (11th Cir. 1986); *United States v. Hollman,* 541 F.2d 196, 198 (8th Cir. 1976); *cf. United States v. Villamonte-Marquez,* 462 U.S. 579, 103 S. Ct. 2573, 77 L. Ed. 2d 22 (1983). But where objective evidence establishes "that a police agency has embarked upon an officially sanctioned or *de facto* policy of targeting minorities for investigation and arrest," any evidence seized will be suppressed to deter future insolence in office by those charged with enforcement of the law and to maintain judicial integrity. *State v. Kennedy,* 247 N.J. Super. 21, 588 A.2d 834 (App.Div. 1991).

Statistics may be used to make out a case of targeting minorities for prosecution of traffic offenses provided the comparison is between the racial composition of the motorist population violating the traffic laws and the racial composition of those arrested for traffic infractions on the relevant roadway patrolled by the police agency. *Wards Cove Packing Co. v. Atonio, supra; State v. Kennedy,* 247 N.J. Super, at 33–34, 588 A.2d 834. While defendants have the burden of proving "the existence of purposeful discrimination," discriminatory intent may be inferred from statistical proof presenting a stark pattern or an even less extreme pattern in certain limited contexts. *McCleskey v. Kemp,* 481 U.S. 279, 107 S. Ct. 1756, 95 L. Ed. 2d 262 (1987). *Kennedy, supra,* implies that discriminatory intent may be inferred from statistical proof in a traffic stop context probably because only uniform variables (Title 39 violations) are relevant to the challenged stops and the State has an opportunity to explain the statistical disparity. "[A] selection procedure that is susceptible of abuse . . . supports the presumption of discrimination raised by the statistical showing." *Castaneda v. Partida,* 430 U.S. 482, 494, 97 S. Ct. 1272, 51 L. Ed. 2d 498 (1977).

Once defendants expose a *prima facie* case of selective enforcement, the State generally cannot rebut it by merely calling attention to possible flaws or unmeasured variables in defendants' statistics. Rather, the State must introduce specific evidence showing that either there actually are defects which bias the results or the missing factors, when properly organized and accounted for, eliminate or explain the disparity. *Bazemore v. Friday,* 478 U.S. 385, 106 S. Ct. 3000, 92 L. Ed. 2d 315 (1986); *EEOC v. General Telephone Co. of Northwest, Inc.,* 885 F.2d 575 (9th Cir. 1989). Nor will mere denials or reliance on the good faith of the officers suffice. *Castaneda v. Partida,* 430 U.S. at 498 n. 19, 97 S. Ct. 1272, 51 L. Ed. 2d 498.

Here, defendants have proven at least a *de facto* policy on the part of the State Police out of the Moorestown Station of targeting blacks for investigation and arrest between April 1988 and May 1991 both south of exit 3 and between exits 1 and 7A of the Turnpike. Their surveys satisfy *Wards Cove, supra.* The statistical disparities and standard deviations revealed are indeed stark. The discretion devolved upon general road troopers to stop any car they want as long as Title 39 is used evinces a selection process that is susceptible of abuse. The utter failure of the State Police hierarchy to monitor and control a crackdown program like DITU or investigate the many claims of institutional discrimination manifests its indifference if not acceptance. Against all this, the State submits only denials and the conjecture and flawed studies of Dr. Cupingood.

The eradication of illegal drugs from our State is an obviously worthy goal, but not at the expense of individual rights. As Justice Brandeis so wisely said dissenting in *Olmstead v. United* States, 277 U.S. 438, 479, 48 S. Ct. 564, 72 L. Ed. 944 (1928):

> "Experience should teach us to be most on our guard to protect liberty when the government's purposes are beneficent. Men born to freedom are naturally alert to repel invasion of their liberty by evil-minded rulers. The greatest dangers to liberty lurk in insidious encroachment by men of zeal, well-meaning but without understanding."

Motions granted.

2

MISSING DATA IN THE FORENSIC CONTEXT

Joseph B. Kadane and Norma Terrin

1. INTRODUCTION

Data collection efforts, whether by survey, administrative records or other methods, often suffer from missing data (Rubin, 1987; Little and Rubin, 1987). Two classical methods to deal with such missing data are extreme: either

(a) ignore the fact that some data are missing or
(b) declare the data set uninterpretable because of the missing data problem.

Neither of these reactions is likely to be faithful to the uncertainty introduced by missing data: it adds some uncertainty, but there is usually some inferential content to the survey even with some data missing.

These extreme reactions are likely to lead to unenlightening courtroom battles among statisticians. Whichever side wishes to use the data will choose (a) above, whereas the other side will find an equal and opposite statistician to argue for (b). This in turn may lead to a search for "statistical rules," saying what proportion of the data must be missing to shift the appropriate treatment from (a) to (b).

Rather than presuming complete knowledge of the missing data, as in (a), or no knowledge at all, as in (b), we may consider various alternatives for the distribution of the missing observations, eliminating only the implausible ones.

This paper discusses the presentation in court of a Bayesian analysis of this type. The next section describes the case, the analysis offered in court and the response of the court. Some general conclusions are given in the third section.

© 1997 Royal Statistical Society. Reprinted from the Journal of the Royal Statistical Society, Series A, *160*, Part 2, pp 351–357.

2. NEW JERSEY TURNPIKE CASE

In the case (re: State *versus* Pedro Soto, *et al.*) the New Jersey Public Defender's Office moved to suppress evidence against 17 black defendants (principally on charges of transporting illegal drugs) on the grounds of selective enforcement, i.e. that blacks were more likely to be stopped by the police than were others, on the southern end of the New Jersey turnpike.

US law on discrimination is most active in the areas of jury participation and employment issues. In both these areas, the law distinguishes between discriminatory *intent* and discriminatory *impact*.

Discriminatory intent is concerned with an expressed intention to discriminate. If a state had a law forbidding blacks to serve on juries, or if an employer had a policy (expressed orally or in writing) stating that it did not hire blacks, nothing further would be required before the law would respond. The analogue here would be if there were written or oral instructions to the state police to target law enforcement against blacks. The main evidence in this case is not of this nature.

The other main way of showing discrimination is by showing discriminatory impact. In such a case, what is to be shown is that similarly situated people have a different probability of gaining something good (jury service or employment) or avoiding something bad (traffic stopping and/or arrest), and that those probabilities differ systematically by race. However, "similarly situated" requires explication.

In employment discrimination cases, a substantial paper trail is generated in the usual processes of hiring and promotion. Generally people apply in writing for jobs, and their applications state various facts about them, including their qualifications. In promotion cases, the employer generally has records relating to the employee's performance on the job. These written materials mean that substantial evidence can be brought to bear on issues of "similar situation" in employment cases. In traffic cases, the analogue to an application is the appearance of an automobile and its behaviour, how it is driven. The "applicants" for stopping and/or arrest do not leave a paper trail; in particular, it is very difficult to compare the "qualifications" of "successful applicants" (those stopped and/or arrested) with those of "unsuccessful" applicants (those not stopped and not arrested).

Two studies were undertaken by the defense: one designed to estimate the proportion of cars on the southern end of the New Jersey turnpike carrying black occupants and the other to estimate the fraction of traffic law violators who were black or carrying black passengers. The traffic study involved 21 randomly selected 2½-hour sessions at four sites on the turnpike in daylight hours of June 1993. About 42000 vehicles were observed, of which 13.5% had a black occupant.

In the violator study, an observer drove at 60 miles per hour (the speed limit is 55 miles per hour) and observed how many cars passed him and how many he passed. Of 2096 cars, 2062 passed him (98.1%), and of these 15% had a black occupant. Because all these 2062 represent cars that can be stopped, we took 15% as a reasonable number for the proportion of cars containing blacks to be expected in randomly chosen stopped cars. Since virtually everyone on the turnpike was driving faster than the speed limit, the state police could legally stop virtually anyone they chose. The broad racial consistency between the two data sets (13.5%, 15%) lent credibility to both, in our opinion.

Finally, data collected on stops by the state police were analysed for 35 randomly selected days between April 1988 and May 1991. Limited to the area between exits 1 and 3 on the turnpike (which is where the traffic study and violator study were done), this showed 892 stops being made, of which 127 were of blacks, 148 of whites and others, and 617 of people of unknown race.

Concentrating first on just the racially identified stops, 46.2% were of blacks. To appreciate this number, a simple application of Bayes theorem yields

$$\Theta = \frac{P(\text{stop}|\text{black})}{P(\text{stop}|\text{white})} = \frac{P(\text{black}|\text{stop})P(\text{stop})/P(\text{black})}{P(\text{white}|\text{stop})P(\text{stop})/P(\text{white})}$$

$$= \frac{0.462/0.15}{0.538/0.85} = 4.86. \tag{1}$$

For brevity, the term "white" refers to anyone who would not be identified as "black" by the state police.

Thus, by this calculation, a black driver was 4.86 times as likely to be stopped on the New Jersey turnpike, strongly suggesting a racially non-neutral policy. The difficulty with this result is that 0.462 in equation (1) is actually P(black | stop and race identified), not P(black | stop).

We need to be concerned about whether the large amount of missing data (69.1 %) might disturb the analysis. That such a large portion of the data is missing is due to the routine destruction of the radio logs for 10 of the 35 days, and to the failure of the state police on the other days to follow its own rules to radio the race of occupants of any car stopped. Nonetheless, such a large amount of missing data, in the absence of further analysis, might create a legitimate doubt about the conclusion.

To advise the court about the likely effect of the missing data, we created the following model. The key question is the possible bias in race reporting. Let

r_1 = P(race reported | black and stopped),

r_2 = P(race reported | white and stopped),

t = P(black | stopped),

$1 - t$ = P(white | stopped),

n_1 = number of blacks reported as stopped,

n_2 = number of whites reported as stopped,

n_3 = number of people stopped whose race is not reported.

Three events may occur with a stop: the person stopped is black and the race is reported, the person stopped is white and the race is reported or the person who is stopped does not have their race reported. These events have respective probabilities $r_1 t$, $r_2 (1 - t)$ and $(1 - r_1)t + (1 - r_2)(1 - t)$. Since, given these parameters, the stops are regarded as independent and identically distributed, the likelihood function is trinomial:

$$(r_1 t)^{n_1} \left\{ r_2 (1-t) \right\}^{n_2} \left\{ (1-r_1)t + (1-r_2)(1-t) \right\}^{n_3}. \tag{2}$$

Treating the parameters as t, r_1 and r_2, the goal is a distribution for Θ, as in equation (1), which in this notation is

$$\Theta = \frac{t/0.15}{(1-t)/0.85} = \frac{0.85t}{0.15(1-t)}. \tag{3}$$

It turns out that $\log \Theta$ is more convenient, as the posterior distributions for it are closer to symmetric.

Observe that the trinomial distribution has two dimensions of information, whereas the parameter space has three: r_1, r_2 and t. This consequence of incomplete information can be viewed as a problem of lack of identifiability. However, lack of identifiability is not a great difficulty for subjective Bayesians (Kadane, 1975).

Finally, it is necessary to choose the prior distribution for the parameters r_1, r_2 and t. What is reasonable to assume about them?

Using the violator study as an anchor, we centre our prior for t at 0.15. This is an estimate of the proportion of cars violating the speed limit on the New Jersey turnpike that contain blacks, which is larger than the proportion of violating cars driven by blacks. Hence, using 0.15 as an estimate of the proportion of violating cars driven by blacks tends to exonerate the police. The beta family of distributions with mean 0.15 is a logical choice, as the distributions are concentrated on (0, 1). The parameter in this beta family is a scale parameter, which can be discussed in terms of the standard deviation.

We present two rather different choices for the standard deviation of t. In court, we presented an analysis based on a standard deviation of 0.064, which corresponds to a beta (4.5, 25.5) distribution. We later explored a much larger standard deviation, 0.30, which corresponds to a beta (0.06, 0.35). This latter choice can be criticized on the grounds that it forces most of the probability for t to be close to 0 or 1. These specifications are referred to below as "small" and "large" standard errors respectively. The analysis to come shows that the results of these two priors are indistinguishable.

Among the possible relationships between r_1 and r_2, which are plausible? Perhaps those whose race was reported are representative of all those stopped. In that case, $r_1 = r_2$, or equivalently

$$\text{odds}(r_1) = \frac{r_1}{1-r_1} = \frac{r_2}{1-r_2} = \text{odds}(r_2).$$

Equivalently, this assumption means that being black and having one's race recorded are independent, given that one is stopped. It is possible, however, that the race of the occupants of a stopped car affects the probability that the officer will report race. We do not presume to know in which direction the probability would be altered, if at all. Giving the benefit of the doubt to the prosecution, we explored the consequence of blacks having twice, and even three times, the odds of whites for having their race reported. Let r be the ratio of the odds of having a stopped driver's race reported if black to that if white, i.e.

$$r = \frac{r_1/(1-r_1)}{r_2/(1-r_2)}.$$

To complete the model, a uniform distribution is taken on r_2 (other beta distributions for r_2 could be used; again the results are robust to this choice as long as var(r_2) is appreciable).

Once the likelihood and prior are specified, the joint posterior distribution of the parameters r_1, r_2 and t is proportional to their product.

The calculations were conducted for the three cases mentioned above: odds of race reporting of a stop given black as one, two or three times that of whites, i.e. $r = 1$, 2 and 3. The upshot is that, even with odds of race reporting for blacks at three times that for whites, the probability of blacks being stopped more often than whites (i.e. $P(\log \Theta > 0)$) is over 0.99. The posterior distribution of log Θ for each case is shown in Figs. 1 and 2. A comparison of these figures shows that the results are virtually identical for the two standard error cases.

We used the Laplace approximation (Kass et al., 1988) as implemented in X-LISP-STAT (Tierney, 1990) to calculate the posterior marginal distribution of log Θ. Fig. 1 was produced from S-PLUS (Spector, 1994).

There are several things to notice from Figs 1 and 2. When $r = 1$, so that the odds of having one's race reported if stopped does not depend on race, there is a distribution for log Θ. This is a result of the fact that data are missing, which introduces uncertainty. As r increases, the curve of log Θ shifts to the left. This is because, as r increases, it becomes increasingly more likely that each person stopped whose race is not reported was white; hence increasingly more of the races of missing persons are imputed to be white. Only if all the stopped drivers with missing race data are assumed to be white does the proportion of blacks stopped, 14.2%, approximate the 15% found in the violator study. But this corresponds to $r = \infty$, which is not plausible.

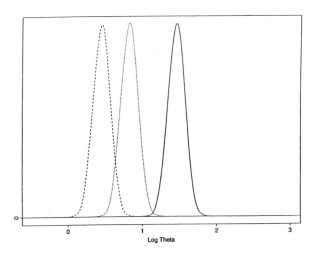

FIGURE 1. Posterior distribution of $L = \log \Theta$ as a function of r, the ratio of the odds of race reporting if stopped and black to that of whites, when the prior standard deviation of t is small: ———, $r = 1$; ...., $r = 2$; -----, $r = 3$

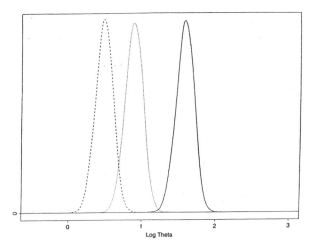

FIGURE 2. Posterior distribution of $L = \log \Theta$ as a function of r, the ratio of the odds of race reporting if stopped and black to that of whites, when the prior standard deviation of t is large: ———, $r = 1$; ...., $r = 2$; -----, $r = 3$

A referee suggested the following alternative formulation: since $L = \log \Theta$ is a function of t, attention can be focused on the posterior distribution of t:

$$p(t \mid n_1, n_2, n_3) = \sum_m p(t \mid n_1, n_2, n_3, m) \, p(m \mid n_1, n_2, n_3),$$

where m is the number of blacks stopped but not recorded. The first factor is binomial with parameters t and $n_1 + n_2 + n_3$, with $n_1 + m$ 'successes' and $n_2 + n_3 - m$ 'failures.' Then attention focuses on the second factor. Although we did not use this method of decomposing the problem, we agree that it may prove to be a useful alternative.

Judge Robert E. Francis of the Superior Court of New Jersey, in an unpublished opinion, accepted our argument and found for the defense. The case is on appeal.

3. PRESENTATION TO COURT OF SURVEYS WITH MISSING DATA

What has Bayesian analysis contributed in this context?

It has brought to light more of the real uncertainty engendered by the missing data. For example, taking classical solution (a) (i.e. ignoring the fact that some data are missing) yields an estimate for Θ of 4.86 (or $\log \Theta = 1.58$) as shown in equation (1). By contrast, if Θ is viewed as a random variable, the posterior distribution for $\log \Theta$ shown in Fig. 2.1 for the case of equal odds of non-reporting for the two racial groups has substantial variance.

Presenting the evidence this way, rather than as an estimator with a standard error, makes it more accessible to the court. Even those without statistical training can understand the conclusion of our analysis, that it is almost certain that blacks are more likely to be stopped than whites, and that we arrive at this conclusion even when our initial assumptions favour the prosecution. By contrast, a detailed explanation of a confidence interval would be much more difficult, i.e. an interval with stochastic end points having the property that before the data are observed 95% of the times that such an interval is used it will contain the true parameter value, the interval being evaluated after the data have been observed.

The Bayesian method by itself will not cure the problem of equal and opposing statisticians. Each side could propose its own models (likelihoods and priors), which might lead to opposite conclusions. Then the matter could come down to rhetoric, of persuading the trier of fact that one's assumptions and conclusions are the most reasonable.

ACKNOWLEDGMENTS

We thank the New Jersey Public Defender's Office in Gloucester County, New Jersey, for involving us in the case: Jeffrey Wintner, Fred Last and Justin Louchry. We also thank Professor John Lamberth of Temple University for his advice and help and Dr. C. Aitken of Edinburgh University for very useful comments.

REFERENCES

Kadane, J. B. (1975) The role of identification in Bayesian theory. In *L. J. Savage Memorial Volume: Studies in Bayesian Statistics and Econometrics* (eds S. E. Fienberg and A. Zellner), pp. 175–191. Amsterdam: North-Holland.
Kass, R. E., Tierney, L. and Kadane, J. B. (1988) Asymptotics in Bayesian computation. In *Bayesian Statistics 3* (eds J. M. Bernardo, M. H. DeGroot, D. V. Lindley and A. M. F. Smith), pp. 261–278. Oxford: Oxford University Press.
Little, R. J. A. and Rubin, D. (1987) *Statistical Analysis with Missing Data*. New York: Wiley.
Rubin, D. (1987) *Multiple Imputation for Nonresponse in Surveys*. New York: Wiley.
Spector, P. (1994) *An Introduction to S and S-Plus*. Belmont: Duxbury.
Tierney, L. (1990) *Lisp-Stat: an Object-oriented Environment for Statistical Computing and Dynamic Graphics*. New York: Wiley.

3

COUNTING CARS IN A LEGAL CASE INVOLVING DIFFERENTIAL ENFORCEMENT

Norma Terrin and Joseph B. Kadane

INTRODUCTION

Many African-Americans perceive the police as having a negative bias toward them, a proposition denied by many white Americans (Moore and Saad 1995). These perceptions are often driven by anecdotes or personal experience, leaving the factual issue in doubt.

The criminal case discussed here serves as a test of the social perception that police are racially biased. Nineteen African-American defendants, all of whom were stopped for motor vehicle violations on the southern end of the New Jersey Turnpike, were then searched and charged with carrying contraband. The Public Defender's office in Woodbury, New Jersey (to whom we were consultants), asked the judge to bar the prosecutors from presenting evidence obtained during the searches, alleging that the state police in the region of the arrests were stopping black motorists at extraordinarily high rates. There are legal limitations on the power of the police. To stop a motorist and demand identification, the police must have credible reasons to believe that the motorist has violated the law. Because of the vagueness of this criterion, it has been very difficult to show unfair enforcement of the law against individuals. In the New Jersey case, the defense collected data to prove that black motorists are more likely to be stopped and searched than white motorists driving in the same manner.

The public defenders conducted two traffic surveys to assess whether blacks and whites violate traffic laws at the same rate. In the first, stationary observers counted and recorded race data on the drivers and occupants of cars traveling in one direction past a designated milepost. About 42,000 cars were counted in 21 randomly selected 2 1/2 hour sessions at four selected sites on the turnpike during daylight hours of June 1993. Roughly 13.5% of these cars had a black occupant.

In the other survey, referred to here as *the moving observer* survey, an observer (call him "Fred") traveled continuously between Exits 1 and 3 at 60 mph (the posted speed limit was 55) and counted the number of cars that passed and were passed by him during several timed sessions and the race of the driver. No car was counted more than once during a session. The speeds of the cars are not known. One purpose of the study by the moving observer was to estimate the proportion of drivers who exceed the speed limit. Because Fred did not observe a simple random sample of cars, the question arises as to how much information can be extracted from the data collected.

The statisticians consulted by the prosecution used the data from these two surveys to calculate that, on average, 833 cars per hour passed the stationary observation point, whereas only 61.5 cars per hour were counted in the turnpike violation study. Because these figures differ by a factor of 13, they claimed that "at least one or both of these studies cannot be relied upon for inferences about the characteristics of the traffic or of the drivers in traffic" (Siskin and Cupingood 1994). They also claimed that the design of the violator study is biased in that it overestimates the percent of cars traveling in excess of 60 mph. The violator study counted a total of 1,768 cars, of which 1,734, or 98.1%, overtook the observer.

Let us examine these two objections. First, one would not expect the two surveys to count the same number of cars per hour. For example, suppose all cars on the turnpike travel at exactly 60 mph. Then the observer would neither overtake nor be overtaken by any car. Computations discussed later demonstrate that 13 is a plausible ratio. As to the second point, Siskin and Cupingood are correct in that the greater the difference in speed between Fred and another driver, the more likely Fred is to see that driver. This point was discussed by Haight (1963). For example, a driver 15 miles behind Fred traveling at 90 mph would catch up to Fred in a half hour, whereas it would take Fred three hours to catch up with a driver 15 miles ahead of him traveling at 55 mph. The relevant question, however, is not whether Fred sees a disproportionate number of high-speed travelers but whether the information he records can be used to estimate the proportion of drivers who are speeding. We show that, in spite of the inherent bias in the moving observer's sample and the lack of velocity data, it is possible to obtain a lower bound on the fraction of drivers who speed.

ANALYSIS

We make a few simplifying assumptions in our model. We assume that each car travels at a constant rate of speed. This assumption is somewhat unrealistic because of speeders who are forced to reduce their speed due to slower traffic. The resulting bias, however, would be against recording speeders and hence would reduce the estimated proportion of drivers who exceed the limit. We ignore Exit 2. This is equivalent to assuming that the volume and speed distribution for those leaving at Exit 2 are the same for those entering. We assume that cars enter at Exit 1 traveling northbound according to a process stationary in time with rate $\lambda = 833$ cars per hour and that cars enter at Exit 3 traveling southbound with the same rate.

Suppose that Fred travels continuously at 60 mph between Exit 1 in the south and Exit 3 in the north. This scenario is very close to his actual experience, because he used cruise control and stopped the clock each time he exited to change direction. We say a car is "seen" or "observed" by Fred if either he overtakes it or it overtakes him. First we check whether it is plausible that Fred sees, on average, 61.5 cars per hour. Without the distribution of speeds, we cannot precisely calculate what Fred's observation rate should be. We can, however, judge whether a rate of 61.5 cars per hour is reasonable by postulating a plausible distribution. Suppose that, on roads with a 55-mph speed limit, 5% of the traffic travels between 55 and 60 mph, 60% between 60 and 65, 25% between 65 and 70, 7% between 70 and 75, 2% between 75 and 80, .8% between 80 and 85, and .2% between 85 and 90. Then, approximating the velocities with the interval midpoints, it can be calculated that Fred would see about 61.2 cars per hour on average and that he would be overtaken by about 97.1% of those cars. These two numbers (61.2 and 97.1%), computed from a simple, hypothetical distribution, are close enough to the numbers computed from the actual data (61.5 and 98.1%) to make the result of the violator survey plausible.

Next, without postulating a distribution of speeds, we use the data collected in the defendants' study to approximate the percent of drivers exceeding the speed limit. The upper bound on this percent is clearly 100% because even those drivers overtaken by Fred might be speeding. To obtain a lower bound on the fraction of speeders, we compute the following three probabilities.

The probability that a car is traveling at a speed s that is 55 mph or below, given that the car is among those observed by Fred, is no larger than the number of cars Fred overtook divided by the total number he observed, which is $34/1768 = .019$. That is,

$$P(s \leq 55 \mid observed) \leq .019.$$

The probability that a car traveling between Exits 1 and 3 in either direction is observed by Fred is the number of cars Fred sees per hour divided by the total number entering either the northbound or southbound lanes. That is,

$$P(observed) = 61.5/1666 = .037.$$

Now we compute the probability that a car is observed by Fred given it is traveling at 55 mph. If a car enters at Exit 1, heading north, and the length of the road between Exits 1 and 3 is m miles, then the car will pass Exit 3 in $m(60/55)$ minutes. To overtake the car, Fred has to reach Exit 1 in $m(60/55) - m$ minutes, because he travels one mile per minute. The probability he's in a position to do that is $m(60/55 - m)/2m = 1/22 = .0455$, because his position is uniformly distributed on the $2m$ miles of roadway. That is,

$$P(observed \mid s = 55) = .0455$$

and hence

$$P(observed \mid s \leq 55) \geq .0455$$

because cars with velocity further from that of the observer are more likely to be observed.

We are now ready to estimate an upper bound on the fraction of cars on the road traveling at or below 55 mph:

$$P(s \leq 55) = \frac{P(s \leq 55 \mid \text{observed}) P(\text{observed})}{P(\text{observed} \mid s \leq 55)} \leq \frac{(.019)(.037)}{.0455} = .015.$$

Although Siskin and Cupingood are correct that the 98.1% of the drivers observed exceeding 60 mph is likely to be an overestimate, our calculations show, in spite of this bias, that nearly all drivers on the southern section of the New Jersey Turnpike (more than 98.5%) drive faster than the speed limit of 55 mph. Therefore the police can practically stop whomever they choose.

The stationary study found that 13.5% of the cars had black drivers; the moving observer study found that 15% of the cars were driven by blacks. Combined with (incomplete) data indicating 46.2% of the state police stops were of blacks, this suggested that the odds of being stopped if one were black is about 4.86 times the odds of being stopped if one were white. The principal source of uncertainty in these figures is the large amount of missing race data in the police records of stops. We presented a Bayesian analysis to the court, supposing that the race data is not missing at random. Even if the odds of a stopped motorist having his race recorded were two or three times higher for blacks than for whites, we still conclude that blacks had a greater chance than whites of being stopped on the turnpike (Kadane and Terrin 1997). In an unpublished decision, Judge Francis found for the defense. His decision is now up on appeal.

CONCLUSION

From a social viewpoint, the importance of this case is that it addresses the belief of a majority of blacks about racism in police behavior. At least in this place and time, they appear to be correct. From a legal view, the case overcame the difficulty of proof that had made other such cases so problematic: showing that "similarly situated" blacks and whites were treated differently. Technically, it is an example showing the care that must be exercised in understanding and composing results from differing sampling frames.

[We thank the New Jersey Public Defenders office in Gloucester County, New Jersey, for involving us in the case—Jeffery Wintner, Fred Last, and Justin Louchry. We also thank John Lamberth of Temple University for his advice and help.]

REFERENCES AND FURTHER READING

Haight, F. A. (1963), *Mathematical Theories of Traffic Flow,* New York: Academic Press.
Jaynes, G. D., and Williams, R. M. (1989), *A Common Destiny: Blacks and American Society,* Washington, DC: National Academy Press.

Kadane, J. B., and Terrin, N. (1997), "Missing Data in the Forensic Context." *Journal of the Royal Statistical Society, Ser. A,* 160, 351–357.

Moore, D. W., and Saad, L. (1995), "No Immediate Signs That Simpson Trial Intensified Racial Animosity," *The Gallup Poll Monthly,* October, 2–9.

Siskin, B. R., and Cupingood, L. A. (1994), "Critique of Defendant's Statistical Analysis" in the matter of State of New Jersey v. Kevin Jackson et al., March.

4

SELECTED HIGHLIGHTS OF THE INTERIM REPORT OF THE STATE POLICE REVIEW TEAM REGARDING ALLEGATIONS OF RACIAL PROFILING

This Interim Report is limited to the examination of the practice commonly referred to as racial profiling. The Report specifically focuses on activities of state troopers assigned to patrol the New Jersey Turnpike, which is considered to be a major drug corridor. This circumstance provides the incentive and opportunity for the State Police to use drug interdiction tactics that appear to be closely linked to the national racial profiling controversy. (p. 2)

Although this is only an Interim Report and is not the final material that will be developed on this subject, it represents a major step, signaling a recognition of the problem and proposing significant changes in State Police practices and procedures. (p. 3)

The Review Team believes that the great majority of state troopers are honest, dedicated professionals who are committed to enforcing the laws fairly and impartially. The Review Team has determined that the State Police has not issued or embraced an official policy to engage in racial profiling or any other discriminatory enforcement practices. In fact, the State Police has undertaken a number of steps to prohibit racial profiling, including issuing Standard Operating Procedures banning such practices; providing in-service training programs and bulletins; requiring state troopers to have reasonable suspicion before requesting permission to search (thereby imposing a prerequisite to consent searches that goes beyond the requirements of state or federal caselaw); and prohibiting the patrol tactic of spotlighting the occupants of motor vehicles at night before deciding whether to initiate a stop. (pp. 3–4)

Despite these official policies and preventative steps, the Interim Report concludes that the problem of racial profiling is real and that minority motorists have been treated differently than non-minority motorists during the course of traffic stops on the New Jersey Turnpike. The problem is more complex and subtle than has generally been reported. (p. 4)

The Interim Report recognizes that to a large extent, conclusions concerning the nature and scope of the problem will depend on the definitions that are used. The Review Team has chosen to define the problem of disparate treatment to include the reliance by a state trooper on a person's race, ethnicity, or national origin in conjunction with other factors in selecting vehicles to be stopped from among the universe of vehicles being operated in violation of the law or in making any discretionary decision during the course of a traffic stop, such as ordering the driver or passengers to step out; subjecting the occupants to questions that are not directly related to the motor vehicle violation that gave rise to the stop; summoning a drug-detection canine to the scene; or requesting permission to conduct a consent search of the vehicle and its contents. (p. 5)

The Interim Report reveals two interrelated problems that may be influenced by the goal of interdicting illicit drugs: (1) willful misconduct by a small number of State Police members, and (2) more common instances of possible *de facto* discrimination by officers who may be influenced by stereotypes and thus may tend to treat minority motorists differently during the course of routine traffic stops, subjecting minority motorists more routinely to investigative tactics and techniques that are designed to ferret out illicit drugs and weapons. (p. 7)

The issues and problems addressed in the Interim Report are not limited to the New Jersey State Police. Because this Interim Report embraces a broad definition of the problem of racial profiling and disparate treatment, the specific remedial action steps described in this Interim Report are offered as a guide to other state and local jurisdictions where the racial profiling controversy has surfaced. This Interim Report goes further than any other jurisdiction to date in facing up to this national problem and in proposing the establishment of multi-faceted systems to ensure that laws are enforced impartially by State Police members assigned to patrol duties. (p. 9)

The Review Team recommends that a clear policy for the New Jersey State Police be announced providing that race, ethnicity, and national origin may not be considered at all by State Police members in selecting vehicles to be stopped and in exercising police discretion during the course of a traffic stop, other than in determining whether a person matches the general description of one or more known suspects. This proposed policy goes beyond the requirements of federal law. (pp. 12, 52–56)

The Interim Report describes the sequence of steps that may occur during a typical traffic stop on the New Jersey Turnpike. This is done to demonstrate the decision points that can arise during a traffic stop where a state trooper must exercise reasoned discretion. (pp. 13–22)

The Interim Report describes compiled statistics for stops, arrests, and consent searches conducted by State Police members assigned to patrol the New Jersey Turnpike.

- These data show that 59.4% of stops that were examined involved whites, slightly more than one out of every four (27.0%) stops involved a black person, 6.9% involved a Hispanic individual, 3.9% involved an Asian person, and 2.8% were identified as other. (pp. 25–26)
- The data reveal that very few stops (0.7%) result in the search of a motor vehicle. The available data indicate that the overwhelming majority of these

searches (77.2%) involved black or Hispanic persons. Specifically, 21.4% of these searches involved a white person, more than one-half (53.1%) involved a black person, and one of every four (24.1%) involved a Hispanic person. (pp. 26–27)
- 32.5% of arrests involved white persons, 61.7% involved African-Americans, and 5.8% involved persons of other races. (pp. 29–30)

Based upon the foregoing statistical information, the Review Team made several observations:

- Minority motorists were disproportionately subject to consent searches. The data concerning consent searches were deemed to be especially instructive because the decision by a trooper to ask for permission to conduct a search is a discretionary one. Given the concerns engendered by this data, the Review Team proposed that the State Police undertake a case-by-case review of every consent search that was conducted on the Turnpike in 1997 and 1998 to determine whether the searches were conducted in accordance with all applicable State Police Standard Operating Procedures and the requirements of law. (pp. 30–31)
- The Review Team expressed concern about the extent of missing information concerning the racial characteristics of detained motorists in previously-kept manual records. This situation has already been addressed to a large extent through remedial efforts taken by the State Police. (pp. 31–32)
- The Review Team expressed concern with the lack of automation and the inherent problems associated with the existing manual system for recording information, which makes it difficult for supervisors throughout the chain of command to monitor the activities of officers assigned to patrol. The State Police has already begun to implement the Computer-Aided Dispatch/Records Management System (CAD/RMS) that will help to rectify this problem. (pp. 32–33)
- The Review Team expressed concern that where state troopers were afforded more discretion by virtue of their duty assignment, they tended to focus more on minority motorists. This analysis is consistent with the notion that officers who had more time to devote to drug interdiction were more likely to rely upon racial or ethnic stereotypes than those officers whose principal concern was to enforce specific motor vehicle laws or to respond to calls for service. (pp. 33–34)
- The Review Team noted that the significance of the stop statistics could not be determined in the absence of a reliable study of the racial and ethnic characteristics of the persons who travel on the Turnpike to serve as a benchmark. The Review Team therefore proposes to undertake a Turnpike population survey in consultation with the Civil Rights Division of the United States Department of Justice. (pp. 34–35)
- The Interim Report concludes that arrest statistics should not be cited for the proposition that minorities are more likely than whites to be engaged in drug trafficking activities. The fact that the arrest rates for whites are comparatively low does not mean that white motorists are less likely to be transporting drugs, but rather that they are less likely to be suspected of being drug traffickers

in the first place and, thus, less likely to be subjected to probing investigative tactics designed to confirm suspicions of criminal activity such as, notably, being asked to consent to a search. (pp. 35–36)

The Interim Report discusses a number of conditions that might foster disparate treatment of minorities, recognizing that one need not be a racist to be influenced by stereotypes that might lead an officer to treat minority motorists differently during the course of a traffic stop. The Interim Report concludes that the potential for the disparate treatment of minorities during traffic stops may be the product of an accumulation of circumstances that created and reinforced the message that the best way to catch drug traffickers is to focus on minorities, which may have undermined other messages in both official and unofficial policies prohibiting any form of disparate treatment. These circumstances include:

- Ambiguities and misunderstandings about the law;
- Ambiguities, imprecision, and omissions in Standard Operating Procedures;
- Conflicting, subtle messages in otherwise *bona fide* drug-interdiction and gang-recognition training programs;
- The tautological use of statistics to tacitly validate pre-existing stereotypes;
- Formal and informal reward systems that encourage troopers to be aggressive in searching for illicit drugs, thereby providing practical incentives to act upon these stereotypes;
- The inherent difficulties in supervising the day-to-day activities of troopers assigned to patrol; and,
- The procedures used to identify and remediate problems and to investigate allegations of disparate treatment. (pp. 37–44)

The Interim Report includes a detailed discussion of law and policy on racial profiling and the disparate treatment of minorities. This portion of the Report describes the negative effects of stereotyping on minority communities, which can leave persons of color with a sense of powerlessness, hostility, and anger directed toward the law enforcement community. Notably, the Interim Report concludes that disparate treatment of minorities reinforces a sense of mistrust, leaving minority citizens less willing to serve as jurors, less likely to report crime, and less appreciative of the efforts of the vast majority of the law enforcement officers who serve the public with honesty and integrity. (pp. 45–48)

The Interim Report explains in detail the critical distinction between legitimate crime trend analysis and inappropriate racial profiling, recognizing that sophisticated, race-neutral crime analysis is sorely needed if police agencies are to remain responsive to emerging new threats and enforcement opportunities. (pp. 49–52)

The Interim Report recognizes that while the phenomenon of racial profiling and other forms of disparate treatment of minorities is real and not just a matter of perception, perceptions concerning the magnitude and impact of the problem are important, and that these perceptions vary widely in that minority and non-minority citizens in this State have markedly different views regarding the nature and scope of the problem. (pp. 56–59)

The Interim Report recognizes that the racial profiling controversy is by no means limited to the New Jersey State Police, but rather is a truly national problem, as reflected in the number of bills pending in Congress and state legislatures across the country. (pp. 60–65)

The Interim Report describes at length why it would be inappropriate as a matter of policy for officers on patrol to rely upon crime trend analysis that, at first blush, suggests that racial or ethnic characteristics could serve as reliable risk factors in predicting and responding to criminal activity. The Report explains that many of the arrest and conviction numbers relied upon by some police executives across the nation are tautological and, thus, inherently misleading. Notably, these arrest statistics only refer to persons who were found to be involved in criminal activity and do not show the number of persons who were detained or investigated who, as it turned out, were not found to be trafficking drugs or carrying weapons. In fact, when one considers all of the stops conducted by State Police, searches are quite rare, and searches that reveal evidence of crime are rarer still. To the extent that law enforcement agencies arrest minority motorists more frequently based on stereotypes, these events, in turn, generate statistics that confirm higher crime rates among minorities which, in turn, reinforces the underpinnings of the very stereotypes that gave rise to the initial arrests. (pp. 65–75)

The Interim Report recognizes that one of the glaring problems with many forms of profiling is that the characteristics that are typically compiled tend to describe a very large category of presumably innocent motorists. Consequently, these profile characteristics may be no better in terms of predicting criminal behavior than allowing individual officers to rely on inchoate and unparticularized hunches, which is clearly not permitted under Fourth Amendment caselaw. To prove this point, the Interim Report discusses certain kinds of intelligence information provided by the Federal Government to show that this information may provide very little help to state troopers patrolling the Turnpike in identifying major drug couriers from among the universe of innocent motorists. (pp. 72–75)

The Interim Report concludes that while there is no doubt that federal, regional, state, and local intelligence reports reliably indicate that a large number of minority narcotics and weapons offenders are traveling between urban areas in and through New Jersey, so too are innocent minority motorists engaged in such travels and in far, far greater numbers. (p. 72)

The Interim Report describes in detail the legal and policy difficulties in relying on suspected gang membership or other types of group associations to establish suspicion of criminal activity. The Interim Report makes clear that while police officers are permitted under the law to consider, for example, gang membership in determining whether there is reasonable, articulable suspicion to initiate a stop or to conduct a protective frisk for weapons, an officer should not be permitted to use the person's race, ethnicity, or national origin in first determining the likelihood that a person is, in fact, a member of any such criminal organization. While many gangs tend to be exclusionary and are comprised of persons of similar racial or ethnic characteristics, the fact remains that the percentage of young minority males who are members of street gangs is so small that no officer could reasonably suspect that a motorist is a member of any such gang based upon the person's race or ethnicity. To do otherwise would be to practice a form of legal bootstrapping, drawing inferences from a fact that has not yet been established. For this reason, the Report recommends that State Police be trained as to the *objective* criteria and indicia of criminal group associations, so that a state trooper would be prepared to articulate why he or she reasonably suspected that a person is a gang member, going beyond the mere fact that the person was not excluded from the possibility of being a member of a particular criminal organization by virtue of his race or ethnic background. (pp. 75–80)

The Interim Report recognizes that the findings of the Review Team may be cited by some defendants who will seek to overturn or preclude their convictions by claiming selective enforcement. While the Review Team cannot prevent defendants from raising these issues in future motions to suppress, it recommends that the State be prepared to fully and fairly litigate the question whether any particular defendant was a victim of unconstitutional conduct warranting the suppression of evidence. The county prosecutors will be asked to examine closely any case involving a State Police member in which the defendant claims selective enforcement, and prosecutors will be asked to recommend to the Division of Criminal Justice how these cases should be handled, considering the individual facts and circumstances of each case. (pp. 80–82)

The Interim Report makes clear that the Review Team is by no means suggesting an abandonment or repudiation of New Jersey's drug enforcement efforts and suggests that the enforcement of our drug laws must remain an urgent priority of the State Police and law enforcement agencies. The Interim Report explains the necessity for taking decisive steps to ensure strict compliance with all search and seizure and equal protection rules, and the need to make clear to the New Jersey State Police and all other law enforcement agencies of the need to embrace the notion that the so-called war on drugs must be waged with not against the communities that the New Jersey State Police and other law enforcement agencies are sworn to protect. (pp. 82–85)

The Interim Report recognizes that highway interdiction constitutes only one small facet of this State's efforts to address the so-called supply side of the drug problem and recommends that a revised drug enforcement strategy closely examine these issues so as to ensure that drug enforcement resources and efforts are focused so as to have the greatest possible impact on the problem while at the same time ensuring that the tactics employed by the New Jersey State Police do not alienate minority communities, since this would only deny other law enforcement agencies opportunities to enlist the support of these communities and thereby to gain access to information necessary to identify, apprehend, and successfully prosecute those drug profiteers who prey upon minority communities. (p. 85)

The Interim Report recommends a series of detailed remedial steps that should be initiated to ensure that all routine traffic stops made by the State Police are conducted in an impartial, even-handed manner. Some of the policies and procedures described in these action steps are new, while others represent a reaffirmation or clarification of existing State Police policies and practices. The Review Team expects that all well-intentioned troopers will understand that procedures of the type recommended in the Interim Report will serve many purposes and will actually help to protect constitutionally-compliant officers, insulating them from unfair and unfounded allegations of selective enforcement. Notably, the Interim Report would establish a comprehensive and multi-faceted early warning system that would serve not only to detect potential problems, but that would serve to deter violations from occurring in the first place. (pp. 86–90)

The Interim Report recognizes that ultimately, the cornerstone of this comprehensive system is to enhance *professionalism* through enhanced *accountability*. The comprehensive system proposed in the Interim Report would send a strong message that racial profiling and other forms of disparate treatment of minorities will not be tolerated but, as importantly, will provide an opportunity to demonstrate conclusively that the overwhelming majority of state troopers are, indeed, dedicated professionals who perform their sworn duties with integrity and honor. (pp. 90–91)

The Interim Report spells out the goals and objectives of this comprehensive early warning system. (pp. 91–92)

The Interim Report recommends the following specific action steps:

- Recommends that the Attorney General issue an updated statewide drug enforcement strategy to ensure the most efficient, effective, and coordinated use of resources by focusing drug enforcement efforts on carefully-identified impact cases and by making certain that the drug enforcement tactics used by one agency do not unwittingly interfere with or undermine the enforcement efforts of other agencies. The updated strategy would evaluate the effectiveness of the use of highway interdiction tactics as part of New Jersey's comprehensive drug enforcement efforts and would review the effectiveness of the use by state troopers of the consent-to-search doctrine. (pp. 92–94)
- Recommends that the Department of Law and Public Safety publish on a quarterly basis aggregate statistics detailing by State Police station the proportion of minority and non-minority citizens who are subject to various actions taken by State Police members during the course of traffic stops. (p. 94)
- Recommends the establishment of a comprehensive and automated early warning system and enhancement of the computerization of records to ensure the prompt identification of individual troopers whose performance suggests a need for further review by supervisory personnel. (pp. 94–96)
- Recommends the development of a comprehensive new Standard Operating Procedure spelling out all of the steps and criteria to be used by State Police members in initiating and conducting traffic stops. (pp. 96–100)
- Recommends the development of a comprehensive new Standard Operating Procedure spelling out the procedures and criteria for requesting permission to search and in conducting consent searches. (pp. 100–102)
- Recommends that in light of the concerns raised by the consent search data examined by the Review Team, the State Police conduct a case-by-case review of all consent searches made by State Police members assigned to the Turnpike in 1997–1998 to determine whether all reporting requirements and Standard Operating Procedures were complied with. (p. 102)
- Recommends that the State Police enhance and modify their training programs to make certain that the policies regarding racial profiling and the disparate treatment of minorities proposed in this Interim Report are understood by all State Police troopers who are assigned to patrol, their supervisors, and dispatchers. (pp. 102–104)
- Recommends that the State Police develop specific criteria for summoning drug-detection canines or equipment to the scene of a traffic stop that would recognize the psychological impact on persons who are subjected to this

procedure and that would ensure that canines are dispatched quickly so as not to violate the rule that requires that investigative detentions be brief. (p. 104)

- Recommends that a policy be instituted that would require a state trooper assigned to patrol duties to inform the dispatcher when feasible of the trooper's intention to conduct a probable cause search. (pp. 104–105)

- Recommends that the State Police establish specific criteria explaining when and under what circumstances a State Police member should make a custodial arrest rather than issue a summons. (p. 105)

- Recommends that the Division of Criminal Justice and the county prosecutors make available deputy attorneys general and assistant prosecutors to serve as police legal advisors on a 24-hour, 7-day per week basis to answer search and seizure, custodial interrogation, and other legal questions raised by State Police members assigned to patrol duties. (p. 106)

- Recommends that the Director of the Division of Criminal Justice in consultation with the county prosecutors establish a comprehensive reporting system whereby the State Police are notified whenever evidence seized during the course of a patrol stop by a State Police member is suppressed by a court or would likely be suppressed by a court were the matter to be prosecuted. (pp. 106–107)

- Recommends that the State Police develop an inventory and impoundment policy explaining when and under what circumstances State Police members may inspect the contents of a disabled vehicle. (pp. 107–108)

- Recommends interim procedures concerning the handling of internal affairs investigations of selective enforcement allegations, requiring that all allegations of discriminatory practices by State Police members be reported to the Review Team and further requiring that no internal investigation into selective enforcement allegations be concluded until the results have been reviewed by the Division of Criminal Justice. (pp. 108–109)

- Recommends that the Division of Criminal Justice, in consultation with the county prosecutors, develop uniform procedures and criteria for handling selective enforcement litigation involving State Police members. (p. 109)

- Recommends the development of a legislative initiative to create new official misconduct offenses to deal specifically with the use of police authority to knowingly or purposely violate a citizen's civil rights. (pp. 109–110)

- Recommends that the Attorney General's Office in consultation with the Civil Rights Division of the United States Department of Justice undertake a population survey of the persons who travel on the New Jersey Turnpike to serve as a benchmark that will be integrated into the early warning system that can be used to trigger heightened scrutiny and supervision of the exercise of police discretion where an automated audit suggests that an individual trooper or group of troopers have stopped a disproportionate percentage of minority motorists. (pp. 110–112)

C. RACIAL STEERING

1

ALLEGED RACIAL STEERING IN AN APARTMENT COMPLEX

Jason T. Connor and Joseph B. Kadane

R acial discrimination may occur in many ways. Blatant refusal to provide goods, services, employment, or housing on grounds of race is most familiar, but many more subtle forms of discrimination are common.

Racial steering is one such subtle form of racial discrimination involving housing. Racial steering typically occurs as real estate agents and brokers act to preserve and encourage patterns of racial segregation. When an individual or family comes seeking housing, an agent may illegally offer only options in a neighborhood already dominated by the individual's or family's race. The customer is not offered housing in neighborhoods dominated by other races. Racial steering violates section 1982 of the Civil Rights Act of 1966 and the Fair Housing Act of 1968.

In a case we were involved in as consultants, a suit was filed by a group of African-American tenants at an apartment complex. The complex is divided into two parts, "Section A" and "Section B." The plaintiffs allege that potential black renters were steered to Section B while potential white renters were steered toward Section A. The racial breakdown of new tenants for the time period in question was 65% white, 21% black, and 14% other races, primarily Asian and Hispanic. Because the suit alleged that whites and others were steered to Section A and blacks were steered to Section B, data for all non-African-American individuals was subsumed into that for whites. Table 1 provides a breakdown of the new renters by race and the section they moved into. Additional details about these data are provided in the next section.

If the alleged racial steering were occurring, one would expect to find a larger proportion of black tenants in Section B and a larger proportion of whites in Section A.

TABLE I. New Renters

Location	White	Black	Total
Section A	87	8	95
Section B	83	34	117
Total	170	42	212

At first glance, choosing a random black tenant who moved in during the relevant time period shows a .81 probability he lived in Section B. The probability of a white tenant who entered the complex in the same two years living in Section B is .49.

This table may, at first glance, lend some credence to the plaintiffs' allegation. Black tenants do seem more likely to acquire apartments in Section B. White tenants, however, seem equally likely to acquire apartments in Sections A and B.

With large sample sizes in both groups, a confidence interval for the difference of proportions can be used to see whether sampling variability alone might account for the apparent disparity. The symmetric 95% confidence interval for the difference of proportions is .18 to .46, indicating that blacks do have a higher probability of renting in Section B than do whites. This result does not indicate why white tenants disproportionately rented in Section A or black tenants in Section B.

DATA

An assumption made using the preceding confidence interval is that every white person has a probability of renting in Section A equal to every other white person and every African-American has a probability of renting in Section A equal to every other African-American. This initial approach also does not consider many events surrounding the choosing of an apartment such as the tenant's desired apartment type or how many apartments in Sections A and B were available to each tenant when their lease was signed.

For each tenant who moved into the complex in the relevant period, the following is known: the tenant's race, the week he or she signed a lease, the location of the apartment, the apartment type and the date of the move-in. The number of available apartments of the chosen type in both sections of the complex at the date of move-in that were known to be available at the date of the lease signing is also known. These two counts indicate the options the agent had to choose from when he or she was selecting which apartments to show a prospective renter. The rental agents were instructed to show only a few apartments to any renter. By including all these variables, it may be possible to construct a better model for the observed rental patterns.

A rental was excluded from the dataset when the number of available apartments in either section equaled 0 for a particular tenant. For example, if a tenant rented a three-bedroom apartment and at the time of their lease signing there were two three-bedroom apartments available in Section A and none in Section B, this situation does not allow for choice of which section the new tenant is to live in by anyone and was therefore excluded. These exclusions yielded the 212 new renters

TABLE 2. Maximum Likelihood Estimates of
Logistic Model Parameters

Parameter	Estimate	Standard Error	p-value
θ	1.1	.2	$< 10^{-6}$
ϕ	$-.3$	.4	.0013

described in Table 1. Typically, there were numerous apartments of any given type available at all times. On average, there were three available apartments of each type in Section A and nearly nine in Section B.

METHODOLOGY

A model that fully describes the probability that a tenant of a particular race acquires an apartment in Section A or B must be influenced by the number of available apartments in Section A of the complex, m_A, and the number of available apartments in Section B, m_B. The model we construct also allows for this probability to depend on the tenant's race, R, scored as 1 for blacks and 0 for all others. Finally the model allows for greater market attractiveness of either Section A or Section B without regard to race.

Let Y be the event that a tenant rents an apartment in Section A. The model incorporating the possibility of race sensitivity is

$$\text{logit} \Pr\left(Y \mid R, m_A, m_B, \theta, \phi\right) = \log\left(\frac{m_A}{m_B}\right) + \theta + \phi R$$

where ϕ is the race parameter that will indicate a tendency for black renters to obtain apartments in Section A or B of the complex and θ measures the degree to which one section is more attractive regardless of race. In terms of interpretation, $\phi > 0$ indicates blacks going disproportionately to Section A, and $\phi < 0$ indicates black tenants going disproportionately to Section B. The allegation, of course, is that $\phi < 0$. If $\theta > 0$, then Section A is more popular regardless of race, whereas $\theta < 0$ indicates a preference for Section B.

This model is equivalent to

$$\Pr\left(Y \mid R, m_A, m_B, \theta, \phi\right) = \frac{m_A e^{\theta + \phi R}}{m_A e^{\theta + \phi R} + m_B}.$$

Notice that, when $\theta = \phi = 0$, the probability of getting an apartment in Section A is simply $m_A/(m_A + m_B)$, the number of available apartments in A divided by the total number of available apartments. This is the situation in which there is no steering and no underlying tendency to go to either section disproportionately.

Performing a logistic regression with the intercept determined by the number of available apartments in each section at each tenant's lease signing yields the estimates in Table 2. The resulting estimates overwhelmingly indicate that, although Section A was generally more attractive (the estimate of θ is significantly greater than 0), black renters are more likely than white renters to acquire apartments in Section B of the complex (the estimate of ϕ is significantly less than 0). For example, if an African-American desired to rent a two-bedroom apartment at a time when there was an equal number of two-bedroom units in each section, her probability of getting an apartment in Section A is estimated to be .44. A white renter seeking a similar two-bedroom apartment at the same time has an estimated probability of .74 of renting an apartment in Section A.

BAYESIAN LOGISTIC REGRESSION

We might stop the statistical analysis with the logistic regression of the previous section. The key piece of evidence obtained is that the estimate of ϕ is -1.3 and the corresponding p-value is sufficiently low to indicate that black tenants were significantly more likely to obtain apartments in Section B. The low p-value explicitly means that if there were truly no difference in the proportions of black tenants and white tenants getting apartments in the two sections of the complex, there is very low probability we would see such extreme data. This is a difficult statement to make sense of even for statisticians; one can imagine how it might play in court. In the courtroom when the judge says, "Tell me the probability steering occurred," we have no answer so far. We prefer to perform a Bayesian analysis so the judge's question can be answered.

Unlike traditional statistics in which hypotheses are accepted or rejected by comparing the p-value to some predetermined value (.05 for instance), Bayesian analyses assign probabilities to each competing hypothesis and let each person interpret those probabilities for himself. Each hypothesis needs to be assigned a probability of being true before the analysis is performed, a prior probability. Here the two competing hypotheses are $\phi < 0$, the plaintiffs' allegation that black tenants go disproportionately to Section B, and $\phi \geq 0$, that contrary to the plaintiffs' allegation there is no difference or black tenants are more likely to get apartments in Section A.

For each new tenant, the likelihood he or she rents an apartment in Section A is given by the equation of the previous section. Putting a prior distribution on θ and ϕ (i.e., specifying a priori plausible values and how likely they are) permits finding a posterior distribution for them (i.e., an assessment of how likely different values are after examining the data). Once the posterior distribution is known, the posterior probability that $\phi < 0$ will answer the judge's question.

Before analyzing the data one has no knowledge of whether black or white tenants are more likely to acquire apartments in specific sections. Therefore, a reasonable first choice of prior probabilities should reflect this prior lack of knowledge so we choose a flat prior distribution over θ and φ, which takes each possible value to be equally likely. Note that this implies our two hypotheses are equally likely *a priori*.

Figure 1 shows the posterior distributions of θ and ϕ and their joint distribution. The mean of θ is 1.07 with standard deviation .16, the same as the estimates in

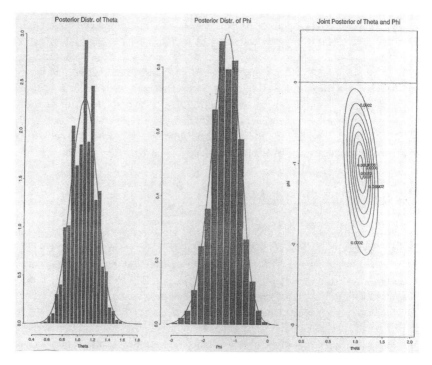

FIGURE 1. Results from Bayesian logistic regression analysis. First plot is posterior distribution of θ; second plot is posterior distribution of ϕ; third plot is contour plot of joint posterior distribution of ϕ and θ.

Table 2. The posterior mean of ϕ is -1.4, while the standard deviation is .4, also very close to the estimates in Table 2. More relevant, the posterior probability of the event $\phi < 0$, indicating black tenants are more likely to rent in Section B, is .9995. An important part of any Bayesian analysis is an assessment of the sensitivity of results to the prior distribution assumed. In this case other reasonable prior distributions had little effect on the posterior probability.

CONCLUSION

Although the statistical analysis indicates that, after taking apartment availability into account, blacks are more likely to go to Section B and whites and others to Section A, the model does contain a few problems. For example, it does not account for tenants specifically requesting particular locations. Moreover, apartment type was broken down by one-bedroom, two-bedroom, three-bedroom, and a few special or deluxe apartment types without regard to floor. This may not account for tenants specifically requesting either first- or second-floor apartments. The final source of error exists when there are a small number of one type of apartment available. Because the available apartments and lease signing data were on a weekly basis, if there were three one-bedroom apartments available, two in Section A and one in Section B, and two get rented that week, the model uses $m_A = 2$ and $m_B = 1$ for both tenants although this assumption is true for only the first tenant. This last issue was not terribly important in practice because, as noted earlier, with the exception of specialty apartments, there were typically numerous apartments available within the entire complex.

The major difficulty in separating the effects of legally benign consumer choices from illegal racial steering by the rental agents is that, on the basis of the data, we cannot say why prospective tenants lease the apartments that they do. If tenants were as a matter of practice made aware of all the vacancies available in the type of apartment of interest to them, one could attribute the flow of tenants solely to their choices. However, tenants were made aware of only a few vacancies chosen by the rental agent, perhaps with the thought of not revealing the extent of the apartment complex's vacancies. Records were not kept about exactly which vacancies were offered to each tenant. Thus the result of the analysis, although strongly indicating the possibility of steering, remains somewhat ambiguous about whether racial steering occurred.

The case settled before it went to trial.

ACKNOWLEDGMENT

The authors would like to thank Tom Wolfe for his legal advice.

Part 3

JURY DISCRIMINATION

Juries play a vital role in American law. The Sixth and Seventh Amendments to the Constitution (part of the Bill of Rights) guarantee the right to a trial by jury in criminal and civil cases, respectively. Most civil cases settle, and most criminal cases lead to plea bargains or are heard, by consent of the defendant, by a judge alone. If either party in a civil case disagrees with a proposal for a waiver of a jury trial, the case must be tried to a jury. Nonetheless, what the parties agree to is influenced by what they believe a jury would decide about the case in question.

The law concerning what is an adequate jury system has developed over time. A series of new laws and Supreme Court decisions required expansion of the group of people called for jury service, and require that the system of jury selection not discriminate on the basis of race, sex, etc. This history is reviewed in Section A in the paper of Kairys, Kadane and Lehoczky (1977). In that paper, we were especially concerned with how jurisdictions get the basic list of names from which potential jurors are selected. We make a case for the use of other lists to supplement the voter list as sources of potential jurors.

Section B addresses the special issues that revolve around juror selection in death penalty cases. The possibility of imposing the death penalty is among the most contentious aspects of American criminal law. As a result, the process by which jurors are chosen for such cases is among the most heavily scrutinized of all jury procedures. A series of Supreme Court decisions require a two-part consideration of a case in which the death penalty is a possibility, and in which the prosecution has asked for it. Such a trial is divided into two parts. In the first, the jury decides whether to find the defendant guilty. If the jury finds the defendant guilty, a second proceeding is held, before the same jury, to weigh aggravating and mitigating circumstances, and to decide whether to impose the death penalty or an alternative punishment, such as life imprisonment without the possibility of parole.

All jurors are asked whether they could decide the case in question fairly and impartially. A jury hearing a case in which the prosecution is asking for the death penalty is also asked questions about their attitudes toward the death penalty. Those who could NEVER impose the death penalty are excluded, as are those who would ALWAYS impose it, every chance they got. The remainder, those who SOMETIMES AND SOMETIMES NOT would impose the death penalty are called "death qualified" jurors, and are the only jurors permitted to serve on juries in cases in which death is a possible penalty.

The question addressed in this section is whether the practice of having the same jury hear both phases biases the jury against the defense by excluding fair and impartial jurors who are not death qualified from hearing the first phase, concerning guilt or innocence.

The first item in this section, Kadane (1983), addresses this question by putting together results from a number of previously conducted studies. Roughly the answer is "yes," the jurors who are fair and impartial but not death qualified are numerically dominated by jurors who would NEVER impose the death penalty, and those jurors are less prone to find defendants guilty than are the SOMETIMES AND SOMETIMES NOT group.

The second item is the Supreme Court decision in Lockhart v. McCree, which ruled that the procedure of having a single jury hear both phases of a trial passes constitutional muster, even if those excluded from hearing the first phase are less conviction-prone.

In addition to progress mandated by legislation, or administrative reform, the principal way in which progress is made in ensuring that juries are a representative cross-section of the community is by a legal challenge. Challenges to the jury system are expensive and difficult. As a result, most challenges occur in the context of cases involving the death penalty. While one is put in the position of challenging the conviction or sentence of someone who has done something egregiously wrong, the result can lead to an improvement of the entire justice system, by finding and correcting sources of misrepresentation in the jury system. The second paper in this section, Kadane (2002), records the evidence in such a jury challenge. The fourth item is the decision of the Indiana Supreme Court on that challenge.

There is one aspect of that decision that deserves special comment. In part I.C. of the majority opinion of the Indiana Supreme Court, they remark "we reach our holding today under Indiana code." The Indiana Supreme Court is the last stop for interpretation of the Indiana Code. Had they based their decision on federal law, their decision could have been appealed to the federal courts, and ultimately the US Supreme Court.

An excellent general work on juries is National Jury Project (2004).

REFERENCE

National Jury Project (2004). Jurywork: Systematic Techniques, E. Krauss, ed. (2 vols.) West, Eagan, Minnesota.

A. JURY REPRESENTATIVENESS

1

JURY REPRESENTATIVENESS: A MANDATE FOR MULTIPLE SOURCE LISTS

David Kairys, Joseph B. Kadane, and John P. Lehoczky

Challenges to jury systems that do not represent the community have increased significantly in the last several years.[1] This phenomenon has occurred not only in the South, but across the country, and, though reforms were instituted by the Federal Jury Selection and Service Act of 1968,[2] federal as well as state jury systems have been challenged for their unrepresentativeness.[3]

Analysis of the evidence presented in support of these challenges reveals that most of the proven unrepresentativeness is attributable to the source list.[4] Indeed,

© 1977 by the California Law Review, Inc Reprinted from the California Law Review, Volume 65, by permission of the California Law Review.

1. REPORT OF THE COMMITTEE ON THE OPERATION OF THE JURY SYSTEM OF THE JUDICIAL CONFERENCE OF THE UNITED STATES 9 (April 7, 1976) [hereinafter cited as REPORT]. For a general discussion of the standards applicable to jury challenges, see NATIONAL JURY PROJECT, THE JURY SYSTEM: NEW METHODS FOR REDUCING PREJUDICE 5–24 (1975) (available from the National Jury Project, 2054 University Ave., Berkeley, California 94704) [hereinafter cited as JURY SYSTEM]; Foster v. Sparks, 506 F.2d 805, 811–37 (5th Cir. 1975) (appendix to the opinion by Judge Walter P. Gewin) [hereinafter cited as Foster Appendix]; Kairys, *Juror Selection: The Law, a Mathematical Method of Analysis, and a Case Study,* 10 AM. CRIM. L. REV. 771 (1972) [hereinafter cited as *Juror Selection*].

2. 28 U.S.C. § 1861, *et seq.* [hereinafter cited as the Federal Act].

3. REPORT, *supra* note 1, at 9.

4. See Table 2 and the text accompanying notes 146–57 *infra;* J. VAN DYKE, JURY SELECTION PROCEDURES 89 (1977) [hereinafter cited as JURY SELECTION PROCEDURES]. The only major exception is in cases where the selection process is systematically discriminatory, which was not the basis of the overwhelming majority of challenges. Cases in which the process is systematically discriminatory fall into two categories, both of which commonly involve claims of discrimination against women. In the first, selection officials have actively discriminated against a cognizable class: for example, in People v. Attica Brothers, 79 Misc. 2d 492, 359 N.Y.S.2d 699 (1974), selection officials admitted that they picked more men than women from the source list (the pool was reconstituted using a nondiscriminatory procedure). Second, qualifications, exclusions, exemptions, and excuses can be discriminatory, either facially or as applied. For example, women are systematically discriminated against and discouraged from serving by various facially neutral but sex-based exemptions nominally linked to child care. *See* Taylor v. Louisiana, 419 U.S. 522 (1975). *See generally* Copelon, Schneider & Stearns, *Sex Discrimination in Jury Selection,* 2 WOMEN'S RTS. L. RPTR. 3 (1975) (the authors discuss the constitutional issues raised by such exemptions and point out that most people who

statistical data indicates that no single available list, including voter registration lists, the most widely used source,[5] adequately represents a cross section of our communities.[6] In recognition of this problem, the Federal Act provides that while voter lists are to be the primary source, federal courts "shall prescribe some other source or sources... where necessary to foster the policy [of representation of a cross section],"[7] and the Uniform Jury Selection and Service Act, adopted in five states,[8] makes the use of multiple lists mandatory.[9]

At present only two federal district courts[10] and several state courts[11] utilize multiple lists, and no court has either required multiple lists or supplemented a primary list on constitutional or statutory grounds.[12] Most decisions[13] require proof of purposeful discrimination in jury selection, even though constitutional[14] and statutory[15] authority indicates that proof of a significant disparity between the composition of the population and the source or pool constitutes a prima facie case of invalidity. Other cases require proof that the underrepresentation resulted in a "substantial impact" on the absolute number of minority members serving

---

are offered an exemption take it and that a large proportion of the women with children who claim the exemption are employed outside the home).

5. The Federal Act, *supra* note 2, provides for use of voter registration or actual voter lists. 28 U.S.C. § 1863(b) (2). Most state courts also use voter registration or actual voter lists. *See* National Center for State Courts, *Facets of the Jury System: A Survey* (1976), Table 3 (available from NCSC, 1660 Lincoln St., Suite 200, Denver, Colorado 80203) [hereinafter cited as *Facets of the Jury System*]. Sixteen states in New England and the South still use the "key man" system or some variation, where selection officials have discretion to choose people they know or hear about. JURY SELECTION PROCEDURES, *supra* note 4, at 86–87.

6. This conclusion is based on data concerning the entire nation and many states and urban areas. See text accompanying notes 149–57 *infra*; JURY SELECTION PROCEDURES, *supra* note 4, at 85–106. It is possible that in some areas voter registration lists, or some other single list, are sufficient, and in such areas it would not be necessary to use multiple lists.

7. 28 U.S.C. § 1863(b) (2).

8. COLO. REV. STAT. ANN. § 13–71-101 *et seq.* (1973); IDAHO CODE § 2–201 *et seq.* (1948); IND. STAT. ANN. § 33–4–5.5–1 *et seq.* (Burns 1975); MISS. CODE § 13–52 *et seq.* (Cum. Supp. 1976); No. DAK. CENT. CODE § 27–09.1–01 *et seq.* (1974).

9. UNIFORM JURY SELECTION AND SERVICE ACT [hereinafter cited as UNIFORM ACT] § 5. Mississippi and Indiana have modified § 5 of the Uniform Act. MISS. CODE, § 13–5–8(1) provides for exclusive use of voter registration lists. IND. STAT. ANN. § 33–4–5.5–7 (Burns 1975) provides for multiple lists but applies only to counties with a population of from 500,000 to 600,000 according to the 1970 census.

10. The two federal district courts are the United States District Court for the District of Colorado and the United States District Court for the District of Columbia. See note 199 *infra*.

11. See Table 5 *infra*.

12. *But see* Ford v. Hollowell, 385 F. Supp. 1392 (N.D. Miss. 1974); United States v. Grant, 471 F.2d 648 (4th Cir. 1973), *rehearing denied*, 475 F.2d 581 (1973) (Winter, J., dissenting). One court has invalidated use of voter registration lists as the sole source, but the decision was quickly reversed by an appellate court. People v. Taylor, No. A-277–425 (Super. Ct. for Los Angeles County, Oct. 11, 1974), *vacated*, Civ. No. 45230 (Cal. Ct. App., 2d Dist., Nov. 26, 1974).

13. *E.g.*, United States v. Test, 550 F.2d 577 (10th Cir. 1976); United States v. Lewis, 472 F.2d 252 (3d Cir. 1973); United States v. Gordon, 455 F.2d 398 (8th Cir. 1972); United States v. Ross, 468 F.2d 1213 (9th Cir. 1972); United States v. Dangler, 422 F.2d 344 (5th Cir. 1970). For an analysis of the legislative history of the federal supplementation provision, see note 198 *infra*.

14. Taylor v. Louisiana, 419 U.S. 522 (1975); Turner v. Fouche, 396 U.S. 346 (1970); Thiel v. Southern Pacific Co., 328 U.S. 217 (1946).

15. 28 U.S.C. § 1861; H.R. REP. No. 1076, 90th Cong., 2d Sess. 8 (1968); S. REP. No. 891, 90th Cong., 1st Sess. 9 (1967). *See* Taylor v. Louisiana, 419 U.S. 522, 529 (1975); United States v. Jenkins, 496 F.2d 57, 65 (2d Cir. 1974), *cert. denied*, 420 U.S. 925 (1975); United States v. McDaniels, 370 F. Supp. 298, 301 (E.D. La. 1973), *aff'd* 509 F.2d 825, *cert. denied*, 423 U.S. 857 (1975).

on a panel,[16] rendering challenges based on the underrepresentation of small or medium-sized minorities impossible.[17] The Committee on the Operation of the Jury System of the Judicial Conference of the United States recently suggested, in response to the increase in challenges, that the Federal Act be amended to "establish a presumption that names of prospective jurors contained in voter lists represent a fair cross-section of the community."[18] There is, however, no factual basis for such a presumption.[19]

A number of factors are responsible for the failure of the courts to require use of multiple lists to correct unrepresentative selection systems. First, there is confusion over the constitutional standard to be applied. Since the 1940's the courts have required that jury pools and source lists be representative of a cross section of the community.[20] The Federal Act[21] and many state statutes[22] contain the same or a similar requirement. Nevertheless, courts have tended to analyze disparities between the composition of the population and the source list under a purposeful discrimination test, rather than a representativeness test.[23] Second, there are no accepted standards for evaluating the representativeness of source lists or pools. Neither courts nor legislatures have established criteria for distinguishing allowable from impermissible deviations from the cross-sectional ideal.[24] Third, though inadequate source lists cause most of the unrepresentativeness, they have received insufficient attention in challenges and court decisions because of a lack of available data[25] and the unquestioned notion that voters are the "best" or "most concerned" citizens.[26] Last, there has been no available methodology for implementing multiple list systems with reasonable costs and effort.[27]

---

16. United States v. Goff, 509 F.2d 825 (5th Cir. 1975), *cert. denied,* 423 U.S. 857 (1975); United States v. Jenkins, 496 F.2d 57, 65 (2d Cir. 1974), *cert. denied,* 420 U.S. 925 (1975).

17. United States v. Test, 550 F.2d 577 (10th Cir. 1976); United States v. Freeman, 514 F.2d 171 (8th Cir. 1975).

18. REPORT, *supra* note 1, at 9. The proposal also prohibits use of multiple lists unless there is a specific finding that voter lists are not representative. Since the two federal districts now using multiple lists have explicitly provided in their plans that their use of multiple lists was not based on such a finding, the proposal would have the effect of invalidating the use of multiple lists in these districts.

19. The Committee's report does not state or refer to any factual basis; rather, the Committee said: "This proposal is a response to the increasing number of cases challenging the process of jury selection." *Id.*

20. Taylor v. Louisiana, 419 U.S. 522 (1975); Turner v. Fouche, 396 U.S. 346 (1970); Thiel v. Southern Pacific Co., 328 U.S. 217 (1946).

21. 28 U.S.C. § 1861.

22. *E.g.,* COLO. REV. STAT. ANN. § 13–71-102 (1973).

23. Castaneda v. Partida, 97 S. Ct. 1272 (1977); Alexander v. Louisiana, 405 U.S. 625 (1972).

24. For a general discussion, see *Juror Selection, supra* note 1, at 772–79; Foster Appendix, *supra* note 1, at 818–19, 833–35. The Federal Act sets "substantial deviation" as the standard, 28 U.S.C. § 1861, but the definition and content of this provision was left to judicial determination. See S. REP. No. 981, 90th Cong., 1st Sess., 11 (1967); H.R. REP. No. 1076, 90th Cong., 2d Sess., 5 (1968).

25. *See Foster Appendix, supra* note 1, at 817.

26. *See* United States v. Test, 550 F.2d 577 (10th Cir. 1976); Simmons v. United States, 406 F.2d 456 (5th Cir. 1969). This notion is directly contrary to the cross-sectional principle and is of questionable factual validity. See text accompanying notes 160–65 *infra.*

27. Judge Walter P. Gewin, although recognizing the need for more representative sources, has stated that "there is no facile way to supplement voter registration lists." Gewin, *The Jury Selection and Service Act of 1968; Implementation in the Fifth Circuit Court of Appeals,* 20 MERCER L. REV. 349, 383 (1969) [hereinafter cited as *Implementation in the Fifth Circuit*]. Later, Judge Gewin proposed use of the old "key man" system as the only practical method of supplementation. Gewin, *Should Guidelines be Established for Determining When District Courts Should Use Other Sources of Names of Prospective Jurors in Addition to Voter Registration Lists or Lists of Actual Voters; and, if so, What Guidelines Should be Used,* COMMITTEE ON THE OPERATION OF THE JURY SYSTEM

These obstacles to representative source lists and jury pools should no longer prevent implementation of the cross section of the community principle. Although there is still considerable controversy concerning the theoretical basis of the jury selection cases, the Supreme Court recently confirmed[28] that a significant under-representation establishes a prima facie case of invalidity. Various measures of representativeness have been proposed, and there is a firm legal and factual basis for adoption of a definitive standard. The necessary data is available.[29] The methodology for using multiple lists at minimal added cost or effort has been developed,[30] and is being utilized in several jurisdictions.[31] All that remains is a traditional judicial task: standards must be formulated and adopted, and the constitutional and statutory mandates must be enforced.

This chapter traces the theoretical framework of the jury selection cases and the various standards that have been proposed as measures of representativeness. Next the guiding constitutional and statutory principles are applied to the problem of source lists and the available data concerning voter registration and other lists is presented. Finally, available methodologies for the easy and inexpensive use of multiple lists are discussed.

THE CONSTITUTIONAL MANDATE OF REPRESENTATIVENESS

Representativeness as a constitutional requirement for jury selection has been derived piecemeal following ratification of the fourteenth amendment.[32] In *Strauder v. West Virginia*,[33] the first successful challenge to a jury selection system, the Supreme Court invalidated a state statute that prohibited blacks from serving on grand or petit juries. Subsequent decisions vindicated challenges to de facto total exclusion of blacks[34] and underrepresentation of blacks[35] and of other "cognizable" groups.[36] The

---

OF THE JUDICIAL CONFERENCE OF THE UNITED STATES, THE JURY SYSTEM IN THE FEDERAL COURTS, 1966–1973 at 109 (1974) [hereinafter cited as WORKS].

28. Castaneda v. Partida, 97 S. Ct. 1272 (1977).

29. See text accompanying notes 145–60.

30. *See* Kadane & Lehoczky, *Random Juror Selection from Multiple Lists*, 24 OPERATIONS RESEARCH 207 (1976) [hereinafter cited as *Multiple Lists*].

31. See Table 5 *infra*.

32. On the historical development of the jury selection cases, see generally *Juror Selection, supra* note 1, at 772–77.

33. 100 U.S. 303 (1880).

34. Norris v. Alabama, 294 U.S. 587 (1935).

35. Smith v. Texas, 311 U.S. 128 (1940).

36. Thiel v. Southern Pacific Co., 328 U.S. 217 (1946); Hernandez v. Texas, 347 U.S. 475 (1954). For a general discussion of the standards for determining cognizable classes and the various groups recognized, see JURY SYSTEM, *supra* note 1, at 10–14. Under either the representativeness or discrimination theories discussed here, it is appropriate to require relief concerning only particular categories in the population, although less stringent standards of cognizability are more consistent with the representativeness principle. For example, underrepresentation of or discrimination against left-handed people would not raise a constitutional question under either theory, but underrepresentation or discrimination on the following bases clearly should: race, ethnicity or ancestry, economic, occupational, social or class status, religious beliefs, sex, age, geography and political beliefs or values. *See* JURY SYSTEM, *supra* note 1, at 10–13.

Court first articulated the affirmative principle[37] that juries be drawn from a "cross-section of the community" in *Thiel v. Southern Pacific Co.,*[38] decided in 1946. Although *Thiel* concerned a federal jury and the decision rested upon the Court's supervisory power,[39] the cross section requirement is now firmly established as a constitutional principle applicable to state as well as federal selection systems.[40]

The representativeness principle furthers important societal interests in addition to the right of litigants to a fair trial and the right of citizens to serve on juries.[41] The concept of the jury as representative of a cross section of the community has long been linked both to notions of representative government and democracy and to the constitutional guarantees of due process, equal protection, and trial by an impartial jury.[42] The jury provides a vehicle for direct citizen participation in an

---

The decisions concerning whether young people are a cognizable class are inconsistent. *Compare* United States v. Butera, 420 F.2d 564 (1st Cir. 1970) (age is a cognizable class) *and* Simmons v. Jones, 317 F. Supp. 397 (S.D. Ga. 1970) (indicating age is a cognizable class) *with* United States v. Allen, 445 F.2d 849 (5th Cir. 1971) (age is not a cognizable class) *and* United States v. Kuhn, 441 F.2d 179 (5th Cir. 1971) (age is not a cognizable class). Application of traditional standards of cognizability to the available evidence leads to the conclusion that young people constitute a cognizable class. *See* JURY SYSTEM, *supra* note 1, at 12–13; JURY SELECTION PROCEDURES, *supra* note 4, at 35–39.

37. The principle was suggested in Smith v. Texas, 311 U.S. 128 (1940). It has been traced back as far as Tudor England. *See* I. HOLDSWORTHY, A HISTORY OF ENGLISH LAW 339–47 (1956); Note, *The "Blue-Ribbon" Jury,* 60 HARV. L. REV. 613 (1947). The Magna Carta, Chp. 39, guaranteed a jury of one's peers, which meant a jury of persons from the same class, legal status, or caste as the accused. *See* W. MCKECHNIE, MAGNA CARTA 378 (2nd ed. 1914) (the "peers of a Crown tenant were his fellow Crown tenants"). *See also* Labat v. Bennett, 365 F.2d 698, 711 (5th Cir.) (en banc), *cert. denied,* 386 U.S. 991 (1966); 4 BLACKSTONE COMMENTARIES 349 (Tucker 1803). The *Strauder* Court clearly had this tradition in mind:

> The very idea of a jury is a body of men composed of the peers or equals of the persons whose rights it is selected or summoned to determine: that is, of his neighbors, fellows, associates, persons having the same legal status in society as that which he holds.

100 U.S. at 308.

38. 328 U.S. 217, 220 (1946). Unless the jury represents a cross-section, the Court noted, there is a danger that it will become "the instrument of the economically and socially privileged" and representative of "narrow class interests." *Id.* at 223–24.

39. *Id.* at 219.

40. Taylor v. Louisiana, 419 U.S. 522 (1975); Alexander v. Louisiana, 405 U.S. 625 (1972); Turner v. Fouche, 396 U.S. 346 (1970); Carter v. Jury Comm'n, 396 U.S. 320 (1970); Witherspoon v. Illinois, 391 U.S. 510 (1968); Glasser v. United States, 315 U.S. 60 (1942) (federal jury); United States v. Zirpolo, 450 F.2d 424 (3d Cir. 1971); Broadway v. Culpepper, 439 F.2d 1253 (5th Cir. 1971); Salary v. Wilson, 415 F.2d 467 (5th Cir. 1969); Witcher v. Peyton, 405 F.2d 725 (4th Cir. 1969) ("[t]here is a constitutional right to a jury drawn from a group which represents a cross-section of the community. And a cross-section of the community includes persons with varying degrees of training and intelligence and with varying economic and social positions....It is a democratic institution...." *Id.* at 727); Labat v. Bennett, 365 F.2d 698 (5th Cir.) (en banc), *cert. denied,* 386 U.S. 991 (1966); Dow v. Carnegie-Illinois Steel Corp., 224 F.2d 414 (3d Cir. 1955) (federal jury); King v. Cook, 298 F. Supp. 584 (N.D. Miss. 1969); Love v. McGee, 297 F. Supp. 1314 (S.D. Miss. 1968); Allen v. State, 110 Ga. App. 56, 137 S.E.2d 711 (1964); Michigan v. Viera, Nos. A-152–598 and A-152–697, Recorders Ct. for the City of Detroit (April 30, 1970).

41. The Court has approved affirmative civil suits by members of an underrepresented group to vindicate their right to serve as jurors. Turner v. Fouche, 396 U.S. 346 (1970); Carter v. Jury Comm'n, 396 U.S. 320 (1970). Other affirmative actions have been successful. Salary v. Wilson, 415 F.2d 467 (5th Cir. 1969); Love v. McGee, 297 F. Supp. 1314 (S.D. Miss. 1968); and White v. Crook, 251 F. Supp. 401 (M.D. Ala. 1966) (three-judge court).

42. Taylor v. Louisiana, 419 U.S. 522 (1975); Thiel v. Southern Pacific Co., 328 U.S. 217 (1946). In Smith v. Texas, 311 U.S. 128 (1940) the Court said:

> For racial discrimination to result in the exclusion from jury service of otherwise qualified groups not only violates our Constitution and the laws enacted under it but is at war with our basic concepts of democratic society and a representative government.

Id. at 130. *See also* Fay v. New York, 332 U.S. 261, 299 (1947) (Murphy, J., dissenting); Glasser v. United States, 315 U.S. 60 (1942).

arena otherwise dominated by professional advocates and government officials. In criminal cases, the jury performs a protective function, interposing a group of citizens between an accused and the punitive mechanism of the state.[43] Thus the representative, popular character of the jury lends legitimacy, integrity, and impartiality[44] to the judicial process.[45] A lack of representativeness tends to compromise the jury as an institution and to undermine the judicial process.[46] The Court has uniformly ruled since 1940 that the right to representativeness is fundamental,[47] comparable in importance to the right to vote,[48] and cannot be "overcome on merely rational grounds."[49]

Nevertheless, standards governing challenges to jury selection systems have usually been formulated in terms of prohibiting discrimination[50] rather than requiring representativeness. This theoretical choice has significant consequences. If discrimination is the focus, the actions, intent, and perhaps even motives of selection officials are crucial in determining the validity of the selection system. If representativeness is the guiding principle, state and federal governments have a duty to provide representative sources and pools, and the presence or absence of discriminatory intent is irrelevant.

43. *E.g.,* Duncan v. Louisiana, 391 U.S. 145, 155–56 (1968).

44. Cross-sectional jury systems yield impartiality in the only sense that concept can have real meaning concerning the pool from which the jury is to be drawn. No one is without attitudes and preferences concerning various social, political, economic, cultural and religious issues, and such attitudes and preferences affect one's judgment and perception regarding factual and legal questions and the credibility of witnesses. Cross-sectionality yields impartiality in the sense that, at least before cause and peremptory challenges, the wide variety of attitudes and life experience in the community are represented and the particular perspectives of any particular group will be prevalent in proportion to that group's numbers in the population.

45. *E.g.,* Taylor v. Louisiana, 419 U.S. 522 (1975); Thiel v. Southern Pacific Co., 328 U.S. 217 (1946). *See* Note, *The Case for Black Juries,* 79 YALE L.J. 531 (1970).

46. Taylor v. Louisiana, 419 U.S. 522 (1975); Thiel v. Southern Pacific Co., 328 U.S. 217 (1946). In *Taylor,* the Court said:

> The purpose of a jury is to guard against the exercise of arbitrary power—to make available the common-sense judgment of the community as a hedge against the overzealous or mistaken prosecutor and in preference to the professional or perhaps over-conditioned or biased response of a judge. Duncan v. Louisiana, 391 U.S. at 155–156.... This prophylactic vehicle is not provided if the jury pool is made up of only special segments of the populace or if large, distinctive groups are excluded from the pool. Community participation in the administration of the criminal law, moreover, is not only consistent with our democratic heritage but is also critical to public confidence in the fairness of the criminal justice system. Restricting jury service to only special groups or excluding identifiable segments playing major roles in the community cannot be squared with the constitutional concept of jury trial. 'Trial by jury presupposes a jury drawn from a pool broadly representative of the community as well as impartial in a specific case.... [T]he broad representative character of the jury should be maintained, partly as assurance of a diffused impartiality and partly because sharing in the administration of justice is a phase of civil responsibility.' Thiel v. Southern Pacific Co. 328 U.S. 217, 227...(Frankfurter, J., dissenting).

419 U.S. at 530–31.

47. Taylor v. Louisiana, 419 U.S. 522 (1975); Carter v. Jury Comm'n, 396 U.S. 320 (1970); Turner v. Fouche, 396 U.S. 346 (1970); Smith v. Texas, 311 U.S. 128 (1940). In *Taylor,* the Court said:

> The unmistakable import of this Court's opinions, at least since 1940, Smith v. Texas,... and not repudiated by intervening decisions, is that the selection of a petit jury from a representative cross section of the community is an essential component of the Sixth Amendment right to a jury trial.

419 U.S. at 528.

48. Carter v. Jury Comm'n, 396 U.S. 320, 330 (1970).

49. Taylor v. Louisiana, 419 U.S. 522, 534 (1975).

50. *E.g.,* Castaneda v. Partida, 97 S. Ct. 1272 (1977); Alexander v. Louisiana, 405 U.S. 625 (1972); Turner v. Fouche, 396 U.S. 346 (1970); Whitus v. Georgia, 385 U.S. 545 (1967); Hill v. Texas, 316 U.S. 400 (1942).

This distinction is unimportant when the challenger proves actual discrimination, for then a violation of both principles has been established. When the challenger has proved "systematic" or "intentional" discrimination by selection officials at any stage, the jury system is presumptively invalid even if the resulting disparity between the composition of the pool and the population is minimal[51] or, in some cases, even if there is no disparity.[52] Thus in *Cassell v. Texas*[53] the Court invalidated a jury system in which there was a proportional limit on the number of blacks although the resulting disparity was not considered large and could be justified. In such cases, good faith or a nondiscriminatory purpose on the part of the selection officials does not save the jury system.[54] For example, where selection officials have intentionally chosen two men for every woman chosen, the selection system is invalid even though selection officials chose more men than women for administrative convenience since women usually ask to be excused.[55] These rules are considered to be well established.

There has been considerable confusion, however, when the challenger relies solely or mainly on proof of a significant disparity between the composition of the population and that of the source or pool.[56] Such proof establishes that the representativeness principle has been violated, but does not explain the reason for the unrepresentativeness or directly establish discriminatory actions, intent or motive on the part of selection officials.

In *Turner v. Fouche,* after finding that there was a substantial disparity, the Court stated that the challengers had "further demonstrated that the disparity originated, at least in part, at the one point in the selection process where jury commissioners invoked their subjective judgment rather than objective criteria."[57] In *Alexander v. Louisiana,* the Court stated that the challenger's prima facie case rested on proof of a substantial disparity and a "clear and easy opportunity for racial discrimination."[58] These decisions led to confusion as to whether proof of an opportunity to discriminate is a necessary element of a prima facie case based on a

---

51. *E.g.,* Cassell v. Texas, 339 U.S. 282 (1950); Avery v. Georgia, 345 U.S. 559 (1953); Arnold v. North Carolina, 376 U.S. 773 (1964); People v. Attica Brothers, 79 Misc. 2d 492, 359 N.Y.S.2d 699 (1974) (selection officials admitted that they discriminated against women in picking names from voter lists; the selection system was invalidated even though their purpose was to minimize administrative tasks since state law provided for a women's exemption which women often claimed). *See also* Brooks v. Beto, 366 F.2d 1, 22 n.40 (5th Cir. 1966); Bell v. Southwell, 376 F.2d 659 (5th Cir. 1967).

52. In Williams v. Georgia, 349 U.S. 375 (1955), the Court invalidated a jury system based on "the system of selection and the resulting danger of abuse...and not an actual showing of discrimination on the basis of comparative numbers of Negroes and whites on the jury lists." *Id.* at 382.

53. 339 U.S. 282 (1950). Three independent grounds were urged by the challenger: the disparity, the imposition of a proportional limit on the number of blacks, and the failure of the "key men" to familiarize themselves with black people. The Court found the disparity insubstantial and justifiable but ruled that each of the remaining claims was sufficient to invalidate the jury system.

54. Taylor v. Louisiana, 419 U.S. 522 (1975); Alexander v. Louisiana, 405 U.S. 625 (1972); Arnold v. North Carolina, 376 U.S. 773 (1964); Avery v. Georgia, 345 U.S. 559 (1953); Thiel v. Southern Pacific Co., 328 U.S. 217 (1946).

55. United States v. Zirpolo, 450 F.2d 424 (3d Cir. 1971); People v. Attica Brothers, 79 Misc. 2d 492, 359 N.Y.S.2d 699 (1974).

56. *See* the majority and dissenting opinions in Castaneda v. Partida, 97 S. Ct. 1272 (1977); Taylor v. Louisiana, 419 U.S. 522 (1975); United States v. Test, 550 F.2d 577 (10th Cir. 1976); United States v. Jenkins, 496 F.2d 57 (2d Cir. 1974), *cert. denied,* 420 U.S. 925 (1975); Black v. Curb, 422 F.2d 656 (5th Cir. 1970).

57. 396 U.S. 346, 360 (1970).

58. 405 U.S. 625, 630 (1972).

substantial disparity.[59] The most recent decision, *Castaneda v. Partida*,[60] indicates that it is not. The Court, while discussing the opportunities for discrimination inherent in the selection system, stated that proof of such opportunity merely "supports" the prima facie case established by proof of a substantial disparity.[61] In *Partida,* the challenger proved that Mexican-Americans constituted 79.1 % of the population and only 45.5% of the grand jurors.[62] The source list for grand jurors was compiled by the "key man" system, in which "key" people select jurors from persons they know or hear about. This system presents officials with an obvious opportunity to discriminate. Nevertheless, the Court explicitly stated that the substantial disparity alone established a prima facie case:

> [A] selection procedure that is susceptible to abuse or not racially neutral supports the presumption of discrimination raised by the statistical showing.... Once the defendant has shown substantial underrepresentation of his group, he has made out a *prima facie* case of discriminatory purpose, and the burden then shifts to the State to rebut that case.[63]

Although *Partida* makes it clear that proof of a significant disparity establishes a prima facie case,[64] the decision is based on the discrimination principle. The Supreme Court has consistently analyzed the substantial disparity cases in terms of

59. In several early decisions, the opportunity to discriminate was an independent basis for invalidating a jury system and was usually analyzed in the framework of the systematic exclusion line of cases. *See, e.g.,* Williams v. Georgia, 349 U.S. 375 (1955); Avery v. Georgia, 345 U.S. 559 (1953). Courts then began to view a prima facie case based on a substantial disparity as bolstered by proof of an opportunity to discriminate. Alexander v. Louisiana, 405 U.S. 625 (1972); Turner v. Fouche, 396 U.S. 346 (1970); Sims v. Georgia, 389 U.S. 404 (1967); Whitus v. Georgia, 385 U.S. 545 (1967); Witcher v. Peyton, 405 F.2d 725 (4th Cir. 1969); Lampkin v. Smith, 309 F. Supp. 1325 (N.D. Ga. 1970); Love v. McGee, 297 F. Supp. 1314 (S.D. Miss. 1968); Bonds v. State, 220 Tenn. 555, 421 S.E.2d 87 (1967).

If the disparity is substantial, there would seem to be little reason for requiring the challenger to prove that the opportunity to discriminate exists before the government must explain the disparity, for proof concerning the workings of the selection system would be presented as part of the government's rebuttal case. Moreover, where there is a substantial disparity that the government cannot explain, the selection system should not be validated simply because the challenger could not obtain proof of the defect or "opportunity" that led to the disparity. In any event, proof of an opportunity to discriminate, as discussed in the later cases, means only proof that the selection system affords selection officials the opportunity, at some stage, to exercise discretion in the selection of jurors.

60. 97 S. Ct. 1272 (1977).

61. *Id.* at 1280.

62. *Id.* at 1276.

63. *Id.* at 1280 (citations omitted). Justice Powell, dissenting, joined by Chief Justice Burger and Justice Rehnquist, would have held that proof of intent, not just a significant disparity, is necessary to a prima facie case and would have applied the reasoning of Washington v. Davis, 426 U.S. 229 (1976) and Arlington Heights v. Metropolitan Hous. Dev. Corp., 97 S. Ct. 555 (1977). *Id.* at 1287–92. See note 64 *infra.* Justice Burger, dissenting, joined by Justices Powell and Rehnquist, would have required the challenger to prove a significant disparity based on figures for the *eligible* population. 97 S. Ct at 1285–86. This would have had the effect of undercutting virtually all challenges to jury selection systems, since eligibility can at present be based on such vague standards as "good character" and since eligible population figures are almost impossible to obtain. *See Juror Selection, supra* note I, at 798–800.

64. It may be thought that recent decisions of the Supreme Court emphasizing the importance of intent as opposed to impact in discrimination cases tend to undercut these jury selection cases. In Arlington Heights v. Metropolitan Hous. Dev. Corp., 97 S. Ct. 555 (1977), and Washington v. Davis, 426 U.S. 229 (1976), the Court, considering, respectively, zoning regulations and police hiring practices that had discriminatory impacts, emphasized the requirement of proof of a discriminatory intent or purpose. Unlike these cases, discrimination in jury selection involves not only the rights of people to serve as jurors but also the due process, equal protection and jury trial rights of litigants and the societal interest in representativeness. The Court has regarded the rights to serve on juries and to have cross-sectional juries as fundamental and compared them in importance to the right to vote. Taylor v. Louisiana, 419 U.S. 522 (1975); Turner v. Fouche, 396 U.S. 346 (1970); Carter v. Jury Comm'n, 396 U.S. 320 (1970). In Taylor v. Louisiana, 419 U.S. at 534, the Court said, "[t]he right to a proper jury cannot be overcome on merely rational grounds."

the "rule of exclusion,"[65] which amounts to an amalgam of the discrimination and representativeness principles. A "significant" or "substantial" disparity is viewed as creating an inference of "systematic," "intentional" or "purposeful" discrimination.[66] But the challenger need not prove actual discrimination, lack of good faith or actual prejudice,[67] and the inference of discrimination is not defeated by proof of a nondiscriminatory intent or purpose, but only by proof that the underrepresented group is less eligible or available for jury duty.[68] Thus, a substantial underrepresentation is viewed as proof of intentional discrimination, but proof that the discrimination was unintentional or was based on administrative feasibility[69] or some other nondiscriminatory purpose does not defeat the inference of intentional discrimination or save the jury system. The analysis is described in terms appropriate to the discrimination principle, but reaches the same result, indirectly, that an analysis based on the representativeness principle would yield directly.[70]

In *Arlington Heights,* the Court noted that "[s]ometimes a clear pattern, unexplainable on grounds other than race, emerges from the effect of the state action even when the governing legislation appears neutral on its face," citing, *inter alia,* Yick Wo v. Hopkins, 118 U.S. 356 (1886) and Gomillion v. Lightfoot, 364 U.S. 339 (1960). A footnote to this statement notes that several jury selection cases fall into this category and says:

> Because of the nature of the jury selection task, however, we have permitted a finding of constitutional violation even when the statistical pattern does not approach the extremes of *Yick Wo* or *Gomillion.* See, e.g., Turner v. Fouche, 396 U.S. 346, 359 (1970); Sims v. Georgia, 389 U.S. 404, 407 (1967).

Id. at 5647 n.13. *See also* Washington v. Davis, 426 U.S. 229, 241. In *Partida,* the Court confirmed that jury selection cases are to be distinguished from the *Arlington Heights-Washington v. Davis* line of cases. 97 S. Ct. 1272, 1279.

65. *See, e.g.,* Castaneda v. Partida, 97 S. Ct. 1272 (1977); Alexander v. Louisiana, 405 U.S. 625 (1972); Turner v. Fouche, 396 U.S. 346 (1970).

66. The idea behind the rule of exclusion is not at all complex. If a disparity is sufficiently large, then it is unlikely that it is due solely to chance or accident, and, in the absence of evidence to the contrary, one must conclude that racial or other class-related factors entered in the selection process.
Castaneda v. Partida, 97 S. Ct. 1272, 1280 n.13.

67. Castaneda v. Partida, 97 S. Ct. 1272 (1977); Taylor v. Louisiana, 419 U.S. 522 (1975); Alexander v. Louisiana, 405 U.S. 625 (1972); Turner v. Fouche, 396 U.S. 346 (1970); Sims v. Georgia, 389 U.S. 404 (1967); Whitus v. Georgia, 385 U.S. 545 (1967); Cassell v. Texas, 339 U.S. 282 (1950); Smith v. Texas, 311 U.S. 128 (1940); Smith v. Yeager, 465 F.2d 272 (3d Cir. 1972); United States v. Zirpolo, 450 F.2d 424 (3d Cir. 1971); Black v. Curb, 422 F.2d 656 (5th Cir. 1970); Salary v. Wilson, 415 F.2d 467 (5th Cir. 1969); Witcher v. Peyton, 405 F.2d 725 (4th Cir. 1969); Pullum v. Greene, 396 F.2d 251 (5th Cir. 1968); Labat v. Bennett, 365 F.2d 698 (5th Cir. 1966), *cert. denied,* 386 U.S. 991 (1967); King v. Cook, 298 F. Supp. 584 (N.D. Miss. 1969); Love v. McGee, 297 F. Supp. 1314 (S.D. Miss. 1968); White v. Crook, 251 F. Supp. 401 (M.D. Ala. 1966) (three-judge court).

68. Evidence that the underrepresented group is less eligible or available can be used to rebut a prima facie case, but a nondiscriminatory motive and denials of discrimination or intent to discriminate are insufficient. Taylor v. Louisiana, 419 U.S. 522 (1975); Alexander v. Louisiana, 405 U.S. 625, 631–32 (1972); Turner v. Fouche, 396 U.S. 346 (1970); Whitus v. Georgia, 385 U.S. 545 (1967); Sims v. Georgia, 389 U.S. 404 (1967); Hernandez v. Texas, 347 U.S. 475 (1954); Cassell v. Texas, 339 U.S. 282 (1950); Norris v. Alabama, 294 U.S. 587 (1935). For a general discussion see *Juror Selection, supra* note 1, at 779–80.
Rebuttal evidence of lower eligibility is only relevant at a stage in the process before which eligibility criteria have been applied; this usually is not the case concerning source lists. *See* Castaneda v. Partida, 97 S. Ct. 1272, 1276 n.8.

69. *See* Taylor v. Louisiana, 419 U.S. 522 (1975); Thiel v. Southern Pacific Co., 328 U.S. 217 (1946); United States v. Zirpolo, 450 F.2d 424 (3d Cir. 1971).

70. One possible difference is that the representativeness principle can mean that selection officials have a duty to find eligible people in an underrepresented cognizable class even though they are less eligible proportionally than people not in that class. Aside from the possibility that there may not be enough eligible people in the class to provide them representation in proportion to their numbers in the population, this would eliminate the basis and need for a rebuttal case and the sole question would be representativeness.

Even if the results are the same, the use of discrimination terminology is undesirable because it obscures the true interests at stake, diverts attention to misleading issues such as intent, motive, and imputed malice, and provides a basis for validation of unrepresentative selection systems.[71] The choice of the appropriate theoretical basis for judicial decisions should be guided by an analysis of the underlying constitutional rights and societal interests. Though lack of access to governmental institutions by citizens in underrepresented groups has been analyzed in terms of discrimination in other contexts,[72] the fundamental rights of litigants and the societal interests in the legitimacy, integrity and impartiality of the judicial process are inextricably tied to the principle of representativeness, regardless of the intent, purpose, or actions of selection officials.[73] The representativeness approach, by placing an affirmative duty on selection officials to provide representative jury pools, focuses on the actual interests underlying the jury system and is therefore preferable from an analytical viewpoint.

The apparent hesitance of the courts to frame standards directly based on the representativeness principle while, at the same time, repeatedly recognizing representativeness as the underlying constitutional requirement, is due in large part to the lack of concrete standards for determining representativeness. Courts are understandably reluctant to embrace an explicit representativeness standard when there is no accepted measure of representativeness or any clear definition of a substantial disparity. Parts II and III demonstrate that these practical objections to the representativeness principle can be met.

71. The *Partida* decision is a good example of the problems that arise from use of the discrimination theory. The Court held that the substantial disparity established a prima facie case, but because the holding was based on an inference of discrimination, the intent and motives of selection officials were brought into question. The government argued, based on the discrimination theory, that the facts that Mexican-Americans were a "governing majority" in the community and there were many Mexican-Americans in official positions created a presumption that there was no discrimination, which rebutted or negated the presumption created by the substantial disparity inference. The majority rejected this argument, stating that "it would be unwise to presume as a matter of law that human beings of one definable group will not discriminate against other members of the group." 97 S. Ct. 1272, 1282. Justice Marshall, concurring, addressed this issue in detail, citing several studies on the matter, *id.* at 1283–85, and the four dissenters discussed the question at length and adopted the government's position, *id.* at 1285–92. Although there would seem to be no basis for the "governing majority" presumption, and it explains, at most, why one might expect that Mexican-Americans would be fully represented, not why they were grossly underrepresented, surely this theoretical debate of assumptions about racial behavior should be extraneous to the fundamental constitutional question.

72. *See, e.g.,* Arlington Heights v. Metropolitan Hous. Dev. Auth., 97 S. Ct. 555 (1977); Washington v. Davis, 426 U.S. 229 (1976); Dunn v. Blumstein, 405 U.S. 330 (1972). When such access involves a "fundamental right," an infringement is unconstitutional unless it is justified by a "compelling governmental interest" and there are no reasonable alternative methods for implementing the government's interest which do not infringe upon the right. *Id.;* Shapiro v. Thompson, 394 U.S. 618 (1969); Harper v. Virginia State Bd. of Elections, 383 U.S. 663 (1966); DeGregory v. Attorney Gen., 383 U.S. 825 (1966); NAACP v. Alabama, 377 U.S. 228 (1964); Shelton v. Tucker, 364 U.S. 479 (1960). The Court has held that both the rights of citizens to access or participation, Turner v. Fouche, 396 U.S. 346 (1970); Carter v. Jury Comm'n, 396 U.S. 320 (1970); and the rights of litigants to representativeness, Taylor v. Louisiana, 419 U.S. 522 (1975), are fundamental, but the Court has never explicitly analyzed the jury cases in terms of the compelling governmental interest standard. This analysis would also lead to the conclusion that a substantial disparity creates a prima facie case that cannot be rebutted by proof of a nondiscriminatory intent or purpose or administrative needs.

73. See text accompanying notes 41–49 *supra.*

DETERMINING REPRESENTATIVENESS AND INCLUSIVENESS

Standards of Representativeness

Whether derived directly from the representativeness principle or from the rule of exclusion, the guiding concept of cross-sectionality[74] is, by its nature, mathematically based. Concrete, appropriate standards must therefore reflect mathematical as well as legal principles. This does not mean that mathematics will yield one correct standard or replace legal analysis; there are a variety of mathematical formulations, and each can and should be evaluated by legal as well as mathematical principles. Indeed, mathematics can only formulate and translate into quantitative terms the guiding principles and assumptions determined by legal principles and analysis.

Establishment of concrete standards of representativeness requires that two distinct questions be resolved. First, a measure of representativeness must be adopted. Then, using that measure, a maximum allowable deviation must be established.

Four methods of measurement have most frequently been proposed.[75] The equations for these standards and various mathematical relationships are presented in the footnotes.[76] For the purpose of discussion and evaluation, each standard will be applied to a hypothetical jurisdiction[77] in which the source is the voter registration list and 30% of the 18 and over population is black, 20% of the voter registration list is black, 70% of the 18 and over whites are registered to vote, and 41% of the 18 and over blacks are registered.[78]

1. Absolute Disparity Standard

This standard measures representativeness by the difference between the proportion of the population and the source or pool that is in the underrepresented

---

74. We have referred to this concept herein interchangeably as representativeness or cross-sectionality, although this analysis concerns the substantial disparity rules whether based on the representativeness or discrimination principle.

75. See generally JURY SELECTION PROCEDURES, supra note 4, at 95–98; Foster Appendix, supra note 1; Juror Selection, supra note 1, at 785–97; Implementation in the Fifth Circuit, supra note 27; WORKS, supra note 27; Finkelstein, The Application of Statistical Decision Theory to Jury Discrimination Cases, 80 HARV. L. REV. 338 (1966).

76. The following notation has been used in presenting the relevant equations:

R_1, R_2, etc. are used for the various standards of representativeness,

P = proportion of the population in the underrepresented category,

L = proportion of the source or pool in the underrepresented category,

A = proportion of the people in the overrepresented category on the source or pool, and

C = proportion of the people in the underrepresented category on the source or pool.

77. The data used in this hypothetical is fairly typical of some large urban areas. The four variables are interrelated as follows:

$$\frac{A}{C} = \frac{P(1-L)}{L(1-P)}.$$

Once three of these variables are specified, the fourth is mathematically determined. Thus, in the hypothetical, once it is specified that P = .3, L = .2, and A = .7, C must be .41.

78. The focus of this article is source lists, but the discussion and evaluation of these standards applies as well to jury pools.

category.[79] Thus, in the hypothetical, the absolute disparity for blacks is 30% minus 20%, or 10%. The absolute disparity standard has been formally suggested with a maximum allowable disparity of 10–15%;[80] it has been used, without discussion or any specific maximum, by most courts.[81]

2. Comparative Disparity Standard

An elementary mathematical statement of the cross-sectional legal principle is that in a fair, cross-sectional system, the probability of any eligible person being included in the source (or in the final pool) would be the same for every eligible person, regardless of race, ethnic background, sex, age, or socio-economic status.[82] The comparative disparity standard measures representativeness by the percentage by which the probability of serving is reduced for people in a particular category or cognizable class. This percentage is determined by the following calculation:

$$\frac{\text{Proportion of the population that is in the specified category} - \text{Proportion of the source that is in the specified category}}{\text{Proportion of the population that is in the specified category}} \times 100. ^{[83]}$$

The comparative disparity is the same as the absolute disparity divided by the proportion of the population that is in the specified category. In the hypothetical, the comparative disparity is

$$\frac{30\% - 20\%}{30\%} \times 100 = 33\%.$$

79. The absolute disparity, R_1, is defined as

$$R_1 = P - L$$

80. Henry D. Moore, Professor of Economics and Director of the Center for Business and Economics Research, University of Alabama, has made the suggestion of 10%. WORKS, *supra* note 27, at 108.

81. *See, e.g.,* Turner v. Fouche, 396 U.S. 346 (1970); Swain v. Alabama, 380 U.S. 202 (1965); Smith v. Yeager, 465 F.2d 272 (3d Cir. 1972); Black v. Curb, 464 F.2d 165 (5th Cir. 1972); Sanford v. Hutto, 394 F. Supp. 1278 (E.D. Ark.), *aff'd.,* 523 F.2d 1383 (8th Cir. 1975). An absolute disparity of 10% was referred to in *Swain* as a minimal showing to support a challenge, 380 U.S. at 208–09. Later decisions of the Court do not apply or mention this requirement. See note 181 *infra.*

82. This probability would be

$$\frac{1}{(\text{eligible population})}$$

83. The comparative disparity, R_2, is defined as

$$R_2 = \frac{P - L}{P}.$$

This means that an eligible black person has 33%, or one-third, less chance of being included than the average person.[84]

The comparative disparity or reduced probability of serving standard has been used, without a specified maximum, by several courts.[85] The courts using this measure of representativeness have not discussed their basis for adopting it and appear to have done so intuitively.

3. Proportion of Eligibles Standard

The proportion of eligibles standard is calculated as follows:

$$\frac{\begin{array}{c}\text{Proportion of the eligibles in}\\ \text{the overrepresented category included}\end{array} - \begin{array}{c}\text{Proportion of eligibles in}\\ \text{the underrepresented category included}\end{array}}{\begin{array}{c}\text{Proportion of the eligibles in the overrepresented}\\ \text{category included}\end{array}} \times 100.\ [86]$$

In the hypothetical, this is

$$\frac{70\% - 41\%}{70\%} \times 100 = 41.3\%.$$

This standard has been used by one court,[87] and the U.S. Civil Rights Commission has recommended its use with an allowable maximum of 20%.[88]

The comparative disparity and proportion of eligibles standards are directly related.[89] Because several courts have used the comparative disparity standard and

---

84. The relationship between the comparative disparity for a group (R_2) and the total proportion of the group not included (D) is

$$(1 - D) = \frac{M}{N}(1 - R_2),$$

where M is the number of people in the source or pool, N is the number of people in the population, and M/N is what we have defined as the inclusiveness of the source or pool.

85. *E.g.,* Alexander v. Louisiana, 405 U.S. 625, 629–30 (1972) (referring to both the absolute and comparative disparities without comment); United States v. Goff, 509 F.2d 825 (5th Cir. 1975), *cert. denied,* 423 U.S. 857 (1975); Stephens v. Cox, 449 F.2d 657 (4th Cir. 1971); Quadra v. Superior Court, 403 F. Supp. 486 (N.D. Cal. 1975); Ford v. Hollowell, 385 F. Supp. 1392 (N.D. Miss. 1974).

86. The proportion of eligibles, R_3, is defined as

$$R_3 = \frac{A - C}{A} = \frac{P - L}{P(1 - L)}.$$

87. United States v. McDaniels, 370 F. Supp. 298 (E.D. La. 1973), *aff'd,* 509 F.2d 825, *cert. denied,* 423 U.S. 857 (1975).

88. Staff Memorandum, Office of General Counsel, U.S. Commission on Civil Rights, *Assuring a Fair Racial Cross Section in the Selection of Jurors Under the Jury Selection and Service Act of 1968* (1969), at 3, cited in WORKS, *supra* note 27, at 105 n. 12.

89. The proportion of eligibles standard is equal to the comparative disparity divided by the proportion of the source that is in the overrepresented category, or

$$R_3 = \frac{R_2}{(1 - L)}.$$

found it intutitively understandable and applicable, and because the comparative disparity is a direct measure of the reduced probability of serving, the proportion of eligibles standard is not further considered in this analysis.

4. Statistical Significance Test

The statistical significance test measures representativeness by calculating the probability of the disparity occurring by chance in a random drawing from the population.[90] If that probability is very low, the conclusion is drawn that the disparity is not the result of chance but results from bias or discrimination. The cutoff probability used in most industrial and scientific applications is 5%.[91]

The statistical significance test depends on the number of people in the sample drawn from the source or pool as well as its proportional makeup, and it has been criticized in the scientific literature on this ground.[92] In the hypothetical, with a sample size of 500, the probability or odds of obtaining a sample of 500 that is 20% black by picking randomly from a population that is 30% black is less than 1 out of 1,000,000.[93] Since a probability of less than 1 out of 1,000,000 is less than 5% (or 1 out of 20), the conclusion is drawn that the selection system discriminates against blacks.

The statistical significance test has been adopted by the courts in employment discrimination cases[94] and has been suggested as a standard for representativeness of jury sources and pools.[95] Although no court has adopted it in the jury selection

---

If the overrepresented group is a very large proportion of the source on pool, the two standards are almost equal. However, if the overrepresented group is one-half of the source or pool, the proportion of eligibles would be double the comparative disparity.

90. The statistical significance test probability, R_4, is determined from a normal distribution table from z, where

$$z = \frac{(P - L)\sqrt{n}}{\sqrt{P(1-P)}}.$$

n is the size of the sample from the source or pool.

91. M. DeGroot, Probability and Statistics 380 (1975).

92. *See, e.g.,* D. Morrison & R. Henkel, The Significance Test Controversy: A Reader (1970); H. Raiffa & R. Schlaifer, Applied Statistical Decision Theory vii (1961) ("In most statistical practice, consequences and performance characteristics receive mere lip service while decisions are actually made by treating the numbers .05 and .95 with the same superstitious awe that is usually reserved for the number 13."); M. DeGroot, Probability and Statistics 380–81 (1975); Kadane, Book Review, 68 J. Am. Statistical Assoc. 1025 (1973).

93.

$$z = \frac{(.3 - .2)\sqrt{500}}{\sqrt{.3(1-.3)}} = 4.88$$

Using National Bureau of Standards, The Tables of Probability Functions, Applied Mathematics Series, No. 23 (1953), this translates into a probability of 5.5×10^{-7}, or less than one in a million.

94. *E.g.,* Pennsylvania v. O'Neill, 473 F.2d 1029 (3d Cir. 1973); Chance v. Board of Examiners, 330 F. Supp. 203 (S.D.N.Y. 1971), *aff'd.,* 458 F.2d 1167 (2d Cir. 1972).

95. Finkelstein, *The Application of Statistical Decision Theory to Jury Discrimination Cases,* 80 Harv. L. Rev. 338 (1966); *Juror Selection, supra* note 1, at 785–97; Report of B.R. Stauber to the Federal Judicial Center (1973), discussed in Foster Appendix, *supra* note 1, at 818.

context, the Supreme Court has referred to the probability of obtaining disparities by chance.[96]

Selecting a Standard of Representativeness

Careful analysis of the proposed standards from both a mathematical and a legal standpoint shows that the comparative disparity method is clearly superior to the others.

The absolute disparity standard is objectionable on both legal and mathematical grounds, because it fails to account for the range at which the disparity occurs.[97] For example, an absolute disparity of 10% in a jurisdiction that is 30% black is quite different from the same absolute disparity in a jurisdiction that is 11% black. In the jurisdiction that is 30% black, a 10% absolute disparity means that the eligible black person has 33% less chance of serving than the average eligible person, while in the jurisdiction that is 11% black, the same absolute disparity means that the eligible black person has 91% less chance of serving. This difference is, of course, legally significant as well: in the 11% black jurisdiction, the 10% absolute disparity amounts to almost total exclusion of black people.[98] The absolute disparity standard yields the same result in situations in which the results clearly should differ.[99]

The statistical significance test, while it accounts for the range of the disparity, involves complicated calculations resulting in answers that are difficult to visualize and to evaluate.[100] Moreover, since the calculated probability is greatly reduced by increasing the size of the sample from the source or pool, the result is significantly

96. Castaneda v. Partida, 97 S. Ct. 1272, 1281 n.17 (1977); Alexander v. Louisiana, 405 U.S. 625, 630 n.9 (1972); Whitus v. Georgia, 385 U.S. 545, 552 n.2 (1967). Every court that has been presented with evidence concerning this test has admitted and considered the evidence. *E.g.,* State v. Little, No. 74 Cr. 4167 (Super. Ct. of Beauford County, N.C. 1975); People v. Attica Brothers, 79 Misc. 2d 492, 359 N.Y.S.2d 699 (1974); United States v. Saxe, Crim. No. 75-236 U.S.D.C. (E.D. Pa. 1976); United States v. Briggs, 366 F. Supp. 1365 (N.D. Fla. 1973).

97. *See Juror Selection, supra* note 1, at 786.

98. One court has noted this deficiency of the absolute disparity standard:

[The comparative disparity] is useful because the importance of a difference of a given amount, for example, 10%, varies depending upon the magnitude of the group's representation in the population. [An absolute] disparity of 10% constitutes a far more significant underrepresentation when the group comprises 12% of the population but only 2% of the grand jury (-83% [comparative disparity]) than when the group comprises 60% of the population but only 50% of the grand jury (-20% [comparative disparity]) [*sic;* this should be -17%].

Quadra v. Superior Court, 403 F. Supp. 486, 495 n.9 (N.D. Cal. 1975).

99. With the suggested 10% absolute disparity standard, an 11% black population and a source or pool that is 1% black would not be impermissible.

100. See cases cited note 96 *supra.* For example, the three probabilities referred to by the Supreme Court were 1 out of 10^{140} in *Partida,* 1 out of 20,000 in *Alexander,* and .000006 (6 out of 1,000,000) in *Whitus.* It is clear that all of these probabilities are quite small, and that provides useful information, but evaluation and differentiation between them is difficult. The choice of a cutoff point in terms of this probability would be difficult to evaluate or visualize.

affected by the choice of sample size.[101] Even with moderate sample sizes, small disparities result in very low probabilities.[102]

The comparative disparity standard is not subject to the difficulties of the other two standards. Because it measures the reduced probability of serving for prospective jurors in a particular category, its results are not affected by the size of the sample or the proportion of the population in the specified category. Moreover, the comparative disparity standard can be easily calculated, readily understood, and consistently interpreted.[103]

---

101. The statistical significance test calculates the probability of randomly drawing a sample of size n that is x% (or less) black from a source or pool that is y% black, not the probability of obtaining the disparity. Thus, in the hypothetical, it calculates the probability of randomly drawing n people, 20% of whom are black, from a source or pool that is 30% black (if that probability is very low, the hypothesis that the source or pool is 30% black is rejected). As n is increased, it is less likely that a random drawing would yield only 20% blacks. For example, if n = 50, the probability is about 6 out of 100; if n = 500, the probability is about 1 out of 1,000,000; and if n = 5,000, the probability is about 8 out of 10^{54}. If all of the source or pool is sampled, the probability of obtaining 20% blacks (or any percentage other than 30%) is zero. In this sense, the significance test tells us that there is a disparity without really providing any information concerning how big or substantial the disparity is and suffers from the contradiction that its results can be less revealing the more information (from a larger sample) we have.

The problem of sample size is not as acute in employment discrimination cases. There, tests of significance have been used to assess whether a group of individuals, such as those hired by a particular company, represent a random sample of people qualified for employment. If the probability of obtaining the composition actually hired by the company is very low, the conclusion drawn that the company discriminated. See authorities cited note 94 *supra*. The dependence of this method on the sample size is less important because the sample size is the number of people hired and is typically small. This use of significance tests is inappropriate for assessing jury sources or pools. The source is not the result of a drawing from the population but a list or combination of lists compiled for other purposes and by other means, and the pool is the result of a drawing from the source. Moreover, if the composition of the source or pool is known exactly, as it must be for this type of significance test, the relevant question is whether the source or pool provides an adequate cross section of the community, not whether it is a random sample of the community. A standard of cross-sectionality, like the comparative disparity, is needed to address and resolve this question appropriately.

102. For example, the probability of randomly drawing a sample of 2,000 that is 28% black from a source or pool that is 30% black is .026 (about half of the commonly used 5% cutoff).

103. In using the comparative disparity standard, although the sample size does not affect the result, care should be taken that the sample size is large enough so the comparative disparity is reliably determined. The sample size necessary to insure any given degree of accuracy can be calculated. This is done by first stipulating an error, e, and a probability of that error occurring, p. Then, assuming the comparative disparity, R_2 has a normal distribution (an assumption supported in statistical theory by the central limit theorem), the sample size, n, necessary to have a probability of (1 — p) that R_2 is between R_2 + e and R_2 — e, is calculated as follows:

$$n \geqq \frac{\left[\frac{z}{e}\right]^2 \left[\frac{L(1-L)}{P^2}\right]}{1 + \frac{\left[\frac{z}{e}\right]^2 \left[\frac{L(1-L)}{P^2}\right]}{M}}.$$

where,

L = proportion of the source or pool with the specified characteristic,
P = proportion of the population with the specified characteristic,
M = size of the source or pool, and
z = a constant calculated from the normal distribution and dependent on p (specifically, it is the (1 — p/2) × 100 percentile of the standard normal distribution).

Since L is not known before the sample is taken, it must be estimated based on an initial sample. An initial sample of about 100 will provide a sufficiently accurate estimate of L. The following table presents some sample calculations of n for an example where $\frac{L(1-L)}{P^2} = 1$ and p = .10:

Judge Walter P. Gewin of the Fifth Circuit Court of Appeals, who has written extensively on this matter,[104] recognizes the deficiencies of the absolute disparity standard, the problems involved with use of statistical significance tests, and the advisability of the comparative disparity standard.[105] He does not, however, believe that the comparative disparity standard should be applied where the cognizable class is a very small percentage of the population.[106] For example, where blacks are 4% of the population and 2% of the source or pool, the comparative disparity is 50%. This is a large comparative disparity, although the resulting number of blacks who will not appear on jury panels because of this disparity is small. On a panel of 50 jurors, there should be 4%, or 2, blacks, but because of the disparity, there is 2%, or 1, black. Since the disparity results in a difference of only one black person out of 50 on the panel, this argument goes, it is insignificant and harmless.[107] Judge Gewin would apply the comparative disparity standard except when the cognizable class is a very small percentage of the population, in which case the absolute disparity standard would be applied.[108]

This suggestion would have the effect of applying the absolute disparity standard in the range where it is least revealing or appropriate. Moreover, application of the absolute disparity standard where the cognizable class is small means that almost all underrepresentations of small and medium-sized minorities—even total exclusions—are validated,[109] for by definition, a small minority can never have a large absolute disparity. If the maximum allowable absolute disparity is 10%, total exclusion of a 9% minority is permissible.[110] Even if the Constitution will tolerate discrimination against or underrepresentation of small or medium-sized minorities as opposed to large minorities,[111] the absolute disparity standard is inappropriate.

	M (size of source or pool)		
e (error)	5,000	100,000	∞
±.01	4,220	21,297	27,060
±.015	3,532	10,736	12,027
±.02	2,875	6,337	6,765
±.05	890	1,071	1,082

This means, for example, that with a source or pool of 100,000 and a sample of 21,297 that results in $R_2 = 15\%$, we are 90% sure that R_2 is between 14% and 16%.

104. See Foster Appendix, *supra* note 1; WORKS, *supra* note 27; *Implementation in the Fifth Circuit, supra* note 27.

105. See Foster Appendix, *supra* note 1, at 818–19, 834–45; WORKS, *supra* note 27, at 107–9. *See also* JURY SELECTION PROCEDURES, *supra* note 4, at 92–98 (referring to the comparative disparity as the "rate of error"; "the comparative method is more revealing of the success of a selection scheme in achieving—or failing to achieve—the goal of producing a representative jury").

106. Foster Appendix, *supra* note 1, at 834–45.

107. Even only one minority person on the panel can hardly be considered insignificant, in practical as well as constitutional terms, particularly where one of the litigants is also in that minority. See note 133 *infra*.

108. Foster Appendix, *supra* note 1, at 835. Judge Gewin does not suggest where the line between large and small cognizable classes should be drawn.

109. See JURY SELECTION PROCEDURES, *supra* note 4, at 96–98.

110. Use of the absolute disparity standard for small minorities means that total exclusion of any minority that is smaller than the maximum allowable absolute disparity would be permissible.

111. See text accompanying notes 132–35 *infra*.

A preferable test would be to apply a larger maximum comparative disparity to the small minority.[112]

The comparative disparity standard is a clear, easily understood and used, and conceptually valid measure of representativeness. Its adoption would fill a long standing vacuum and provide the courts with a workable standard for determining representativeness. Adoption of a standard for determining representativeness will not, however, alone insure the rights of litigants or define the duties of selection officials. A maximum allowable underrepresentation, a numerical cutoff beyond which a jury system loses its presumption of validity, must be established. Otherwise, inconsistent application of the comparative disparity standard could further erode the fundamental constitutional (and statutory) principle of representativeness.

The occasions on which establishment of workable standards for safeguarding constitutional rights requires the adoption of a precise numerical cutoff point are not frequent. Yet the Court has established precise numerical standards when defining constitutional rights concerning jury trials,[113] voting requirements[114] and privacy.[115] In each instance, the Court has analyzed the conflicting rights and interests involved and used available expertise to draw as accurate a line as possible. The right to a representative jury embodies a cross-sectional principle that is itself mathematically based, making it impossible to protect that right or define its limits without establishing a maximum allowable underrepresentation.

---

112. A 15% maximum comparative disparity has been advocated here. If the harmless error reasoning is adopted as to small minorities, a 25% maximum comparative disparity could be applied to cognizable classes that are less than 10% of the population. *See* Jury Selection Procedures, *supra* note 4, at 98.

A continuous but not as easily comprehensible measure of representativeness that gives any given comparative disparity less significance the smaller the underrepresented group's proportion of the population is

$$R_s = \frac{L(1-P)}{P(1-L)}.$$

R_s is the ratio of the odds of drawing a person in the underrepresented group from the source or pool, $\frac{L}{(1-L)}$, to the odds of drawing a person in the underrepresented group from the population, $\frac{P}{(1-P)}$.   Also, $R_s = A/C$.

113. In Duncan v. Louisiana, 391 U.S. 145 (1968), the Court ruled that there is a right to a jury trial where the accused is charged with a "serious" as opposed to a "petty" offense. The accused in *Duncan* was charged with assault, which had a two-year maximum sentence. The Court ruled that this was a "serious" offense, but no further guidance was provided as to the line between "serious" and "petty." In Baldwin v. New York, 399 U.S. 66 (1970), the Court adopted the length of the maximum authorized penalty as the controlling criterion in drawing a line between petty and serious for purposes of the sixth amendment, setting six months as the maximum allowable penalty for nonjury trials. In drawing this precise line, the Court examined the practice in the various states, analyzed, as precisely as possible, the constitutional right and state interest involved, and adopted a numerical cutoff. *See also* Bloom v. Illinois, 391 U.S. 194 (1968) (an accused charged with criminal contempt has a right to a jury trial if the sentence imposed is greater than six months); Codispoti v. Pennsylvania, 418 U.S. 506 (1974) (consecutive sentences totalling more than six months for multiple criminal contempts arising from one trial require a right to a jury trial).

114. Residency requirements for voting pose a conflict between the citizen's constitutional right to vote and the state's interest in preparing and maintaining appropriate records to protect against fraud. The Court established a maximum allowable residency requirement of 50 days. Burns v. Forsten, 410 U.S. 686 (1973); Marston v. Lewis, 410 U.S. 679 (1973). *See also* Dunn v. Blumstein, 405 U.S. 330 (1972).

115. In the abortion cases, Roe v. Wade, 410 U.S. 113 (1973), the Court found a conflict between a woman's right to privacy concerning the abortion decision and the state's interest in protecting the health of pregnant women. The Court analyzed these rights and interests, examined the available medical evidence, and concluded that a woman has a right to an abortion during the first trimester of pregnancy.

Determining how large an underrepresentation should be allowed before a prima facie case of invalidity is established involves determining a level at which the government is required to justify the disparity in representation of a cognizable group.[116] For example, if the selected figure were 15%, the government would have to justify the comparative disparity of 33% in the hypothetical. If the selected figure were 35%, no justification need be made.

There is, of course, no legitimate governmental interest in unrepresentativeness, but the ease and costs of increasing representativeness should be considered[117] in setting a standard.[118] The representativeness of sources can be increased by the use of multiple lists, which is the method already provided for in the Federal Act[119] and the Uniform Act.[120] Easy and inexpensive methods for use of multiple lists are presented in this article;[121] they are presently used in several jurisdictions.[122] These methods can provide, at small cost, sources that are within a 5% comparative disparity concerning every cognizable class. The available nationwide data indicates that most systems are within a 10–15% comparative disparity concerning all or most cognizable classes without using multiple lists.[123]

Since the available methodology for increasing representativeness is widely used and not prohibitively expensive, there seems to be no legitimate reason to allow more than a 15% comparative disparity for any cognizable class before the government must at least justify the disparity.[124]

---

116. Under the rule of exclusion, the basis for rebuttal is a showing that the underrepresented group is less eligible or available. The representativeness principle may mean that there is no basis for rebuttal. See notes 68 & 70 *supra*.

117. In Taylor v. Louisiana, the Court said,

...the administrative convenience in dealing with women as a class is insufficient justification for diluting the quality of community judgment represented by the jury in criminal trials.

419 U.S. 522, 535 (1975).

118. The ease and expense involved were seen as important considerations by the framers of the Federal Act. Foster Appendix, *supra* note 1, at 816; *Implementation in the Fifth Circuit, supra* note 27, at 383. Judge Gewin, who was a member of the Committee on the Operation of the Jury System of the Judicial Conference of the United States, has said:

[S]ubstantiality with regard to relatively small disparities had to be defined in terms of the practicalities of the circumstances. Thus a disparity which could be eliminated easily and effectively would tend to appear more substantial; and, conversely, a disparity which would be difficult to eliminate would tend to appear less substantial. This working definition of substantiality would, of course, have to be applied so as to place any error on the side of safety.

Id. (footnotes omitted).

119. 28 U.S.C. § 1863(b)(2) (1968).

120. UNIFORM ACT, *supra* note 9, § 5.

121. See text accompanying notes 218–49 *infra*.

122. See Table 5 *infra*.

123. See Table 3 *infra*. For example, the nationwide data indicates that there was a comparative disparity concerning black people of 11.7% in 1974 and 9.4% in 1972.

124. The Civil Rights Commission has suggested a 20% maximum in terms of the proportion of eligibles standard. WORKS, *supra* note 24, at 105 n.12. This is directly related to the comparative disparity standard, and the values are close where the proportion of the source or pool in the overrepresented group is large. See note 89 *supra*. In JURY SELECTION PROCEDURES, *supra* note 4, at 98, a maximum allowable comparative disparity of 20% is recommended based on the Civil Rights Commission suggestion (without noting the difference between the comparative disparity and proportion of eligibles standards).

"The essential if not wholly satisfactory [task] of determining the line," Baldwin v. New York, 399 U.S. 66, 68 (1970), involves a value judgment and a balancing of interests that cannot be accomplished by mathematics. We have placed a high value on representativeness while allowing leeway for administrative feasibility. Selection systems within a 15% comparative disparity as to all cognizable classes are not difficult or expensive to accomplish.

The "Substantial Impact" Test: A Source of Judicial Confusion

Recently, a new rule, the "substantial impact" test, has been adopted by the Tenth Circuit in *United States v. Test*,[125] the Second Circuit in *United States v. Jenkins*,[126] and the Fifth Circuit in *United States v. Goff*.[127] In *Jenkins*, blacks were approximately 5.5% of the voting age population and 3.3% of the voter registration list.[128] The court referred to both the absolute disparity, 2.2%, and the comparative disparity, 40%, which the court noted was "substantial indeed."[129] Nevertheless, the court held that this disparity was insubstantial in terms of the number of blacks it eliminated from a panel of 60 jurors. Thus, if there were 5.5% blacks instead of 3.3%, a panel of 60 jurors would have 2.2% of 60, or 1.32, additional black jurors. The court then held that "a difference of one (1) Negro in a panel of 60 jurors is not substantial," and ruled, on this basis, that a prima facie case had not been established.[130] The same calculations were made in *Goff* (concerning a minority that constituted over a quarter of the population) and *Test*, with similar results and the same conclusion. The data concerning all three cases are presented in Table 1.

Although the "substantial impact" rule purports to shed some light beyond the substantiality of the absolute and comparative disparities, it is in fact just another way of expressing the absolute disparity. The "impact" is just the absolute disparity times the size of the panel. Thus, if a panel has 60 jurors, an absolute disparity of 10% is the same as an "impact" of 6, an absolute disparity of 2.2% is the same as an "impact" of 1.32, and so on. And since it adds another variable, the size of the panel, it confuses rather than enlightens. For example, the *Jenkins* court found insubstantial an "impact" of one juror on a panel of 60 and the *Goff* court found insubstantial an "impact" of one juror on a grand jury of 23. The smaller the panel, the smaller the "impact" of a given absolute disparity.[131] The "substantial impact" rule adds nothing to the absolute disparity standard and is subject to the same mathematical and legal criticisms.[132] It further confuses an already confused area of the law by introducing another variable, the panel size.

The absolute disparity standard and the substantial impact rule also raise serious constitutional questions. The unavoidable effect of both is to validate discrimination against small and medium-sized minorities. The use of voter registration lists resulted in a reduced probability of serving of 33% for eligible Chicanos in *Test*, 40% for eligible blacks in *Jenkins*, and 20% for eligible blacks in *Goff*—and yet all three courts failed to require the government to explain or justify these serious underrepresentations. None of these opinions refers to any authority for the proposition that the Constitution only prohibits underrepresentation of or discrimination

---

125. 550 F.2d 577 (10th Cir. 1976).

126. 496 F.2d 57 (2d Cir. 1974).

127. 509 F.2d 825 (5th Cir. 1975).

128. 496 F.2d 57, 64 (2d Cir. 1974).

129. *Id.* at 64–65. The court did not calculate the comparative disparity, 40%, but referred to a "ratio of roughly 5 to 3," which is another way of saying the same thing.

130. *Id.* at 66.

131. The maximum "impact" for any underrepresented group is the group's proportion of the population times the panel size.

132. See text accompanying notes 97–99 *supra*.

TABLE I. Disparities in *Jenkins, Test* and *Goff* Cases

Case and group underrepresented	Proportion of population (%)	Proportion of voter registration list (source) (%)	Absolute disparity (%)	"Impact"	Comparative disparity of source (%)
Jenkins (blacks)	5.5	3.3	2.2	1 (of 60)	40.0
Test (Chicanos)	10.4	6.9	3.5	2 (of 50)	33.7
Goff (blacks)	26.3	21.0	5.3	1 (of 23)	20.2

against large as opposed to small or medium-sized minorities, or that the exclusion of only one minority juror is insignificant.[133]

On the contrary, discrimination against and underrepresentation of all cognizable classes—which surely include blacks and Chicanos, no matter what proportion of the population they may be—have always been the constitutional focus of the jury selection cases.[134] If any special rule is to be formulated for small minorities, our history and our constitutional tradition would seem to require that discrimination against them calls for a "more searching judicial inquiry,"[135] not a less stringent standard.

Standards of Inclusiveness

A source list can be fully representative of all cognizable classes but comprise only a small portion of the eligible population in a community.[136] For example, in a county of 1,000 eligible people of whom 25% are black, a source list of 100 people, 25 of whom are black, would be fully representative of blacks but only 10% inclusive, thereby excluding 90% of the eligible population. The principles underlying the jury selection cases require that source lists[137] be sufficiently inclusive that a significant proportion of the eligible population is not excluded, although the courts have not explicitly recognized an inclusiveness requirement.[138]

---

133. Even only one minority person on the panel can hardly be considered insignificant, in practical as well as constitutional terms, particularly if one of the litigants is a member of a minority or if racial or ethnic issues are involved in the case. The presence of one minority person can affect the tenor of a case and counsel's approach to the issues. *See* JURY SELECTION PROCEDURES, *supra* note 4, at 33; Note, *The Case for Black Juries,* 79 YALE L.J. 531 (1970). Although one person can be removed with a peremptory challenge, that in itself can affect the other jurors, and the number of peremptory challenges is small and the use of each one significant. In this regard, the Supreme Court has recently ruled that the exclusion of "merely one" juror in violation of Witherspoon v. Illinois, 391 U.S. 510 (1969) ("death-qualified" jury unconstitutional) requires a new trial. Davis v. Georgia, 97 S. Ct. 399 (1976).

134. *E.g.,* Taylor v. Louisiana, 419 U.S. 522 (1975); Hernandez v. Texas, 347 U.S. 475 (1954); Thiel v. Southern Pacific Co., 328 U.S. 217 (1946).

135. Justice Stone, in United States v. Carolene Products Co., 304 U.S. 144, 152 n.4 (1938), said:

Discrete and insular minorities may be a special condition, which tends seriously to curtail the operation of those political processes ordinarily to be relied upon to protect minorities, and which may call for a correspondingly more searching judicial inquiry.

136. The converse is not true; a source that is fully inclusive is, by definition, fully representative.

137. Unlike the standards of representativeness, see text accompanying notes 74–96 *supra,* the inclusiveness requirement and the standards of inclusiveness discussed here are applicable only to the source and not to the pool.

138. *But see* Broadway v. Culpepper, 439 F.2d 1253 (5th Cir. 1971); Simmons v. Jones, 317 F. Supp. 397 (S.D. Ga. 1970); United States v. Hunt, 265 F. Supp. 178 (W.D. Tex. 1967).

A substantially underinclusive source, though sufficiently representative as to every cognizable class, can be in contradiction to the rights of citizens and litigants and to the societal interests in the legitimacy, integrity and impartiality of the judicial process. For example, the list of real property owners in a jurisdiction may be sufficiently representative but include less than half of the eligible people. The use of such a source list compromises the societal interest in broad based citizen participation.[139] Moreover, the people on such a source list may well have considerably different values, attitudes and experience from the rest of the eligible population, which would affect the rights of litigants and undermine the impartiality, legitimacy and integrity of the judicial process.[140] One court has suggested that a source list that contains only half of the eligible people in a community may be unconstitutional.[141]

If some degree of inclusiveness is deemed to be constitutionally required, the courts will have to establish standards. Experience with multiple list systems has shown that use of one or two well chosen supplemental lists can double the number of names on voter registration lists.[142] In the context of easy and inexpensive methods for obtaining 95% inclusiveness,[143] perhaps 80% should be required.[144]

APPLICATION OF CONSTITUTIONAL AND STATUTORY PRINCIPLES TO SOURCE LISTS

Representativeness and Inclusiveness of Voter Registration and Other Single Source Lists

Analysis of the evidence presented in recent challenges to jury systems reveals that most of the proven unrepresentativeness is attributable to the unrepresentativeness of the source list.[145]

---

139. See text accompanying notes 41–49 *supra.*

140. In Thiel v. Southern Pacific Co., 328 U.S. 217, 227 (1946) the Court said:

   [T]he broad representative character of the jury should be maintained, partly as assurance of diffused impartiality and partly because sharing in the administration of justice is a phase of civic responsibility,

quoted in Taylor v. Louisiana, 419 U.S. 522, 530 (1974). In White v. Crook, 251 F. Supp. 401, 408 (M.D. Ala. 1966), the court said:

   Jury service on the part of the citizens of the United States is considered under our law in this country as one of the basic rights and obligations of citizenship [and] a form of participation in the processes of government, a responsibility and a right that should be shared by all citizens....

See also UNIFORM ACT, *supra* note 9 and text accompanying notes 210–17 *infra,* which suggests the need for inclusiveness as well as representativeness, and Modified Plan for the United States District Court for the District of Columbia for the Random Selection of Grand and Petit Jurors (March 16, 1976), at 3, which recognizes the need to make a "greater number of citizens...eligible."

141. United States v. Hunt, 265 F. Supp. 178 (W.D. Tex. 1967).

142. See text accompanying notes 236–49 *infra; see* JURY SELECTION PROCEDURES, *supra* note 4, at 103–4. After Colorado adopted the Uniform Act, *supra* note 9, which makes use of multiple lists mandatory, "most Colorado counties had almost twice as many names for potential jurors as they had from the voters' list alone." *Id.* at 103. In North Dakota, which also adopted the Uniform Act, use of the lists of licensed drivers with actual voters' lists increased the sources by from 80% to 100%. *Id.* at 104.

143. See text accompanying notes 217–49 *infra.*

144. An additional standard, analogous to the representativeness standard based on the systematic exclusion principle, discussed *infra,* could provide that a prima facie case of underinclusiveness is established by proof that a significantly more inclusive (perhaps 10%) list or combination of lists is available with reasonable costs and effort.

145. See note 4 *supra.*

Table 2 summarizes the statistical evidence presented in four typical challenges.[146] For each court and category listed, the first three columns present the specified category's percentage of the population, source list and final pool, respectively. The fourth column presents the comparative disparity of the final pool, and the fifth column presents the percentage of the comparative disparity of the final pool that is attributable to the source. For example, in the challenge to the jury system in the United States District Court for the Northern District of Florida, evidence indicated that 22.8% of the population was black while 16.0% of the final pool was black. Since the source was only 16.3% black, 95.6% of the total resulting comparative disparity was attributable to the source and only 4.4% to the process.[147]

Except for the underrepresentation of women in the challenge to the jury system in the Supreme Court of New York, Erie County, which officials admitted resulted from intentional discrimination against women in choosing names from the source, all or most of the unrepresentativeness proved in these cases was attributable to the unrepresentativeness of the sources utilized.

The overwhelming majority of jury systems throughout the country use voter registration lists as the source,[148] although all available studies and data indicate that voter registration lists do not represent a cross section of our communities and are substantially underinclusive.[149] The Census Bureau does a thorough study of voting and registration after each election. Since 1960, these studies have reached the same conclusions concerning the representativeness and inclusiveness of voter registration lists. The study of the 1972 election stated:

> [H]igher levels of registration and voting were associated with persons who were male, white, those in the middle age group (35–64), those persons with at least a high school diploma, those in families with incomes greater than $10,000, and those in white collar occupations. Conversely, females, Negroes, persons of Spanish ethnic origin, the youngest (18–34) and oldest age groups (65 or older), those who did not complete elementary school education, those in families with incomes less than $5,000, and those in unskilled occupations, such as laborers and private household workers, were less likely to be registered and vote.[150]

---

146. One of the authors, David Kairys, has been counsel in several challenges, including the second and fourth challenges listed in Table 2 and the Philadelphia state court challenge discussed in *Juror Selection, supra* note 1, at 789–801, and works with the National Jury Project, which collects data concerning jury challenges. The examples presented in Table 2 are typical of challenges over the last several years. *See also* JURY SELECTION PROCEDURES, *supra* note 4, at 85–106 and Appendices.

147. This was calculated as follows: The disparity between the proportion of the pool and population that is black is 22.8 minus 16.0, or 6.8. The disparity attributable to the source is 22.8 minus 16.3, or 6.5, which is 95.6% of the total disparity. The result is the same whether the absolute or comparative disparities are used.

148. See note 5 *supra*.

149. In addition to the Census Bureau studies discussed in the text, see JURY SELECTION PROCEDURES, *supra* note 4, at 88–93, App. F-I: COMMITTEE FOR THE STUDY OF THE AMERICAN ELECTORATE, NON-VOTER STUDY 1976 (Sept. 5, 1976) (available from the Committee at 421 New Jersey Ave., S.E., Washington, D.C. 20003) [hereinafter cited as NON-VOTER STUDY 1976]; ROSENSTONE AND WOLFINGER, THE EFFECT OF REGISTRATION LAWS ON VOTER TURNOUT (1976) (presented to the 1976 Annual Meeting of the American Political Science Assoc.); W. FLANIGAN & N. ZINGALE, POLITICAL BEHAVIOR OF THE AMERICAN ELECTORATE 9 (1975); Note, *Voter Registration Lists: Do They Represent a Cross-Section of the Community?*, 5 J.L. REF. 385 (1971–72).

150. BUREAU OF THE CENSUS, U.S. DEPT. OF COMMERCE, VOTING AND REGISTRATION IN THE ELECTION OF NOV., 1972, SERIES P-20, No. 253, at 1 (1973). *See also* JURY SELECTION PROCEDURES, *supra* note 4, at 88–90; BUREAU OF THE CENSUS, U.S. DEPT. OF COMMERCE, VOTING AND REGISTRATION IN THE ELECTION OF NOV., 1974, SERIES P-20,

TABLE 2. Proportion of Unrepresentativeness of Various Jury Systems That Is Attributable to Source Lists[151]

Court and Category Underrepresented	% of Population	% of Source	% of Final Pool	Comparative Disparity of Final Pool (%)	% of Comparative Disparity of Final Pool Attributable to Source
1. Superior Court of Beaufort County, North Carolina Race					
Blacks	30.2	18.9	17.0	−43.7	85.6
2. U.S. District Court for the Eastern District of Pennsylvania Age					
Under 30	25.5	17.2	18.6	−27.1	100.0
Under 40	41.8	32.2	33.3	−20.3	100.0
Race					
Nonwhites	15.7	11.7	12.8	−18.5	100.0
3. U.S. District Court for the Northern District of Florida Race					
Blacks	22.8	16.3	16.0	−29.8	95.6
4. Supreme Court of New York, Erie County Race					
Blacks	8.4	6.5	5.1	−39.3	57.6
Age					
21 to 29	20.7	4.5	3.4	−83.6	93.6
Sex					
Women	53.0	53.0	16.7	−70.4	0.0

No. 293 (1976); Voting and Registration in the Election of Nov., 1970, Series P-20, No. 225 (1971); Voting and Registration in the Election of Nov., 1964, Series P-20, No. 143 (1965). The Census Bureau studies are hereinafter cited as Voting and Registration in the Election of [Year].

151. The data presented in the first three columns of Table 2 was submitted to the courts listed in challenges to the jury selection systems in the following cases:

1) State v. Little, No. 74 Cr. 4176 (Super. Ct. of Beaufort County, N.C. 1975). The source was voter registration and tax lists (with no rules or established procedures for using the two lists together; see text accompanying notes 218–49 *infra*). The figure presented for the proportion of the source that is black, 18.9%, was for the voter registration list alone; the figure for the tax list was not known, but testimony established that it was lower than the corresponding figure for the voter registration list. The proportion of underrepresentation of blacks attributable to the source was therefore actually greater than the table indicates. Venue was changed before the challenge was resolved. *See generally* Michael, Mullin, O'Reilly & Rowan, *Challenges to Jury Composition in North Carolina*, 7 N. Carolina Cent. L.J. 1 (1976).

The proportion of people not participating in the election process, a measure of the underinclusiveness of voter registration lists as a source, has been steadily increasing since 1960. In the 1972 presidential election, 27% of those eligible did not register, and 37% did not vote.[152] In the 1974 nonpresidential election, 38% of those eligible did not register and 55% did not vote.[153] In the 1976 presidential election, 33% of those eligible did not register and 41% did not vote.[154]

The data provided by the Census Bureau studies are summarized in Tables 3 and 4. Table 3 presents, from nationwide data for the 1972 and 1973 elections, the underinclusiveness (in columns 1 and 3) and unrepresentativeness (in columns 2 and 4), measured by the comparative disparity standard, of voter registration lists regarding the various categories of the population specified. The proportion of the whole population not appearing on voter registration lists was 27.7% in 1972 and 37.8% in 1974. These large segments of the population have no opportunity to serve as jurors; they are excluded from jury service by the use of voter registration lists as the source.

This large underinclusiveness is not evenly distributed among the various races, age groups, occupations, or income levels in the population. The proportions excluded are considerably higher for nonwhites, people under 40 years old,[155] people with less formal education, people employed in blue collar positions or unemployed, and people with annual incomes of less than $10,000. Columns 2 and 4 indicate that the people in these categories[156] are significantly underrepresented.[157]

Table 4 presents the underinclusiveness data for various states and metropolitan areas from the 1974 election study,[158] which indicate that the inadequacy of voter registration lists as the source is not limited to any particular area but is nationwide.

---

2) United States v. Saxe, Crim. No. 75–236 (E.D. Pa. 1976). The source was voter registration lists. The figures indicate that the process served to make the system more representative. The defendant pled guilty before a decision on the jury challenge was rendered.

3) United States v. Briggs, 366 F. Supp. 1365 (N.D. Fla. 1973). The source was voter registration lists. The challenge was denied; defendants were acquitted, so there was no appeal.

4) People v. Attica Brothers. 79 Misc. 2d 492, 359 N.Y.S.2d 699 (1974). The source was a permanent pool supplemented with voter registration lists. The challenge was granted based on discrimination against women and students, and a new pool was constituted.

152. VOTING AND REGISTRATION IN THE ELECTION OF 1972, *supra* note 151, at 22.

153. VOTING AND REGISTRATION IN THE ELECTION OF 1974, *supra* note 151, at 11.

154. BUREAU OF THE CENSUS, U.S. DEPT. OF COMMERCE, VOTER PARTICIPATION IN THE ELECTION OF NOV., 1976, *Advance Report,* SERIES P-20, No. 304 (Dec., 1976) (based on a preliminary study).

155. Elderly people are also seriously underrepresented. *See* JURY SELECTION PROCEDURES, *supra* note 4, at 35–39.

156. The underrepresentation of people in more than one of these categories is even more severe. For example, young blacks are underrepresented more than either young people or black people. *E.g.,* VOTING AND REGISTRATION IN THE ELECTION OF 1974, *supra* note 151, Table 1.

157. The Census Bureau has noted in its studies that these underrepresentations are actually significantly larger than their studies indicate, due to biases inherent in their procedures that cause an undercount of racial minorities, young people, and low income people. *See* VOTING AND REGISTRATION IN THE ELECTION OF 1972, *supra* note 151, at 7–8; SHRYOCK AND SIEGEL, THE MATERIALS AND METHODS OF DEMOGRAPHY (U.S. Government Printing Office 1973); Siegel, *Estimates of Coverage of the Population by Sex, Race, and Age in the 1970 Census,* 11 DEMOGRAPHY 1 (1974).

158. Data concerning representativeness is not available.

TABLE 3. Representativeness and Inclusiveness of Voter Registration Lists Nationwide[159]

Category	1972 Election		1974 Election	
	% Not on Registration Lists	% Over- or Under-represented	% Not on Registration Lists	% Over- or Under-represented
Whole population	27.7	–	37.8	–
Sex				
Men	26.9	+1.0	37.2	+1.0
Women	28.4	–1.0	38.3	–0.8
Race				
White	26.6	+1.5	36.5	+2.1
Black	34.5	–9.4	45.1	–11.7
Spanish origin	55.6	–38.6	65.1	–43.9
Age				
18 to 20	41.9	–19.6	63.6	–41.5
21 to 24	40.5	–17.7	54.7	–27.2
25 to 29	33.9	–8.6	48.6	–17.4
30 to 34	28.8	–1.5	41.4	–5.8
35 to 44	25.2	+3.5	33.3	+7.2
45 to 54	20.7	+9.7	27.5	+16.6
55 to 64	19.8	+10.9	24.9	+20.7
65 to 74	21.5	+8.6	27.0	+17.4
75 and over	19.3	+11.6	34.8	+4.8
Education				
Years completed				
Elementary				
0 to 4	51.8	–33.3	60.7	–36.8
5 to 7	40.5	–17.7	48.2	–16.7
8	32.0	–5.9	38.9	–1.8
High school				
1 to 3	37.0	–12.9	45.7	–12.7
4	26.0	+2.4	38.1	+0.5
College				
1 to 3	18.3	+13.0	33.1	+7.6
4	12.9	+20.5	25.2	+20.3
5 or more	11.2	+22.8	22.1	+25.2

159. All of the data presented in Tables 3 and 4 is derived from the Census Bureau's Voting and Registration in the Election of 1972 and Voting and Registration in the Election of 1974, *supra* note 151, and the categories listed were established and defined by the Census Bureau. "Percentage Not on Registration Lists" was calculated by subtracting the percentage that is registered, which appears in the Census Bureau studies, from 100%. "Percentage Over- or Underrepresented," on Table 3, is measured by the comparative disparity standard, which reduces to

$$\left[1 - \frac{\text{Proportion of those in the specified category that are registered}}{\text{Proportion of the population that is registered}} \right] \times 100.$$

The proportion of those in the specified category that are registered and the proportion of the population that is registered appear in the Census Bureau studies. Underrepresentations are shown as negative and overrepresentations as positive. *See also* Jury Selection Procedures, *supra* note 4, at App. F-I.

TABLE 3. (Continued)

Category	1972 Election		1974 Election	
	% Not on Registration Lists	% Over- or Under-represented	% Not on Registration Lists	% Over- or Under-represented
Not enrolled in school				
18 to 20	50.3	−31.3	–	–
21 to 24	44.1	−22.7	–	–
Employment status agriculture				
Wage and salary	47.9	−27.9	50.2	−19.9
Self-employed	13.4	+19.8	19.1	+30.1
Nonagriculture industry				
Wage and salary	38.8	−15.4	40.0	−3.5
Self-employed	20.7	+9.7	28.7	+14.6
Government employed	15.6	+16.7	24.0	+22.2
Unemployed	42.7	−20.7	55.7	−28.8
Occupational groupings				
White collar workers	17.6	+14.0	28.9	+14.3
Professional & technical	13.3	+19.9	25.3	+20.1
Managers & administrators	16.9	+ 14.9	25.9	+19.1
Sales	19.8	+10.9	31.2	+10.6
Clerical	21.0	+9.3	32.2	+9.0
Blue collar workers	35.1	−10.2	45.7	−12.7
Craftsmen	29.8	−2.9	40.8	−4.8
Operatives	39.3	−16.0	48.1	−16.6
Transport eqt. operatives	34.0	−8.7	39.7	−3.1
Laborers	40.1	−17.2	51.9	−22.7
Service workers	31.8	−5.7	41.9	−6.6
Private household workers	37.1	−13.0	44.6 (only women)	−10.9
Other	31.0	−4.6	41.7	−6.3
Farm workers	24.0	+5.1	29.6	+13.2
Farmers & farm managers	11.2	+22.8	16.0 (only men)	+35.1
Farm laborers & foremen	43.4	−21.7	43.0	−8.4
Income				
Under $3,000	38.8	−15.4	–	–
$3,000 to $4,999	35.9	−11.3	–	–
Under $5,000	–	–	47.8	−16.1
$5,000 to $7,499	34.3	−9.1	–	–
$7,500 to $9,999	29.1	−1.9	–	–
$5,000 to $9,999	–	–	43.1	−8.5
$10,000 to $14,999	22.3	+7.5	35.3	+4.0
$15,000 and over	15.0	+17.6	–	–
$15,000 to $19,999	–	–	28.6	+14.8
$20,000 to $24,999	–	–	26.4	+18.3
$25,000 and over	–	–	22.4	+24.8

TABLE 4. Inclusiveness of Voter Registration Lists: States and Metropolitan Areas[160]

State	Inclusiveness	% Not on Registration Lists	Metropolitan Area	Inclusiveness	% Not on Registration Lists
California	58.7	41.3	Atlanta, Ga.	62.3	37.7
Florida	56.1	43.9	Baltimore, Md.	61.8	38.2
Georgia	61.2	38.8	Boston, Mass.	68.7	31.3
Illinois	66.9	33.1	Chicago, Ill.	65.0	35.0
Indiana	69.6	30.4	Cleveland, Ohio	66.2	33.8
Massachusetts	69.2	30.8	Dallas, Texas	52.5	47.5
Michigan	63.1	36.9	Denver, Colo.	67.2	32.8
Missouri	63.9	36.1	Detroit, Mich.	65.0	35.0
New Jersey	61.9	38.1	Houston, Texas	57.0	43.0
New York	57.8	42.2	Kansas City,		
North Carolina	55.9	44.1	Kan./Mo.	63.3	36.7
Ohio	60.6	39.4	Los Angeles, Ca.	58.4	41.6
Pennsylvania	70.6	29.4	Miami, Fla.	49.9	50.1
Texas	56.6	43.4	Milwaukee, Wis.	64.7	35.3
Virginia	54.0	46.0	Minneapolis-St. Paul, Minn.	81.3	18.7
			Newark, N.J.	64.0	36.0
			New York, N.Y.	51.1	48.9
			Philadelphia, Pa.	67.7	32.3
			St. Louis, Mo.	64.3	35.7
			San Francisco, Ca.	60.4	39.6
			Washington, D.C.	54.8	45.2

Thus, voter registration lists are substantially unrepresentative and underinclusive. It has been suggested, however, that registration or actual voter lists are the most appropriate source because they include the "most concerned," "most competent," or "best" citizens. This argument reflects subjective, unsupported judgments about voters and nonvoters and a theoretical stance that directly contradicts basic constitutional and democratic principles.

There is no factual support for the notion that the almost one-half of our people who did not vote (or the third not registered) in the 1976 election are unconcerned, incompetent,[161] or unintelligent.[162] Moreover, the search for the "best" jurors,

160. *See* note 159, *supra.*

161. There is substantial evidence that the expert, "blue ribbon" or elite jury in fact yields no more quality or integrity of judgment than the representative jury. *See* JURY SELECTION PROCEDURES, *supra* note 4, at 15–19, 88–93; Note, *The Case for Black Juries,* 79 YALE L.J. 531 (1970); *Implementation in the Fifth Circuit, supra,* note 27, at 349–50; S. REP. No. 891, 90th Cong., 1st Sess., 18–23 (1967).

162. Rather, the available evidence indicates that nonvoters view elections, the electoral or political process, the issues presented or ignored, and/or the choice presented by the candidates and major parties differently than voters. NON-VOTER STUDY 1976, *supra* note 150, at 3, concluded, based on a survey of nonvoters, that nonvoters have a "distrust of, and disaffection from major political and economic institutions, political leadership, and the media," not a lack of interest or concern. *See also* VOTING AND REGISTRATION IN THE ELECTION OF 1972, *supra* note 151, at 6;

whether one believes them to be on voter registration lists or elsewhere, necessarily involves subjective judgments and intentional discrimination.[163] It is not a legitimate basis for constructing or evaluating a jury system.[164] From a societal and constitutional perspective, subjective notions of quality have no place in a jury selection system, and the best jury system is one that is representative of a cross section of the community.

Implementation in the Fifth Circuit, supra note 27, at 350. The U.S. Commission on Civil Rights has concluded that people vote when and if the issues or candidates interest or concern them, citing the following example:

> In Alabama, the Negro turnout for the May 3, 1966 primary was estimated at 74% of the total Negro registration of just under 250,000; in the general election, faced with a choice between two segregationists who were the major candidates in the Governor's race, less than half the registered Negroes voted.

Id. at 350 n.85. Registration procedures also serve as a significant deterrent to registration and voting. *See* ROSENSTONE AND WOLFINGER, THE EFFECTS OF REGISTRATION LAWS ON VOTER TURNOUT (1976) (presented to the 1976 Annual Meeting of the American Political Science Assoc.); JURY SELECTION PROCEDURES, *supra* note 4, at 91–93.

One aspect of our electoral system, when compared to the systems in other countries, helps explain the lack of voter interest more than any negative judgments about nonvoters. Ours is a two-party, winner-take-all system, where compromises must be made and coalitions built in advance of elections since the loser, no matter how large his or her minority is, loses all. One result of this is a tendency toward moderation and a hesitance to address questions of principle on the part of our major candidates and parties. In a parliamentary or proportional, multi-party system, each party and candidate has power in proportion to the votes they receive. Since compromises and coalitions are made after the election, voters are provided with wider issue-oriented and ideological choices. It should not be surprising that voter turnout is significantly higher in countries with proportional, multi-party systems than in the United States. For example, in recent elections, the voter turnout was 94% in Australia and 80% in England. JURY SELECTION PROCEDURES, *supra* note 4 at 91–92. That voter turnout is directly related to this structural difference was dramatically demonstrated in Switzerland in 1919, when a change to proportional representation doubled the voter turnout in many districts (from 40% to 80%). *See* S. LIPSET, THE FIRST NEW NATION 310 (1963).

163. Selection systems that provide officials broad discretion to select jurors have consistently resulted in under-representation of racial and ethnic minorities, people of lower economic and social status, women and young people. *See* JURY SELECTION PROCEDURES, *supra* note 4, at 15–16, 23–44, App. F-I; *Juror Selection, supra* note 1, at 806; S. REP. No. 891, 90th Cong., 1st Sess., 10, 19, 23 (1967); *Federal Jury Selection: Hearings Before the Subcomm. on Improvements in Judicial Machinery of the Senate Comm. on the Judiciary,* 90th Cong., 1st Sess., at 48–49, 56 (Attorney General Ramsey Clark), 255 (Judge Irving Kaufman) (1967). In Witcher v. Peyton, 405 F.2d 725, 727 (4th Cir. 1969), the court said, "It is a simple truth of human nature that we usually find the 'best' people in our own image, including, unfortunately, our own pigmentation." During the evidentiary hearings concerning a jury composition challenge in the state court in Philadelphia, jury selection officials were asked to define "good character," a standard they used for selection. The responses, from the transcript, included: "His mannerisms, the way he speaks"; "belief in an almighty"; "moral or ethical values"; no "personality faults." Concerning the standard "antagonism to our form of government," their responses included "no respect for government" and "doesn't believe in the jury system, law and order, judges, or anything else." *Juror Selection, supra* note 1, at 806. At the initial stage of this process, in which judges, law clerks, tipstaffs, secretaries and a sheriff were asked to select names, without provision of any standards, from a source list on which each person's occupation appeared, they selected "people with 'better' jobs who were not 'too busy' to serve. Unemployed persons were totally excluded by some and at least disfavored by the rest." *Id.* at 799.

164. See authorities cited note 40 *supra.* In Glasser v. United States, 315 U.S. 60, 86 (1942), the Court said selection officials "must not allow the desire for competent jurors to lead them into selections which do not comport with the concept of the jury as a cross-section of the community." Professor Van Dyke has noted that

> [e]xcluding certain people from participation in [the jury] process because, for one reason or another, they have not voted or do not want to vote contradicts the system itself. It further alienates those who may already be alienated, instead of providing them with an opportunity to participate and thus, ideally, to become more involved in society. Serving on a jury is not meant to be a reward for good citizenship; it is an opportunity and a responsibility that derives from citizenship itself. (JURY SELECTION PROCEDURES, *supra* note 4, at 91.)

It has also been suggested that voter registration lists are the best *single* available list.[165] They may well be;[166] but there is no legitimate reason for limiting jury sources to one list.[167]

Application of Constitutional Principles

The constitutional principle that jury systems be representative of a cross section of the community applies to source lists,[168] so that a substantial disparity between the representation of a cognizable class in the source and in the population establishes a prima facie case of invalidity.[169] No particular type of source list or method of compiling the source list has been invalidated. In this regard, all the courts have said is that a source that is itself discriminatory, such as a list maintained on a segregated basis,[170] or one that is clearly a subterfuge for discrimination,[171] is invalid.[172] Tax lists, property lists, actual voter lists, and even the "key man" system, where "key"

165. *See, e.g.,* S. REP. No. 891, 90th Cong., 1st Sess. 16–17; *Implementation in the Fifth Circuit, supra* note 27, at 368.

166. The Federal Act provides:

> State, local, and Federal officials have custody, possession, or control of voter registration lists, lists of actual voters, or other appropriate records shall make such lists and records available to the jury commission or clerks for inspection, reproduction, and copying at all reasonable times as the commission or clerk may deem necessary and proper for the performance of duties under this title. The district courts shall have jurisdiction upon application of the Attorney General of the U.S. to compel compliance with this subsection by appropriate process.

28 U.S.C. § 1863(d). Nevertheless, the two lists that are generally more representative than voter registration, Census and Social Security, have been traditionally regarded as confidential. 13 U.S.C. §§ 8, 9 and 15 C.F.R. § 60 (1977) (census lists); 42 U.S.C. § 1306 and 20 C.F.R. § 401 (1976) (Social Security lists). For a general discussion, see JURY SELECTION PROCEDURES, *supra* note 4, at 98–100, suggesting that these lists be made available.

167. The Report of the President's Commission on Registration and Voting Participation (Nov., 1963) recommended that registration lists not be used for jury selection since such use deters people from registering. This consideration was cited in the Commissioner's Comment to § 5 of the Uniform Jury Selection and Service Act, which requires use of multiple lists. *See also* the comments of Sen. Edward Kennedy in this regard, 121 CONG. REC. S.5985–87 (daily ed. April 15, 1975).

168. Taylor v. Louisiana, 419 U.S. 522, 538 (1975) ("the jury wheels [source lists], pools of names, panels or venires from which juries are drawn must not systematically exclude distinctive groups or the community and thereby fail to be reasonably representative thereof").

169. Taylor v. Louisiana, 419 U.S. 522 (1975); Broadway v. Culpepper, 439 F.2d 1253 (5th Cir. 1971); Simmons v. United States, 406 F.2d 456 (5th Cir. 1969), *cert. denied,* 395 U.S. 982 (1969); King v. Cook, 298 F. Supp. 584 (N.D. Miss. 1969).

170. Sims v. Georgia, 389 U.S. 404 (1967); Whitus v. Georgia, 385 U.S. 545 (1967); Arnold v. North Carolina, 376 U.S. 773 (1964).

171. Simmons v. United States, 406 F.2d 456 (5th Cir. 1964), *cert. denied,* 395 U.S. 982 (1969); King v. Cook, 298 F. Supp. 584 (N.D. Miss. 1969); United States v. Van Allen, 208 F. Supp. 331 (S.D. N.Y. 1962), *modified,* 349 F.2d 720 (2d Cir. 1965), *cert. denied,* 384 U.S. 947 (1966).

172. The courts have not considered whether selection officials may deliberately overrepresent a particular group in the source in order to assure that, even though they are disproportionally eliminated in the selection process, they will be proportionally represented in the pool. For example, in a jurisdiction that is 50% women and in which it is known that women request and receive excuses more than men, it may be constitutional or even constitutionally required for selection officials deliberately to compile a source that is more than 50% women in order to assure that the pool is 50% women. This is different from purposely *under*representing women as a matter of administrative convenience because they ask to be excused anyway, which results in additional underrepresentation of women since some are excluded who would not request or receive an excuse. Pursuant to the representativeness principle, the primary concern is that the pool be representative of a cross-section of the community and that the selection system be designed to yield such a pool. In the example, women are 50% of the population, and they should be fully represented in the pool although many cannot serve.

people select names for the source from people they know or have heard about, have not been *per se* invalidated.[173] Voter registration lists have been uniformly upheld as a valid single source list,[174] although several courts have expressed substantial doubts about their representativeness.[175]

Several recent lower court decisions,[176] decided before *Partida,* seem to require considerably more than proof of a substantial disparity to establish a prima facie case. These cases are best exemplified by *United States v. Test,*[177] recently decided by the Tenth Circuit.[178] In *Test* the defendants challenged use of the voter registration list as the single source list in the District of Colorado. They introduced evidence that Chicanos comprise approximately 10.4% of the voting age population but only 6.9% of the registered voters.[179] The court held that this proof failed to "establish a prima facie case of systematic exclusion"[180] since the absolute disparity, of approximately 4%, was less than the 16% absolute disparity referred to in *Swain v. Alabama.*[181] The comparative disparity, not referred to in the opinion, was 33%.

Of course, there could never have been an absolute disparity of 16% concerning Chicanos, since they comprised only 10.4% of the population; even total exclusion

---

A harder question is presented when a group is underrepresented in the pool because of eligibility problems. For example, should selection officials overrepresent Mexican-Americans in the source to assure that they are fully represented in the pool even though they will be disproportionally disqualified, because many cannot meet the language requirement? Again, the fact that many in this group are not eligible does not mean that the group should be underrepresented or have a smaller voice in the jury system. If selection officials are able to find sufficient numbers of people in such a group who are eligible without an undue burden, it would seem to be consistent with constitutional and democratic principles to assure their representation in and the cross-sectionality of the pool.

173. Castaneda v. Partida, 97 S. Ct. 1272 (1977); United States v. Freeman, 514 F.2d 171 (8th Cir. 1971); United States v. Grant, 475 F.2d 581 (4th Cir. 1973), *cert. denied,* 414 U.S. 868 (1973); United States v. Butera, 420 F.2d 564 (1st Cir. 1970); Rabinowitz v. United States, 366 F.2d 34 (5th Cir. 1966); United States v. Hunt, 265 F. Supp. 178 (W.D. Tex. 1967), *aff'd,* 400 F.2d 306 (5th Cir. 1968), *cert. denied,* 393 U.S. 1021 (1969). Although the key man system seems to be invalid pursuant to the "systematic exclusion" line of cases, see text accompanying notes 51–55 *supra,* the courts, while disfavoring this system, have not ruled it invalid per se.

174. Hallman v. United States, 490 F.2d 1088 (8th Cir. 1973); Simmons v. United States, 406 F.2d 456 (5th Cir. 1969), *cert. denied,* 395 U.S. 982 (1969); Chance v. United States, 322 F.2d 201 (5th Cir. 1963), *cert. denied,* 379 U.S. 823 (1964).

175. Ford v. Hollowell, 385 F. Supp. 1392 (N.D. Miss. 1974); Pullum v. Greene, 396 F.2d 251, 255 (5th Cir. 1968); Dow v. Carnegie-Illinois Steel Corp., 224 F.2d 414, 427 (3d Cir. 1955); King v. Cook, 298 F. Supp. 584 (N.D. Miss. 1969); United States v. Hunt, 265 F. Supp. 178 (W.D. Tex. 1967), *aff'd,* 400 F.2d 306 (5th Cir. 1968), *cert. denied,* 393 U.S. 1021 (1969). *See also* United States v. Grant, 475 F.2d 581 (4th Cir. 1973) (Winter, J., dissenting), *cert. denied,* 414 U.S. 868 (1973); United States v. Andrews, 342 F. Supp. 1261 (D. Mass.), *rev'd on procedural grounds,* 462 F.2d 914 (1st Cir. 1972). One court has held that use of voter registration lists as the sole source was unconstitutional, but the decision was quickly reversed on appeal. People v. Taylor, No. A-277–425 (Super. Ct. for Los Angeles County, Oct. 11, 1974), *vacated,* Civ. No. 45230 (Cal. Ct. App., 2d Dist., Nov. 26, 1974).

176. United States v. Test, 550 F.2d 577 (10th Cir. 1976); United States v. Goff, 509 F.2d 825 (5th Cir. 1975), *cert. denied,* 423 U.S. 857 (1975); United States v. Jenkins, 496 F.2d 57 (2d Cir. 1974), *cert. denied,* 420 U.S. 925 (1975).

177. 550 F.2d 577 (10th Cir. 1976).

178. The *Test* court also adopted the "substantial impact" test, discussed at text accompanying notes 125–35 *supra.*

179. 550 F.2d 577, 582–83 (10th Cir. 1976).

180. *Id.* at 587.

181. 380 U.S. 202 (1965). In *Swain,* the Court referred to an absolute disparity of 10–16% as a minimal showing for a prima facie case, but this has never been regarded by the Court as a minimum standard or even referred to in later opinions of the Court. See note 40 *supra.* The *Test* court adopted and mechanically applied the outside limit referred to in *Swain,* 16%, as a necessary requirement for proof of a substantial disparity, stating that in *Swain* the Court "implicitly held that a prima facie case of systematic exclusion was not established by demonstrating a disparity of as much as 16%." 550 F.2d at 577, 587 (10th Cir. 1976). See note 81 *supra.*

of Chicanos would not have met the court's substantiality requirement.[182] The court explicitly rejected a statistical analysis and praised the use of "subjective" rather than "objective" criteria.[183] The court also seemed to require proof of some discrimination in addition to the discrimination inherent in the use of the voter registration list. First, citing Supreme Court decisions discussing the "opportunity to discriminate," a concept derived from early systematic discrimination cases that invalidated jury systems solely on the basis of such opportunity,[184] the court required proof of some opportunity to discriminate beyond the opportunity inherent in the choice of the source list or lists. In addition, the court ruled that the defendants could not establish a prima facie case "simply because an identifiable group votes in a proportion lower than the rest of the population" and seemed to require proof of electoral discrimination,[185] as if registration to vote were a qualification for jury service and the unrepresentativeness of the voter registration list were irrelevant. This reasoning is tautological, since it allows voter registration lists to be upheld by their bootstraps: all disparities between a voter registration list and the population result from an identifiable group voting in a proportion lower than the rest of the population,[186] and to validate the use of voter registration list on this basis is to validate all voter registration lists, no matter how unrepresentative.[187]

From both a legal and mathematical perspective, a prima facie case of invalidity of a source or pool should be established by proof of a comparative disparity of

---

182. The court did observe that "token" representation may establish a prima facie case, 550 F.2d 577, 586, but no line or concrete exception to the 16% absolute disparity requirement was suggested.

183. The court said:

Defendants' 'standardized' approaches merely present alternative methods of measuring departures from a statistically ideal cross section of the community. Irrespective of the analytical approach selected, the process of characterizing the 'substantiality' of the data derived therefrom remains a subjective function.... What defendants in effect urge is that we reject the collective experience of the courts... and instead allow ourselves to be 'led' by defendants' visceral reactions to the substantiality of the disparities demonstrated below.

550 F.2d 577, 589. The defendants had offered a statistical analysis, not a visceral reaction; the court applied its "subjective" and visceral reaction and a mechanical rule that legitimizes discrimination against small and medium sized minorities.

184. See note 59 supra.

185. 550 F.2d 577, 586 & 586 n.8. See also United States v. Freeman, 514 F.2d 171 (8th Cir. 1975); United States v. Guzman, 468 F.2d 1245 (2d Cir.), cert. denied, 410 U.S. 937 (1972); United States v. James, 453 F.2d 27 (9th Cir. 1971).

186. In United States v. Burkett, 342 F. Supp. 1264, 1265 (D. Mass. 1972), Judge Wyzanski said:

The defendant has a right... not to have the pool diminished at the start by the actions or inaction of public officials, nor by the inertia, indifference, or inconvenience of any substantial group or class who do not choose to vote or to serve on juries. From the viewpoint of a black, or young, or poor, or rich defendant, his interest is in having a pool with a fair proportion of blacks, young, poor, and rich. To him it is a matter of indifference as to whether a diminished pool is due to action or inaction or third persons, whether public or private. In substance, the defendant is entitled to require that the public officials charged with jury selection, including judges who excuse jurors, proceed in such a way as to compel the calling of all eligible for jury duty who do not have socially valid excuses. In this connection jury duty is an obligation owed to the defendant, not a privilege which at the juror's pleasure the juror may choose to exercise or forgo.

187. "The use of voter lists is not the end sought. Rather, that is the principal source. If the source is deficient or infected its use alone will not suffice," Broadway v. Culpepper, 439 F.2d 1253, 1257 (5th Cir. 1971). Similarly, to reject a challenge to the use of voter lists on the grounds that nonvoters are not a cognizable class, as many courts have done, see, e.g., United States v. Lewis, 472 F.2d 252 (3d Cir. 1973); United States v. Dangler, 422 F.2d 344 (5th Cir. 1970); United States v. Van Allen, 208 F. Supp. 331 (S.D. N.Y. 1962), modified, 349 F.2d 720 (2d Cir. 1965), cert. denied, 384 U.S. 947 (1966), is to base the decision on a nonsequitur. The question is whether the list is representative of cognizable classes.

at least 15%. This standard would have had the effect of requiring the government to justify the disparities in *Test, Jenkins* and *Goff*.[188] Furthermore, the "systematic exclusion" line of cases[189] yields an additional standard for evaluating source lists. Selection officials should not be required to go door to door to find every eligible person in a community. However, under the systematic exclusion standard, which requires proof of only a minimal disparity,[190] where a challenger can prove that a significantly[191] more representative list or combination of lists is available with reasonable cost and effort, the burden should shift to selection officials to justify their failure to use such lists. The choice of which list or lists will be utilized, whether made by a legislature, by a court, or by selection officials, is an intentional act, and any resulting discrimination is systematic.[192] The "systematic exclusion" cases have never required proof of an improper motive or a lack of good faith. If selection officials cannot, pursuant to the systematic exclusion standard, choose women from the source less frequently than men because men are thought to be better jurors or because women usually ask to be excused anyway,[193] surely they cannot choose to utilize a source list that underrepresents women or any other cognizable class, for the same or other purposes, when another list or combination of lists is available that would remedy the imbalance with reasonable costs and effort.

In sum, based on analysis of the interests underlying the representativeness principle (and the rule of exclusion), it is suggested that a constitutional violation has occurred if the comparative disparity between a source list and the population is greater than 15% concerning any cognizable class or if a source list is less than 80% inclusive,[194] unless the unrepresentativeness or underinclusiveness can be appropriately justified in terms of eligibility or availability.[195] If use of a single source list, such as a voter registration list, or a multiple source list results in a violation of these standards, supplementation by additional lists to correct the violation should be constitutionally required.[196]

---

188. Since these courts held that the challengers failed to establish a prima facie case, we do not know whether the disparities proved were justifiable. Following the initiation of the challenge in *Test*, the district court adopted a multiple list system. See note 199 *infra*.

189. "Systematic exclusion cases" refers to the line of cases in which proof of actual systematic or intentional discrimination was presented, as opposed to proof only of a disparity. See text accompanying notes 51–55 *supra*.

190. See note 51 *supra*.

191. "Significantly more representative" is used to connote a standard that is less stringent than the "substantial disparity" or "substantial deviation" standards. If the comparative disparity standard is adopted, it would be appropriate to consider an additional list or combination of lists significantly more representative if it will lower the comparative disparity by 10%.

192. *See* Foster Appendix, *supra* note 1, at 823, 828.

193. *See* United States v. Zirpolo, 450 F.2d 424 (3d Cir. 1971); People of New York v. Attica Brothers, 79 Misc. 2d 492, 359 N.Y.S.2d 699 (1974).

194. Pursuant to the intentional or "systematic exclusion" line of cases, it is also concluded that a prima facie constitutional violation is established by proof that a significantly more representative (resulting in a reduction of 10% in the comparative disparity) or significantly more inclusive (resulting in an addition of at least 10% to the inclusiveness) list or combination of lists is available.

195. A justification based on eligibility is appropriate only concerning a stage in the process before which eligibility criteria have been applied. See notes 68 & 172 *supra*. The representativeness principle places an affirmative duty on selection officials and limits the availability justification to situations in which, even with affirmative efforts, sufficient numbers of persons in specific categories cannot be made available. See notes 70 & 172 *supra*.

196. With a multiple source list, it is possible that the disparity can best be remedied by eliminating one or more of the lists used.

The failure of the courts to formulate and enforce appropriate standards for source lists has confused the issues and eroded the constitutionally mandated representativeness principle. The rules advocated here would provide the courts with understandable and workable standards that could secure and safeguard the representativeness principle while providing due consideration to administrative feasibility.

Statutory Mandate for Multiple Lists

Although the legislatures have left the task of formulating standards of representativeness to the courts, federal and state statutes presently in effect require use of multiple source lists when single lists are not representative.

1. The Federal Act

The Federal Act provides that "voter registration lists or lists of actual voters" are to be the primary source and that each district "shall prescribe some other source or sources of names...where necessary to foster the policy [of representation of a cross section of the community]."[197] Although the language of this provision clearly requires supplementation where voter lists are not representative, and the legislative history supports that interpretation,[198] only two districts supplement voter lists,[199] and no court has as yet ordered supplementation.[200]

The failure to implement this provision is not because of a lack of challenges.[201] Most of the decisions have required proof of a constitutional violation and have

---

197. 28 U.S.C. § 1863 (b) (2).

198. The Report of the Committee on the Operation of the Jury System of the Judicial Conference of the United States, 42 F.R.D. 353 (1967), which drafted the Act, states that, although supplementation was not envisioned for most districts, "the committee's draft bill requires the use of other lists in addition to the voter lists to obtain the representative cross-section." *Id.* at 362. The Senate Report recognizes the particular importance of the source stage in terms of the purposes of the Act, and states:

> The bill *requires* that voter lists be supplemented by other sources whenever they do not adequately reflect a cross-section of the community....
>
> The voting list requirement, together with the provisions for supplementation...is therefore the primary technique for implementing the cross-sectional goal of this legislation..... [A]*ny substantial percentage deviations must be corrected by the use of supplemental sources....*
>
> [T]he bill recognizes that in some areas voter lists of all kinds may be insufficient to implement the policies of the Act, by reason of local voting practices. Where that is true, the plan *must* prescribe other sources to supplement the voter lists.

S. REP. No. 891, 90th Cong., 1st Sess. 17, 27 (1967) (emphasis added). The House "adopt[ed] for its report the Senate committee report." H.R. REP. No. 1076, 90th Cong., 2d Sess. 3 (1968). In a brief submitted to the Supreme Court, Solicitor General Robert Bork stated that "Congress intended that voter lists be supplemented by other sources where the use of voter lists results in substantial failure to achieve the cross-sectional goal." Brief for the United States, Test v. United States, No. 73–5993, at 15. See also Foster Appendix, *supra* note 1; WORKS, *supra* note 27, at 64.

199. *See* Modified Plan for the U.S. District Court for the District of Columbia for the Random Selection of Grand and Petit Jurors (March 16, 1976) and Amended Plan for the Random Selection of Grand and Petit Jurors in the U.S. District Court for the District of Colorado (August 18, 1975), on file with the Administrative Office of the U.S. Courts, Washington, D.C., both of which provide for supplementation of voter registration lists. These are the only two districts, out of a total of 94, that have used supplemental lists. Letter of September 23, 1976 from William R. Burchill, Jr., Associate General Counsel of the Administrative Office of the U.S. Courts, to David Kairys, on file at the *California Law Review.*

200. *But see* Ford v. Hollowell, 385 F. Supp. 1392 (N.D. Miss. 1974); United States v. Grant, 475 F.2d 581 (4th Cir. 1973) (Winter, J., dissenting), *cert. denied,* 414 U.S. 868 (1973).

201. See cases cited notes 12–17 *supra* & 202 *infra.*

adopted the substantial impact and electoral discrimination standards discussed above,[202] even though the Act clearly establishes less stringent standards.[203] The Act requires representation of a cross section of the community and mandates that relief (supplementation) be granted where there is a "substantial" deviation.[204]

The Committee on the Operation of the Jury System of the Judicial Conference of the United States, which drafted the Act, has noted "the increasing number of cases challenging the process of jury selection" and has responded with proposals regarding source lists.[205] Unfortunately, rather than encouraging enforcement of the supplementation provision or investigating the representativeness of voter lists, the Committee has proposed an amendment to the Act that would "establish a presumption that names of prospective jurors contained in voter lists represent a fair cross-section of the community."[206] The basis for this presumption is surely not factual,[207] since all available data directly opposes it.[208] The Committee's proposal would define away a serious problem, and sacrifice the basic principles of the Act.[209]

The Act provides the basis for representative and inclusive source lists in the federal courts; all that is required is enforcement by the courts.

2. The Uniform Act

The Uniform Act, adopted in five states,[210] provides that:

"The jury commission for each [county] [district] shall compile and maintain a master list consisting of all [voter registration lists] [lists of actual voters] for the [county] [district] supplemented with names from other lists of persons resident therein, such as lists of utility customers, property [and income] taxpayers, motor vehicle registrations, and drivers' licenses, which the [Supreme Court] [Attorney General] from time to time designates."[211]

202. *E.g.*, United States v. Test, 550 F.2d 577 (10th Cir. 1976); United States v. Jenkins, 496 F.2d 57 (2d Cir. 1974), *cert. denied*, 420 U.S. 925 (1975); United States v. Jones, 480 F.2d 1135 (2d Cir. 1973); United States v. Lewis, 472 F.2d 252 (3d Cir. 1973); United States v. Dangler, 422 F.2d 345 (5th Cir. 1970); United States v. Gordon, 455 F.2d 398 (8th Cir. 1972); United States v. Ross, 468 F.2d 1213 (9th Cir. 1972). In a brief presented to the Supreme Court, Solicitor General Robert Bork stated that, pursuant to the Act, "substantial underrepresentation of a cognizable group...would necessitate resort to supplementary sources...even though no history of voter discrimination against the group is proved." Brief of the United States, Test v. United States, No. 73–5993, at 13–14.

203. Indeed, these cases also misconstrue the actual constitutional standard, since, although the courts use the term "systematic exclusion," they only require an inference of discrimination from proof of a "substantial disparity." See text accompanying notes 32–73 *supra*.

204. 28 U.S.C. § 1867(d). See authorities cited notes 15 & 198 *supra*.

205. REPORT, *supra* note 1, at 9.

206. *Id.*

207. The Committee's report does not state or refer to any factual basis; rather, the Committee said: "This proposal is a response to the increasing number of cases challenging the process of jury selection." *Id.*

208. See text accompanying notes 145–60 *supra*. The Uniform Act creates the opposite presumption and makes it irrebuttable, requiring use of multiple lists. See text accompanying notes 215–17 *infra*.

209. This proposal seems to suggest an implicit recognition of the deficiencies in the decisions construing the present supplementation provision. However, contrary to the apparent belief of its proponents, it would lead to more, not less, litigation, since the provision itself and the unrepresentative sources and pools it would spawn would be subject to constitutional attack.

210. See note 8 *supra*.

211. UNIFORM ACT, *supra* note 9, § 5.

The Commissioners' Comment to this provision states that supplementation is "mandatory."[212]

The Uniform Act embodies provisions designed to maximize the representativeness and inclusiveness of sources and the entire jury system. There are no exemptions,[213] and all requests for excuse are resolved on an individual basis by the court "only upon a showing of undue hardship, extreme inconvenience, or public necessity, for a period the court deems necessary..."[214] Jurors are reimbursed for travel expenses and paid at a "more adequate [rate] than has commonly been provided" to compensate them adequately and to "reduce the occasions for excusing prospective jurors...because of financial hardship."[215] Length and frequency of service is limited,[216] and employers are prohibited from punitive actions because of absence from work.[217] Presumably, states adopting the Uniform Act will have representative and inclusive jury source lists and pools, making challenges rare.

METHODOLOGY FOR USE OF MULTIPLE LISTS

In order to achieve the representative source lists and pools required by the Constitution, relevant statutes, and societal considerations, multiple lists must be used. This section deals with the theoretical and practical aspects of implementing multiple-list systems,[218] and discusses various techniques for combining multiple lists and the choice of the lists to be used.

Methods for Combining Multiple Lists

Any multiple list plan must be governed by the fundamental principle that each person whose name appears on at least one of the lists used should have an equal chance of being selected, independent of the total number of lists on which the name appears. For example, if four lists are used and one individual is on all four lists while another individual is on only one list, then a procedure must be used that results in both having the same chance of selection. Otherwise, the result will be no better and may even be worse than with single list procedures.

At first glance, it may appear that implementation of a nonduplicating, multiple list procedure is both costly and time consuming.[219] However, simple methods have

212. UNIFORM ACT, *supra* note 9, Commissioners' Comment to § 5.

213. UNIFORM ACT, *supra* note 9, § 10.

214. *Id.* at § 11. Experience with various excuses and exemptions has shown that any group offered a basis for not serving will take it. See note 4 *supra*. The provisions eliminating exemptions and narrowing excuses should result in a broader cross-section and widespread participation.

215. *Id.* at § 14, and Commissioners' Comment to § 14.

216. UNIFORM ACT, *supra* note 9, § 15.

217. *Id.* at § 17.

218. For a general discussion see *Multiple Lists, supra* note 30, in which the authors present the theoretical basis and relevant equations concerning the procedures outlined here.

219. *See, e.g., Facets of the Jury System, supra* note 5, at 13, where, without benefit of the techniques discussed here and based upon the experience in one jurisdiction, use of multiple lists is said to be "desirable" but "expensive." The problem of duplications has been cited as one of the primary reasons that multiple list systems have been

recently become available that can eliminate the effect of duplications and require only slight additional cost or time.[220]

The most straightforward procedure consists of combining the lists into a single master list on which all duplicates are eliminated and each individual is listed only once. In most jurisdictions only a small proportion of the people on a source would actually be called for jury duty, so that much of the effort of compiling such a master list is unnecessary. A better approach involves not combining the lists initially, but selecting random samples from each and then using a procedure regarding the names selected that insures each individual has an equal chance of being selected independent of the number of lists on which he or she appears.[221]

Such a procedure begins by ordering the lists to be used. For example, the voter registration list may be first, the driver's license list second, the public assistance list third, and the telephone list fourth. The particular ordering used does not influence the probability of any particular individual being selected, but it does affect the cost.[222] Once the lists have been ordered, they are still kept separate, but, for purposes of this procedure, they are regarded collectively as one long list containing many duplications. If a person's name appears on more than one list, it is regarded as "good" on the first list on which it appears, while all further appearances are regarded as duplicates or "blanks."[223] Using this definition, each person listed at least once has one good listing and possibly several duplicate listings or blanks.

In order to insure that each individual has an equal chance of being selected, only good listings can result in a person being selected, and all blank listings are ignored. Names are selected from each of the lists, and each selection is then checked to determine if it is good or blank. For example, a name selected from list 1 is automatically good. A name selected from list 2 must be checked against list 1. If it is on list 1, the selection is blank, and if it is not on list 1, it is good. In general, a name from a particular list is good provided that it is not to be found on any of the lists preceding it in the order described above. The concept of good and blank names will give each individual an equal chance of selection if an appropriate number of names is selected randomly from each list. This can be accomplished in a number of ways, two of which are described here.[224]

One plan requires that a fixed sampling fraction of the names on each list be selected and checked. The total number of good names that will result from this process is not known in advance; it will be approximately the sampling fraction multiplied by the total number of good names on all lists.[225] Since all names on the

viewed as impractical. *Implementation in the Fifth Circuit, supra* note 27, at 383. The author was not aware of the methodology, discussed here.

220. On the costs, see text accompanying notes 236–40 *infra.*

221. These procedures also allow an updated version of any one or all of the lists to be used without requiring a re-compilation of an entire master list.

222. *See Multiple Lists, supra* note 30, at 212–18.

223. For example, if an individual were on lists 3 and 4, the listing on list 3 would be good and the one on list 4 would be a blank.

224. Five plans are presented and discussed in *Multiple Lists, supra* note 30, at 208–11.

225. For example, suppose the lists utilized are voter registration, licensed drivers, telephone and public assistance and they contain, respectively, 1,000,000, 800,000, 500,000 and 40,000 names. If the sampling fraction is 5%, then 50,000, 40,000, 25,000 and 2,000 names would be selected and checked from the voters, drivers, telephone and public assistance lists respectively. Of these 2,340,000 names, if 1,500,000 were good and 834,000 were blanks, then

first list are good and therefore require no checking, time and costs will often be minmized by ordering the lists so that the longest is first.[226]

A second plan consists of two stages. First, after the lists are ordered, one of them is selected with a probability proportional to the total number of names on that list. A name is then chosen at random and checked as described above. If it proves to be a duplicate, it is discarded; otherwise, it is kept as a good selection. Then a list, perhaps the same, perhaps a different one, is again chosen at random and the checking procedure is followed. This continues until the required number of names has been selected. This method is quite flexible and is particularly useful when only a small number of names is required.

The costs and effort involved in use of multiple lists will depend on many factors, including the number of lists used, their lengths, the number to be selected, and whether or not the procedure is computerized. The most costly aspect is the checking procedure, implementation of which requires a method for determining whether a given name is on a particular list. There are three basic alternatives.[227] First, checking can be accomplished manually.[228] The cost and time involved with this method are significant, but it has been used without undue expenditures.[229]

Second, computer programs are available that will check lists for duplications at a minimal cost.[230] These programs may not be able to detect all the duplications because of spelling mistakes and differing conventions for names, addresses, and districts. Failure to detect these duplications results in certain individuals having a slightly larger chance of selection; however, this bias should not be serious, for it is unlikely to affect a significant number of names, and no particular group should have a special concentration of these errors.

Third, checking can be accomplished by questionnaire.[231] Once a name is selected, a qualification questionnaire is usually sent to determine whether or not the individual meets the established requirements for jury service. In addition to the usual questions, each person can be asked to identify the lists on which his or her name appears. It is then a simple matter for selection officials to determine whether a selection was good or a blank. With such a procedure, the only added burden is the cost of mailing additional questionnaires.[232]

---

the yield of good names would be approximately 5% of 1,500,000 or 75,000. The sampling fraction can be varied to yield the required good names.

226. This is so if the cost of a check for each name and on each list is uniform throughout, which is not always the case.

227. Equations for computation of costs are presented in *Multiple Lists, supra* note 30, at 212–15.

228. See Table 5 *infra* for jurisdictions in which a manual checking procedure has been utilized.

229. See text accompanying notes 236–40 *infra* concerning the costs.

230. See Table 5, *infra,* for jurisdictions in which a computerized checking procedure has been used. Information concerning computer programs and methods for formatting the lists is available from Judicial Department, State of Colorado, 323 State Capital, Denver, Colorado 80203 (Attention: Rayma Jordon); Office of Court Administrator, 29th Judicial District, Kansas Courthouse, Kansas City, Kansas 66101 (Attention: Jerry Larson); Administrative Office of the Courts, 303 K St., Anchorage, Alaska 99501 (Attention: M. Martin).

231. No jurisdiction of which we are aware has utilized a questionnaire checking procedure.

232. The use of questionnaires raises the possibility that the information received could be inaccurate or unreliable, but experiences in similar situations have demonstrated that this is not a serious problem. The Census Bureau used this technique in its studies of voting and registration, determining who was registered and who voted by asking the respondents, and a check of the procedure revealed minimal inaccuracies and unreliability. *See Voting and Registration in the Election of 1972, supra* note 151, at 7–8.

TABLE 5. Jurisdictions That Presently Use Multiple Lists[233]

Jurisdiction	Lists	Method of Combination	Implementation Technique	Present Status
U.S. District Court for the District of Colo.	Voter registration Licensed drivers	1	Computer	In use
U.S. District Court for the District of Columbia	Voter registration Licensed drivers	1	Computer	In use
Alaska (statewide)	Voter registration Fish and game Income tax	1	Computer	In use
California:				
San Diego Co.	Voter registration Licensed drivers	3	Under study	Under study
San Joaquin Co.	Voter registration Licensed drivers	1	Computer	In use
San Mateo Co.	Voter registration Licensed drivers	1	Computer	In use
Colorado (statewide)	Voter registration Licensed drivers City directory	1	Computer	In use
Idaho:				
Ada Co.	Voter registration Licensed drivers	1	Computer	In use
All other Counties (separately)	Voter registration Licensed drivers	1	Manual	In use
Wyandotte Co., Kansas	State census Voter registration	1	Computer	In use
Jefferson Co., Kentucky	Voter registration Property tax	*	Computer	In use
Hennepin Co., Minnesota	Voter registration Licensed drivers	3	Under study	Under study
St. Louis Co., Missouri	Voter registration Licensed drivers	1	Computer	In use shortly
North Dakota:				
Burleigh Co.	Actual voters Licensed drivers	3	Manual	In use

233. Table 5 was compiled by Bird Engineering—Research Associates (P.O. Box 37, Vienna, Virginia 22180), a firm with extensive experience in developing techniques and advising courts concerning multiple list methods for jury selection (the authors appreciate the assistance given by Bird, particularly by Chester Mount, Thomas Munsterman and William Pabst). The plan numbers listed under "Method of Combination" indicate the following: Plan 1 is the master list plan discussed *infra* at text accompanying notes 219–20 (*see Multiple Lists, supra* note 30, at 209, where this plan is also referred to as plan 1); Plan 2 is the proportional list selection plan discussed *infra* at text accompanying notes 224–26 (*see Multiple Lists, supra* note 30, at 210, where this plan is also referred to as plan 2); Plan 3 is the random list selection plan discussed *infra*, at text accompanying notes 226–27 (*see Multiple Lists, supra* note 30, at 210, where this plan is referred to as plan 4). An asterisk (*) means the lists are used without combination method to eliminate the effect of duplications.

TABLE 5. (Continued)

Jurisdiction	Lists	Method of Combination	Implementation Technique	Present Status
Canyon Co.	Actual voters Licensed drivers Utilities	1	Manual	In use
Ward Co.	Actual voters Licensed drivers	1	Manual	In use
New York: Bronx Co.	Voter registration Licensed drivers City income tax	*	Computer	Under study
Kings Co.	Voter registration Licensed drivers	*	Computer	In use
New York Co.	Voter registration City income tax	1	Computer	In use
Queens Co.	Voter registration Licensed drivers City income tax	*	Computer	Under study
Allegheny Co., Pennsylvania	Voter registration Telephone book Welfare	3	Manual	In use (Plan 3 system under study)

Finally, perhaps the best checking procedure from the standpoint of costs and complete accuracy is a combination of the computer and questionnaire techniques. A computer will eliminate almost all of the blanks. Any remaining blanks can be detected by use of the questionnaire technique. The additional costs of such a hybrid procedure over a single list procedure would be minimal.[234]

The methods for multiple list systems utilized in several jurisdictions are presented in Table 5. Most of these jurisdictions have used the master list method and computerized checking procedures. However, three jurisdictions are using or are about to use the new methods discussed here for minimizing the number of names to be checked.

Application of these techniques and the cost and effort involved can be illustrated by a recent example. Allegheny County, Pennsylvania (which includes Pittsburgh) has recently adopted the fixed fraction multiple list system (Plan 2) using three lists: voter registration, telephone and public assistance.

The total number of names, the number of "good" names, and the number of "checks" on each list are presented in Table 6. Once the overlap has been determined, the sampling fraction, or the proportion of names to be selected from the lists, can be determined based on the total number of jurors needed.[235] In this instance, 50,000

---

234. This procedure has been recommended in San Diego County, California. See Bird Engineering—Research Associates. *The Use of Multiple Lists for Jury Selection, A Report to the Superior Court of San Diego County* (May 2, 1977).

235. The sampling fraction is 4.81%. It is determined from the equation $N_0 = (N_1 + d_2N_2 + d_3N_3)f$, where

N_0 = number of names needed,

N_1 = number of names on list 1,

TABLE 6. Allegheny County, Pa. Multiple List System[236]

List	Total names	"Good" (non-duplicating) Names	Number of "Checks" Required (for 50,000 "good" names)
Voter registration	921,000	921,000	0
Telephone	559,000	112,694	26,890
Public Assistance	35,000	5,705	2,757
Total	1,515,000	1,039,399	29,647

names are needed, and 72,880 must be drawn from the lists in order to meet this requirement. In addition, a total of 29,647 checks must be made. The added tasks attributable to the use of multiple lists, in addition to conceptual and developmental tasks,[237] are drawing 22,880 additional names and checking 29,647 names. This has not yet been done in Allegheny County,[238] but cost figures for the same tasks are available from the experience in San Mateo County, California. If all the lists are available on computer tapes and therefore no keypunching is necessary,[239] the cost in computer time would be less than $30.00.[240] Compilation of a master list, which would require approximately 600,000 checks, would cost approximately $600.00 in computer time.[241]

Choice of Lists

Multiple list procedures are necessary to overcome the biases and exclusiveness inherent in available single lists. As more lists are employed, the total number of unique

---

N_2 = number of names on list 2,

N_3 = number of names on list 3,

d_2 = proportion of names on list 2 that are not on list 1,

d_3 = proportion of names on list 3 that are not on list 1 or 2, and

f = sampling fraction.

In this example,

$$50,000 = [921,000 + (.2016)559,000 + (.1630)35,000]f$$
$$f = 4.81\%$$

d_2 and d_3 are determined by a sample survey. The checks required for the telephone and public assistance lists are determined as follows. A total of (.0481)559,000 = 26,980 names will be selected from the phone book, all of which must be checked against the voters registration list. A total of (.0481)35,000 = 1684 names must be selected from the public assistance list. All of these names must be checked against the telephone book. Of the 1684 names, 611 will be found in the telephone book and be discarded as bad names. The remaining 1073 must be checked against the voters registration list which gives a total of 1684 + 1073 = 2757 checks for the public assistance names.

236. These data are from Report of Dr. William Brinckloe to the Jury Commission of Allegheny County (May 29, 1974).

237. The Federal Judicial Center and other agencies could reduce the costs and effort of development and programming by developing programs for using the various plans and making them generally available, as they have already done concerning the task of picking randomly from source lists.

238. Allegheny County plans to do the drawing and checking manually, and it is not known how long this will take or how much it will cost.

239. If the lists are not available on tapes and have to be keypunched, there is a significant additional cost.

240. In San Mateo County, computer time costs $63.00 per hour. The cost of computer time and the amount of time necessary vary with various machines and locations.

241. 600,000 checks would require approximately 20 times as much computer time, which would cost about $600.00.

names increases, and the overall list becomes more inclusive. However, care must be exercised in choosing the lists to insure that no biases result. If a group is underrepresented on all the lists, it will be underrepresented in the combined source no matter how many lists are used. Moreover, supplementation with some lists can increase a group's underrepresentation or create an underrepresentation of another group.[242]

This observation indicates that if the purpose of using multiple lists is to be achieved, the additional lists utilized must overrepresent the groups underrepresented on the primary list.[243] The most widely used primary list is the voter registration list, which significantly underrepresents racial minorities, people under 40, those with lower incomes and less education, blue collar workers, and the unemployed. The lists chosen to supplement the voter registration list should therefore compensate for these specific deficiencies. Some commonly available lists, including lists of telephone and utility customers, city directories and tax lists, generally underrepresent the same groups as voter registration and actual voter lists.[244] However, public assistance[245] and unemployment lists generally give strong representation to people with lower socioeconomic status and minority groups.[246] Licensed drivers' lists are typically a good source for young people and provide substantially increased inclusiveness.[247]

The advisability of various combinations will, of course, vary in different areas. Data from various jurisdictions indicate that generally voter registration lists combined with licensed drivers, public assistance and unemployment lists will provide representative and inclusive sources.[248] These and other lists should be examined and evaluated before determining which lists to use.[249]

242. Care should always be taken that the use of lists that compensate for the underrepresentation of one group and/or increase inclusiveness does not result in underrepresentation of another group. For example, city directory and telephone customer lists will increase inclusiveness, but they seriously overrepresent men and higher income people. See note 244 *infra*.

243. *See* WORKS, *supra* note 27, at 109; JURY SELECTION PROCEDURES, *supra* note 4, at 98–104.

244. JURY SELECTION PROCEDURES, *supra* note 4, at 100–2.

245. In *Implementation in the Fifth Circuit, supra* note 27, at 384, Judge Gewin states, without citing any authority or indicating his basis, that public assistance lists are not useful because people on them are mostly excusable or disabled. He does not discuss unemployment lists. *See* JURY SELECTION PROCEDURES, *supra* note 4, at 102, recommending public assistance lists.

246. A 1975 study by the Department of Health, Education and Welfare determined that of the people receiving aid to families with dependent children, 50.2% were white, 44.3% were black, 1.1% were American Indian, .5% were Asian, and 3.9% were in some other minority or unknown. Of those classified in the study as white, about 13% were from a Spanish speaking background, mostly Mexican and Puerto Rican. NATIONAL CENTER FOR SOCIAL STATISTICS, POPULATION SURVEY (December 21, 1976). *See also* NATIONAL CENTER FOR SOCIAL SATISTICS, FINDINGS OF THE 1973 AFDC STUDY, DHEW PUBLICATION No. (SRS) 74–03764 (1974) (provides detailed demographic data on public assistance recipients, some of which is broken down by states). For a demographic breakdown of unemployed people (not all of whom are on unemployment lists), see U.S. DEPT. OF LABOR, BUREAU OF LABOR STATISTICS, EMPLOYMENT AND EARNINGS, Vol. 24, No. 4 (April, 1977) (published monthly); U.S. DEPT. OF LABOR, BUREAU OF LABOR STATISTICS, GEOGRAPHIC PROFILE OF EMPLOYMENT AND UNEMPLOYMENT, Report 481 (1975).

247. JURY SELECTION PROCEDURES, *supra* note 4, at 102–4. In California, there were about 12 million licensed drivers in 1973, while as of 1970 there were only 8.7 million registered voters. *Id.* at 102. See CAL. CIV. PROC. CODE, § 204e (West Supp. 1977) which provides for use of licensed drivers lists with voter registration lists. In some counties in North Dakota, use of licensed drivers lists with lists of actual voters increased the number of people included by from 80% to 100% and greatly increased the number of young people. JURY SELECTION PROCEDURES, *supra* note 4, at 104.

248. These lists are typically computerized and are therefore easily used with computerized selection and checking procedures.

249. It is advisable to include demographic questions on jury questionnaires so the demographic composition of the source and pool will be known and appropriate changes can be made. Such data is also useful regarding possible challenges to the jury system.

CONCLUSION

This chapter advocates the adoption of concrete standards of representativeness or cross-sectionality for jury source lists and pools pursuant to the representativeness principle and the rule of exclusion. The representativeness principle embodies the fundamental right of litigants to a fair trial, the right of citizens to serve on juries, and the societal interest in the legitimacy, integrity and impartiality of the judicial process. Adoption of concrete standards directly based upon the representativeness principle is analytically appropriate and would ensure that jury selection systems are evaluated in terms of the underlying constitutional concerns.

Most of the unrepresentativeness in our jury systems is directly attributable to the unrepresentativeness of the source lists. Multiple lists are in use in several jurisdictions, and the methodology for easy and inexpensive implementation of multiple list systems is now available. Constitutional and statutory mandates concerning source lists must be enforced, or representativeness will be an empty generalization rather than a constitutional cornerstone.

B. JURIES IN DEATH PENALTY CASES

1

JURIES HEARING DEATH PENALTY CASES: STATISTICAL ANALYSIS OF A LEGAL PROCEDURE

Joseph B. Kadane

This article considers a question of current legal interest: whether the present jury selection procedure used in cases in which capital punishment is a possibility is biased against the defense. Section 1 gives the legal background of the question addressed here. Section 2 describes two major studies on jurors in capital cases that were recently presented to the California Supreme Court in *Hovey v. Superior Court* (1980): the Fitzgerald and Ellsworth (1984) work on the attitudes of jurors favoring either the defense or the prosecution, and the Cowan, Thompson, and Ellsworth (1984) experiment on juror behavior. The response of the California Supreme Court in *Hovey* is then discussed. Section 3 introduces the basic measures used in this article, and applies them to the Fitzgerald and Ellsworth, and the Cowan, Thompson, and Ellsworth data. Section 4 reviews two surveys on attitudes toward the death penalty conducted since *Hovey:* a Field Research Corporation survey of Californians and a Harris national survey. Then Section 5 gives the assumptions needed to justify the likelihood function. Section 6 gives the numerical results and discusses their implications. Substantial evidence of bias against the defense is found.

1. LEGAL BACKGROUND

Some 90 percent of the criminal cases prosecuted in America are settled by plea bargaining without a trial. Of the remainder, fewer than half are heard by juries. And of those heard by juries, in only a portion is the outcome truly in doubt. Hence in terms of frequency of invocation, a trial by jury is the exception rather than the rule. Nonetheless, juries play a vital role in American criminal law. Because every defendant has the right to a trial by jury, the likely outcome were the case to go to a jury forms the background to a plea bargain or an agreement to let a judge decide the case without a jury. It is precisely the unusual case in which the jury has to work hard that sets the tone for the system as a whole. As but one example, the refusal of a jury in New York City to convict John Peter Zenger of sedition for printing criticism of British Colonial authorities established practical freedom of the press in that era and context even before the American Revolution (Alexander 1963).

In each state and in each federal district, law and court rules establish the basic lists from which juries are to be chosen (see Van Dyke 1977, Appendix A). Generally the list of those registered to vote is used, although recently proposals have been made to supplement the voters list with others (Kadane and Lehoczky 1976; Kairys, Kadane, and Lehoczky 1977). These people are then summoned to appear on a particular day, to serve for a set length of time. (To avoid large numbers of excusals, some jurisdictions have reduced the duration of service to one day or one trial.)

When a case is ready to be heard by a jury, a group of summoned potential jurors are examined—usually by the judge, often by the attorneys as well—to see if they should be excused from the case for cause. The reasons a person may be excused for cause are restricted to those that would directly impair their ability to hear the case impartially, for example, being a close relative of the defendant, the victim, or one of the attorneys, or having a fixed opinion on the defendant's guilt. In addition, the law provides for peremptory challenges for each side (different numbers in each state, see Van Dyke 1977, Appendix D), which can be used by the defense or the prosecution to eliminate potential jurors without having to explain their reasons.

In capital cases jurors are also questioned on their views about the death penalty, and some are excused because of their attitudes toward the death penalty. At one time all jurors who opposed the death penalty were excluded from capital cases. In *Witherspoon v. Illinois* (1968) the Supreme Court limited this practice. Under *Witherspoon* the only jurors who may be constitutionally excused from capital cases for cause because of their opposition to the death penalty are those who say either (a) that they would automatically vote against the death penalty in every capital case, regardless of the evidence, or (b) that they could not be fair and impartial in deciding a capital defendant's guilt or innocence. In addition, in some states—notably California—jurors are also excused for cause from capital cases if they state that they would automatically vote *for* the death penalty in every capital case in which they made a finding of first degree murder and aggravated circumstances, regardless of the evidence. This article analyzes the California procedure.

Commonly in capital cases a single jury decides both (a) the guilt or the innocence of the defendant and (b) if the defendant is convicted of a capital offense, the penalty. Under current practice jurors who would never vote for the death penalty and jurors who would always vote for the death penalty are excluded from the guilt or innocence phase of a capital trial even though they say they could fairly and impartially try that issue. The Supreme Court recognized in *Witherspoon* that this practice might be prejudicial to the capital defendant in the determination of guilt or innocence (*Witherspoon v. Illinois,* p. 520, fn. 18). This article examines whether the practice is in fact prejudicial.

2. RECENT STUDIES AND THE HOVEY DECISION

For our analysis, consider each potential juror to be in one and only one of the following four groups:

A_1: those who say they would not decide the question of guilt or innocence in a fair and impartial manner;

A_2: those who say they are fair and impartial, and say they would ALWAYS vote for the death penalty, regardless of the facts, if the defendant is found guilty;

A_3: those who say they are fair and impartial, and say they would NEVER vote for the death penalty, regardless of the facts, if the defendant is found guilty;

A_4: those who say they are fair and impartial, who say they would consider the death penalty and would SOMETIMES AND SOMETIMES NOT vote for it, depending on the facts, if the defendant is found guilty.

There are two other groups for which it is useful to have notation:

A_5: those who say they are fair and impartial, who say they would consider the death penalty and would AT LEAST SOMETIMES vote for it, if the defendant is found guilty ($A_5 = A_2 \cup A_4$);

A_6: those who say they are fair and impartial, and say they would ALWAYS OR NEVER vote for the death penalty ($A_6 = A_2 \cup A_3$), if the defendant is found guilty.

Under the procedure in effect in California and various other states, the question that the Supreme Court left open in *Witherspoon* can be restated as follows: does the exclusion from the pool of jurors who could fairly and impartially try a capital defendant's guilt or innocence of two groups of jurors—the automatic death penalty (or ALWAYS death penalty) group, and the automatic life imprisonment (and hence NEVER death penalty) group—bias the jury pool against the defendant?

In 1980, a major presentation citing many studies was made to the California Supreme Court on this question. Two of these studies will be reviewed here in detail because they are the most recent and most thorough of their respective types. The first, by Fitzgerald and Ellsworth (1984), studies the relationship between attitudes toward capital punishment and attitudes on other relevant criminal justice issues. A questionnaire was administered to 811 persons eligible for jury duty in Alameda County, California at random by random digit dialing by the Field Research Corporation of San Francisco, CA. The questionnaire asked whether they could decide the question of guilt or innocence fairly and impartially; 717 respondents indicated that they could. (The remaining analyses are limited to those self-designated fair and impartial respondents.) They were also asked whether their views on capital punishment were such that they could never vote to impose it, or whether they would vote to impose it in some cases. The results are shown in Table 1.

Fitzgerald and Ellsworth summarize these data by finding that those who would AT LEAST SOMETIMES impose the death penalty "were consistently more prone to favor the point of view of the prosecution, to mistrust the criminal defendant and his counsel, to take a punitive attitude toward criminals, and to be more concerned with crime control than with due process." By contrast, those who would NEVER

TABLE I. The Relationship Between Attitudes of Jurors on the Death Penalty and Defense Bias Items (from Fitzgerald and Ellsworth). Percent of Death Penalty Groups NEVER (A3) and AT LEAST SOMETIMES (A5) Giving Each Answer and the Number Answering

	Group of Respondent	Agree Strongly	Agree Somewhat	Disagree Somewhat	Disagree Strongly	Number Answering
1. Better some guilty go free	NEVER	32.5	30.0	20.8	16.7	120
	ALS	16.1	27.9	27.2	28.9	585
2. Failure to testify indicates guilt	NEVER	10.9	12.6	31.9	44.5	119
	ALS	16.0	16.3	39.5	28.2	582
3. Consider worst criminal for mercy	NEVER	40.2	37.6	10.3	12.0	117
	ALS	15.0	29.0	15.5	40.5	575
4. District attorneys must be watched	NEVER	21.2	31.9	32.7	14.2	113
	ALS	24.0	25.0	26.4	24.6	568
5. Enforce all laws strictly	NEVER	22.3	24.0	25.6	28.1	121
	ALS	38.1	19.0	25.3	17.6	585
6. Guilty if brought to trial	NEVER	14.9	11.6	17.4	56.2	121
	ALS	17.2	15.1	17.7	49.9	581
7. Exclude illegally obtained evidence	NEVER	50.0	13.9	17.2	18.9	122
	ALS	38.4	18.1	24.0	19.6	576
8. Insanity plea is a loophole	NEVER	27.5	31.7	22.5	18.3	120
	ALS	51.5	26.5	13.7	8.3	581
9. Harsher treatment not solution to crime	NEVER	55.0	25.0	14.2	5.8	120
	ALS	32.7	26.3	17.9	23.1	571
10. Defense attorneys must be watched	NEVER	21.0	43.7	23.5	11.8	119
	ALS	38.9	34.6	17.4	9.1	581
		Unemployment	Crime	Number Answering		
11. More serious problem: Unemployment, crime	NEVER	50.4	49.6	117		
	ALS	37.5	62.5	581		
		Would Not Consider	Would Consider			
12. Consider confession reported by news media	NEVER	60.2	39.8	118		
	ALS	49.1	50.9	581		
		Should not infer	Should infer			
13. Infer guilt from silence	NEVER	86.0	14.0	118		
	ALS	76.0	24.0	588		

NOTE: NEVER = NEVER vote for Death Penalty. ALS = AT LEAST SOMETIMES vote for death penalty.

impose the death penalty "tended to be more concerned with mercy, more oriented toward due process, and less mistrustful of the defendant and his legal representative." These findings confirm those of Bronson (1970) and of Vidmar and Ellsworth (1974) that jurors' attitudes with respect to whether they would ever impose the death penalty are related to many other juror attitudes of concern to a defendant seeking a fair trial.

The second study of major interest here, by Cowan, Thompson, and Ellsworth (1984), reports an experiment relating attitudes toward capital punishment to simulated juror behavior. In this study, 288 adults eligible for jury service in California were shown a videotaped reenactment of an actual murder trial that occurred in Boston. Thirty-seven subjects were recruited from venire lists of the Santa Clara Superior Court after completing their terms, 218 responded to a newspaper advertisement for volunteers for a study of "how juries make decisions," and 33 were referred by friends who had seen the advertisement.

The tape included appropriate instructions to the jury on California law. All the subjects in this study said they would be fair and impartial in deciding guilt or innocence; 30 would NEVER impose capital punishment, the other 258 would impose it AT LEAST SOMETIMES. After having seen the videotape, the jurors were asked whether they would vote for conviction of first-degree murder, second-degree murder, or manslaughter, or for acquittal. The data collected suggested that the only real issue in the case was the choice between manslaughter and acquittal. Over 70 percent of the subjects in each category voted for one of these verdicts on the initial ballot; in a second ballot taken after an hour's deliberations this figure rose to over 80 percent for each group. Accordingly, the focus of the analysis by Cowan, Thompson, and Ellsworth was on the dichotomy in the initial ballot between conviction and acquittal, rather than on the possible levels of conviction. The results of the study are given in Table 2.

Cowan, Thompson, and Ellsworth conclude from the data in Table 2 that jurors in the AT LEAST SOMETIMES group are more likely to convict than are jurors in the NEVER group. Similar experiments are reported by Goldberg (1970), Jurow (1971), and Wilson (1964).

The California Supreme Court in its *Hovey* decision reviewed both the Fitzgerald and Ellsworth attitude study and the Cowan, Thompson, and Ellsworth study of juror behavior in a simulated trial. The court found the evidence relevant and persuasive except for one fact: both studies included the automatic death penalty group (ALWAYS) with those who would SOMETIMES AND SOMETIMES NOT impose the death penalty, rather than with the automatic life imprisonment group (NEVER). Thus the court remained unconvinced that the exclusion of both the ALWAYS and the NEVER groups results in a bias against the defense, and consequently it did not find the form of death qualification in use in California to be legally flawed. In the remainder of this article the Fitzgerald and Ellsworth study and the Cowan, Thompson, and Ellsworth study are extended to take into account the effect of moving the automatic death penalty group from the AT LEAST SOMETIMES to join the NEVER group in a new group called NEVER OR ALWAYS (A_6).

3. STATISTICAL MEASURES

That groups of people that differ in one respect (here views on the death penalty) differ also in others (attitude on the Fitzgerald-Ellsworth survey items and behavior in the Cowan-Thompson-Ellsworth experiment) is unsurprising. It is apparent in the data from Tables 1 and 2 that those who would consider voting for the death penalty are consistently more favorable to the prosecution than those who would never

TABLE 2. Juror Attitudes toward Capital Punishment and Behavior
in a Simulated Trial (from Cowan, Thompson, and Ellsworth).
Percent of Death Penalty Groups NEVER (A3) and AT LEAST
SOMETIMES (A5) Giving Each Verdict, and Their Number

	Acquit	Convict	Number
NEVER	46.7	53.3	30
ALS	22.1	77.9	258

NOTE: NEVER = NEVER vote for death penalty. ALS = AT LEAST SOMETIMES
vote for death penalty.

do so. It is not obvious, however, how strongly each item reveals a more favorable
attitude toward the prosecution among the AT LEAST SOMETIMES (A_5) group
than among the NEVER (A_3) group. To be most helpful to the court, I looked for
measures to quantify this difference, so that the court could judge whether the differences shown are sufficiently large to require the imposition of a new procedure
for jury selection.

There are several sources in the statistical literature for measures of association:
the measures proposed and studied by Goodman and Kruskal (1954, 1959, 1963,
1972), the literature on log-linear models, and the nonparametric literature. In this
section, I first reorient the data in Tables 1 and 2 so that answers favorable to the
defense would be early, and answers favorable to the prosecution late, in the num-
bering system. Then I introduce the measure I chose, and explain its interpretation
and relation to others not chosen. Finally I disuss the statistical model appropriate
to the data of Tables 1 and 2 and apply the chosen measure to these data.

Suppose items are indexed with an integer subscript g. Thus in considering the
Fitzgerald and Ellsworth data, $g = 1, 2, \ldots, 13$. Also, suppose that item g has n_g
responses. Then for the Fitzgerald and Ellsworth data, $n_1 = n_2 = \ldots = n_{10} = 4$, and
$n_{11} = n_{12} = n_{13} = 2$. The n_g responses can then be reordered if necessary so that
response 1 is most favorable to the defense, and response n_g most favorable to the
prosecution. This requires reversal of items 2, 5, 6, 8, and 10 in Table 1. The data
in Table 2 do not require this reordering.

Recall that groups holding the views A_3 (NEVER) and A_5 (AT LEAST
SOMETIMES) are of interest here. Suppose for the moment that the numbers $p_{i,l}$
were known, where $p_{i,l}$ is the proportion of jurors holding views A_l on capital punishment ($l = 3$ or 5), taking position i ($1 \leq i \leq n_g$) on an item g. How might these
numbers be summarized in a single number representing how much more favorble
to the defense is a typical juror in group A_3 by comparison with a juror in group
A_5 on item g?

One convenient notion, borrowed from nonparametric statistics, is the probability
that a random juror from group A_3 has a view less favorable to the defense than
does a random juror from group A_5 (with ties being regarded as equally likely to
go in either direction). The Mann-Whitney Statistic (1947) is an unbiased estimate
of this probability (Fraser 1957). Expressed symbolically,

$$W_g = \sum_{i=1}^{n_g} p_{i,5} \left(\sum_{i \geq i} p'_{i,3} + \tfrac{1}{2} p_{i,3} \right). \tag{1}$$

As an example, we do the calculations for question 12 or Table 1, interpreting the proportions from the data as if they were probabilities in the entire population.

$$W_{12} = p_{1,5}(p_{2,3} + p_{1,3}/2) + p_{2,5}(p_{2,3}/2)$$
$$= .491 (.398 + .602/2) + .509(.398/2)$$
$$= .4445.$$

Thus, for this example, the probability is less than 45 percent that a random NEVER juror is less favorable to the defense than is a random AT LEAST SOMETIMES juror on question 12, which relates to consideration of a confession reported in the news media.

It is convenient to reexpress W_g in odds form as

$$\phi = (1 - W_g)/W_g.$$

In the example $\phi_{12} = .5555/.4445 = 1.25$. Thus the odds are 1.25 to 1, or 5 to 4, that a random juror in group A_3 (NEVER) is more favorable to the defense on item 12 than is a random juror in group A_5 (AT LEAST SOMETIMES). If ϕ_g is greater than 1, this would indicate bias against the defense in the AT LEAST SOMETIMES group compared with the NEVER group; if ϕ_g is less than 1, this would indicate bias against the prosecution; finally, if ϕ_g is exactly 1, this would indicate no bias on item g. (This is why the reorientation of the data is convenient.)

The measures studied by Goodman and Kruskal (1963), particularly their measures γ and Somers's (1962) Δ are possible alternatives. Somers's measure and the Mann-Whitney W are related by

$$\Delta = 1 - 2W_g$$

so that $\phi = (1 + \Delta)/(1 - \Delta)$. Another alternative source of inspiration is the log-linear model literature, in which the work of Haberman (1974) specifically addresses tables with ordered classifications. Somers pointed out that his and the Goodman-Kruskal measures differ in their denominators. I found the Somers measure with the Mann-Whitney explanation convenient because its urn interpretation seemed natural to me for this problem. The log-linear approach would have imposed a particular parametric form on the data, and then would have interpreted the extent to which one group is more favorable to the defense than is the other in terms of a parameter in this distribution. This seemed to me to require extra assumptions that are really extraneous to the essential problem. For these reasons, I am satisfied with the Mann-Whitney-Somers Statistic transformed to odds (ϕ).

The next issue is an appropriate statistical model for the data reported in Tables 1 and 2, now taking into account that the numbers $p_{i,l}$ are not known with certainty. Two important alternatives are to consider the data as jointly multinomial, or to consider them independently multinomial, row by row. If one believed that the Fitzgerald-Ellsworth sample of Alameda County residents were typical of California potential jurors in the proportion of NEVER, and AT LEAST SOMETIMES views among them, the former would be appropriate. Since Alameda County, containing Berkeley and Oakland, has the reputation of being more liberal than California

TABLE 3. (Fitzgerald and Ellsworth 1983). How Much More Likely to Favor the Defense is the NEVER Group Than is the AT LEAST SOMETIMES Group?

Item	$\hat{\phi}$	$SD(\hat{\phi})$	$(\hat{\phi}-1)/SD(\hat{\phi})$	$Pr(\phi \leq 1)$
1. Better some guilty go free	1.642	.124	5.16	1.3×10^{-7}
2. Failure to testify indicates guilt	1.427	.081	5.26	8×10^{-8}
3. Consider worst criminal for mercy	2.456	.160	9.12	7.5×10^{-20}
4. District attorneys must be watched	1.131	.085	1.55	6.1×10^{-2}
5. Enforce all laws strictly	1.448	.091	4.93	4.2×10^{-7}
6. Guilty if brought to trial	1.149	.063	2.38	8.8×10^{-3}
7. Exclude illegally obtained evidence	1.224	.085	2.62	4.4×10^{-3}
8. Insanity plea is a loophole	1.796	.134	5.94	1.5×10^{-9}
9. Harsher treatment not solution to crime	1.836	.122	6.84	4.0×10^{-12}
10. Defense attorneys must be watched	1.451	.115	3.93	4.3×10^{-5}
11. More serious problem: unemployment, crime	1.296	.093	3.20	6.9×10^{-4}
12. Consider confession reported by news media	1.250	.078	3.19	7.1×10^{-4}
13. Infer guilt from silence	1.220	.047	4.67	1.5×10^{-6}

as a whole, this assumption would strain credibility. The same assumption for the Cowan-Thomspon-Ellsworth experiment, involving volunteers, would be even harder to justify. However, assuming a row-by-row multinomial distribution means assuming only that the views of the NEVER (AT LEAST SOMETIMES) jurors in the Fitzgerald-Ellsworth study and their behavior in the Cowan-Thompson-Ellsworth experiment are typical of those groups in California. This is a more reasonable assumption, and consequently I adopt it.

Under this model, maximum likelihood estimates for ϕ and an asymptotic standard error for it can be calculated. The former consists simply of substituting sample proportions for $p_{i,5}$ and $p_{i,3}$ in (1). Since the latter is a special case of a more complicated computation described later, I defer discussion of it.

Table 3 reports maximum likelihood estimates and asymptotic standard errors for each of the Fitzgerald-Ellsworth questions reported in Table 1. It shows that on each item, the odds are that a juror who would consider imposing the death penalty AT LEAST SOMETIMES (ALS) will be more favorable to the prosecution than one who would NEVER consider imposing the death penalty. The magnitudes of the odds of bias range from a modest finding of 1.131 on item 4, to very sizable odds on several other items. This analysis confirms and strengthens the Fitzgerald-Ellsworth conclusions about Table 1.

A discussion of the philosophy behind the computation of the fourth column is necessary. In this problem, the parameters are the $p_{i,l}$'s. There is a function of them, ϕ, of particular interest and a special value of ϕ, namely 1, of substantive concern. One classical method would be to test $\phi = 1$ against $\phi > 1$ at, say, the .05 level. If $(\hat{\phi} - 1)/SD(\hat{\phi})$ is greater than 1.645, significance at the .05 level is achieved. To say that $\hat{\phi}$ is significantly different from 1 at the .05 level says that if the system were exactly nondiscriminatory, the probability is less than 5 percent that $\hat{\phi}$ would be as large, or larger, than the value observed.

TABLE 4. (Cowan, Thompson, and Ellsworth 1983).
How Much More Likely to Vote for the Defense Is the
NEVER Group Than Is the AT LEAST SOMETIMES
Group?

$$\hat{\phi} = 1.652$$
$$SD(\hat{\phi}) = .265$$
$$(\hat{\phi} - 1)/SD(\hat{\phi}) = 2.45$$
$$Pr\{\phi \leq 1\} = 7.1 \times 10^{-3}$$

A Bayesian approach can be based on a theorem of Walker (1962), showing that the asymptotic posterior distribution of ϕ is normal, with mean $\hat{\phi}$ and standard deviation $SD(\hat{\phi})$. This result does not depend on the particular prior probability distribution in the parameter space chosen for the analysis, as long as it is smooth and has positive probability everywhere. The probability that ϕ is greater than 1 can now be calculated in a straightforward way.

The fourth column of Table 3 can thus be interpreted either as the significance level of a one-tailed test of $\phi = 1$ against the alternative $\phi > 1$, or as the probability that $\phi \leq 1$. I find the Bayesian interpretation more responsive to the legal question of the probability that the currently used juror selection procedure discriminates against the defense: the Bayesian analysis attempts to answer the question at hand, while the classical testing approach does not.

Table 4 gives a similar analysis of the Cowan-Thompson-Ellsworth experiment. Again the results indicate those who would NEVER impose capital punishment to be more lenient to the defense than are those who would impose it AT LEAST SOMETIMES, confirming the Cowan-Thompson-Ellsworth conclusions.

4. SURVEYS MEASURING THE SIZE OF THE DEATH PENALTY GROUPS

In order to conduct a further analysis of the Fitzgerald and Ellsworth attitude survey and the Cowan-Thompson and Ellsworth experiment, taking into account the effect of the exclusion of the automatic death penalty group, it is necessary to have data on the proportion of people in the population holding various attitudes toward the death penalty. In particular, it is necessary to have data on the size of the automatic death penalty group (ALWAYS). Two recent surveys, which contain the needed data, form the basis for the analysis presented in this article.

The first survey was conducted by the Field Research Corporation in March 1981 for the National Council on Crime and Delinquency. In this survey some 1,014 face-to-face interviews were conducted in a cluster sample of households. The respondents were asked about their general views on the death penalty, whether they would fairly and impartially decide the question of guilt or innocence, and whether in the guilt phrase they would NEVER favor the death penalty, or whether in AT LEAST SOME cases they would consider voting for it.

Field describes its sample as follows: "The sample presently consists of 200 primary sampling-point clusters. These primary sampling units enter the sample

TABLE 5. Attitudes of Californians on the Death
Penalty (Field Survey)

Would be fair and impartial in deciding guilt or innocence	89.8%
NEVER	11.3%
AT LEAST SOMETIMES	78.2%
Don't know/no answer/refused	.4%
Would not be fair and impartial	7.9%
Don't know/no answer/refused	2.2%

with a probability of selection in proportion to the population of California coun-ties. Specific cluster locations are determined by random selection of key addresses, using current telephone directories as the initial sampling frame within counties. Households in a given cluster are consecutively listed with a procedure to assure that the interviewers exert no influence on the selection of households. This procedure also draws non-telephone homes into the sample and permits telephone-density bias to be removed."

The results of the interviews are then weighted by giving each cluster of inter-views a weight inversely proportional to the density of listed telephones found in that cluster. A second-stage weighting by age and sex within geographic area is done using Census data. The weighted results are given in Table 5. The data indicate that 11.3 percent of the total adult population of California—which amounts to 12.6 percent of those in the adult population who could fairly and impartially try the guilt or innocence of a capital defendant—would be excluded from jury service in capital cases because they would never consider voting for the death penalty.

The Field survey does not permit disaggregation of the AT LEAST SOMETIMES imposers of the death penalty into the ALWAYS group and the SOMETIMES AND SOMETIMES NOT group. For this, I turn to a Harris survey conducted in January 1981 by a national telephone poll, with 100 clusters (telephone exchanges) stratified by geographic region and metropolitan/nonmetropolitan residence. The respondents were asked whether they strongly favored capital punishment, or whether there were some (or many, or most, or all) cases in which it is legally possible but not appropriate. Since only those who had said they were strongly in favor of the death penalty had given answers consistent with the ALWAYS position, only they were asked whether they would be fair and impartial in determining guilt or innocence in a capital case. Those who said they would be fair and impartial were asked whether they would ALWAYS vote to impose the death penalty, or whether they would con-sider the evidence, and hence be in the SOMETIMES AND SOMETIMES NOT category. Of the 1,499 people questioned, 13 were in the ALWAYS group. After reweighting for age, sex, and race, the data support an estimate that 1 percent of the adult American population falls into the automatic death penalty (ALWAYS) group. The Harris results are analyzed by region (East, Midwest, South, West), and area (city, suburbs, towns, rural), as well as by age, education, sex, race, income, type of work, religion, and union membership. No substantial regional differences in general attitudes toward the death penalty are revealed by the data.

There are two other sources cited in the *Hovey* decision (fn. 111) for indications as to the size of the automatic death penalty (ALWAYS) group. Jurow (1971) studied the views of 211 volunteer subjects who were employees of Sperry Rand. His subjects were more conservative than the population of New York City at the time. Of these, five answered that in a jury vote to determine the penalty for a serious crime, "I would always vote for the death penalty in a case where the law allows me to." This finding of 2.37 percent is not inconsistent with the 1981 Harris survey finding of 1 percent ALWAYS jurors since (a) those not fair and impartial were not eliminated and (b) a more conservative than average, volunteer population at one company cannot be taken to be representative of the national population.

The second study cited by the California Supreme Court in *Hovey* is Smith (1976), which relies on a 1973 Harris survey. Here 28 percent of the respondents answered that all persons convicted of first-degree murder should receive the death penalty. However, this question does not distinguish between people's views on what the law should be and how they would vote as jurors. Jurow (1971), for one, found 21 percent of his population strongly favored the death penalty, while only 2.37 percent would impose it in every case. Additionally, the Harris 1973 question—unlike the 1981 question—does not tell the respondents that it would be their obligation as jurors to follow the law, and that the judge would instruct them that the law requires them to consider all penalties and to weigh all the evidence before making up their minds. Thus the 1973 Harris survey cited by Smith did not ask the relevant question to determine the size of the ALWAYS group. Consequently, we use the 1981 Harris survey number of 1 percent because, unlike Jurow. it was based on a national probability sample, and, unlike the 1973 Harris survey, it asked the relevant set of questions.

5. ASSUMPTIONS AND A MODEL

It is entirely possible that an analysis will reveal that on some questions the ALWAYS OR NEVER group is more favorable to the defense than the SOMETIMES AND SOMETIMES NOT group, and that on other questions the reverse might be true. In order to learn whether this is the case, the findings of the Cowan, Thompson, and Ellsworth experiment and of each of the 13 attitude questions in the Fitzgerald and Ellsworth study are considered separately.

The first assumption I make is that the Field survey and the Harris survey are random samples from the population of jurors in question. Thus if $p_l = \Pr(A_l)$, the Field survey contributes a likelihood function

$$p_1^{f_1} \, p_3^{f_3} \, p_5^{f_5}, \tag{2}$$

where f_l is the weighted number of persons found in the Field survey to be in category A_l ($l = 1, 3, 5$). Similarly the Harris survey contributes a likelihood

$$p_2^{h_2} (1 - p_2)^{h_7}, \tag{3}$$

where h_2 is the weighted number of persons in the Harris survey to be in category A_2, and h_7 are the others in the Harris survey. In making this assumption, I am relying on the geographical homogeneity found in both surveys to indicate that Californians are very similar to other Americans in their views, and on the expertise of the Field and Harris organizations in doing their surveying properly.

My second assumption is that the attitude survey and the experiment constitute random samples from the populations studied (A_3 and A_5). Thus if j_i persons in group A_3 take position i on item g, and k_i persons in group A_5 take the position $i(1 \leq i \leq n_g)$, the contribution to the likelihood is

$$\prod_{i=1}^{n_k} p_{i,3}^{ji} p_{i,5}^{ki} \tag{4}$$

One need not assume that the number of subjects in groups A_3 and A_5 in the Cowan, Thompson, and Ellsworth and the Fitzgerald and Ellsworth studies is typical of the population, but only that the views of the subjects in groups A_3 and A_5 are typical of the views of those groups in the population at large.

Finally, some assumption is necessary about the views of the group A_2, since these were not directly measured. The harsher these views are to the defense, the more difficult the defense's task of showing that exclusion of the whole group ($A_6 = A_2 \cup A_3$) is disadvantageous to them. Accordingly I take the most conservative stance on this issue. Those in group A_2 are assumed, on each item g, to take the position n_g most opposed to the defense's interests. Inevitably, this assumption is, to some extent, false; to the extent that it is false, my estimates ϕ will be too low. To the extent that bias against the defense is shown using this assumption, the real extent of the bias against the defense is greater by some amount.

In summary, the likelihood function for the data is

$$\mathcal{L} = p_1^{f1} p_3^{f3} p_5^{f5} p_2^{h1} (1-p_2)^{h_2} \prod_{i=1}^{n_g} p_{i,3}^{ji} p_{i,5}^{ki}, \tag{5}$$

where $p_{i,5} = (p_2 p_{i,2} + p_4 p_{i,4})/(p_2 + p_4)$, and where $p_{i,2} = 0$ $(i \neq n_g)$ and $p_{ng,2} = 1$.

The maximum likelihood estimates are found in closed form by straightforward maximization of (5). The computation of the asymptotic standard deviation is equally straightforward, but considerably more tedious. The principal danger is that one might forget the Goodman-Kruskal (1972) admonition not to change parameterization in midcomputation. An Appendix giving the details of both computations is available on request from the author.

6. IMPLEMENTATION AND RESULTS

Before implementing the model in the last section, several details must be discussed about how to relate the model of Section 5 to the surveys of Section 4. First, the "don't know/no answer/refused" groups were distributed among the other categories in proportion to their size. After doing this, the Field survey numbers f_1, f_3, and f_5 are taken to be proportional to .0808, .1162, and .8040, respectively. The Harris numbers h_2 and h_7 are taken from the weighted analysis to be proportional to .01 and .99, respectively.

The question of effective sample size arises in both studies, because both used cluster sampling (see Kish 1965). Hence the actual sample sizes, 1,014 for the Field survey and 1,499 for the Harris survey, may overstate the amount of information available.

The Field survey reports the results of six replicates of approximately equal size, weighted as were the results reported in Table 4. For NEVER, the replicate standard error is .011, compared with the standard error computed as if simple random sampling had occurred, which is $[(.113)(.887)/1014]^{1/2} = .0099$. Thus the ratio is $.011/.0099 \cong 1.1$. Similarly from the AT LEAST SOMETIMES group, the reported standard error is .010, which gives a ratio of 1.3. Finally the NOT FAIR AND IMPARTIAL question has a standard error of .088, yielding a ratio of 1.06. I take the median of these numbers, 1.1, as representative of them, and hence use the effective sample size for the Field survey of $1,014/(1.1)^2 = 838$.

In the Harris survey, each of the 13 in the ALWAYS group was found in a different cluster. This cannot be taken as evidence of a correlation due to being in the same cluster, so I use the unadjusted sample size, 1,499, as the effective sample size for the Harris survey.

6.1 The Fitzgerald and Ellsworth Study

Using these effective sample sizes, values for $\hat{\phi}$, $SD(\hat{\phi})$, $(\hat{\phi}-1)/SD(\hat{\phi})$, and $\Pr\{\phi \le 1\}$ were calculated for each item in the Fitzgerald and Ellsworth study, and are recorded in Table 6. In analyzing the results of Table 6, recall the assumption that the ALWAYS group takes the worst position on each question from the defense viewpoint. To the extent that this extreme assumption is wrong, the numbers of $\hat{\phi}$ recorded in Table 6 are too low.

I propose a working assumption that estimated odds ($\hat{\phi}$) closer to 1 than .1 reveal an item close enough to nondiscriminatory that it need not be worried about. These items are those closer to even than 11 to 10 on the high side, and 9 to 10 on the low side. This eliminates four items, numbers 4, 6, 7, and 13. All nine of the others display bias against the defense, rising to estimated odds of 2 to 1 for item 3. I conclude from Table 6 a showing of substantial bias against the defense from the current procedure.

6.2 The Cowan, Thompson, and Ellsworth Study

The showing of substantial bias in the Fitzgerald and Ellsworth study is confirmed by the reanalysis of the Cowan, Thompson, and Ellsworth experiment. As shown in Table 7, $\hat{\phi}$ is 1.519 so the estimated odds of a NEVER or ALWAYS juror being more favorable to the defense than a SOMETIMES AND SOMETIMES NOT juror are more than 3 to 2. Hence the probability of neutrality or bias against the prosecution ($\phi \le 1$) is 1.3 percent. Again we have a finding of substantial bias against the defense.

What makes all the results true is that the ALWAYS group is so small (1% of the population) that the ALWAYS or NEVER group is dominated by the NEVER part, and the AT LEAST SOMETIMES group is dominated by the SOMETIMES AND SOMETIMES NOT majority. Even attributing the least favorable views to the

TABLE 6. (Fitzgerald and Ellsworth 1983). How Much More Favorable to the Defense Is the ALWAYS OR NEVER Group (A_6) Than Is the SOMETIMES AND SOMETIMES NOT (A_4) Group?

Item	$\hat{\phi}$	SD($\hat{\phi}$)	$(\hat{\phi} - 1)/\mathrm{SD}(\hat{\phi})$	Pr($\phi \leq 1$)
1. Better some guilty go free	1.380	.132	2.89	1.9×10^{-3}
2. Failure to testify indicates guilt	1.189	.108	1.75	4.0×10^{-2}
3. Consider worst criminal for mercy	2.006	.174	5.78	4.0×10^{-9}
4. District Attorneys must be watched	.974	.101	−.26	6.0×10^{-1}
5. Enforce all laws strictly	1.252	.106	2.38	8.7×10^{-3}
6. Guilty if brought to trial	.977	.089	−.26	6.0×10^{-1}
7. Exclude illegally obtained evidence	1.040	.100	.40	3.4×10^{-1}
8. Insanity plea is a loophole	1.558	.133	4.18	1.6×10^{-5}
9. Harsher treatment not crime solution	1.506	.143	3.54	2.0×10^{-4}
10. Defense Attorneys must be watched	1.256	.120	2.13	1.7×10^{-2}
11. More serious problem: unemployment, crime	1.184	.091	2.02	2.2×10^{-2}
12. Consider confession reported by news media	1.121	.086	1.42	7.8×10^{-2}
13. Infer guilty from silence	1.044	.085	.52	3.0×10^{-1}

TABLE 7. (Cowan, Thompson, and Ellsworth 1983). How Much More Favorable to the Defense are the Votes of the ALWAYS OR NEVER Group (A6) Compared to the SOMETIMES AND SOMETIMES NOT (A4) Group?

$$\hat{\phi} = 1.519$$
$$\mathrm{SD}(\hat{\phi}) = .229$$
$$(\hat{\phi} - 1)/\mathrm{SD}(\hat{\phi}) = 2.27$$
$$\mathrm{Pr}(\phi \leq 1) = 1.3\%$$

ALWAYS group does not disturb the finding of substantial bias against the defense. Of course, this extreme assumption does reduce the estimated odds in Table 6 compared with those in Table 3, and in Table 7 compared with those in Table 4, as arithmetically it must.

7. CONCLUSION

The Cowan-Thompson-Ellsworth experiment, because it deals with a simulated juror vote, is the evidence most closely related to actual juror behavior. Consequently, I would pay most attention to the odds (about 3 to 2) of a more proprosecution vote in that experiment among the SOMETIMES AND SOMETIMES NOT group than among the ALWAYS OR NEVER group. This conclusion is strengthened by the finding that 9 of the 13 attitude questions from the Fitzgerald-Ellsworth study show the same favoritism to the prosecution among the SOMETIMES AND SOMETIMES NOT group compared with the ALWAYS or NEVER group. (The other four attitude questions show essentially no difference.) This conclusion is further strengthened by the fact these calculations are conducted under the extreme assumption that those who would ALWAYS impose the death penalty, but would reach the question of guilt fairly and impartially, would unanimously vote to convict in the experiment, and would unanimously take the most proprosecution view of each attitude question. Therefore there appears to be a distinct and substantial antidefense bias in the exclusion for cause of the ALWAYS OR NEVER group from the jury deciding guilt or innocence.

[*Received November 1981. Revised April 1983.*]

REFERENCES

ALEXANDER, JAMES (ed.) (1963), *A Brief Narration of the Case and Trial of John Peter Zenger,* Cambridge, Mass.: Harvard University Press.

BRONSON, E.J. (1970), "On the Conviction-Proneness and Representativeness of the Death-Qualified Jury: An Empirical Study of Colorado Veriremen," 42 U. Colo. L Rev. 1.

COWAN, CLAUDIA, THOMPSON, WILLIAM, and ELLSWORTH, PHOEBE C. (1984), "The Effects of Death Qualification on Jurors' Predisposition to Convict and on the Quality of Deliberation," *Law and Human Behavior,* to appear.

FITZGERALD, ROBERT, and ELLSWORTH, PHOEBE C. (1984), "Due Process vs. Crime Control: Death Qualification and Jury Attitudes," *Law and Human Behavior,* to appear.

FRASER, D.S. (1957), *Nonparametric Methods in Statistics,* New York: John Wiley, 268, 269.

GOLDBERG, F. (1970), "Toward Expansion of Witherspoon: Capital Scruples, Jury Bias, and the Use of Psychological Data to Raise Presumption in the Law," *Harvard Civil Rights and Civil Liberties Law Review,* 5, 53.

GOODMAN, L.A., and KRUSKAL, W.H. (1954), "Measures of Association for Cross Classifications," *Journal of the American Statistical Association,* 49, 732–764.

—— (1959), "Measures of Association for Cross Classifications II, Further Discussion and References," *Journal of the American Statistical Association,* 54, 123–163.

—— (1963), "Measures of Association for Cross Classifications III. Approximate Sampling Theory," *Journal of the American Statistical Association,* 58, 310–364.

—— (1972), "Measures of Association for Cross Classifications IV. Simplification of Asymptotic Variances," *Journal of the American Statistical Association,* 67, 415–521.

HABERMAN, S.J. (1974), "Log-Linear Models for Frequency Tables with Ordered Classifications," *Biometrics,* 30, 589–600.

HOVEY V. SUPERIOR COURT (1980). 28 Cal. 3d1, 616 P.2d 1301.

JUROW, G.L. (1971), "New Data on the Effect of a Death-Qualified Jury on the Guilt Determination Process," *Harvard Law Review,* 84, 567.

KADANE, J.B., and LEHOCZKY, J.P. (1976). "Random Juror Selection from Multiple Lists," *Operations Research,* 24, 207–219.

KAIRYS, D., KADANE, J.B., and LEHOCZKY, J.P. (1977), "Jury Representativeness: A Mandate for Multiple Source Lists," *California Law Review,* 65, 776–827.

KISH, L. (1965), *Survey Sampling,* New York: John Wiley.

MANN, H.B., and WHITNEY, D.R. (1947), "On a Test of Whether One of Two Random Variables Is Stochastically Larger than the Other," *Annals of Mathematical Statistics.* 18, 50–60.

SMITH, T.W. (1976), "A Trend Analysis of Attitudes Toward Capital Punishment, 1936–1974," in *Studies in Social Change Since 1948,* ed. James A. Davis. NORC Report 127B, Chicago: National Opinion Research Center.

SOMERS, R.H. (1962), "A New Asymmetric Measure of Association for Ordinal Variables," *American Sociological Review,* 27, 799–811.

VAN DYKE, J.M. (1977), *Jury Selection Procedures,* Cambridge, Mass.: Ballinger.

VIDMAR, N., and ELLSWORTH, PHOEBE (1974), "Public Opinion and the Death Penalty," *Stanford Law Review,* 26.

WALKER, A.M. (1962), "On the Asymptotic Behavior of Posterior Distributions," *Journal of the Royal Statistical Society,* Ser. B, 31, 80–89.

WILSON, W.C. (1964), "Belief in Capital Punishment and Jury Performance," University of Texas, unpublished.

WITHERSPOON V. ILLINOIS (1968), 391 U.S. 510.

2

LOCKHART V. McCREE

LOCKHART, DIRECTOR, ARKANSAS DEPARTMENT OF CORRECTIONS v. **McCREE**

No. 84–1865

SUPREME COURT OF THE UNITED STATES

476 U. S. 162; 106 S. Ct. 1758; 90 L. Ed. 2d 137; 54 U.S.L.W. 4449

January 13, 1986, Argued

May 5, 1986, Decided

At respondent's Arkansas state-court trial for capital felony murder, the judge at *voir dire* removed for cause, over respondent's objections, those prospective jurors who stated that they could not under any circumstances vote for the imposition of the death penalty—that is, so-called "*Witherspoon*-excludables" under the principles of *Witherspoon* v. *Illinois,* 391 U.S. 510. The jury convicted respondent, but at the sentencing phase of the trial it rejected the State's request for the death penalty and set punishment at life imprisonment without parole. The conviction was affirmed on appeal, and respondent's petition for state postconviction relief was denied. He then sought federal habeas corpus relief, contending that the "death qualification" of the jury by the removal for cause of the "*Witherspoon*-excludables" violated his rights under the Sixth and Fourteenth Amendments to have his guilt or innocence determined by an impartial jury selected from a representative cross section of the community. The District Court ruled that "death qualification" of the jury prior to the guilt phase of the bifurcated trial violated both the fair-cross-section and the impartiality requirements of the Constitution. The Court of Appeals affirmed on the ground that removal for cause of "*Witherspoon*-excludables" violated respondent's Sixth Amendment right to a jury selected from a fair cross section of the community.

Held: The Constitution does not prohibit the removal for cause, prior to the guilt phase of a bifurcated capital trial, of prospective jurors whose opposition to the death penalty is so strong that it would prevent or substantially impair the performance of their duties as jurors at the sentencing phase of the trial. This is so even assuming, *arguendo*, that the social science studies introduced in the courts below were adequate to establish that "death qualification" in fact produces juries somewhat more "conviction-prone" than "non-death-qualified" juries. Pp. 173–183.

(a) "Death qualification" of a jury does not violate the fair-cross-section requirement of the Sixth Amendment, which applies to jury panels or venires but does not require that petit juries actually chosen reflect the composition of the community at large. Even if the requirement were extended to petit juries, the essence of a fair-cross-section claim is the systematic exclusion of a "distinctive group" in the community—such as blacks, women, and Mexican-Americans—for reasons completely unrelated to the ability of members of the group to serve as jurors in a particular case. Groups defined solely in terms of shared attitudes that would prevent or substantially impair members of the group from performing one of their duties as jurors, such as the "*Witherspoon*-excludables" at issue here, are not "distinctive groups" for fair-cross-section purposes. "Death qualification" is carefully designed to serve the State's legitimate interest in obtaining a single jury that can properly and impartially apply the law to the facts of the case at both the guilt and sentencing phases of a capital trial. Pp. 173–177.

(b) Nor does "death qualification" of a jury violate the constitutional right to an impartial jury on the theory asserted by respondent that, because all individual jurors are to some extent predisposed towards one result or another, a constitutionally impartial jury can be constructed only by "balancing" the various predispositions of the individual jurors, and when the State "tips the scales" by excluding prospective jurors with a particular viewpoint, an impermissibly partial jury results. An impartial jury consists of nothing more than jurors who will conscientiously apply the law and find the facts. Respondent's view of jury impartiality is both illogical and impractical.

Neither *Witherspoon, supra,* nor *Adams* v. *Texas,* 448 U.S. 38, supports respondent's contention that a State violates the Constitution whenever it "slants" the jury by excluding a group of individuals more likely than the population at large to favor the defendant. Here, the removal for cause of "*Witherspoon*-excludables" serves the State's entirely proper interest in obtaining a single jury (as required by Arkansas law) that could impartially decide all of the issues at both the guilt and the penalty phases of respondent's trial. Moreover, both *Witherspoon* and *Adams* deal with the special context of capital sentencing, where the range of jury discretion necessarily gave rise to far greater concern over the effects of an "imbalanced" jury. The case at bar, by contrast, deals not with capital sentencing, but with the jury's more traditional role of finding the facts and determining the guilt or innocence of a criminal defendant, where jury discretion is more channeled. Pp. 177–183.

COUNSEL: John Steven Clark, Attorney General of Arkansas, argued the cause for petitioner. With him on the briefs were Jack Gillean, Assistant Attorney General, Victra L. Fewell, and Leslie M. Powell.

Samuel R. Gross argued the cause for respondent. With him on the brief were John Charles Boger, James S. Liebman, William R. Wilson, Jr., and Anthony G. Amsterdam.*

Briefs of amici curiae urging affirmance were filed for the National Center on Institutions and Alternatives by Allan Blumstein and Eric M. Freedman; for Robert Popper et al. by Robert Popper, pro se; and for Billy Junior Woodward by Reed E. Hundt and Thomas M. Carpenter.

Donald N. Bersoff filed a brief for the American Psychological Association as amicus curiae.

JUDGES: REHNQUIST, J., delivered the opinion of the Court, in which BURGER, C. J., and WHITE, POWELL, and O'CONNOR, JJ., joined. BLACKMUN, J., concurred in the result. MARSHALL, J., filed a dissenting opinion, in which BRENNAN and STEVENS, JJ., joined,

OPINION BY: REHNQUIST

OPINION: JUSTICE REHNQUIST DELIVERED THE OPINION OF THE COURT

In this case we address the question left open by our decision nearly 18 years ago in *Witherspoon* v. *Illinois,* 391 U.S. 510 (1968): Does the Constitution prohibit the removal for cause, prior to the guilt phase of a bifurcated capital trial, of prospective jurors whose opposition to the death penalty is so strong that it would prevent or substantially impair the performance of their duties as jurors at the sentencing phase of the trial? See *id.,* at 520, n. 18; *Bumper* v. *North Carolina,* 391 U.S. 543, 545 (1968). We hold that it does not.

Respondent Ardia McCree filed a habeas corpus petition in the United States District Court for the Eastern District of Arkansas claiming that such removal for cause violated the Sixth and Fourteenth Amendments and, after McCree's case was consolidated with another habeas case involving the same claim on remand from the Court of Appeals for the Eighth Circuit, the District Court ruled in McCree's favor

---

* Briefs of amici curiae urging reversal were filed for the State of Alabama et al. by Susan Crump, David Crump, Charles K. Graddick, Attorney General of Alabama, John J. Kelly, Chief State's Attorney of Connecticut, Jim Smith, Attorney General of Florida, Michael J. Bowers, Attorney General of Georgia, James Thomas Jones, Attorney General of Idaho, Neil F. Hartigan, Attorney General of Illinois, Linley E. Pearson, Attorney General of Indiana, Edwin Lloyd Pittman, Attorney General of Mississippi, William L. Webster, Attorney General of Missouri, Irwin I. Kimmelman, Attorney General of New Jersey, Paul Bardacke, Attorney General of New Mexico, David B. Frohnmayer, Attorney General of Oregon, Mark V. Meierhenry, Attorney General of South Dakota, Jim Mattox, Attorney General of Texas, Kenneth O. Eikenberry, Attorney General of Washington, and Stephen E. Merrill, Attorney General of New Hampshire; and for the State of Arizona et al. by Michael C. Turpen, Attorney General of Oklahoma, and David W. Lee, Hugh A. Manning, Tomilou Gentry Liddell, Robert A. Nance, and Jean M. LeBlanc, Assistant Attorneys General, Robert K. Corbin, Attorney General of Arizona, John Van de Kamp, Attorney General of California, Charles M. Oberly, Attorney General of Delaware, David L. Armstrong, Attorney General of Kentucky, William J. Guste, Jr., Attorney General of Louisiana, Stephen H. Sachs, Attorney General of Maryland, Robert M. Spire, Attorney General of Nebraska, Brian McKay, Attorney General of Nevada, Stephen E. Merrill, Attorney General of New Hampshire, Anthony J. Celebrezze, Jr., Attorney General of Ohio, LeRoy S. Zimmerman, Attorney General of Pennsylvania, Travis Medlock, Attorney General of South Carolina, W. J. Michael Cody, Attorney General of Tennessee, David L. Wilkinson, Attorney General of Utah, William G. Broaddus, Attorney General of Virginia, and Archie G. McClintock, Attorney General of Wyoming.

and granted habeas relief. *Grigsby* v. *Mabry,* 569 F.Supp. 1273 (1983). A sharply divided Eighth Circuit affirmed, *Grigsby* v. *Mabry,* 758 F.2d 226 (1985) (en banc), creating a conflict with recent decisions of the Fourth, Fifth, Seventh, and Eleventh Circuits. See *Keeten* v. *Garrison,* 742 F.2d 129, 133–135 (CA4 1984), cert. pending, No. 84–6187; *Smith* v. *Balkcom,* 660 F.2d 573, 576–578 (CA5 1981), modified on other grounds, 671 F.2d 858, cert. denied *sub nom. Tison* v. *Arizona,* 459 U.S. 882 (1982); *Spinkellink* v. *Wainwright,* 578 F.2d 582, 594 (CA5 1978), cert. denied, 440 U.S. 976 (1979); *United States ex rel. Clark* v. *Fike,* 538 F.2d 750, 761–762 (CA7 1976), cert. denied, 429 U.S. 1064 (1977); and *Corn* v. *Zant,* 708 F.2d 549, 564 (CA.11 1983), cert. denied, 467 U.S. 1220 (1984). We granted certiorari to resolve the conflict, 474 U.S. 816 (1985), and now reverse the judgment of the Eighth Circuit.

On the morning of February 14, 1978, a combination gift shop and service station in Camden, Arkansas, was robbed and Evelyn Boughton, the owner, was shot and killed. That afternoon, Ardia McCree was arrested in Hot Springs, Arkansas, after a police officer saw him driving a maroon and white Lincoln Continental matching an eyewitness' description of the getaway car used by Boughton's killer. The next evening, McCree admitted to police that he had been at Boughton's shop at the time of the murder. He claimed, however, that a tall black stranger wearing an overcoat first asked him for a ride, then took McCree's rifle out of the back of the car and used it to kill Boughton. McCree also claimed that, after the murder, the stranger rode with McCree to a nearby dirt road, got out of the car, and walked away with the rifle. McCree's story was contradicted by two eyewitnesses who saw McCree's car between the time of the murder and the time when McCree said the stranger got out and walked away, and who stated that they saw only one person in the car. The police found McCree's rifle and a bank bag from Boughton's shop alongside the dirt road. Based on ballistics tests, a Federal Bureau of Investigation officer testified that the bullet that killed Boughton had been fired from McCree's rifle.

McCree was charged with capital felony murder in violation of Ark. Stat. Ann. § 41–1501(1)(a) (1977). In accordance with Arkansas law, see *Neal* v. *State,* 259 Ark. 27, 31, 531 S. W. 2d 17, 21 (1975), the trial judge at *voir dire* removed for cause, over McCree's objections, those prospective jurors who stated that they could not under any circumstances vote for the imposition of the death penalty. Eight prospective jurors were excluded for this reason. The jury convicted McCree of capital felony murder, but rejected the State's request for the death penalty, instead setting McCree's punishment at life imprisonment without parole. McCree's conviction was affirmed on direct appeal, *McCree* v. *State,* 266 Ark. 465, 585 S. W. 2d 938 (1979), and his petition for state post-conviction relief was denied.

McCree then filed a federal habeas corpus petition raising, *inter alia,* the claim that "death qualification," or the removal for cause of the so-called "*Witherspoon-*excludable" prospective jurors,[1] violated his right under the Sixth and Fourteenth

---

1. In *Wainwright* v. *Witt,* 469 U.S. 412 (1985), this Court emphasized that the Constitution does not require "ritualistic adherence" to the "talismanic" standard for juror exclusion set forth in footnote 21 of the *Witherspoon* opinion. 469 U.S., at 419, 423. Rather, the proper constitutional standard is simply whether a prospective juror's views would "prevent or substantially impair the performance of his duties as a juror in accordance with his

Amendments to have his guilt or innocence determined by an impartial jury selected from a representative cross section of the community. By stipulation of the parties, this claim was consolidated with another pending habeas case involving the same claim, which had been remanded by the Eighth Circuit for an evidentiary hearing in the District Court. App. 9–11; *Grigsby* v. *Mabry,* 637 F.2d 525 (1980). The District Court denied the remainder of McCree's petition, and the Eighth Circuit affirmed. *McCree* v. *Housewright,* 689 F.2d 797 (1982), cert. denied, 460 U.S. 1088 (1983).

The District Court held a hearing on the "death qualification" issue in July 1981, receiving in evidence numerous social science studies concerning the attitudes and beliefs of "*Witherspoon*-excludables," along with the potential effects of excluding them from the jury prior to the guilt phase of a bifurcated capital trial. In August 1983, the court concluded, based on the social science evidence, that "death quali-fication" produced juries that "were more prone to convict" capital defendants than "non-death-qualified" juries. *Grigsby* v. *Mabry,* 569 F.Supp., at 1323. The court ruled that "death qualification" thus violated both the fair-cross-section and impar-tiality requirements of the Sixth and Fourteenth Amendments, and granted McCree habeas relief. *Id.,* at 1324.[2]

The Eighth Circuit found "substantial evidentiary support" for the District Court's conclusion that the removal for cause of "*Witherspoon*-excludables" resulted in "conviction-prone" juries, and affirmed the grant of habeas relief on the ground that such removal for cause violated McCree's constitutional right to a jury selected from a fair cross section of the community. *Grigsby* v. *Mabry,* 758 F.2d, at 229. The Eighth Circuit did not address McCree's impartiality claim. *Ibid.* The Eighth Circuit left it up to the discretion of the State "to construct a fair process" for future capital trials that would comply with the Sixth Amendment. *Id.,* at 242–243. Four judges dissented. *Id.,* at 243–251.

Before turning to the legal issues in the case, we are constrained to point out what we believe to be several serious flaws in the evidence upon which the courts below reached the conclusion that "death qualification" produces "conviction-prone" juries.[3]

---

instructions and his oath." *Id.,* at 433, quoting *Adams* v. *Texas,* 448 U.S. 38, 45 (1980). Thus, the term *"Witherspoon*-excludable" is something of a misnomer. Nevertheless, because the parties and the courts below have used the term *"Witherspoon*-excludables" to identify the group of prospective jurors at issue in this case, we will use the same term in this opinion.

2. James Grigsby, the habeas petitioner with whose case McCree's had been consolidated, died prior to the District Court's decision, so his case became moot. *Grigsby* v. *Mabry,* 569 F.Supp., at 1277, n. 2. Dewayne Hulsey, a third habeas petitioner whose "death qualification" claim was consolidated with Grigsby's and McCree's, was found to be procedurally barred, under *Wainwright* v. *Sykes,* 433 U.S. 72 (1977), from asserting the claim. *Hulsey* v. *Sargent,* 550 F.Supp. 179 (ED Ark. 1981).

3. McCree argues that the "factual" findings of the District Court and the Eighth Circuit on the effects of "death qualification" may be reviewed by this Court only under the "clearly erroneous" standard of Federal Rule of Civil Procedure 52(a). Because we do not ultimately base our decision today on the invalidity of the lower courts' "factual" findings, we need not decide the "standard of review" issue. We are far from persuaded, however, that the "clearly erroneous" standard of Rule 52(a) applies to the kind of "legislative" facts at issue here. See generally *Dunagin* v. *City of Oxford, Mississippi,* 718 F.2d 738, 748, n. 8 (CA5 1983) (en banc) (plurality opinion of Reavley, J.). The dif-ficulty with applying such a standard to "legislative" facts is evidenced here by the fact that at least one other Court of Appeals, reviewing the same social science studies as introduced by McCree, has reached a conclusion contrary to that of the Eighth Circuit. See *Keeten* v. *Garrison,* 742 F.2d 129, 133, n. 7 (CA4 1984) (disagreeing that studies show relationship between generalized attitudes and behavior as jurors), cert. pending, No. 84–6187.

McCree introduced into evidence some 15 social science studies in support of his constitutional claims, but only 6 of the studies even purported to measure the potential effects on the guilt-innocence determination of the removal from the jury of "*Witherspoon*-excludables."[4] Eight of the remaining nine studies dealt solely with generalized attitudes and beliefs about the death penalty and other aspects of the criminal justice system, and were thus, at best, only marginally relevant to the constitutionality of McCree's conviction.[5] The 15th and final study dealt with the effects on prospective jurors of *voir dire* questioning about their attitudes toward the death penalty,[6] an issue McCree raised in his brief to this Court but that counsel for McCree admitted at oral argument would not, standing alone, give rise to a constitutional violation.[7]

Of the six studies introduced by McCree that at least purported to deal with the central issue in this case, namely, the potential effects on the determination of guilt or innocence of excluding "*Witherspoon*-excludables" from the jury, three were also before this Court when it decided *Witherspoon*.[8] There, this Court reviewed the studies and concluded:

"The data adduced by the petitioner...are too tentative and fragmentary to establish that jurors not opposed to the death penalty tend to favor the prosecution in the determination of guilt. We simply cannot conclude, either on the basis of the record now before us or as a matter of judicial notice, that the exclusion of jurors opposed to capital punishment results in an unrepresentative jury on the issue of guilt or substantially increases the risk of conviction. In light of the presently available

4. The Court of Appeals described the following studies as "conviction-proneness surveys": H. Zeisel, Some Data on Juror Attitudes Toward Capital Punishment (University of Chicago Monograph 1968) (Zeisel); W. Wilson, Belief in Capital Punishment and Jury Performance (unpublished manuscript, University of Texas, 1964) (Wilson); Goldberg, Toward Expansion of *Witherspoon:* Capital Scruples, Jury Bias, and Use of Psychological Data to Raise Presumptions in the Law, 5 Harv. Civ. Rights-Civ. Lib. L. Rev. 53 (1970) (Goldberg); Jurow, New Data on the Effect of a "Death Qualified" Jury on the Guilt Determination Process, 84 Harv. L. Rev. 567 (1971) (Jurow); and Cowan, Thompson, & Ellsworth, The Effects of Death Qualification on Jurors' Predisposition to Convict and on the Quality of Deliberation, 8 Law & Hum. Behav. 53 (1984) (Cowan-Deliberation). In addition, McCree introduced evidence on this issue from a Harris Survey conducted in 1971. Louis Harris & Associates, Inc., Study No. 2016 (1971) (Harris-1971).

5. The Court of Appeals described the following studies as "attitudinal and demographic surveys": Bronson, On the Conviction Proneness and Representativeness of the Death-Qualified Jury: An Empirical Study of Colorado Veniremen, 42 U. Colo. L. Rev. 1 (1970); Bronson, Does the Exclusion of Scrupled Jurors in Capital Cases Make the Jury More Likely to Convict? Some Evidence from California, 3 Woodrow Wilson L. J. 11 (1980); Fitzgerald & Ellsworth, Due Process vs. Crime Control: Death Qualification and Jury Attitudes, 8 Law & Hum. Behav. 31 (1984); and Precision Research, Inc., Survey No. 1286 (1981). In addition, McCree introduced evidence on these issues from Thompson, Cowan, Ellsworth, & Harrington, Death Penalty Attitudes and Conviction Proneness, 8 Law & Hum. Behav. 95 (1984); Ellsworth, Bukaty, Cowan, & Thompson, The Death-Qualified Jury and the Defense of Insanity, 8 Law & Hum. Behav. 81 (1984); A. Young, Arkansas Archival Study (unpublished, 1981); and various Harris, Gallup, and National Opinion Research Center polls conducted between 1953 and 1981.

6. McCree introduced evidence on this issue from Haney, On the Selection of Capital Juries: The Biasing Effects of the Death-Qualification Process, 8 Law & Hum. Behav. 121 (1984).

7. We would in any event reject the argument that the very process of questioning prospective jurors at *voir dire* about their views of the death penalty violates the Constitution. McCree concedes that the State may challenge for cause prospective jurors whose opposition to the death penalty is so strong that it would prevent them from impartially determining a capital defendant's guilt or innocence. *Ipso facto*, the State must be given the opportunity to identify such prospective jurors by questioning them at *voir dire* about their views of the death penalty.

8. The petitioner in *Witherspoon* cited the Wilser and Goldberg studies, and a prepublication draft of the Zeisel study. 391 U.S., at 517, n. 10; see n. 4, *supra*.

information, we are not prepared to announce a *per se* constitutional rule requiring the reversal of every conviction returned by a jury selected as this one was." 391 U.S. at 517–518 (footnote omitted).

It goes almost without saying that if these studies were "too tentative and fragmentary" to make out a claim of constitutional error in 1968, the same studies, unchanged but for having aged some 18 years, are still insufficient to make out such a claim in this case.

Nor do the three post-*Witherspoon* studies introduced by McCree on the "death qualification" issue provide substantial support for the *"per se* constitutional rule" McCree asks this Court to adopt. All three of the "new" studies were based on the responses of individuals randomly selected from some segment of the population, but who were not actual jurors sworn under oath to apply the law to the facts of an actual case involving the fate of an actual capital defendant.[9] We have serious doubts about the value of these studies in predicting the behavior of actual jurors. See *Grigsby v. Mabry,* 758 F.2d, at 248, n. 7 (J. Gibson, J., dissenting). In addition, two of the three "new" studies did not even attempt to simulate the process of jury deliberation,[10] and none of the "new" studies was able to predict to what extent, if any, the presence of one or more *"Witherspoon*-excludables" on a guilt-phase jury would have altered the outcome of the guilt determination.[11]

Finally, and most importantly, only one of the six "death qualification" studies introduced by McCree even attempted to identify and account for the presence of so-called "nullifiers," or individuals who, because of their deep-seated opposition to the death penalty, would be unable to decide a capital defendant's guilt or innocence fairly and impartially.[12] McCree concedes, as he must, that "nullifiers" may properly be excluded from the guilt-phase jury, and studies that fail to take into account the presence of such "nullifiers" thus are fatally flawed.[13] Surely a *"per se* constitutional

---

9. The Harris-1971 study polled 2,068 adults from throughout the United States, the Cowan-Deliberation study involved 288 jury-eligible residents of San Mateo and Santa Clara Counties in California, and the Jurow study was based on the responses of 211 employees of the Sperry Rand Corporation in New York.

10. The Harris-1971 and Jurow studies did not allow for group deliberation, but rather measured only individual responses.

11. JUSTICE MARSHALL'S dissent refers to an "essential unanimity" of support among social science researchers and other academics for McCree's assertion that "death qualification" has a significant effect on the outcome of jury deliberations at the guilt phase of capital trials. See *post,* at 189. At least one of the articles relied upon by the dissent candidly acknowledges, however, that its conclusions ultimately must rest on "[a] certain amount of...conjecture" and a willingness "to transform behavioral suspicions into doctrine." Finch & Ferraro, The Empirical Challenge to Death-Qualified Juries: On Further Examination, 65 Neb. L. Rev. 21, 67 (1986). As the authors of the article explain:

"[Uncertainty] inheres in every aspect of the capital jury's operation, whether one focuses on the method of identifying excludable jurors or the deliberative process through which verdicts are reached. So it is that, some seventeen years after *Witherspoon,* no definitive conclusions can be stated as to the frequency or the magnitude of the effects of death qualification.

....

"Nor is it likely that further empirical research can add significantly to the current understanding of death qualification. The true magnitude of the phenomenon of conviction proneness is probably unmeasurable, given the complexity of capital cases and capital adjudication." *Id.,* at 66–67 (footnote omitted).

12. Only the Cowan-Deliberation study attempted to take into account the presence of "nullifiers."

13. The effect of this flaw on the outcome of a particular study is likely to be significant. The Cowan-Deliberation study revealed that approximately 37% of the *"Witherspoon*-excludables" identified in the study were also "nullifiers."

rule" as far-reaching as the one McCree proposes should not be based on the results of the lone study that avoids this fundamental flaw.

Having identified some of the more serious problems with McCree's studies, however, we will assume for purposes of this opinion that the studies are both methodologically valid and adequate to establish that "death qualification" in fact produces juries somewhat more "conviction-prone" than "non-death-qualified" juries. We hold, nonetheless, that the Constitution does not prohibit the States from "death qualifying" juries in capital cases.

The Eighth Circuit ruled that "death qualification" violated McCree's right under the Sixth Amendment, as applied to the States via incorporation through the Fourteenth Amendment, see *Duncan* v. *Louisiana,* 391 U.S. 145, 148–158 (1968), to a jury selected from a representative cross section of the community. But we do not believe that the fair-cross-section requirement can, or should, be applied as broadly as that court attempted to apply it. We have never invoked the fair-cross-section principle to invalidate the use of either for-cause or peremptory challenges to prospective jurors, or to require petit juries, as opposed to jury panels or venires, to reflect the composition of the community at large. See *Duren* v. *Missouri,* 439 U.S. 357, 363–364 (1979); *Taylor* v. *Louisiana,* 419 U.S. 522, 538 (1975) ("[We] impose no requirement that petit juries actually chosen must mirror the community and reflect the various distinctive groups in the population"); cf. *Batson* v. *Kentucky, ante,* at 84–85, n. 4 (expressly declining to address "fair-cross-section" challenge to discriminatory use of peremptory challenges).[14] The limited scope of the fair-cross-section requirement is a direct and inevitable consequence of the practical impossibility of providing each criminal defendant with a truly "representative" petit jury, see *ante,* at 85–86, n. 6, a basic truth that the Court of Appeals itself acknowledged for many years prior to its decision in the instant case. See *United States* v. *Childress,* 715 F.2d 1313 (CA8 1983) (en banc), cert. denied, 464 U.S. 1063 (1984); *Pope* v. *United States,* 372 F.2d 710, 725 (CA8 1967) (Blackmun, J.) ("The point at which an accused is entitled to a fair cross-section of the community is when the names are put in the box from which the panels are drawn"), vacated on other grounds, 392 U.S. 651 (1968). We remain convinced that an extension of the fair-cross-section requirement to petit juries would be unworkable and unsound, and we decline McCree's invitation to adopt such an extension.

But even if we were willing to extend the fair-cross-section requirement to petit juries, we would still reject the Eighth Circuit's conclusion that "death qualification" violates that requirement. The essence of a "fair-cross-section" claim is the systematic exclusion of "a 'distinctive' group in the community." *Duren, supra,* at 364. In our view, groups defined solely in terms of shared attitudes that would prevent or substantially impair members of the group from performing one of their duties

---

14. The only case in which we have intimated that the fair-cross-section requirement might apply outside the context of jury panels or venires, *Ballew* v. *Georgia,* 435 U.S. 223 (1978) (opinion of BLACKMUN, J.), did not involve jury *selection* at all, but rather the *size* of the petit jury. JUSTICE BLACKMUN's opinion announcing the judgment, and the opinions concurring in the judgment which agreed with him, expressed the view that Georgia's limitation of the size of juries to five "prevents juries from truly representing their communities," *id.,* at 239.

as jurors, such as the *"Witherspoon-excludables"* at issue here, are not "distinctive groups" for fair-cross-section purposes.

We have never attempted to precisely define the term "distinctive group," and we do not undertake to do so today. But we think it obvious that the concept of "distinctiveness" must be linked to the purposes of the fair-cross-section requirement. In *Taylor, supra,* we identified those purposes as (1) "[guarding] against the exercise of arbitrary power" and ensuring that the "commonsense judgment of the community" will act as "a hedge against the overzealous or mistaken prosecutor," (2) preserving "public confidence in the fairness of the criminal justice system," and (3) implementing our belief that sharing in the administration of justice is a phase of civic responsibility." *Id.,* at 530–531.

Our prior jury-representativeness cases, whether based on the fair-cross-section component of the Sixth Amendment or the Equal Protection Clause of the Fourteenth Amendment, have involved such groups as blacks, see *Peters* v. *Kiff,* 407 U.S. 493 (1972) (opinion of MARSHALL, J.) (equal protection); women, see *Duren, supra* (fair cross section); *Taylor, supra* (same); and Mexican-Americans, see *Castaneda* v. *Partida,* 430 U.S. 482 (1977) (equal protection). The wholesale exclusion of these large groups from jury service clearly contravened all three of the aforementioned purposes of the fair-cross-section requirement. Because these groups were excluded for reasons completely unrelated to the ability of members of the group to serve as jurors in a particular case, the exclusion raised at least the possibility that the composition of juries would be arbitrarily skewed in such a way as to deny criminal defendants the benefit of the common-sense judgment of the community. In addition, the exclusion from jury service of large groups of individuals not on the basis of their inability to serve as jurors, but on the basis of some immutable characteristic such as race, gender, or ethnic background, undeniably gave rise to an "appearance of unfairness." Finally, such exclusion improperly deprived members of these often historically disadvantaged groups of their right as citizens to serve on juries in criminal cases.

The group of *"Witherspoon*-excludables" involved in the case at bar differs significantly from the groups we have previously recognized as "distinctive." "Death qualification," unlike the wholesale exclusion of blacks, women, or Mexican-Americans from jury service, is carefully designed to serve the State's concededly legitimate interest in obtaining a single jury that can properly and impartially apply the law to the facts of the case at both the guilt and sentencing phases of a capital trial.[15] There is very little danger, therefore, and McCree does not even argue, that "death qualification" was instituted as a means for the State to arbitrarily skew the composition of capital–case juries.[16]

15. See *Rector* v. *State,* 280 Ark. 385, 396–397, 659 S. W. 2d 168, 173–174 (1983), cert. denied, 466 U.S. 988 (1984). McCree does not dispute the existence of this interest, but merely contends that it is not substantial. See Brief for Respondent 74–79.

16. McCree asserts that the State often will request the death penalty in particular cases solely for the purpose of "death qualifying" the jury, with the intent ultimately to "waive" the death penalty after a conviction is obtained. We need not consider the implications of this assertion, since the State did not "waive" the death penalty in McCree's case.

Furthermore, unlike blacks, women, and Mexican-Americans, "*Witherspoon*-excludables" are singled out for exclusion in capital cases on the basis of an attribute that is within the individual's control. It is important to remember that not all who oppose the death penalty are subject to removal for cause in capital cases; those who firmly believe that the death penalty is unjust may nevertheless serve as jurors in capital cases so long as they state clearly that they are willing to temporarily set aside their own beliefs in deference to the rule of law. Because the group of "*Witherspoon*-excludables" includes only those who cannot and will not conscientiously obey the law with respect to one of the issues in a capital case, "death qualification" hardly can be said to create an "appearance of unfairness."

Finally, the removal for cause of "*Witherspoon*-excludables" in capital cases does not prevent them from serving as jurors in other criminal cases, and thus leads to no substantial deprivation of their basic rights of citizenship. They are treated no differently than any juror who expresses the view that he would be unable to follow the law in a particular case.

In sum, "*Witherspoon*-excludables," or for that matter any other group defined solely in terms of shared attitudes that render members of the group unable to serve as jurors in a particular case, may be excluded from jury service without contravening any of the basic objectives of the fair-cross-section requirement. See *Lockett* v. *Ohio,* 438 U.S. 586, 597 (1978) ("Nothing in *Taylor,* however, suggests that the right to a representative jury includes the right to be tried by jurors who have explicitly indicated an inability to follow the law and instructions of the trial judge"). It is for this reason that we conclude that "*Witherspoon-excludables*" do not constitute a "distinctive group" for fair-cross-section purposes, and hold that "death qualification" does not violate the fair-cross-section requirement.

McCree argues that, even if we reject the Eighth Circuit's fair-cross-section holding, we should affirm the judgment below on the alternative ground, adopted by the District Court, that "death qualification" violated his constitutional right to an impartial jury. McCree concedes that the individual jurors who served at his trial were impartial, as that term was defined by this Court in cases such as *Irvin* v. *Dowd,* 366 U.S. 717, 723 (1961) ("It is sufficient if the juror can lay aside his impression or opinion and render a verdict based on the evidence presented in court"), and *Reynolds* v. *United States,* 98 U.S. 145 (1879). He does not claim that pretrial publicity, see *Rideau* v. *Louisiana,* 373 U.S. 723 (1963), *ex parte* communications, see *Remmer* v. *United States,* 347 U.S. 227 (1954), or other undue influence, see *Estes* v. *Texas,* 381 U.S. 532 (1965), affected the jury's deliberations. In short, McCree does not claim that his conviction was tainted by any of the kinds of jury bias or partiality that we have previously recognized as violative of the Constitution. Instead, McCree argues that his jury lacked impartiality because the absence of "*Witherspoon*-excludables" "slanted" the jury in favor of conviction.

We do not agree. McCree's "impartiality" argument apparently is based on the theory that, because all individual jurors are to some extent predisposed towards one result or another, a constitutionally impartial *jury* can be constructed only by "balancing" the various predispositions of the individual *jurors.* Thus, according to McCree, when the State "tips the scales" by excluding prospective jurors with a particular viewpoint, an impermissibly partial jury results. We have consistently rejected this view of jury impartiality, including as recently as last Term when we squarely held that an impartial *jury* consists of nothing more than "*jurors* who will conscientiously apply the law and find the facts." *Wainwright* v. *Witt*, 469 U.S. 412, 423 (1985) (emphasis added); see also *Smith* v. *Phillips*, 455 U.S. 209, 217 (1982) ("Due process means a jury capable and willing to decide the case solely on the evidence before it"); *Irvin* v. *Dowd, supra*, at 722 ("In essence, the right to jury trial guarantees to the criminally accused a fair trial by a panel of impartial, 'indifferent' jurors").

The view of jury impartiality urged upon us by McCree is both illogical and hopelessly impractical. McCree characterizes the jury that convicted him as "slanted" by the process of "death qualification." But McCree admits that exactly the same 12 individuals could have ended up on his jury through the "luck of the draw," without in any way violating the constitutional guarantee of impartiality. Even accepting McCree's position that we should focus on the *jury* rather than the individual *jurors*, it is hard for us to understand the logic of the argument that a given jury is unconstitutionally partial when it results from a state-ordained process, yet impartial when exactly the same jury results from mere chance. On a more practical level, if it were true that the Constitution required a certain mix of individual viewpoints on the jury, then trial judges would be required to undertake the Sisyphean task of "balancing" juries, making sure that each contains the proper number of Democrats and Republicans, young persons and old persons, white-collar executives and blue-collar laborers, and so on. Adopting McCree's concept of jury impartiality would also likely require the elimination of peremptory challenges, which are commonly used by both the State and the defendant to attempt to produce a jury favorable to the challenger.

McCree argues, however, that this Court's decisions in *Witherspoon* and *Adams* v. *Texas*, 448 U.S. 38 (1980), stand for the proposition that a State violates the Constitution whenever it "slants" the jury by excluding a group of individuals more likely than the population at large to favor the criminal defendant. We think McCree overlooks two fundamental differences between *Witherspoon* and *Adams* and the instant case, and therefore misconceives the import and scope of those two decisions.

First, the Court in *Witherspoon* viewed the Illinois system as having been deliberately slanted for the purpose of making the imposition of the death penalty more likely. The Court said:

> "But when it swept from the jury all who expressed conscientious or religious scruples against capital punishment and all who opposed it in principle, the State crossed the line of neutrality. In its quest for a jury capable of imposing the death penalty, the State produced a jury uncommonly willing to condemn a man to die.

"It is, of course, settled that a State may not entrust the determination of whether a man is innocent or guilty to a tribunal 'organized to convict.' *Fay* v. *New York*, 332 U.S. 261, 294 [1947]. See *Tumey* v. *Ohio*, 273 U.S. 510 [1927]. It requires but a short step from that principle to hold, as we do today, that a State may not entrust the determination of whether a man should live or die to a tribunal organized to return a verdict of death." 391 U.S., at 520–521 (footnotes omitted).

In *Adams* v. *Texas, supra,* the Court explained the rationale for *Witherspoon* as follows:

> "In this context, the Court held that a State may not constitutionally execute a death sentence imposed by a jury culled of all those who revealed during *voir dire* examination that they had conscientious scruples against or were otherwise opposed to capital punishment. The State was held to have no valid interest in such a broad-based rule of exclusion, since '[a] man who opposes the death penalty, no less than one who favors it, can make the discretionary judgment entrusted to him...and can thus obey the oath he takes as a juror.' *Witherspoon* v. *Illinois,* 391 U.S., at 519." 448 U.S., at 43.

Adams, in turn, involved a fairly straightforward application of the *Witherspoon* rule to the Texas capital punishment scheme. See *Adams, supra,* at 48 (Texas exclusion statute "focuses the inquiry directly on the prospective juror's beliefs about the death penalty, and hence clearly falls within the scope of the *Witherspoon* doctrine").

Here, on the other hand, the removal for cause of "*Witherspoon*-excludables" serves the State's entirely proper interest in obtaining a single jury that could impartially decide all of the issues in McCree's case. Arkansas by legislative enactment and judicial decision provides for the use of a unitary jury in capital cases. See Ark. Stat. Ann. § 41–1301(3) (1977); *Rector* v. *State,* 280 Ark. 385, 395, 659 S. W. 2d 168, 173 (1983), cert. denied, 466 U.S. 988 (1984). We have upheld against constitutional attack the Georgia capital sentencing plan which provided that the same jury must sit in both phases of a bifurcated capital murder trial, *Gregg* v. *Georgia,* 428 U.S. 153, 158, 160, 163 (1976) (opinion of Stewart, POWELL, and STEVENS, JJ.), and since then have observed that we are "unwilling to say that there is any one right way for a State to set up its capital sentencing scheme." *Spaziano* v. *Florida,* 468 U.S. 447, 464 (1984).

The Arkansas Supreme Court recently explained the State's legislative choice to require unitary juries in capital cases:

> "It has always been the law in Arkansas, except when the punishment is mandatory, that the same jurors who have the responsibility for determining guilt or innocence must also shoulder the burden of fixing the punishment. That is as it should be, for the two questions are necessarily interwoven." *Rector, supra,* at 395, 659 S. W. 2d, at 173.

Another interest identified by the State in support of its system of unitary juries is the possibility that, in at least some capital cases, the defendant might benefit at the sentencing phase of the trial from the jury's "residual doubts" about the evidence presented at the guilt phase. The dissenting opinion in the Court of Appeals also adverted to this interest:

> "[As] several courts have observed, jurors who decide both guilt and penalty are likely to form residual doubts or 'whimsical' doubts...about the evidence so as to bend them to decide against the death penalty. Such residual doubt has been recognized as an

extremely effective argument for defendants in capital cases. To divide the responsibility...to some degree would eliminate the influence of such doubts." 758 F.2d, at 247–248 (J. Gibson, J., dissenting) (citations omitted).

JUSTICE MARSHALL's dissent points out that some States which adhere to the unitary jury system do not allow the defendant to argue "residual doubts" to the jury at sentencing. But while this may justify skepticism as to the extent to which such States are willing to go to allow defendants to capitalize on "residual doubts," it does not wholly vitiate the claimed interest. Finally, it seems obvious to us that in most, if not all, capital cases much of the evidence adduced at the guilt phase of the trial will also have a bearing on the penalty phase; if two different juries were to be required, such testimony would have to be presented twice, once to each jury. As the Arkansas Supreme Court has noted, "[such] repetitive trials could not be consistently fair to the State and perhaps not even to the accused." *Rector, supra,* at 396, 659 S. W. 2d, at 173.

Unlike the Illinois system criticized by the Court in *Witherspoon,* and the Texas system at issue in *Adams,* the Arkansas system excludes from the jury only those who may properly be excluded from the penalty phase of the deliberations under *Witherspoon, Adams,* and *Wainwright* v. *Witt,* 469 U.S. 412 (1985).[17] That State's reasons for adhering to its preference for a single jury to decide both the guilt and penalty phases of a capital trial are sufficient to negate the inference which the Court drew in *Witherspoon* concerning the lack of any neutral justification for the Illinois rule on jury challenges.

Second, and more importantly, both *Witherspoon* and *Adams* dealt with the special context of capital sentencing, where the range of jury discretion necessarily gave rise to far greater concern over the possible effects of an "imbalanced" jury. As we emphasized in *Witherspoon:*

> "[In] Illinois, as in other States, the jury is given broad discretion to decide whether or not death *is* 'the proper penalty' in a given case, and a juror's general views about capital punishment play an inevitable role in any such decision.
> "....Guided by neither rule nor standard, 'free to select or reject as it [sees] fit,' a jury that must choose between life imprisonment and capital punishment can do little more—and must do nothing less—than express the conscience of the community on the ultimate question of life or death." 391 U.S., at 519 (emphasis in original; footnotes omitted).

Because capital sentencing under the Illinois statute involved such an exercise of essentially unfettered discretion, we held that the State violated the Constitution when it "crossed the line of neutrality" and "produced a jury uncommonly willing to condemn a man to die." *Id.,* at 520–521.

In *Adams,* we applied the same basic reasoning to the Texas capital sentencing scheme, which, although purporting to limit the jury's role to answering several "factual"

---

17. The rule applied by Arkansas to exclude these prospective jurors was scarcely a novel one; as long ago as *Logan* v. *United States,* 144 U.S. 263 (1892), this Court approved such a practice in the federal courts, commenting that it was also followed "by the courts of every State in which the question has arisen." *Id.,* at 298.

questions, in reality vested the jury with considerable discretion over the punishment to be imposed on the defendant. See 448 U.S., at 46 ("This process is not an exact science, and the jurors under the Texas bifurcated procedure unavoidably exercise a range of judgment and discretion while remaining true to their instructions and their oaths"); cf. *Jurek v. Texas*, 428 U.S. 262, 273 (1976) (opinion of Stewart, POWELL, and STEVENS, JJ.) ("Texas law essentially requires that...in considering whether to impose a death sentence the jury may be asked to consider whatever evidence of mitigating circumstances the defense can bring before it"). Again, as in *Witherspoon,* the discretionary nature of the jury's task led us to conclude that the State could not "exclude all jurors who would be in the slightest way affected by the prospect of the death penalty or by their views about such a penalty." *Adams,* 448 U.S., at 50.

In the case at bar, by contrast, we deal not with capital sentencing, but with the jury's more traditional role of finding the facts and determining the guilt or innocence of a criminal defendant, where jury discretion is more channeled. We reject McCree's suggestion that *Witherspoon* and *Adams* have broad applicability outside the special context of capital sentencing,[18] and conclude that those two decisions do not support the result reached by the Eighth Circuit here.

In our view, it is simply not possible to define jury impartiality, for constitutional purposes, by reference to some hypothetical mix of individual viewpoints. Prospective jurors come from many different backgrounds, and have many different attitudes and predispositions. But the Constitution presupposes that a jury selected from a fair cross section of the community is impartial, regardless of the mix of individual viewpoints actually represented on the jury, so long as the jurors can conscientiously and properly carry out their sworn duty to apply the law to the facts of the particular case. We hold that McCree's jury satisfied both aspects of this constitutional standard. The judgment of the Court of Appeals is therefore

Reversed.

JUSTICE BLACKMUN concurs in the result.

DISSENT BY: MARSHALL

DISSENT: JUSTICE MARSHALL, with whom JUSTICE BRENNAN and JUSTICE STEVENS join, dissenting.

Eighteen years ago, this Court vacated the sentence of a defendant from whose jury the State had excluded all venirepersons expressing any scruples against capital punishment. Such a practice, the Court held, violated the Constitution by creating a "tribunal organized to return a verdict of death." *Witherspoon v. Illinois,* 391 U.S. 510, 521 (1968). The only venirepersons who could be constitutionally excluded from service in capital cases were those who "made unmistakably clear...that they would *automatically* vote against the imposition of capital punishment" or that they could not assess the defendant's guilt impartially. *Id.,* at 522–523, n. 21.

---

18. The majority in *Adams* rejected the dissent's claim that there was "no plausible distinction between the role of the jury in the guilt/innocence phase of the trial and its role, as defined by the State of Texas, in the sentencing phase." 448 U.S., at 54 (REHNQUIST, J., dissenting).

Respondent contends here that the "death-qualified" jury that convicted him, from which the State, as authorized by *Witherspoon,* had excluded all venirepersons unwilling to consider imposing the death penalty, was in effect "organized to return a verdict" of guilty. In support of this claim, he has presented overwhelming evidence that death-qualified juries are substantially more likely to convict or to convict on more serious charges than juries on which unalterable opponents of capital punishment are permitted to serve. Respondent does not challenge the application of *Witherspoon* to the jury in the sentencing stage of bifurcated capital cases. Neither does he demand that individuals unable to assess culpability impartially ("nullifiers") be permitted to sit on capital juries. All he asks is the chance to have his guilt or innocence determined by a jury like those that sit in noncapital cases—one whose composition has not been tilted in favor of the prosecution by the exclusion of a group of prospective jurors uncommonly aware of an accused's constitutional rights but quite capable of determining his culpability without favor or bias.

With a glib nonchalance ill suited to the gravity of the issue presented and the power of respondent's claims, the Court upholds a practice that allows the State a special advantage in those prosecutions where the charges are the most serious and the possible punishments, the most severe. The State's mere announcement that it intends to seek the death penalty if the defendant is found guilty of a capital offense will, under today's decision, give the prosecution license to empanel a jury especially likely to return that very verdict. Because I believe that such a blatant disregard for the rights of a capital defendant offends logic, fairness, and the Constitution, I dissent.

I

Respondent is not the first to argue that "death qualification" poses a substantial threat to the ability of a capital defendant to receive a fair trial on the issue of his guilt or innocence. In 1961, one scholar observed that "Jurors hesitant to levy the death penalty would...seem more prone to resolve the many doubts as to guilt or innocence in the defendant's favor than would jurors qualified on the 'pound of flesh' approach." Oberer, Does Disqualification of Jurors for Scruples Against Capital Punishment Constitute Denial of Fair Trial on Issue of Guilt?, 39 Texas L. Rev. 545, 549 (1961).

When he claimed that the exclusion of scrupled jurors from his venire had violated his constitutional right to an impartial jury, the petitioner in *Witherspoon* v. *Illinois, supra,* sought to provide empirical evidence to corroborate Oberer's intuition. See Brief for Petitioner in *Witherspoon* v. *Illinois,* O. T. 1967, No. 1015, pp. 28–33. The data on this issue, however, consisted of only three studies and one preliminary summary of a study.[19] Although the data certainly supported the validity of Witherspoon's challenge to his conviction, these studies did not provide the Court with the firmest basis for constitutional adjudication. As a result, while it reversed

---

19. The *Witherspoon* Court had before it only a preliminary summary of the results of H. Zeisel, Some Data on Juror Attitudes Towards Capital Punishment (University of Chicago Monograph (1868), "not the data nor the analysis that underlay his conclusions, nor indeed his final conclusions themselves." *Hovey* v. *Superior Court,* 28 Cal. 3d 1, 30, n. 63, 616 P. 2d 1301, 1317, n. 63 (1980).

Witherspoon's death sentence, the Court was unable to conclude that "the exclusion of jurors opposed to capital punishment results in an unrepresentative jury on the issue of guilt or substantially increases the risk of conviction." 391 U.S., at 518, and declined to reverse Witherspoon's conviction. Nonetheless, the Court was careful to note:

> "[A] defendant convicted by [a properly death-qualified] jury in some future case might still attempt to establish that the jury was less than neutral with respect to *guilt*. If he were to succeed in that effort, the question would then arise whether the State's interest in submitting the penalty issue to a jury capable of imposing capital punishment may be vindicated at the expense of the defendant's interest in a completely fair determination of guilt or innocence—given the possibility of accommodating both interests by means of a bifurcated trial, using one jury to decide guilt and another to fix punishment. That problem is not presented here, however, and we intimate no view as to its proper resolution." *Id.,* at 520, n. 18.

In the wake of *Witherspoon,* a number of researchers set out to supplement the data that the Court had found inadequate in that case. The results of these studies were exhaustively analyzed by the District Court in this case, see *Grigsby* v. *Mabry,* 569 F.Supp. 1273, 1291–1308 (ED Ark. 1983) (*Grigsby II*), and can be only briefly summarized here.[20] The data strongly suggest that death qualification excludes a significantly large subset—at least 11% to 17%—of potential jurors who could be impartial during the guilt phase of trial.[21] Among the members of this excludable class are a disproportionate number of blacks and women. See *id.,* at 1283, 1293–1294.

The perspectives on the criminal justice system of jurors who survive death qualification are systematically different from those of the excluded jurors. Death-qualified jurors are, for example, more likely to believe that a defendant's failure to testify is indicative of his guilt, more hostile to the insanity defense, more mistrustful of defense attorneys, and less concerned about the danger of erroneous convictions. *Id.,* at 1283, 1293, 1304. This pro-prosecution bias is reflected in the greater readiness of death-qualified jurors to convict or to convict on more serious charges. *Id.,* at 1294–1302; *Grigsby* v. *Mabry,* 758 F.2d 226, 233–236 (CA8 1985). And, finally,

20. Most of the studies presented here were also comprehensively summarized in *Hovey v. Superior Court, supra.* Because the California Supreme Court found the studies had not accounted for jurors who could be excluded because they would automatically vote for the death penalty where possible, that court ultimately rejected a defendant's constitutional challenge to death qualification. But see Kadane, After *Hovey:* A Note on Taking Account of the Automatic Death Penalty Jurors, 8 Law & Hum. Behav. 115 (1984).

21. Bronson, On the Conviction Proneness and Representativeness of the Death-Qualified Jury: An Empirical Study of Colorado Veniremen, 12 U. Colo. L. Rev. 1 (1970) (using classification only approximating *Witherspoon* standard, and finding 11% of subjects *Witherspoon*-excludable); Bronson, Does the Exclusion of Scrupled Jurors in Capital Cases Make the Jury More Likely to Convict? Some Evidence from California, 3 Woodrow Wilson L. J. 11 (1980) (using more appropriate *Witherspoon* question and finding 93% overlap of "strongly opposed" group in prior Bronson study with *Witherspoon-excludables*); Jurow, New Data on the Effect of a "Death Qualified Jury" on the Guilt Determination Process, 84 Harv. L. Rev. 567 (1971) (finding only 10% of sample excludable, but likely to have underestimated size of class in general population because sample 99% white and 80% male); Fitzgerald & Ellsworth, Due Process vs. Crime Control: Death Qualification and Jury Attitudes, 8 Law & Hum. Behav. 31 (1984) (random sample with nullifiers screened out finding 17% still excludable under *Witherspoon*); A. Young, Arkansas Archival Study (unpublished, 1981) (14% of jurors questioned in *voir dire* transcripts excludable); Precision Research, Inc., Survey No. 1286 (1981) (11% excludable, not counting nullifiers); see 569 F.Supp., at 1285; *Grigsby v. Mabry,* 758 F.2d 226, 231 (CA8 1985).

the very process of death qualification—which focuses attention on the death penalty before the trial has even begun—has been found to predispose the jurors that survive it to believe that the defendant is guilty. 569 F.Supp., at 1302–1305; 758 F.2d, at 234.

The evidence thus confirms, and is itself corroborated by, the more intuitive judgments of scholars and of so many of the participants in capital trials—judges, defense attorneys, and prosecutors. See 569 F.Supp., at 1322.[22]

II

A

Respondent's case would of course be even stronger were he able to produce data showing the prejudical effects of death qualification upon actual trials. Yet, until a State permits two separate juries to deliberate on the same capital case and return simultaneous verdicts, defendants claiming prejudice from death qualification should not be denied recourse to the only available means of proving their case, recreations of the *voir dire* and trial processes. See *Grigsby* v. *Mabry, supra,* at 237 ("[It] is the courts who have often stood in the way of surveys involving real jurors and we should not now reject a study because of this deficiency").

The chief strength of respondent's evidence lies in the essential unanimity of the results obtained by researchers using diverse subjects and varied methodologies. Even the Court's haphazard jabs cannot obscure the power of the array. Where studies have identified and corrected apparent flaws in prior investigations, the results of the subsequent work have only corroborated the conclusions drawn in the earlier efforts. Thus, for example, some studies might be faulted for failing to distinguish within the class of *Witherspoon*-excludables, between nullifiers (whom respondent concedes may be excluded from the guilt phase) and those who could assess guilt impartially. Yet their results are entirely consistent with those obtained after nullifiers had indeed been excluded. See, *e. g.,* Cowan, Thompson, & Ellsworth, The Effects of Death Qualification on Jurors' Predisposition to Convict and on the Quality of Deliberation, 8 Law & Hum. Behav. 53 (1984). And despite the failure of certain studies to "allow for group deliberations," *ante,* at 171, n. 10, the value of their results is underscored by the discovery that initial verdict preferences, made

---

22. The Court reasons that because the State did not "waive" the death penalty in respondent's case, we "need not consider the implications" of respondent's assertion that "the State often will request the death penalty in particular cases solely for the purpose of 'death qualifying' the jury, with the intent ultimately to 'waive' the death penalty after a conviction is obtained." *Ante,* at 176, n. 16. If, by this, the Court intended to limit the effects of its decision to the case pending before us, I would gladly join that effort. However, I see all too few indications in the Court's opinion that future constitutional challenges to death-qualified juries stand much chance of success here. In view of the sweep of the Court's opinion and the fact that, in any particular case, a defendant will never be able to demonstrate with any certainty that the prosecution's decision to seek the death penalty was merely a tactical ruse, I find the Court's refusal to consider the potential for this abuse rather disingenuous. See *Grigsby* v. *Mabry,* 483 F.Supp. 1372, 1389, n. 24 (ED Ark. 1980) (*Grigsby I*) (suggesting possibility that prosecutor had initially sought death penalty merely to get more conviction-prone, death-qualified jury). Cf. Oberer, Does Disqualification of Jurors for Scruples Against Capital Punishment Constitute Denial of Fair Trial on Issue of Guilt?, 39 Texas L. Rev. 545, 555, n. 45b (1961) (reporting testimony of one prosecutor that there might be other prosecutors who death-qualified a jury "without hope of obtaining a death verdict but in the expectation that a jury so selected would impose a higher penalty than might otherwise be obtained").

prior to group deliberations, are a fair predictor of how a juror will vote when faced with opposition in the jury room. See Cowan, Thompson, & Ellsworth, *supra,* at 68–69; see also R. Hastie, S. Penrod, & N. Pennington, Inside the Jury 66 (1983); H. Kalven & H. Zeisel, The American Jury 488 (1966).

The evidence adduced by respondent is quite different from the "tentative and fragmentary" presentation that failed to move this Court in *Witherspoon,* 391 U.S., at 517. Moreover, in contrast to *Witherspoon,* the record in this case shows respondent's case to have been "subjected to the traditional testing mechanisms of the adversary process," *Ballew* v. *Georgia,* 435 U.S. 223, 246 (1978) (POWELL, J., concurring in judgment). At trial, respondent presented three expert witnesses and one lay witness in his case in chief, and two additional lay witnesses in his rebuttal. Testimony by these witnesses permitted the District Court, and allows this Court, better to understand the methodologies used here and their limitations. Further testing of respondent's empirical case came at the hands of the State's own expert witnesses. Yet even after considering the evidence adduced by the State, the Court of Appeals properly noted: "there are no studies which contradict the studies submitted [by respondent]; in other words, all of the documented studies support the district court's findings." 758 F.2d, at 238.

B

The true impact of death qualification on the fairness of a trial is likely even more devastating than the studies show. *Witherspoon* placed limits on the State's ability to strike scrupled jurors for cause, unless they state "unambiguously that [they] would automatically vote against the imposition of capital punishment no matter what the trial might reveal," 391 U.S., at 516, n. 9. It said nothing, however, about the prosecution's use of peremptory challenges to eliminate jurors who do not meet that standard and would otherwise survive death qualification. See Gillers, Deciding Who Dies, 129 U. Pa. L. Rev. 1, 85, n. 391 (1980). There is no question that peremptories have indeed been used to this end, thereby expanding the class of scrupled jurors excluded as a result of the death-qualifying *voir dire* challenged here. See, *e. g., People* v. *Velasquez,* 26 Cal. 3d 425, 438, n. 9, 606 P. 2d 341, 348, n. 9 (1980) (prosecutor informed court during *voir dire* that if a venireperson expressing scruples about the death penalty "were not a challenge for cause, I would kick her off on a peremptory challenge"). The only study of this practice has concluded: "For the five-year period studied a prima facie case has been demonstrated that prosecutors in Florida's Fourth Judicial Circuit systematically used their peremptory challenges to eliminate from capital juries venirepersons expressing opposition to the death penalty." Winick, Prosecutorial Peremptory Challenge Practices in Capital Cases: An Empirical Study and a Constitutional Analysis, 81 Mich. L. Rev. 1, 39 (1982).[23]

Judicial applications of the *Witherspoon* standard have also expanded the class of jurors excludable for cause. While the studies produced by respondent generally

---

23. At this point, the remedy called for is not the wholesale removal of the prosecution's power to make peremptory challenges, but merely the elimination of death qualification. But cf. *Batson* v. *Kentucky, ante,* p. 102 (MARSHALL, J., concurring). Without the extensive *voir dire* now allowed for death qualification, the prosecution would lack sufficient information to be able to expand the scope of *Witherspoon* through the use of peremptories. See n. 8, *infra.*

classified a subject as a *Witherspoon*-excludable only upon his unambiguous refusal to vote death under any circumstance, the courts have never been so fastidious. Trial and appellate courts have frequently excluded jurors even in the absence of unambiguous expressions of their absolute opposition to capital punishment. Schnapper, Taking *Witherspoon* Seriously: The Search for Death-Qualified Jurors, 62 Texas L. Rev. 977, 993–1032 (1984). And this less demanding approach will surely become more common in the wake of this Court's decision in *Wainwright* v. *Witt*, 469 U.S. 412 (1985). Under *Witt*, a juror who does not make his attitude toward capital punishment "unmistakably clear," *Witherspoon*, 391 U.S., at 522, n. 21, may nonetheless be excluded for cause if the trial court is left with the impression that his attitude will "prevent or substantially impair the performance of his duties as a juror in accordance with his instructions and his oath." *Witt, supra,* at 433 (quoting *Adams* v. *Texas*, 448 U.S. 38, 45 (1980)). It thus "seems likely that *Witt* will lead to more conviction-prone panels" since " 'scrupled' jurors—those who generally oppose the death penalty but do not express an unequivocal refusal to impose it—usually share the pro-defendant perspective of excludable jurors." See Finch & Ferraro, The Empirical Challenge to Death Qualified Juries: On Further Examination, 65 Neb. L. Rev. 21, 63 (1986).

C

Faced with the near unanimity of authority supporting respondent's claim that death qualification gives the prosecution a particular advantage in the guilt phase of capital trials, the majority here makes but a weak effort to contest that proposition. Instead, it merely assumes for the purposes of this opinion "that 'death qualification' in fact produces juries somewhat more 'conviction-prone' than 'non-death-qualified' juries," *ante,* at 173, and then holds that this result does not offend the Constitution. This disregard for the clear import of the evidence tragically misconstrues the settled constitutional principles that guarantee a defendant the right to a fair trial and an impartial jury whose composition is not biased toward the prosecution.

III

In *Witherspoon* the Court observed that a defendant convicted by a jury from which those unalterably opposed to the death penalty had been excluded "might still attempt to establish that the jury was less than neutral with respect to *guilt*." 391 U.S., at 520, n. 18. Respondent has done just that. And I believe he has succeeded in proving that his trial by a jury so constituted violated his right to an impartial jury, guaranteed by both the Sixth Amendment and principles of due process, see *Ristaino* v. *Ross*, 424 U.S. 589, 595, n. 6 (1976). We therefore need not rely on respondent's alternative argument that death qualification deprived him of a jury representing a fair cross section of the community.[24]

---

24. With respect to the Court's discussion of respondent's fair-cross-section claim, however, I must note that there is no basis in either precedent or logic for the suggestion that a state law authorizing the prosecution before trial to exclude from jury service all, or even a substantial portion of, the members of a "distinctive group" would not constitute a clear infringement of a defendant's Sixth Amendment right. "The desired interaction of a cross section of the community does not take place within the venire; it is only effectuated by the jury that is selected and sworn to try the issues." *McCray* v. *New York,* 461 U.S. 961, 968 (1983) (MARSHALL, J., dissenting from denial

A

Respondent does not claim that any individual on the jury that convicted him fell short of the constitutional standard for impartiality. Rather, he contends that, by systematically excluding a class of potential jurors less prone than the population at large to vote for conviction, the State gave itself an unconstitutional advantage at his trial. Thus, according to respondent, even though a nonbiased selection procedure might have left him with a jury composed of the very same individuals that actually sat on his panel, the process by which those 12 individuals were chosen violated the Constitution.

I am puzzled by the difficulty that the majority has in understanding the "logic of the argument." *Ante*, at 178. For the logic is precisely that which carried the day in *Witherspoon*, and which has never been repudiated by this Court—not even today, if the majority is to be taken at its word. There was no question in *Witherspoon* that if the defendant's jury had been chosen by the "luck of the draw," the same 12 jurors who actually sat on his case might have been selected. Nonetheless, because the State had removed from the pool of possible jurors all those expressing general opposition to the death penalty, the Court overturned the defendant's conviction, declaring "that a State may not entrust the determination of whether a man should live or die to a tribunal organized to return a verdict of death." 391 U.S., at 521. Witherspoon had been denied a fair sentencing determination, the Court reasoned, not because any member of his jury lacked the requisite constitutional impartiality, but because the manner in which that jury had been selected "stacked the deck" against him. *Id.*, at 523. Here, respondent adopts the approach of the *Witherspoon* Court and argues simply that the State entrusted the determination of his guilt and the level of his culpability to a tribunal organized to convict.

The Court offers but two arguments to rebut respondent's constitutional claim. First, it asserts that the "State's reasons for adhering to its preference for a single jury to decide both the guilt and penalty phases of a capital trial are sufficient to negate the inference which the Court drew in *Witherspoon* concerning the lack of any neutral justification for the Illinois rule on jury challenges." *Ante*, at 182. This argument, however, does not address the question whether death qualification infringes a defendant's constitutional interest in "a completely fair determination of guilt or innocence," *Witherspoon*, 391 U.S., at 520, n. 18. It merely indicates the state interest that must be considered once an infringement of that constitutional interest is found.

The Court's second reason for rejecting respondent's challenge to the process that produced his jury is that the notion of "neutrality" adumbrated in *Witherspoon* must

---

of certiorari); see *McCray* v. *Abrams*, 750 F.2d 1113, 1128–1129 (CA2 1984), cert. pending, No. 84–1426. The right to have a particular group represented on venires is of absolutely no value if every member of that group will automatically be excluded from service as soon as he is found to be a member of that group. Whether a violation of the fair-cross-section requirement has occurred can hardly turn on *when* the wholesale exclusion of a group takes place. If, for example, blacks were systematically struck from petit juries pursuant to state law, the effect—and the infringement of a defendant's Sixth Amendment rights—would be the same as if they had never been included on venires in the first place.

be confined to "the special context of capital sentencing, where the range of jury discretion necessarily gave rise to far greater concern over the possible effects of an 'imbalanced' jury." *Ante,* at 182. But in the wake of this Court's decision in *Adams v. Texas,* 448 U.S. 38 (1980), this distinction is simply untenable.

B

In *Adams,* this Court applied the principles of *Witherspoon* to the Texas death-penalty scheme. Under that scheme, if a defendant is convicted of a capital offense, a separate sentencing proceeding is held at which additional aggravating or mitigating evidence is admissible. The jury then must answer three questions based on evidence adduced during either phase of the trial:

> "(1) whether the conduct of the defendant that caused the death of the deceased was committed deliberately and with the reasonable expectation that the death of the deceased or another would result;
>
> "(2) whether there is a probability that the defendant would commit criminal acts of violence that would constitute a continuing threat to society; and
>
> "(3) if raised by the evidence, whether the conduct of the defendant in killing the deceased was unreasonable in response to the provocation, if any, by the deceased."
> Tex. Code Crim. Proc. Ann., Art. 37.071(b) (Vernon Supp. 1986).

See *Adams, supra,* at 40–41. If the jury finds beyond a reasonable doubt that the answer to each of these questions is "yes," the court must impose a death sentence; a single "no" answer requires the court to impose a sentence of life imprisonment. With the role of the jury so defined, *Adams* held that Texas could not constitutionally exclude every prospective juror unable to state under oath that "the mandatory penalty of death or imprisonment for life will not affect his deliberations on any issue of fact." Tex. Penal Code Ann. § 12.31(b) (1974); see *Adams, supra,* at 42. The "process" of answering the statutory questions, the Court observed, "is not an exact science, and the jurors under the Texas bifurcated procedure unavoidably exercise a range of judgment and discretion while remaining true to their instructions and their oaths." 448 U.S., at 46. Consequently, while Texas could constitutionally exclude jurors whose scruples against the death penalty left them unable "to answer the statutory questions without conscious distortion or bias," *ibid.,* it could not exclude those

> "[who] aver that they will honestly find the facts and answer the questions in the affirmative if they are convinced beyond reasonable doubt, but not otherwise, yet who frankly concede that the prospects of the death penalty may affect what their honest judgment of the facts will be or what they may deem to be a reasonable doubt. Such assessments and judgments by jurors are inherent in the jury system, and to exclude all jurors who would be in the slightest way affected by the prospect of the death penalty or by their views about such a penalty would be to deprive the defendant of the impartial jury to which he or she is entitled under the law." *Id.,* at 50.

The message of *Adams* is thus that even where the role of the jury at the penalty stage of a capital trial is limited to what is essentially a factfinding role, the right to an impartial jury established in *Witherspoon* bars the State from skewing the composition of its capital juries by excluding scrupled jurors who are nonetheless able to find those facts without distortion or bias. This proposition cannot be limited to the penalty stage of a capital trial, for the services that Adams' jury was called upon to perform at his penalty stage "are nearly indistinguishable" from those required of juries at the culpability phase of capital trials. Gillers, Proving the Prejudice of Death-Qualified Juries after *Adams* v. *Texas,* 47 U. Pitt. L. Rev. 219, 247 (1985). Indeed, JUSTICE REHNQUIST noted in *Adams* that he could "see no plausible distinction between the role of the jury in the guilt/innocence phase of the trial and its role, as defined by the State of Texas, in the sentencing phase." 448 U.S., at 54 (dissenting). Contrary to the majority's suggestion, *ante* at 183, this point was at no time repudiated by the *Adams* Court. And the absence of a reply to JUSTICE REHNQUIST was not an oversight. At the penalty stage of his trial, Adams' jury may have been called upon to do something more than ascertain the existence *vel non* of specific historical facts. Yet the role assigned a jury at a trial's culpability phase is little different, for there the critical task of the jury will frequently be to determine not whether defendant actually inflicted the fatal wound, but rather whether his level of culpability at the time of the murder makes conviction on capital murder charges, as opposed to a lesser count, more appropriate. Representing the conscience of the community, the jurors at both stages "unavoidably exercise a range of judgment and discretion while remaining true to their instructions and their oaths." 448 U.S., at 46.

Adams thus provides clear precedent for applying the analysis of *Witherspoon* to the guilt phase of a criminal trial. Indeed, respondent's case is even stronger than Witherspoon's. The Court in *Witherspoon* merely presumed that the exclusion of scrupled jurors would unacceptably increase the likelihood that the defendant would be condemned to death. Respondent here has gone much further and laid a solid empirical basis to support his claim that the juries produced by death qualification are substantially more likely to convict.

IV

A

One need not rely on the analysis and assumptions of *Adams* and *Witherspoon* to demonstrate that the exclusion of opponents of capital punishment capable of impartially determining culpability infringes a capital defendant's constitutional right to a fair and impartial jury. For the same conclusion is compelled by the analysis that in *Ballew* v. *Georgia,* 435 U.S. 223 (1978), led a majority of this Court[25] to hold that a criminal conviction rendered by a five-person jury violates the Sixth and Fourteenth Amendments.

Faced with an effort by Georgia to reduce the size of the jury in a criminal case beyond the six-member jury approved by this Court in *Williams* v. *Florida,* 399 U.S. 78 (1970), this Court articulated several facets of the inquiry whether the

---

25. Although the opinion of JUSTICE BLACKMUN in *Ballew* was cosigned by only JUSTICE STEVENS, three other Justices specifically joined that opinion "insofar as it holds that the Sixth and Fourteenth Amendments require juries in criminal trials to contain more than five persons." 435 U.S., at 246 (opinion of BRENNAN, J., joined by Stewart and MARSHALL, JJ.).

reduction impermissibly "[inhibited] the functioning of the jury as an institution to a significant degree." *Ballew, supra,* at 231. First, the Court noted that "recent empirical data" had suggested that a five-member jury was "less likely to foster effective group deliberation" and that such a decline in effectiveness would likely lead "to inaccurate factfinding and incorrect application of the common sense of the community to the facts." *Id.,* at 232. The Court advanced several explanations for this phenomenon:

> "As juries decrease in size…, they are less likely to have members who remember each of the important pieces of evidence or argument. Furthermore, the smaller the group, the less likely it is to overcome the biases of its members to obtain an accurate result. When individual and group decisionmaking were compared, it was seen that groups performed better because prejudices of individuals were frequently counterbalanced, and objectivity resulted." *Id.,* at 233 (footnotes omitted).

The Court also cited empirical evidence suggesting "that the verdicts of jury deliberation in criminal cases will vary as juries become smaller, and that the variance amounts to an imbalance to the detriment of one side, the defense." *Id.,* at 236. Lastly, the Court observed that further reductions in jury size would also foretell problems "for the representation of minority groups in the community." *Ibid.*

B

Each of the concerns that led this Court in *Ballew* to find that a misdemeanor defendant had been deprived of his constitutional right to a fair trial by jury is implicated by the process of death qualification, which threatens a defendant's interests to an even greater extent in cases where the stakes are substantially higher. When compared to the juries that sit in all other criminal trials, the death-qualified juries of capital cases are likely to be deficient in the quality of their deliberations, the accuracy of their results, the degree to which they are prone to favor the prosecution, and the extent to which they adequately represent minority groups in the community.

The data considered here, as well as plain common sense, leave little doubt that death qualification upsets the "counter-balancing of various biases" among jurors that *Ballew* identified as being so critical to the effective functioning of juries. *Id.,* at 234. The evidence demonstrates that "a person's attitude toward capital punishment is an important indicator of a whole cluster of attitudes about crime control and due process." Fitzgerald & Ellsworth, Due Process vs. Crime Control: Death Qualification and Jury Attitudes, 8 Law & Hum. Behav. 31, 46 (1984). Members of the excluded group have been shown to be significantly more concerned with the constitutional rights of criminal defendants and more likely to doubt the strength of the prosecution's case. No doubt because diversity promotes controversy, which in turn leads to a closer scrutiny of the evidence, one study found that "the members of mixed juries [composed of *Witherspoon*-excludables as well as death-qualified jurors] remember the evidence better than the members of death-qualified juries." Cowan, Thompson, & Ellsworth, 8 Law & Hum. Behav., at 76. This study found that not only is the recall of evidence by a death-qualified jury likely to be below the standard of ordinary juries, but its testing of that evidence will be less rigorous. *Id.,* at 75. It thus appears that in the most serious criminal cases—those in which the State has expressed an intention to seek the death penalty—the ability of the jury to find historical truth may well be impaired.

The role of a jury in criminal cases, however, is not limited to the determination of historical facts. The task of ascertaining the level of a defendant's culpability requires a jury to decide not only whether the accused committed the acts alleged in the indictment, but also the extent to which he is morally blameworthy. Thus, especially in capital cases, where a defendant invariably will be charged with lesser included offenses having factual predicates similar to those of the capital murder charges, see *Beck* v. *Alabama*, 447 U.S. 625 (1980), it may be difficult to classify a particular verdict as "accurate" or "inaccurate." However, the *Ballew* Court went beyond a concern for simple historical accuracy and questioned any jury procedure that systematically operated to the "detriment of . . . the defense." 435 U.S., at 236.

With even more clarity than the data considered in *Ballew,* the studies adduced by respondent show a broad pattern of "biased decisionmaking," *Ballew, supra,* at 239. It is not merely that the jurors who survive death qualification are more likely to accept the word of the prosecution and be satisfied with a lower standard of proof than those who are excluded. The death-qualified jurors are actually more likely to convict than their peers. Whether the verdict against a particular capital defendant will actually be different depends on the strength of the evidence against him, the likelihood that, absent death qualification, one or more *Witherspoon-excludables* would have sat on his jury, and a host of other factors. See Gillers, 47 U. Pitt. L. Rev., at 232–238. However, *Ballew* points to the importance of considering the effects of a particular jury procedure over a range of cases, and not focusing on the fairness of any single trial. Because it takes only one juror unwilling to find that the prosecution has met its burden for a trial to end in either a mistrial or a compromise verdict, "it can be confidently asserted that, over time, some persons accused of capital crimes will be convicted of offenses—and to a higher degree— who would not be so convicted" had all persons able to assess their guilt impartially been permitted to sit on their juries. *Hovey* v. *Superior Court,* 28 Cal. 3d 1, 25, n. 57, 616 P. 2d 1301, 1314, n. 57 (1980).

Death qualification also implicates the *Ballew* Court's concern for adequate representation of minority groups. Because opposition to capital punishment is significantly more prevalent among blacks than among whites, the evidence suggests that death qualification will disproportionately affect the representation of blacks on capital juries. Though perhaps this effect may not be sufficient to constitute a violation of the Sixth Amendment's fair-cross-section principle, see *Duren* v. *Missouri,* 439 U.S. 357, 363–364 (1979), it is similar in magnitude to the reduction in minority representation that the *Ballew* Court found to be of "constitutional significance" to a defendant's right to a fair jury trial, 435 U.S., at 239. See White, Death-Qualified Juries: The "Prosecution Proneness" Argument Reexamined, 41 U. Pitt. L. Rev. 353, 387–389 (1980).

The principle of "impartiality" invoked in *Witherspoon* is thus not the only basis for assessing whether the exclusion of jurors unwilling to consider the death penalty but able impartially to determine guilt infringes a capital defendant's constitutional interest in a fair trial. By identifying the critical concerns that are subsumed in that interest, the *Ballew* Court pointed to an alternative approach to the issue, drawing on the very sort of empirical data that respondent has presented here. And viewed in light of the concerns articulated in *Ballew,* the evidence is sufficient to establish that death qualification constitutes a substantial threat to a defendant's Sixth and Fourteenth Amendment right to a fair jury trial—a threat constitutionally acceptable only if justified by a sufficient state interest.

C

Respondent's challenge to the impartiality of death-qualified juries focuses upon the imbalance created when jurors particularly likely to look askance at the prosecution's case are systematically excluded from panels. He therefore appears to limit his constitutional claim to only those cases in which jurors have actually been struck for cause because of their opposition to the death penalty. Tr. of Oral Arg. 26. However, this limitation should not blind the Court to prejudice that occurs even in cases in which no juror has in fact been excluded for refusing to consider the death penalty.

There is considerable evidence that the very process of determining whether any potential jurors are excludable for cause under *Witherspoon* predisposes jurors to convict. One study found that exposure to the *voir dire* needed for death qualification "increased subjects' belief in the guilt of the defendant and their estimate that he would be convicted." Haney, On the Selection of Capital Juries: The Biasing Effects of the Death-Qualification Process, 8 Law & Hum. Behav. 121, 128 (1984). See *Hovey, supra,* at 73, 616 P. 2d, at 1349. Even if this prejudice to the accused does not constitute an independent due process violation, it surely should be taken into account in any inquiry into the effects of death qualification. "[The] process effect may function additively to worsen the perspective of an already conviction-prone jury whose composition has been distorted by the outcome of this selection process...." Haney, Examining Death Qualification: Further Analysis of the Process Effect, 8 Law & Hum. Behav. 133, 151 (1984).

The majority contends that any prejudice attributed to the process of death-qualifying jurors is justified by the State's interest in identifying and excluding nullifiers before the guilt stage of trial. *Ante,* at 170, n. 7. It overlooks, however, the ease with which nullifiers could be identified before trial without any extended focus on how jurors would conduct themselves at a capital sentencing proceeding. Potential jurors could be asked, for example, "if there be any reason why any of them could not fairly and impartially try the issue of defendant's guilt in accordance with the evidence presented at the trial and the court's instructions as to the law." *Grigsby II,* 569 F.Supp., at 1310.[26] The prejudice attributable to the current pretrial focus on the death penalty should therefore be considered here and provides but another reason for concluding that death qualification infringes a capital defendant's "interest in

26. Restriction of *voir dire* to this inquiry would limit the ability of the prosecution to use its peremptory challenges to strike those jurors who would have been excluded for cause under the current system of death qualification. And the power of this inquiry to root out *all* prejudices and biases, no matter how great a threat they pose to the fairness of a guilt determination, has only recently been established by this Court as a matter of law. See *Turner* v. *Murray, ante,* p. 28. But see *ante,* at 45 (MARSHALL, J., concurring in judgment in part and dissenting in part).

a completely fair determination of guilt or innocence." *Witherspoon*, 391 U.S., at 520, n. 18.

V

As the *Witherspoon* Court recognized, "the State's interest in submitting the penalty issue to a jury capable of imposing capital punishment" may be accommodated without infringing a capital defendant's interest in a fair determination of his guilt if the State uses "one jury to decide guilt and another to fix punishment." *Ibid.* Any exclusion of death-penalty opponents, the Court reasoned, could await the penalty phase of a trial. The question here is thus whether the State has *other* interests that require the use of a single jury and demand the subordination of a capital defendant's Sixth and Fourteenth Amendment rights.

The only two reasons that the Court invokes to justify the State's use of a single jury are efficient trial management and concern that a defendant at his sentencing proceedings may be able to profit from "residual doubts" troubling jurors who have sat through the guilt phase of his trial. The first of these purported justifications is merely unconvincing. The second is offensive.

In *Ballew*, the Court found that the State's interest in saving "court time and ... financial costs" was insufficient to justify further reductions in jury size. 435 U.S., at 243–244. The same is true here. The additional costs that would be imposed by a system of separate juries are not particularly high.

> "First, capital cases constitute a relatively small number of criminal trials. Moreover, the number of these cases in which a penalty determination will be necessary is even smaller. A penalty determination will occur only where a verdict on guilt has been returned that authorizes the possible imposition of capital punishment, and only where the prosecutor decides that a death sentence should be sought. Even in cases in which a penalty determination will occur, the impaneling of a new penalty jury may not always be necessary. In some cases, it may be possible to have alternate jurors replace any 'automatic life imprisonment' jurors who served at the guilt determination trial." Winick, 81 Mich. L. Rev., at 57.

In a system using separate juries for guilt and penalty phases, time and resources would be saved every time a capital case did not require a penalty phase. The *voir dire* needed to identify nullifiers before the guilt phase is less extensive than the questioning that under the current scheme is conducted before *every* capital trial. The State could, of course, choose to empanel a death-qualified jury at the start of every trial, to be used only if a penalty stage is required. However, if it opted for the cheaper alternative of empaneling a death-qualified jury only in the event that a defendant were convicted of capital charges, the State frequently would be able to avoid retrying the entire guilt phase for the benefit of the penalty jury. Stipulated summaries of prior evidence might, for example, save considerable time. Thus, it cannot fairly be said that the costs of accommodating a defendant's constitutional rights under these circumstances are prohibitive, or even significant.

Even less convincing is the Court's concern that a defendant be able to appeal at sentencing to the "residual doubts" of the jurors who found him guilty. Any suggestion that the current system of death qualification "may be in the defendant's best interests, seems specious unless the state is willing to grant the defendant the option to waive this paternalistic protection in exchange for better odds against conviction." Finch & Ferraro, 65 Neb. L. Rev., at 69. Furthermore, this case will stand as one of the few times in which any legitimacy has been given to the power of a convicted capital defendant facing the possibility of a death sentence to argue as a mitigating factor the chance that he might be innocent. Where a defendant's sentence but not his conviction has been set aside on appeal, States have routinely empaneled juries whose only duty is to assess punishment, thereby depriving defendants of the chance to profit from the "residual doubts" that jurors who had already sat through a guilt phase might bring to the sentencing proceeding. In its statute authorizing resentencing without any reassessment of culpability, Arkansas has noted: "it is a waste of judicial resources to require the retrying of an error-free trial if the State wishes to seek to reimpose the death penalty." 1983 Ark. Gen. Act No. 546, § 3, note following Ark. Rev. Stat. Ann. § 41–1358 (Supp. 1985).

But most importantly, it ill behooves the majority to allude to a defendant's power to appeal to "residual doubts" at his sentencing when this Court has consistently refused to grant certiorari in state cases holding that these doubts cannot properly be considered during capital sentencing proceedings. See *Burr* v. *Florida,* 474 U.S. 879 (1985) (MARSHALL, J., dissenting from denial of certiorari); *Heiney* v. *Florida,* 469 U.S. 920 (1984) (MARSHALL, J., dissenting from denial of certiorari); *Burford* v. *State,* 403 So. 2d 943 (Fla. 1981), cert. denied, 454 U.S. 1164 (1982). Any suggestion that capital defendants will benefit from a single jury thus is more than disingenuous. It is cruel.

VI

On occasion, this Court has declared what I believe should be obvious—that when a State seeks to convict a defendant of the most serious and severely punished offenses in its criminal code, any procedure that "[diminishes] the reliability of the guilt determination" must be struck down. *Beck* v. *Alabama,* 447 U.S., at 638. But in spite of such declarations, I cannot help thinking that respondent here would have stood a far better chance of prevailing on his constitutional claims had he not been challenging a procedure peculiar to the administration of the death penalty. For in no other context would a majority of this Court refuse to find any constitutional violation in a state practice that systematically operates to render juries more likely to convict, and to convict on the more serious charges. I dissent.

3

ANATOMY OF A JURY CHALLENGE

Joseph B. Kadane

The right to a trial by a jury is guaranteed by the U.S. Constitution. The purpose of this right is to ensure that the administration of justice, both civil and criminal, stays close to ordinary people's conception of what is right, and not become solely the province of specialized elites, judges and lawyers. Juries are an important part of American democracy, shared with other countries whose legal systems grew out of England's, but not with democracies whose law developed from the Napoleonic Code.

In a jury trial, the judge explains what the relevant law is, and the jury decides how the facts presented at trial match up with the law. Few cases are decided by trials; most criminal cases are settled by plea bargain and most civil cases by agreement among the parties. Additionally, the defendant in a criminal case and both parties in a civil case can waive their right to a jury trial. But the terms of plea bargains and settlements of civil cases depend on predictions of (and uncertainty about) what a jury would decide.

In view of the critical role of juries in deciding cases, it is important to understand how they are chosen. A jury pool (group of potential jurors) is supposed to be a representative cross section of the community (more on this later). A jury venire is a sample drawn from the jury pool and is the source of jurors for a particular case. There are two different procedures by which members of the venire are not accepted as members of the jury. First, each potential juror is asked questions that could result in him or her being excused by the judge for cause, such as having a relationship with one of the parties or one of the attorneys. Second, the potential jurors are also asked questions (just how varies) that inform the attorneys on both sides about their general attitudes about the case. Each side then has the opportunity to excuse a number (set by law) of jurors using peremptory challenges. The reasons for these challenges do not have to be explained, but are not allowed to be used systematically to eliminate persons of a given race.

Because of the use of challenges for cause and peremptory challenges, as well as the luck of the random draw from the jury pool, an individual defendant does not have the right to a jury that is demographically balanced. However, each side has had some role in shaping membership on the jury through the use of peremptory challenges. Nonetheless, a defendant does have a right to a jury venire that is drawn from a representative cross section of the community.

A jury challenge is a motion to invalidate or forbid a decision by a jury on the grounds that the jury pool was not a representative cross section of the community. The burden of proof is with the maker of the motion—typically the defendant in a criminal case. There are three possible grounds for such a challenge: provisions of the U.S. Constitution, of state constitutions, or of statutes.

Not all violations of randomness are regarded by the courts as sufficiently serious as to warrant sustaining a federal constitutional challenge. For example, if persons whose last name begin with the letter L through Z were underrepresented on a jury venire, the courts would not act, unless there were other evidence. A series of Supreme Court decisions developed the concept of a "cognizable class" of persons who, by membership in that class, have a distinctive viewpoint that adds to the diversity of the jury. Thus demographic underrepresentation by race or sex is cognizable, for example, while underrepresentation by first letter of last name is not.

Court decisions have not resolved the question of how to measure the extent to which a jury system fails cross-sectionality. Two of the most frequently used measures are absolute disparity, the difference between the proportion of the population in the cognizable class and the proportion in the pool, and relative disparity, the absolute disparity divided by the proportion in the population. The former is criticized for failing to protect small minorities, while the latter is criticized for magnifying the effects of small differences in small cognizable classes. Another method sometimes used is a test of significance, with all of its well-known infirmities. The courts have also not agreed upon a critical value for these measures, above which they would sustain a jury challenge and below which they would not.

In addition to the vagueness of the law, there are three other factors that make jury challenges difficult to win. The first is the disruption they cause to the system of justice. Because they attack (as they must) the jury pool, a successful jury challenge impugns many trials. This effect is mitigated by the rule that to be able to launch a jury challenge on appeal, the matter must have been raised when the original case was heard. Nonetheless the prospect of requiring the jury system to be fixed before a trial can go forward, the effect of such a finding on all the courts that use the same system to get a jury venire, and the pall of illegitimacy cast over previous cases heard by juries, even if a challenge is barred by rule, makes judges very reluctant to sustain a jury challenge.

A second factor that makes jury challenges difficult is the unpopularity of the defendant. Often, a jury challenge is brought on behalf of someone who has been sentenced to execution and who typically is accused of doing something awful. While there is a public benefit to examining the jury system and correcting such egregious errors as may be found, the immediate effect of sustaining a jury challenge is to permit another trial or postconviction sentencing hearing for someone hard to regard sympathetically. This factor may weigh especially heavily in states that elect judges.

Finally, many courts have a conflict of interest in hearing a jury challenge. Typically the court, or its presiding judge, has a managerial role in approving a plan for how jury venires are to be chosen, hiring the people to do it, and supervising the process. For a judge to sustain a jury challenge is to criticize his or her own work, that of his or her immediate colleagues and superiors, and that of the administrative staff with whom they have close contact. Sometimes judges are participants in the very decisions being challenged in court. Yet judges rarely recuse themselves from hearing these challenges.

Thus a combination of factors, including vagueness in the law, disruption to the system, unpopular defendants, and conflict of interest, make jury challenges difficult to be heard dispassionately, let alone to win.

In a not atypical case, Zolo Azania was accused of murdering a policeman in the course of a bank robbery in 1981. In 1982 he was convicted and sentenced to death. In 1993, the Indiana Supreme Court reversed the sentence and required a new sentencing hearing. This was held in February 1996 and again resulted in a death sentence. It is this second death sentencing trial that is at issue.

Because this case arises in Indiana, it is necessary to explain a bit about Indiana jury law as well. By state law, Indiana requires that jury venire selection be such that each commissioner's district in a county be represented in proportion to its population. In Allen County, where Azania's sentencing retrial was conducted in 1996, there are three such commissioners' districts.

In 1996 there were complaints in the local press about the nature of the juries being selected in Allen County, where Azania was tried. Defense attorneys speculated that perhaps a specific area of Wayne Township (where most of the African-Americans in Allen County live) was somehow being excluded. This led to a review of the computer programs and correction of them, without public specification of what the problem had been. I was asked to help figure out what consequences the computer error might have had for the jury venire that tried Azania, with a view toward his constitutional and statutory rights.

The computer programs in question turned out to still be available and were written in COBOL, an old mainframe language that would not be used for writing programs now. However, it comes up surprisingly often in jury challenges. The defense team provided me with a COBOL expert. The attorneys also conducted extensive depositions of jury managers and computer personnel to gain a better understanding of the system as it existed in 1996.

One intriguing result of those depositions was that when a request was made to the jury system for a given number of jurors for a year's call, the program would always produce more potential jurors than requested. A request for 10,000 resulted in a draw of 10,528; a request of 14,000 in 14,364. This was called round-off error, but because of changes in personnel over the years, no one was available who could explain technically how this came about.

Careful study of the COBOL program used for the annual draw revealed that the fraction of the voter list to be selected was rounded up to the nearest integer reciprocal, say $1/k$. Then one out of each successive group of k voters in the township voter list would be summoned for jury service. This results in a draw of at least the number requested and can result in a factor of

$$\left(\frac{k+1}{k}\right)$$

more. If the voter list were approximately 150,000 in Allen County in 1995 when the 1996 lists were prepared (exact numbers unavailable), a request of 10,000 leads to a k of 15, and a draw between 10,000 and 10,666, consistent with the observed 10,528. Similarly a request of 14,000 yields a k of 10, and a draw between 14,000 and 15,000, consistent with an observed 14,364 draw.

This same program also ordered the potential jurors alphabetically by township. As it happens, Wayne Township is last, alphabetically, among the townships in Allen County. This alphabetized list is then sent to a second program, which randomly assigns persons on the first list to four groups to be used as quarterly draws. This second program has a tape input of 10,000 random numbers, which has the consequence that only the first 10,000 persons in the tape from the first program are assigned by the second program. Hence the unassigned names all come from one township, Wayne.

In the draw in which 10,000 jurors were requested and 10,528 actually drawn, the excess names unassigned to the second program, 528, would have all been Wayne Township voters. The jury venire that actually heard Azania's second penalty trial had the request of 14,000, with an actual draw of 14,364, so the excess over 10,000, here 4,364, would all have come from Wayne Township. This feature of the jury system went unnoticed when the request was for 10,000 jurors, as it apparently was for years. However, the increase in the number of trials led to a request for 14,000 jurors in 1995 for the 1996 year, which led to the surfacing of the issue. The question I had to address is how this set of circumstances might have affected the various requirements a jury system is intended to meet.

According to the 1990 census, Wayne Township had 88,671 residents 18 years or older, while all of Allen County had 217,332. Thus Wayne Township had 40.8% of the residents of jury-eligible age. In 1995, Allen County had 159,914 registered voters, of whom 55,136, or 34.9%, resided in Wayne Township. Since the Indiana jury system uses only the voter list as a source list for potential jurors, the latter figure, 34.9%, is the proportion of Wayne Township residents one would expect to find in an Allen County jury were the system choosing jurors from the voter list randomly.

In a draw of 14,364, therefore, one would expect $(14,364)(.349) = 5013$ from Wayne Township. Since 4,364 of those jurors would not have been called because of the interaction of the two jury selection programs described above, the call from Wayne County would be expected to be $5013 - 4364 = 649$, or 6.5% of the jury draw. A count of the quarter draw from which Azania's jury was selected showed that 123 of the 2,050 called, or 6.0%, were from Wayne Township. The calculations above show that this underrepresentation of Wayne Township was systematic, that is, that it is not a fluke of a random process, but rather was inherent to the system in place at the time.

The question the courts have to deal with, then, is whether a process expected to yield a jury venire that is 6.5% from Wayne Township and actually produced a venire of 6.0% from Wayne Township, drawn from a voter list that is 34.9% from Wayne Township, while Wayne Township has 40.8% of the residents of jury-eligible age, violates Azania's rights. There are three different ways in which it might have.

First, there is an argument that, having chosen to allocate voters by township, Azania is entitled to a jury venire that, but for random fluctuations, reflects the voter registrations by township. If so, it is hard to argue that the 1996 jury venire from which the jury was chosen that, for the second time, sentenced Azania to die, should pass muster.

A second argument concerns whether the underrepresentation of Wayne County leads to an underrepresentation by race. According to the 1990 census, of the 18,552 African-Americans 18 and older who live in Allen County, 13,937, or 75.1%, live in Wayne Township. Therefore, to underrepresent Wayne Township is almost inevitably to underrepresent African-Americans. Hence the question is whether the extent of the underrepresentation is enough to matter to the courts.

If the Allen County jury system represented African-Americans in accordance with their numbers in the 18 and older population, such venires would have $18,552/217,332 = 8.5\%$ African-Americans. Taking a draw that is 6.5% from Wayne Township and 93.5% from the rest of the county yields a draw expected to be

$$(.065)\left(\frac{13,937}{88,671}\right)+(.935)\left(\frac{4,615}{128,661}\right)=.044=4.4\%$$

African-American. This in turn implies an absolute disparity of $8.5\% - 4.4\% = 4.1\%$ and a relative disparity of $4.1\%/8.5\% = 48.2\%$. This means that African-Americans have slightly more than half the probability of appearing on a jury venire compared with what they would have in a random system.

A third way to understand the import of these numbers is to look at the probability of having at least one African-American on a jury of 12 randomly selected from the jury venire. Absent this computer system error, this probability would be

$$1 - (1 - .085)^{12} = .66 = 66\%,$$

while with the computer glitch, it is

$$1 - (1 - .044)^{12} = .417 = 41.7\%.$$

The question of what consequences these numbers have is both a legal and a statistical question.

The third way in which Azania's rights may have been violated is with respect to the Indiana statute requiring that commissioners' districts be represented in accordance with their populations. Allen County has three such districts, with the following populations from the 1990 census:

Commissioner District	Population	Proportion
1	101,448	33.7%
2	100,545	33.4%
3	98,841	32.9%
Total	300,834	

In order to estimate the impact of the jury system's systematic underrepresentation of Wayne township, it is necessary to have data on the extent to which each of these districts include population from Wayne Township. Those numbers are as follows:

Commissioner District	Wayne	Non-Wayne	Total
1	13,227	88,221	101,448
2	28,969	71,576	100,545
3	73,808	25,033	98,841

Thus District l's proportion of a random jury venire chosen as was Azania's sentencing jury is proportional to

$$13,227(.065) + 88,221(.935) = 83,346.$$

Similarly, for District 2, the draw is proportional to

$$28,969(.065) + 71,576(.935) = 68,807.$$

Finally, for District 3, it is proportional to

$$73,808 \ (.065) + 25,033(.935) = 28,203.$$

Hence the proportions for the three districts are, respectively, 46.2%, 38.2%, and 15.6%. The following table permits the comparison:

District	Proportion of Population	Proportion of Jury Venire
1	33.7%	46.2%
2	33.4%	38.2%
3	32.9%	15.6%

These numbers were the heart of my testimony before Judge Scheibenberger on March 13, 2001. On April 12, he released his ruling, denying Azania a new penalty trial. With respect to race, he adopted an absolute disparity standard of 10%, which means that defendants could be tried in Allen County from jury venires with virtually no African-Americans without violating this interpretation of the law. With respect to the commissioners' districts, he found that "to the extent that the computer bug resulted in the selection of jurors to be slightly out of proportion to the commissioner districts during 1996, the Court finds that the random and impartial nature of the selection process still substantially complies with the statute." This ruling is now being appealed.

Helping to keep the jury system functioning properly is a crucial part of maintaining our democracy. Statisticians have an important role to play in this process. That Azania might get a third sentencing trial as a result is either part of the cost or part of the benefit, depending on your point of view.

ACKNOWLEDGMENTS

I thank the Azania defense team of Erica Thompson, Esq., Michael Deutsch, Esq., Bob Weeks, and Brian Reid for help in preparing this material and David Kairys, Esq., and Erica Thompson for their helpful comments.

REFERENCES AND FURTHER READING

Kairys, D., Kadane, J. B., and Lehoczky, J. (1977), "Jury Representativeness: A Mandate for Multiple Source Lists," *California Law Review,* 65, 776–827. (Reprinted in the record of a hearing before the Subcommittee on Improvements in Judicial Machinery of the Committee on the Judiciary, U.S. Senate, September 28, 1977.)

National Jury Project (1983), *Jurywork: Systematic Techniques,* E. Krauss and B. Bonara, editors (2nd ed.), New York: Clark Boardman.

Van Dyke, J. M. (1977), *Jury Selection Procedures: Our Uncertain Commitment for Representative Panels,* Cambridge, MA: Ballinger Publishing Co.

4

AZANIA V. STATE OF INDIANA

ZOLO AGONA **AZANIA,** Appellant (Petitioner Below), v. STATE OF **INDIANA,** Appellee (Respondent Below).

SUPREME COURT OF INDIANA

778 N.E.2d 1253; 110 A.L.R.5th 725

November 22, 2002, Decided

COUNSEL: ATTORNEY FOR APPELLANT: Jesse A. Cook, Deputy Public Defender, Terre Haute, Indiana, Michael E. Deutsch, Deputy Public Defender, Chicago, Illinois.

BRIEF OF AMICI CURIAE: William Goodman, Jaykumar A. Menon, New York, New York, Monica Foster, Indianapolis, Indiana.

ATTORNEYS FOR APPELLEE: Steve Carter, Attorney General of Indiana, Christopher L. Lafuse, Deputy Attorney General, Indianapolis, Indiana.

JUDGES: BOEHM, Justice. SULLIVAN and RUCKER, JJ., concur. SHEPARD, C.J., dissents with separate opinion, in which DICKSON, J., concurs.

OPINION BY: BOEHM

OPINION

ON APPEAL FROM THE DENIAL OF SUCCESSIVE PETITION FOR POST-CONVICTION RELIEF

BOEHM, Justice.

Zolo Agona Azania, formerly known as Rufus Averhart, was convicted of murder and sentenced to death. In this appeal from the denial of his second petition for post-conviction relief, Azania argues that his death sentence must be vacated because the jury that recommended imposition of the death penalty was the product of a system for jury pool selection that systematically and materially reduced participation of African–American jurors.

In an ordinary lawsuit we would not find the irregularities in the Allen County jury selection process sufficient to require a reversal. The disproportionate reduction of African–Americans in the jury pool was, as the Chief Justice's dissent observes, the result of a "computer glitch," more precisely, a flawed program, not a hardware defect. But computer failures can have serious consequences, and this is an example of that. Because of the heightened need for public confidence in the integrity of a death penalty, we conclude that although the conviction was proper, the jury pool selection process was fundamentally flawed, and reversal of the death penalty and a new penalty phase or resentencing is required.

FACTUAL AND PROCEDURAL BACKGROUND

Azania was convicted of murder and sentenced to death for the 1981 slaying of Gary Police Lieutenant George Yaros in the course of a bank robbery.[1] In 1984, this Court affirmed his conviction and sentence on direct appeal. Averhart v. State, 470 N.E.2d 666 (Ind. 1984). Azania was denied post-conviction relief, and in a 1993 appeal from that ruling, this Court affirmed Azania's conviction but reversed his sentence, citing ineffective assistance of counsel at the sentencing phase and the failure of the prosecution to provide gunshot residue test results to the defense. Averhart v. State, 614 N.E.2d 924, 930 (Ind. 1993).

After remand for a new penalty phase, Azania unsuccessfully moved to strike the entire jury pool on the grounds that it did not represent a reasonable cross section of the community. A new jury was impaneled and it also recommended death. After the trial court again sentenced Azania to death, this Court affirmed the sentence on direct appeal. Azania v. State, 730 N.E.2d 646 (Ind. 2000). Azania was then granted leave to file a successive petition for post-conviction relief on two grounds: newly discovered evidence, and alleged abnormalities in the Allen County jury pool selection system. Azania v. State, 738 N.E.2d 248 (Ind. 2000). The successive post-conviction court denied relief, and this appeal followed.

I. JURY POOL SELECTION

A. The Statutory Standard

The method by which jury pools are selected in Indiana is governed by statute. Indiana Code section 33–4–5–2(c) allows jury commissioners to use a computerized jury selection system, but requires that the system employed "must be fair and may not violate the rights of persons with respect to the impartial and random selection of prospective jurors." This Court long ago held that the purpose of the jury selection statute is to ensure that the method used to select a jury is not arbitrary and does not result in the systematic exclusion of any group. Shack v. State, 259 Ind. 450,

---

1. For a detailed account of the robbery and killing, see Averhart v. State, 470 N.E.2d 666, 673–75 (Ind. 1984).

459–60, 288 N.E.2d 155, 162 (1972). Nevertheless, there is no requirement that any particular segment of the population be represented on every jury, Daniels v. State, 274 Ind. 29, 35, 408 N.E.2d 1244, 1247 (1980), and completely random selection of jurors is not required as long as the system used is impartial and not arbitrary. State ex rel. Burns v. Sharp, 271 Ind. 344, 348, 393 N.E.2d 127, 130 (1979). Minor irregularities will not constitute reversible error unless there is a showing of substantial prejudice to the accused's rights as a result of the irregularities. Porter v. State, 271 Ind. 180, 201, 391 N.E.2d 801, 816 (1979), overruled on other grounds. Despite these somewhat flexible standards, an accused is entitled to a trial by a jury selected in substantial compliance with the statute, and if there is a lack of substantial compliance, the accused need not show actual prejudice. Cross v. State, 272 Ind. 223, 226, 397 N.E.2d 265, 268 (1979); Wireman v. State, 432 N.E.2d 1343, 1354 (Ind. 1982) (Hunter, J., dissenting); Rogers v. State, 428 N.E.2d 70, 72 (Ind. Ct. App. 1981); Bagnell v. State, 413 N.E.2d 1072, 1075 (Ind. Ct. App. 1980).

B. Allen County's System of Pool Selection

The computerized system used to select the jury pool for Azania's 1996 sentencing recommendation hearing was designed in 1980. The successive post-conviction court found that the system had four flaws, the net effect of which was exclusion of a number of jury pool members who resided in Wayne Township from the possibility of being called to serve. Specifically, in 1996, when Azania's penalty phase was retried, these problems excluded 4364 of 5013, or 87%, of Wayne Township voters from jury service. In that year, the countywide jury pool was 14,364.

1. Overview of the Problem The problem in Allen County's jury selection procedures may be readily stated in broad overview. The number of jurors needed for 1996 was first identified as 14,000. The program then selected 14,364 registered voters to be assigned a random number. Only persons assigned a number could be drawn for a panel. The assignment stopped after 10,000 voters had received numbers. Because the program worked through the voter list by township in alphabetical order, all of the excluded 4364 registered voters were Wayne Township residents. As a result, 87% of Wayne Township was excluded. This had a materially disproportionate effect on African-Americans because African-Americans comprised 8.5% of the total population of Allen County, and three fourths of that 8.5% resided in Wayne Township. The remainder of this Part I:B explains the details of how this occurred. Its legal implications are addressed in Part C.

2. Truncation The first problem resulted from a truncation feature embedded in the program since 1980. The program would first read the registered voter list and determine the total number of registered voters in the county and in each township. The program would then determine the percentage of all Allen County registered voters who resided in each township. Before each calendar year, the court administrator determined the desired number of jurors required for all Allen County courts for the entire year. Based upon the requested size of this "master pool," the program then determined the number of jurors it needed to select from each township to ensure proportional representation of that township in the master pool. The total voter list for the township was then to be divided into that number of "selection groups" by dividing the total number of registered voters in the township by the number of jurors needed from the township. One juror was then to be chosen from each group. This division rarely produced an integer (e.g., 21). In almost all cases, it produced a real number (e.g., 21.2439). The program then truncated this real number by eliminating everything after the decimal point and converting the real number (21.2439) into an integer (21). The program then used the integer, rather than the real number, to select groups, identifying the first 21 as group 1, then 22 through 42 as group 2, etc. By using the truncated integer, which was a fraction smaller than the real number, rather than rounding to the nearest integer, the program produced roughly 5% more groups than the requested size of the master pool. A random number was then used to select one juror from each group, producing a response in the range of 10,500 names to a request for 10,000 jurors. Thus, from the outset of the program in 1980, this truncation caused more voters than were requested to be chosen for assignment of a random number.[2]

3. The Effect of Growth in the Requested Number of Jurors Regardless of how many names were included on the master jury pool list, from the outset the program assigned random numbers—necessary for actual selection to serve—to only 10,000 voters. When the list exceeded 10,000 names, the effect of this was to cut the list off at 10,000. From 1980 to 1994, the court administrator requested annual master jury pools of 10,000 people. During that period, the approximately 500 excess jurors produced by the truncation feature were excluded from service, but only those 500 jurors were affected. In 1995, however, the requested number grew to 12,000 jurors, and the truncation feature added another 693, so 12,693 voters were selected. As a result of assigning a random number to only 10,000 jurors, 2693 of those jurors could not be called to serve. In 1996, the year of Azania's resentencing, the requested jury pool was 14,000, and the truncation feature added 364 names. As a result of the limitation to 10,000, 4364 of those did not receive random numbers and could not serve.

4. The Accident of the Alphabet Finally, and importantly, the computer organized the county jury pool by townships in alphabetical order. This placed all Wayne Township jurors at the end of the list of 14,364. Thus, in each year since 1980 all of the excluded jury pool members were Wayne Township residents. The effect of these problems was not unfocused or randomly distributed over the county or over population groups. According to the 1990 census, African-Americans comprised 18,552 or 8.5% of the total age 18 and over of the Allen County population of

---

2. For example, in a hypothetical county comprised of 1100 registered voters evenly distributed across the county's 10 townships, the program would first determine that 10% of the registered voters, or 110 voters, resided in each township. If the requested size of the master jury pool was 200 jurors, the program would next determine that 20 jurors (10% of 200) were needed from each township to ensure proportional representation of that township in the master pool. As to hypothetical Township A, the program would divide the total number of registered voters in Township A (110) by the number of jurors needed from Township A (20), to determine that 5.5 voters should be placed into each of Township A's 20 "selection groups." Next, the program would truncate 5.5 to the integer 5, and then take the first 5 voters on the list and select one, then take the next five voters on the list and select one, and so on. The result would be that 22 voters from Township A would be included on the list, rather than the 20 required for proportional representation.

217,332. In addition, 13,937 (75.1%) of these 18,552 African-Americans resided in Wayne Township. Accordingly, the program excluded 87% of the jury pool members from the township in which 75.1% of Allen County's age 18 and over African-Americans resided.

Azania argues that the result of these problems was that in the quarterly draw from which his jury pool was taken, African-Americans—who in a truly representative system would have comprised 8.5% of the pool—in fact comprised only 4.4% of the pool. The post-conviction court rejected Azania's calculation as unreliable. The court ruled that using 1990 census data "as a proxy for the racial composition of the 1996 voter registration list"—as well as using a mathematical formula to estimate the number of African-Americans in the quarterly draw from which Azania's jury was comprised—was akin to "asking the court to make an inference from an inference, something the court is not allowed to do." The post-conviction court may be correct that African-American citizens do not necessarily register to vote in proportion to their population, but Allen County did not maintain racial information about the voter list and we have nothing to go by except the census. Both the United States Supreme Court and the lower federal courts have repeatedly upheld the use of census figures in constitutional assaults on jury selection procedures. See Duren v. Missouri, 439 U.S. 357, 365, 58 L. Ed. 2d 579, 99 S. Ct. 664 (1979) (upholding the use of six–year–old census data in fair cross–section challenge); Alexander v. Louisiana, 405 U.S. 625, 627, 31 L. Ed. 2d 536, 92 S. Ct. 1221 (1972) (upholding the use of six-year-old census data in equal protection challenge); Davis v. Warden, 867 F.2d 1003, 1014 (7th Cir. 1989); United States v. Osorio, 801 F. Supp. 966, 977–78 (D.Conn. 1992). We agree with the courts that have concluded that under these circumstances a "defendant should not be expected to carry a prohibitive burden in proving underrepresentation." Davis, 867 F.2d at 1014. Similarly, because no statistical data was available regarding the number of African–Americans in the quarterly draw from which Azania's jury was comprised, it was appropriate for Azania's expert witness to use a mathematical formula derived directly from the operation of Allen County's computerized system to estimate that number.

C. The Effect of the Elimination of 87% of Wayne Township from Jury Service

The United States Supreme Court has long held that "the selection of a petit jury from a representative cross section of the community is an essential component of the Sixth Amendment right to a jury trial." Taylor v. Louisiana, 419 U.S. 522, 528, 42 L. Ed. 2d 690, 95 S. Ct. 692 (1975). We think our state statute, in requiring an "impartial and random selection," demands no less. Although we reach our holding today under Indiana Code section 33–4–5–2(c) and not under the Sixth Amendment to the Federal Constitution, we think that the Indiana statute ultimately turns on an issue very similar to Sixth Amendment analysis: whether the flaws in a jury selection system are so minor as to be inconsequential or are material enough that a segment of the population has been materially excluded.

The federal courts have developed two competing tests under the Sixth Amendment to determine if a jury pool adequately represents the community. Under the absolute disparity test, the "disparity" is the difference between the percentage of the distinctive group eligible for jury duty and the percentage represented in the pool. In this case, where the percentage of African–Americans eligible for jury duty in Allen county is 8.5% and the percentage represented in the pool is 4.4%, this amounts to an absolute disparity of 4.1%. Under the comparative disparity test, the "disparity" is calculated by dividing the absolute disparity by the percentage of the group eligible for jury duty. Here, that results in the division of 4.1% by 8.5%, for a comparative disparity of 48.2%. Put differently, as the result of flaws in Allen County's system, African–Americans as a group had roughly half the chance of being included on a jury panel than a truly random system would have produced. Nevertheless, the post-conviction court concluded that in Azania's case the computerized system "impartially and randomly selected citizens to be jurors, and thus substantially complied with [section 33–4–5–2(c)]." We agree this may be true for non–death penalty cases, but we do not agree that the Allen County system in place in 1996 was sufficiently impartial or random to support a jury recommendation of the death penalty.

As the Supreme Court of the United States held in Powers v. Ohio, 499 U.S. 400, 413, 113 L. Ed. 2d 411, 111 S. Ct. 1364 (1991):

> The purpose of the jury system is to impress upon the criminal defendant and the community as a whole that a verdict of conviction or acquittal is given in accordance with the law by persons who are fair. The verdict will not be accepted or understood in these terms if the jury is chosen by unlawful means at the outset.

The Indiana jury selection statute is designed to ensure that the method used to select a jury is not arbitrary and does not result in the systematic exclusion of any group. The United States Supreme Court has long emphasized that "the qualitative difference of death from all other punishments requires a correspondingly greater degree of scrutiny of the capital sentencing determination." California v. Ramos, 463 U.S. 992, 998–99, 77 L. Ed. 2d 1171, 103 S. Ct. 3446 (1983). The Supreme Court has also held, in a death penalty case, that a jury's being chosen from a fair cross section of the community is "critical to public confidence in the fairness of the criminal justice system," and that the systematic exclusion of "identifiable segments playing major roles in the community cannot be squared with the constitutional concept of jury trial." Taylor, 419 U.S. at 530. Widespread concern over the fairness and reliability of death sentences demands that the courts and the public have no significant doubts as to the integrity and fairness of the process. These same considerations require heightened sensitivity in a death penalty case in determining whether a jury selection system is "random" and "impartial" as required by Indiana law.

In this case, Azania properly preserved his right to contest the impartiality of the computerized system by moving to strike the entire jury pool. As the court below noted, the system's programming error excluded 4364 people—roughly one–third of the jury pool—from possible service, and reduced by nearly one–half the odds that an African–American would appear on the jury panel. Every one of the excluded jury pool members was from Wayne Township, the township in which three-fourths of Allen County's African-Americans over age 18 resided. The net result was that the flaws inherent in the selection system materially reduced the probability that

African–Americans would serve on Azania's penalty phase jury. Accordingly, the system did not substantially comply with section 33–4–5–2(c), and a new penalty phase is required.

Finally, as the dissent observes, in 1982 Azania requested a transfer of this case from Lake County, where Officer Yaros was slain. Unlike the dissent, we do not consider that to be relevant here. Azania exercised his right under generally applicable procedures to seek a transfer to another county. In so electing, he did not forfeit his right to a properly selected jury in the new county, whatever its demographic composition.

II. CHANGE OF JUDGE

Prior to his second post-conviction hearing, Azania twice unsuccessfully moved for a change of judge pursuant to Indiana Post-Conviction Rule 1(4)(b). Azania alleged bias on the part of the trial judge, who as a member of the Allen Superior Court Board of Judges had some oversight responsibility for the computerized jury selection system. Rule 1(4)(b) mandates a change of judge when the historical facts recited in the affidavit filed in support of the motion, if taken as true, support a rational inference of bias or prejudice. This Court will presume that a judge is not biased against a party. Lambert v. State, 743 N.E.2d 719, 728 (Ind. 2001). This Court recently held that denial of a motion for change of judge under Criminal Rule 12 is reviewed under a clearly erroneous standard. Sturgeon v. State, 719 N.E.2d 1173, 1182 (Ind. 1999). We think the same standard applies to post-conviction court motions under Rule 1(4)(b). Both rules call for a change of judge if the affidavits support "a rational inference of bias or prejudice."

The pertinent historical facts recited in Azania's affidavits in support of his motions for a change of judge were that (1) the trial judge was personally involved in the investigation of the computer problems and, if not serving as judge, might be called as a witness in the case, (2) because of the destruction of some of the jury selection evidence, the court would be called upon to assess the credibility of some Allen County court employees and judicial officers, and (3) after the judge became aware of the problems, he did not notify Azania of any problems with the computer system, even though the judge had recently presided over Azania's sentencing hearing with a jury selected by the same system. Our holding in Part I of this opinion renders moot Azania's first two historical facts. As to the remaining one, we do not believe the trial judge's failure to notify Azania of problems with the computer system raises a rational inference of bias against Azania. Azania points to no authority requiring—or even suggesting—such a notification by a trial judge to a defendant in a closed matter. Nor are we aware of any. The trial court's denial of Azania's motions for a change of judge was not clearly erroneous.

III. ALLEGEDLY FALSE TESTIMONY

At Azania's 1982 trial, James McGrew identified Azania as the man McGrew saw place a pistol and jacket in some bushes not far from the scene of the robbery. McGrew also testified that when a police officer pursuing Azania approached McGrew, McGrew told the officer, "I believe that the guy you're looking for is over there," and pointed in the direction Azania had gone. McGrew testified that when the officer returned "about a minute later," with Azania now face down in the back of a patrol car, McGrew positively identified Azania as the man who placed the objects in the bushes.

In a 1995 deposition in preparation for Azania's penalty phase retrial, McGrew recanted his earlier testimony and identification. In a 2001 videotaped deposition prepared for Azania's successive post-conviction proceeding, McGrew claimed that he had never been able to identify Azania as the man he saw place the objects in the bushes, and that he told this to police and prosecutors, but that they pressured him to make the identification at trial anyway. McGrew claimed that when he was interviewed at the Gary police station, he saw a photograph of Azania and Azania's name on a bulletin board behind the officer who interviewed him, and that the officer pointed to the picture and told McGrew that Azania had killed a police officer. McGrew also testified that in 1982, while waiting in a room adjacent to the courtroom and preparing to testify, an armed man McGrew assumed to be a bailiff pointed Azania out to McGrew through the room's doorway. McGrew claimed he felt threatened by the armed man's action and the trial atmosphere, and was afraid that if he did not identify Azania his own life would be in jeopardy. McGrew testified that he could not otherwise have identified Azania, since he never saw the face of the man who placed the objects in the bushes.

Former Lake County Deputy Prosecutor James McNew, who assisted in the 1982 prosecution of Azania, testified that he was never aware that McGrew allegedly could not identify Azania. McNew denied directing anyone to coerce McGrew into identifying Azania, and testified that he did not consider Azania's identity a problem at trial in light of the other evidence of Azania's guilt, including security camera photographs from the bank and clothing evidence. McNew testified that to present a complete story to the jury, the State would have asked McGrew on direct examination if he could identify Azania even if the State knew McGrew could not do so.

Captain Michael Nardini interviewed McGrew the day after the murder in an interview room at the Gary police station. Nardini testified that McGrew told him McGrew could identify the man who placed the objects in the bushes. Nardini also testified that there was no bulletin board in the interview room, that he did not remember a photograph of Azania being posted anywhere in the station at the time of the interview, that at the time of the interview he did not know that Azania was a suspect in the case, and that he did not tell McGrew that Azania killed a police officer.

Allen County Deputy Sheriff Jerry Fruchey served as a bailiff during Azania's trial and closely meets McGrew's description of the "armed man" who allegedly pointed Azania out to McGrew. Fruchey testified that he did not remember ever speaking with McGrew, did not tell him to identify Azania, did not point out Azania, and did not in any way threaten McGrew.

The successive post-conviction court considered all this evidence and held that McGrew's deposition testimony was not credible and accordingly did not satisfy the nine-prong test for newly discovered evidence mandating a new trial. See

Carter v. State, 738 N.E.2d 665, 671 (Ind. 2000). Substantial evidence contradicts McGrew's recantation of his trial identification. First, as is true of all recanted testimony, McGrew's 1982 trial testimony directly contradicts his current claims. Second, his current claims contradict his statements to police at the time of Azania's apprehension and to Nardini the next day. Third, his claim that he was intimidated by an armed man is contradicted by Fruchey. This issue turns on credibility of witnesses. The successive post-conviction court viewed McGrew and the other post-conviction witnesses and found that his recantation was not credible. That finding is not clearly erroneous, and is accordingly affirmed.

CONCLUSION

The successive post-conviction court erred when it concluded that Allen County's computerized jury selection system substantially complied with Indiana Code section 33-4-5-2(c) on the facts of this case. The court did not err when it denied Azania's motions for a change of judge and his claim for a new trial based on the prosecution's use of allegedly false testimony. Accordingly, we vacate Azania's death sentence and remand to the trial court for new penalty phase hearings, or, if the prosecution elects not to pursue the death penalty, for sentencing.

SULLIVAN and RUCKER, JJ., concur.

SHEPARD, C.J., dissents with separate opinion, in which DICKSON, J., concurs.

DISSENT BY: DICKSON; SHEPARD

DISSENT: DICKSON, J., dissents, believing that this accidental, inadvertent, and impartial exclusion of jurors did not undermine Allen County's essential substantial compliance with the statutory method for selection of jury pools, and that any error was harmless in light of the overwhelming evidence of aggravators which have previously led two separate penalty phase juries to unanimously recommend the death sentence.

SHEPARD, Chief Justice, dissenting.

Zolo Azania and two cohorts burst into the Gary National Bank in broad daylight with guns drawn. By the time they were ready to depart with the money, the bank's security camera was already recording the robbery and an alarm had summoned the police.

The trio decided to shoot their way out, and exited the bank with guns blazing at the uniformed officer who had arrived on the scene.

They ran past the fallen body of Gary Police Lieutenant George Yaros and headed for the getaway car. Not content merely to take off, Azania went over to the officer, kicked his gun away, then put yet another shot into him at close range. After that, the three perpetrators led police on a chase through the streets of Gary at 80–100 mph, firing back at the pursuing officers. All of this has been largely settled fact for more than a generation.

In the meantime, twenty-four jurors and two different trial judges have unanimously agreed that the State's request for the death penalty was a just one.

In the face of this, the judgment of judge and jury is today set aside on the basis of a computer glitch.

Equally unattractive is Azania's play of the system. He seeks relief on the grounds that a mathematically perfect computer run would enhance his chance to have an African-American on the jury. He asserts that having even one black juror is crucial to his cause. Of course, the reason that even an ideal, random jury pool might still produce an all–white jury is that this litigation was transferred, at Azania's request and with his participation, to a county with a modest minority population. The State filed these charges in Lake County, where Azania would have had the most diverse jury pool Indiana has to offer. He asked to get away from that jury pool, citing reports of his crimes on Chicago television, and he was accommodated.

I would find that the jury was assembled in substantial compliance with the statute and that any error was harmless. Instead, we will now move along to a third decade of judicial effort aimed not at assessing whether Azania is guilty but rather at settling on an appropriate penalty.

DICKSON, J., concurs.

Part 4

OTHER ADVENTURES

OTHER ADVENTURES AS AN EXPERT WITNESS

A reader may at this point have the mis-impression that the only use of statistics in the law is to address matters of discrimination. The purpose of this part is to demonstrate how misleading that impression is. I offer here six diverse cases to illustrate the variety of issues to which statistics is relevant.

The first of these is the question of whether electronic draw poker requires skill to play. Electronic draw poker is played on and against a machine, typically in a bar. The player inserts a coin (typically a quarter), is "dealt" a hand of five cards by the machine, decides which cards to keep and which to return, and gets new cards to replace those discarded. The resulting hand is scored according to a point schedule, where each point won permits the player to play a free game. In some situations, the bartender may pay the player a quarter for each game won but not redeemed for free play. When that is done, both the player and the bartender may have violated gambling laws, which forbid games with three elements: (1) consideration (do you have to pay to play?), (2) reward (do you get something valuable if you win?), and (3) skill (roughly, does the game require skill to play?). Machines that have no other purpose than to be used as gambling devices are illegal per se, and can be seized by the state police from any bar. Two such machines were seized, and I was hired to study the issue of whether skill is a factor in winning electronic draw poker.

The first paper in this section reports my testimony in the matter. I contrasted a "dumb" strategy (discarding nothing) with a "smart" strategy (where I used my knowledge of probability to decide which cards to keep and which to discard). As I did much better with the smart strategy, I concluded that skill is definitely an element in winning. The opposing expert witness was an FBI agent who testified that skill meant manual dexterity, and therefore there was no skill involved.

The paper reports how my argument was received by each of the three courts to consider it.

The second paper in this section is a critique of my paper, written by my colleague John Lehoczky. His argument is that a few simple rules, which a child could learn, are sufficient to do quite well at electronic draw poker, and hence the skill involved is minimal. At the time, we did not know the expected worth (in points) of the optimal strategy for electronic draw poker, and hence it was difficult to evaluate Lehoczky's argument.

Later, software became available that enabled Marc Ware and me to find the expected worth of the optimal strategy. Our conclusion, in the third paper offered here, is that Lehoczky was essentially right. (This illustrates an important point about expert testimony. It is your best opinion at the time. Later, you may learn new facts that change your mind.)

The second section concerns a "case" heard not before an ordinary court, but before an honor court at a university. (It might not have occurred to you before, but each college and university operates its own system of law and courts, which decide whether students (and in some cases, staff and faculty) have violated university standards.) Here a student was accused, on the basis of body-language, of copying from other students in a series of multiple-choice exams. I was given the answers each student gave to each question, and the identity of the exam of the student accused, and those of each alleged copyee. My conclusion was that the student was close to the top of the class and that position was not dependent on copying. However the data were such that I could not exclude the possibility of a small amount of copying.

The next paper in this part has to do with sampling for the purpose of tax auditing, here a state sales tax. Even if done correctly (which was not the case in this audit), the best that a sample can give is a probability distribution for the amount owed. This is a difficult amount to write a check for. So what considerations should go into moving from a probability distribution to an amount owed? Together with Bright and Nagin, we develop a state loss function that balances the loss to the state from overcollecting taxes with the loss to the state of undercollecting tax. This leads to a rule for how much the check should be.

The fourth section here concerns a disputed election, and specifically whether ballot boxes were tampered with between the count on election night, and a recount that took place several weeks later. The paper, written jointly with Ilaria DiMatteo, reports three different ways of looking at the data: a data-analytic method that uses an index of the vulnerability of a prescient to tampering, based on changes in the overvotes and undervotes between the two counts, a formal Bayesian hierarchical model simulated using Markov Chain Monte Carlo methods, and the court's method, which required identifying individual ballots which it believed were tampered with. Fortunately, all three methods came to the same conclusion that tampering had occurred.

The next-to-last section deals with an important and difficult social problem, how to protect the public against re-offense by violent sexual offenders, on the one hand, and how to uphold these offenders' civil rights once they have served their sentence on the other. The state laws of many states permit the civil incarceration of such prisoners once they have served their sentence. This civil incarceration depends on expert testimony that the prisoner is likely to re-offend. This expert testimony is usually based on either the intuition of a psychiatrist or psychologist (found in studies to be quite unreliable), or on the basis of "actuarial" (which means statistical) instruments that are alleged to be predictive of re-offense. George Woodworth and I examine the basis of the claims made for these instruments, and find the claims in general unsubstantiated.

The last case in this part deals with an issue of patent misconduct. When an inventor applies for a patent, there is assumed to be a high level of integrity in the submissions made to the Patent Office. Falsified material submitted can invalidate a patent. In this case, the inventor was accused of adding material to a page of his notebook after it had been witnessed. Thus the issue comes to which ink lies on top, the text in question or the witness line. An initial analysis by the side wishing to invalidate the patent showed that the intersections of the two lines was more like the text than it was like the witness line in the concentration of various rare elements. This, it was alleged, constituted proof of inventor misconduct. A plot of the data, including the raw paper, shows, however, that the paper and the witness line were very similar in these measurements. Hence the witness line is essentially invisible, and the observation that the intersections are similar to the text is as consistent with the text lying under the witness line as over. Hence there no evidence of misconduct in these data. The contention of misconduct was withdrawn.

As you can now see, there are many factual situations that occur in the law in which statistics has a useful role to play.

A. ELECTRONIC DRAW POKER

This section records the history of one of my earliest adventures as an expert witness. Pennsylvania has made it illegal to possess a "gambling device," defined as a machine you have to pay to play, that gives you a reward for winning, and in which chance, not skill, drives the outcome.

Thus a slot machine, into which you put money to play, gives you money if you win, and whose outcome is determined solely by chance, is a gambling device *per se* (i.e., in and of itself). By contrast, a deck of cards could be used for gambling, but could be used solely for amusement, and hence is not a gambling device *per se*.

The issue in this case was whether a certain kind of draw poker machine was a gambling device *per se* (and hence could be seized by the State Police on sight), or was a gambling device only if it was used for gambling (which would require further evidence). This question boiled down to a discussion of skill versus chance.

At the time, there was little guidance to be had, in the legal or statistical literature, on how to think about that issue. It wasn't clear what the legal standard was, and it wasn't clear how to implement such vague guidance as existed. So I conducted a simple experiment that I thought addressed the issue. The first paper in this section gives a summary of my testimony, and the responses of three levels of Pennsylvania State Courts to it. John Lehoczky's comment raises a number of very interesting points challenging my thinking. (He would have made a formidable opposing expert witness in this case.) My brief rejoinder defended my work as best I could at the time.

Both Lehoczky's comment and my paper left many issues unresolved. Later advances permitted some of it to be resolved, as recorded in the paper with Marc Ware. First, computer programs were available to calculate the expected worth of the optimal strategy. Second, a theory had been developed for making some sense of quantification of the relative degree to which chance and skill influence the outcome of a game involving both. Finally, that paper raises, and leaves open, the question of whether the machines I had been given to examine in the original case had been doctored to give a payoff higher than would have been achieved under a random draw of the "cards."

1

DOES ELECTRONIC DRAW POKER REQUIRE SKILL TO PLAY?

Joseph B. Kadane

1. INTRODUCTION

On 11 March 1980, a Pennsylvania State Trooper seized a draw poker machine in Allen's Grill in Neville Island Township in Allegheny County, Pennsylvania, charging it was a gambling device per se. I was hired by the machine's defenders to testify on the question of whether these machines require skill to play. I begin with a description of how a draw poker machine is played, and a brief description of the law of gambling devices per se. I then describe my testimony and the decisions of the courts. Ultimately the Pennsylvania Supreme Court decided that chance predominates over skill in playing draw poker machines.

2. WHAT IS A DRAW POKER MACHINE?

A draw poker machine plays a form of solitaire poker with the customer. It is activated by a coin (typically a quarter), which causes a point to appear in the "points" counter. By pushing a "play credit" button, one may then choose any number of points, up to 10, to be transferred to the "bet" column. Pushing the "deal" button causes five card images to appear on the screen. These are electronic pictures of ordinary playing cards, from ace to ten, and the ordinary four suits. Under each card is a "discard" button. By pushing that button, the image of the card disappears. If one wishes not to discard at all, one may push a "stand" button. Otherwise, when one has finished discarding, one pushes the draw button, which causes new cards to appear in place of those discarded. If one feels one has made a mistake in pushing

TABLE I. Evaluation of Hands in Solitaire
Draw Poker

Type of Hand	Number of Skill Points
Straight flush	50
Four of a kind	25
Full house	10
Flush	8
Straight	5
Three of a kind	3
Two pairs	2
Pair of aces	1

discard buttons, a "cancel" button may be pushed, which restores the initial five cards. "Cancel" will work only if pushed before the "draw" button.

Hands are evaluated in "skill" points according to the system shown in Table 1.

The number of skill points awarded corresponds to the most valuable type of hand it qualifies for. Some machines award 250 skill points for a royal flush. The number of skill points, multiplied by the "bet," is added to the "points" counter, allowing the player to play more games.

All the machines I saw had a counter measuring the number of quarters that had been spent in the machine. In addition, some had a "knock-off button," permitting all remaining points to be removed from the "points" counter (i.e., restoring it to zero). These machines also had a meter recording the number of points knocked off.

3. WHAT IS A GAMBLING DEVICE PER SE?

Nearly anything (including the *World Almanac* and a television set) may be used in the course of making bets. However, some devices (like roulette wheels and slot machines) have no other reasonable uses, and these are called gambling devices per se. A gambling device per se may be confiscated by the Commonwealth of Pennsylvania without any evidence of its being used for gambling.

The common law of Pennsylvania provides three elements that must all be present in order to establish that a device is a gambling device per se: consideration, a result determined by chance rather than skill, and a reward. "Consideration" is whether one must pay to play. This element is conceded by all parties to be present in the Electro-Sport Draw Poker Machine. "Reward" is whether there is a valuable prize for winning. In this case, one can win only further free games, which in other cases (*In re Wigton*, 1943) had been held not to constitute reward. While "reward" has many aspects, the court in this case ruled that an Electro-Sport draw poker machine with a knock-off button and meter satisfies the element of reward, and that one without it does not. A machine with a knock-off button and meter that is used for gambling is easy for absentee owners to monitor, since the difference between the number of quarters that were put into the machine and the number of games knocked-off by the bartender and paid for by him (at 25 cents per game) is the net take of the machine. The Pennsylvania Supreme Court ruled in this case that an Electro-Sport draw poker machine without a knock-off button and meter does not satisfy the element of reward, and hence is not a gambling device per se. For those with the button and meter, however, the remaining question is the issue of chance and skill, which I was asked to address.

4. MY TESTIMONY

Using an Electro-Sport machine like the one that had been seized, I wanted to make dramatic the role that skill plays in the game. To do this, I compared two strategies, a "dumb" strategy consisting of automatically standing pat with every hand, and a "smart" strategy in which I did my best to win, knowing what I do about probability. I played 128 games on the machine, and found that the smart strategy won 159 skill points, whereas the dumb strategy won only 34 skill points. Thus I found a factor of about 4½ between my smart and dumb strategies, and proposed this as a measure of how much skill is involved in the game.

In the hands that I played, I never threw away a winning hand. There are, however, some winning hands I would have thrown away. For example, if I had been dealt the aces of hearts and spades and the king, queen, and jack of spades, I would have discarded the ace of hearts. While this does break up a winning hand (the pair of aces is worth one point), it gives a better expected number of points. Of course, the probability is greater than ½ that this move would have resulted in zero points for the hand. Consequently it was not inevitable (but it was highly likely) that the skill ratio I based my testimony on would come out to be greater than one.

A larger factor could have been found, of course, with a "dumber" strategy that actively sought to minimize the expected number of points scored. It was my decision that such a comparison would not have been useful to the court, and I made no such comparisons. Conversely, a smaller factor would have resulted if I had compared the smart strategy to a "smarter" dumb strategy, a point made in cross-examination in this case, and by an opposing expert witness in a similar case in Michigan.

Another decision I made was whether to report plays of the machine at all. Under the assumption that each hand is equally likely, one could in principle compute the ratio of the expected winnings of the strategy that maximizes expected winnings to the expected winnings of the stand-pat strategy. However, to present such evidence would have highlighted the assumption. Verifying the assumption empirically looked at least as difficult a task as to play the machine and report the results. Thus although I played the machine using my "smart" strategy with the assumption of equally likely hands, my statement about skill did not depend on such an assumption. An electrical engineer who had participated in the design of the Electro-Sport machine testified that it was designed with a pseudorandom number generator that started manipulating a Boolean string when the machine is turned on. Thus to get nonrandom results from the machine would have required timing in microseconds, which is beyond human capability. I found no empirical evidence that suggested the machine's random device had been tampered with.

As the number of games played goes to infinity, the factor I used will presumably settle down to a number that reflects the ratio of the expected winnings of my "smart" strategy to that of the stand-pat strategy. However, I have not specified my "smart" strategy for all possible hands (it is even possible that, by whim, I would play the same hand differently on different occasions). Thus theoretical calculation of this ratio is not possible.

Under the assumption of independent and identically distributed trials, I could have calculated a variance for the estimated ratio. This would have quantified the obvious problem that would have occurred with very small numbers of plays. But in this instance it seemed unnecessary, and no one asked me for it.

One might ask why I played 128 games. As I recall, my playing and recording of hands were interrupted by a telephone call at that point. I decided that I had enough data to analyze, and that became the data set used in the trial. One might argue that because I knew the results when I chose to stop, I might have biased the outcome. In fact, the ratio had been essentially constant for some time, so I doubt that any such bias occurred.

I testified that both skill and chance played a role in solitaire draw poker, and that skill and chance are not necessarily opposed elements in a game. I compared draw poker to the game of matching pennies, which in a single play is entirely a matter of luck; to chess, which is entirely a matter of skill; and to backgammon, which has both chance and skill. In my judgment, draw poker is more like backgammon, as it involves both chance and skill.

Apparently the courts were looking for a statement that goes something like "success at playing electronic draw poker is determined 70% by skill and 30% by chance." What might be the meaning of such numbers? Why would not 30% skill and 70% chance be equally plausible? If one could encode full strategies, specifying what cards one would discard in every possible hand, it would be possible to compute (in principle) the probability, in k hands, that a player with one strategy would win more skill points than a player with another strategy (assuming independent draws from equally likely hands). The strategy with the higher expected number of skill points would be nearly certain to win more skill points as the number of plays k gets large, as is guaranteed by the central limit theorem. But real players find it hard to say how they would discard in each of the $\binom{52}{5}$ possible hands. Even if they could, would these calculations get us any closer to skill versus chance? In fact, skill can mean many things. It might mean whether one can identify players who are particularly good at the game. It might mean whether one can, by practice or by thinking about it, find ways to improve. But none of these theories appears to get one any closer to the desired form of statement. The interpretation I chose emphasized that there are decisions under the control of the player that greatly influence how many skill points the player wins. I had no knowledge of, and no way to measure, how "skillful" typical players in bars are. Hence I was not able to quantify the extent to which skill, as opposed to chance, matters in draw poker.

The opposing expert witness was an FBI agent who testified that solitaire draw poker had no element of skill, apparently defining skill in the sense of manual dexterity. My profession and curriculum vitae were contrary to that position—how could it be that one could be a professor and grant Ph.D.'s in a subject that was not a skill? None of the courts to review this case apparently took the FBI agent's testimony seriously.

A similar problem was faced by Solomon and several of his colleagues in relation to pinball in the early 1960s. Solomon (1966, pp. 322, 323) writes:

> About four years ago I appeared as an expert witness in several trials in San Francisco. Some colleagues at Stanford and Berkeley also appeared there in this role in several similar trials. A number of shopkeepers had been arrested on the charge that they had paid a prize in money to the players (San Francisco policemen in mufti) of pinball machines which the State labeled a game of chance, that is, skill did not predominate in achieving a high score in the game. In those trials in which I participated, some experimentation preceded the day in court. A pinball machine was obtained, of the same model as the one to be produced in court (the machine in question being held in storage by the Police Department until the day of the trial). For one thousand games, the plunger was pulled back and released to propel each of the metal balls (usually five) on their way to drop into various holes and thus produce a score for the game. The ratio of the number of wins—score greater than a critical value—out of 1000 games was assumed to be the probability of winning by chance alone, that is, the probability of win which was built into the machine. On this same machine an expert then tried his prowess. The distance the plunger is pulled back and the manner in which it is released, plus the application of hand-pressure to the body of the machine, influence the score. Too much pressure causes the game to terminate abruptly and this is recorded as a defeat. The expert, who happened to be a house-painter with a neighborhood reputation for proficiency in this activity, ran through 1000 games. As before, the ratio of number of wins to number of games was assumed to be the expert's probability of win. The sample size of 1000 trials is large enough to make sampling error negligible (in these situations a standard deviation less than 0.01), and yet not wear out the expert, or the graduate student who pulled the plunger without ever checking to see the flight of the metallic ball. Over several pinball machines which differed in structure, the expert's performance was just about twice as good as could be attained without the use of any human skill. However, the ratios of winning performance without skill went from 0.05 to 0.10, and correspondingly, the winning performance ratios of the expert from 0.10 to 0.20.

It is notable that Solomon's measure of skill is very similar to mine. See also Solomon (1982). The California pinball cases do not seem to provide a basis for how to decide whether a game is predominantly skill or predominantly chance.

5. COURT RESPONSE

The hearing was held before Judge Dauer of the Allegheny County Court of Common Pleas, who ruled orally at the time that "the court finds as a matter of fact that skill is a definite factor in playing this game and winning this game; certainly the evidence presented [showed that] a knowledge of the use of probabilities and statistics was certainly involved" [transcript at pp. 120–121]. His written opinion repeats this idea: "A player with a high degree of skill will produce a better outcome than a player with a low degree of skill....To constitute a gambling device 'per se,' the game must be purely one of chance with no skill involved in reaching the desired result" (Opinion, p. 6).

The case was appealed by the prosecution to the Pennsylvania Superior Court, which wrote:

A device or game does not involve gambling per se merely because an element of chance is involved in its play or because it may be the subject of a bet. Football, baseball and golf, as well as bridge, ping pong, billiards or, for that matter, tiddledywinks, all involve an element of chance, yet the mere playing thereof is not gambling; betting on them is. *See Commonwealth v. Mihalow*, 142 Pa.Super.Ct. 433, 16 A.2d 656 (1940). Admittedly, someone might attempt to bet on the outcome of games played with the Electro-Sport Draw Poker Machine. This possibility, however, does not make the machine a gambling device per se.

In support of its argument that no skill is involved in the play of draw poker, the Commonwealth compares it to a slot machine where the object of the game, to match a variety of objects on a visual display, depends entirely on a random spin. We find the analogy to be inapplicable. Draw poker involves far more than a mere glance at the face of cards, whether they appear on pieces of pasteboard, or on a viewing screen. Skill is exercised in choosing which cards to hold, in deciding which to discard, in considering whether to stand pat and in weighing the probabilities of drawing the desired card.[1] It is disingenuous to compare the Draw Poker Machine to a game in which the player merely inserts a coin and waits for bells, bars, cherries, oranges, etc., to match up. In playing the Draw Poker Machine, there are a variety of "hands" which can be sought and a varying range of possibilities of obtaining them through a draw. A player faces similar odds to any game of draw poker; he has a measure of control over the outcome and he is similarly dependent upon his individual skill. To successfully play the Draw Poker Machine, as in regular poker, the player must be familiar with the relative value of the various combinations which may be held and must also know the odds against improving any given combination by drawing to it. *See Foster's Complete Hoyle, supra* at 169. An expert witness, testifying in Mr. Allen's behalf, Mr. Kadane, Professor of Statistics at Carnegie Mellon University, found, after over a hundred plays, that it was possible to win four times as many games playing a "smart" strategy as opposed to a "dumb" strategy. Professor Kadane therefore concluded that success with Electro-Sport Draw Poker depends "a great deal" on the player's skill.

The assertion by the Commonwealth that the "maximum score that a player can achieve is preordained upon the deal" is correct, but irrelevant; it is equally applicable to any game of cards. Once the cards are shuffled and cut, they are aligned so that the maximum hand is predetermined, unless someone engages in improper dealing, but the extent to which that hand is subsequently achieved is dependent upon the skill of each player as manifested by his subsequent play. In the Draw Poker Machine, in particular, the element of skill comes into operation when the player calculates the odds and determines whether to draw cards, and if so, how many.

---

1. Illustrations of the kind of choices available to Draw Poker Machine players are found in R. F. Foster, *Foster's Complete Hoyle*, at 174 (1927):

There is no rule to prevent his throwing away a pair of aces and keeping three clubs if he is so inclined; but the general practice is for the player to retain whatever pairs or triplets he may have, and to draw to them. Four cards of a straight or a flush may be drawn to in the same way, and some make a practice of drawing to one or two high cards, such as an ace and a king, when they have no other choice. Some hands offer opportunities to vary the draw. For instance: A player has dealt to him a small pair; but finds he has also four cards of a straight. He can discard three cards and draw to the pair; or one card, and draw to the straight; or two cards, keeping his ace in the hope of making two good pairs, aces up.

The argument that a player may occasionally win against the law of probabilities does not mean that the game does not [differ] in substance from a card game of poker is to ignore contemporary reality.[2]

Thus the Pennsylvania Superior Court also found that the skill element is present in the Electro-Sport Draw Poker Machine, and hence that it is not a gambling device per se. Finally, the case was appealed by the prosecution to the Pennsylvania Supreme Court, which wrote:

Whether the result is determined by chance poses a far more difficult question. Superior Court, citing *Nu-Ken, supra*, and *In re Wigton*, 151 Pa. Superior Ct. 337, 30 A.2d 352 (1943), stated the standard to be that:

> In order to conclude that a machine is a gambling device per se, *it is necessary to find that successful play is* entirely a matter of chance as opposed to skill. *Electro-Sport, supra*, at 58, 443 A.2d at 297 (emphasis added). Initially, we note that the cited cases do not stand for the proposition that success be *entirely* a matter of chance. Rather, they hold that the "mere fact that a machine involves a substantial element of chance is insufficient" to find the machine a gambling device per se. *Nu-Ken, supra*, at 433, 288 A.2d at 920, citing *In re Wigton, supra*. Thus a showing of a large element of chance, without more, is not sufficient. Nor must the outcome of a game be wholly determined by skill in order for the machine to fall outside the per se category. As Superior Court pointed out:

> A peculiar combination of luck and skill is the sine qua non of almost all games common to modern life. It is hard to imagine a competition or a contest which does not depend in part on serendipity. It cannot be disputed that football, baseball and golf require substantial skill, training and finesse, yet the result of each game turns in part upon luck or chance.

> *Electro-sport, supra*, at 60, 443 A.2d at 298. We are thus left with the task of determining in each case the relative amounts of skill and chance present in the play of each machine and the extent to which skill or chance determines the outcome.

> The expert witness who testified on behalf of the tavern owner is a professor of statistics at Carnegie-Mellon University. He stated that he could win at a rate four and one half ($4\frac{1}{2}$) times greater using a "smart" strategy (*i.e.*, employing his knowledge of statistics) than by using a "dumb" strategy (*i.e.*, always "standing pat" on the initial hand dealt by the machine). Based on his play of the Electro-Sport machine, he concluded that skill was a "definite factor" in playing the game. RR-130a. On cross-examination, however, he conceded that chance was also a factor both in determining the initial hand dealt and the cards from which one can draw, concluding that there was a "random element" present. *Id.* He could not say how to apportion the amounts of skill and chance. The expert witness for the Commonwealth testified that no skill was involved in playing the game.

> While appellee has demonstrated that some skill is involved in the playing of Electro-Sport, we believe that the element of chance predominates and the outcome is largely determined by chance. While skill, in the form of knowledge of probabilities, can improve a player's chances of winning and can maximize the size of the winnings, chance ultimately determines the outcome because chance determines the cards

---

2. Appellant also contends that the brief duration of the game, approximately fifteen seconds, is further evidence that the machine is a gambling device. *Laris Enterprises, Inc. Appeal*, 201 Pa. Super. Ct. 28, 192 A.2d 240 (1963). While in a proper case, brevity of play may tend to prove one of the elements required for a finding that a machine is a gambling device per se, we conclude that since there is ample evidence that skill is required to play Electro-Sport Poker, the fact that the playing time is brief provides little, if any, support for Appellant's position.

dealt and the cards from which one can draw—in short, a large random element is always present. *See Commonwealth v. 9 Mills Mechanical Slot Machines*, 62 Pa. Commonwealth Ct. 397, 437 A.2d 67 (1981). That the skill involved in Electro-Sport is not the same skill which can indeed determine the outcome in a game of poker between human players can be appreciated when it is realized that holding, folding, bluffing, and raising have no role to play in Electro-Sport poker. Skill can improve the outcome in Electro-Sport; it cannot determine it.

Thus the Pennsylvania Supreme Court found that skill was not as important as chance in determining the outcome in Electro-Sport Draw Poker, and thus that the chance element is present. As explained above, this led them to determine that draw poker machines equipped with knock-down buttons and meters are gambling devices per se and that those without them are not.

Part of the difference between the decisions of Judge Dauer of the Allegheny County Court of Common Pleas and the Pennsylvania Superior Court, on the one hand, and the Pennsylvania Supreme Court on the other, can be explained by their different understandings of the law. Both Judge Dauer and the Pennsylvania Superior Court interpreted "skill versus chance" to mean that if any skill had been shown, this is sufficient to remove the chance element, and hence sufficient to declare the machines not gambling devices per se. They also both found, on the facts that I had shown, that skill was involved, and thus ruled that these are not gambling devices per se.

The Pennsylvania Supreme Court, however, interpreted the requirement as being whether skill or chance predominates. Thus it was not sufficient for the Supreme Court that I had shown that skill was definitely involved. In its last few words quoted above, the court advances a theory based on "determining the outcome." While those words are vague, no doubt they will be scrutinized by counsel and courts as they argue other cases on this point, until another case goes to the Supreme Court and, perhaps, further clarification is obtained.

CASES CITED

In re Wigton (1943), 151 Pa.Super.Ct. 337, 30 A.2d 352.
Commonwealth v. 9 Mills Mechanical Slot Machines (1981), 62 Pa.Commonwealth Ct. 397, 437 A.2d 67.
Commonwealth v. Mihalow (1940), 142 Pa.Super.Ct. 433, 16 A.2d 656.

REFERENCES

Solomon, H. (1966). Jurimetrics. In F. N. David (ed.), *Research Papers in Statistics*, New York: John Wiley & Sons, pp. 319–350.
Solomon, H. (1982). Measurement and burden of evidence. In J. Tiago de Oliveira and B. Epstein (eds.), *Some Recent Advances in Statistics*, London: Academic, pp. 1–22.

2

COMMENT

John Lehoczky

1. INTRODUCTION

In his chapter, Kadane raises a variety of provocative issues associated with the quantification and measurement of skill in a game involving both skill and chance. Kadane considered these issues in the context of the game of draw poker as played on certain electronic machines. In this game, an individual is "dealt" five cards and can decide which, if any, of the cards to discard. Any discarded cards are replaced. The resulting hand is evaluated according to a table of values, and the player earns a certain number of skill points on each play of the game. The table of values is given by Kadane. Throughout this paper, we assume that a royal flush earns 50 points. The cost of playing is assumed to be $1.00, as is the value of a skill point. The legal issue involves the determination of whether electronic draw poker can be considered to be a game of skill.

Kadane attacks the question of skill in the following way. He considers two strategies, one called "dumb" and one called "smart." The dumb strategy consists of never drawing any cards. It was reported that this strategy earned 34 points in 128 plays. Unfortunately, the smart strategy was never defined by Kadane. Indeed, the only indication of its definition is that when dealt AKQJ of one suit and a second A, the smart strategy discards the second A. The performance of the smart strategy was reported as yielding 159 points in the same 128 hands. Kadane evaluates the performance of smart relative to dumb by computing the skill ratio, $\frac{159}{34} = 4.68$. He feels that this ratio is a measure of the "amount" of skill in the game and in this case is sufficiently large to allow one to consider electronic draw poker to be considered a game of skill.

This comment addresses a variety of issues raised by the Kadane paper including:

1. The assumption of randomness in evaluating draw poker strategies,
2. The number of trials needed to accurately measure the skill ratio,
3. The use of a "skill ratio" as a measure of the amount of skill in a game,
4. The introduction and evaluation of strategies for playing draw poker.

2. THE RANDOMNESS ASSUMPTION

Kadane introduces two strategies for draw poker, smart and dumb, and evaluates the performance of each by playing each over 128 trials. He indicates that his statement of skill did not depend on any assumption of the randomness of machine generated hands. It is likely he was implicitly assuming that all hands are equally likely in developing his smart strategy. For example, he indicates that he would discard an extra ace (thus risking earning 0 points instead of earning 1 or more for certain) in order to draw for a royal flush. This strategy is smart with respect to a model of randomness in which all cards are equally likely to be dealt, but need not be at all smart for other models of randomness. Moreover, in using a skill ratio as the sole measure of skill in the game, he is implicitly assuming that hands are independent of each other and identically distributed. If one did not make this assumption, then far more sophisticated experimentation would be required to determine if the smart strategy really was superior to dumb. Under the standard model of randomness, one can calculate the expected return from a variety of different strategies, although neither the optimal strategy nor its return are known. Throughout the remainder of this comment, we assume a total randomness model and evaluate all strategies with respect to it.

3. THE SAMPLE SIZE REQUIRED

In the Kadane chapter, a skill ratio approach was taken as a basis for evaluation of the degree of skill involved in the game. This ratio was estimated using the empirical skill ratio, namely, the number of points earned under the smart strategy divided by the number of points earned under the dumb strategy, assuming a common number of trials. Generally, owing to the volatility of ratio estimators and the skewness in the point scale, this approach requires a large number of trials to provide sufficient accuracy. Furthermore, in Section 4, we argue that when dealing with unfair games, the average payoffs (excluding entry fees) of strategies are not an adequate measure of their efficacy. Let us assume, however, that we want to estimate the skill ratio, namely, the ratio of the payoff (excluding entry fee) using the smart strategy to the payoff using the dumb strategy. Indeed, one can use standard combinatorial techniques to evaluate the average payoff of the dumb strategy (it is 0.246); thus one need only estimate the average payoff using the smart strategy. If one were to play independent trials and let X denote the average payoff of the smart strategy, then $X/0.246$ would estimate the skill ratio. This statistic would be approximately normally distributed about the true mean and would have variance in the order of $64/n$. The variance of the optimal strategy is not known, nor is it estimable from the data in Kadane's chapter. Nevertheless, theoretical calculations for enhanced unsophisticated strategies suggest that this variance should be at least 4. If one uses the observed average payoff for the dumb strategy in the denominator rather than the known theoretical value, then the variance of the empirical skill ratio increases substantially. One can approximate the situation using asymptotic methods. The empirical skill ratio X/Y is approximately asymptotically normal with mean μ_X/μ_Y and variance $\left(\sigma_x^2 + \sigma_y^2/\mu_y^2\right)/n\mu\gamma^2$. For the draw poker game, we can get a rough order of magnitude by letting $\mu_X = 1$, $\mu_Y = 0.25$, $\sigma_X = 2$, and $\sigma_Y = 1$. This gives a variance of $544/n$ or roughly 8 times the variance arising from using the known value in the denominator. Indeed, with $n = 128$, we find that the empirical skill ratio has variance about 4, surely too large for accurate estimation of the true skill ratio. This analysis is very rough. The numerator and denominator used by Kadane are actually positively correlated, a fact that serves to reduce the variance somewhat. Nevertheless, using the known performance of the dumb strategy (which gives a variance of 0.5 for 128 trials), one can conclude that the skill ratio is at least 3. An approximate 95% confidence interval for the payoff of the smart strategy is (0.900, 1.59). It will be seen in Section 6 that there is an unsophisticated strategy that offers a payoff of 0.857. The skill ratio for this strategy versus the dumb strategy is nearly 3.5, so it is clear that Kadane's smart strategy and the optimal strategy must have skill ratios at least this large.

4. THE USE OF THE SKILL RATIO AS A MEASURE OF SKILL

Section 3 indicated that the number of trials used by Kadane was not adequate to accurately estimate a skill ratio. Suppose, however, one were able to carry out an unlimited number of trials or calculate the skill ratio exactly using theory. The skill ratio itself, even if estimated exactly, cannot determine whether the game is one of sufficient skill to be legal in Pennsylvania. There are two problems:

1. The skill ratio obscures the absolute rates of return of the strategies used.
2. A strategy with a large skill ratio relative to another may still be inferior to the other.

The skill ratio measures the relative rates of return of two strategies. If one strategy (smart) has an expected payoff given by $r\theta$, while another (dumb) has payoff θ, then the ratio is r, independent of θ. As θ decreases, the payoff from the smart strategy decreases. If θ is very small, then the ratio r obscures the fact that the difference in rates of return, $\theta(r-1)$, is small and that the game involves little skill.

A second problem is that of the general evaluation of the performance of one strategy versus another. It is important to restrict our analysis to unfair games (supermartingales). This means that we shall consider only the case in which the expected payoff for any strategy is less than the $1 entry fee. We assume throughout that electronic draw poker forms a supermartingale. This means that the expected payoff of the optimal strategy is less than $1. Kadane's data suggest that his smart strategy may have a positive rate of return; however, we believe this is due solely to sampling variability. When dealing with supermartingales arising from this sort of game (i.e., sums of i.i.d. bounded random variables) a player's fortune will become arbitrarily small if he plays the game a sufficiently long time. That is, he will lose an arbitrarily large amount, limited only by his initial capital. Ruin is certain, although the time until ruin occurs is random and will vary with each distinct strategy.

A simple example may help to clarify these points. Suppose instead of draw poker, we play a game with an entry fee of $1 and the possibility of winning $2, $1, or $0 on each play of the game. Suppose using one strategy (smart) the player wins $1 with probability 0.99 or $0 with probability 0.01. The expected payoff is $0.99 for this strategy. Consider now a second strategy (dumb) which results in winning $2 with probability 0.125 and $0 with probability 0.875. This strategy has a rate of return of 0.25. The resulting skill ratio is approximately 4, thus showing a strong superiority in favor of smart. In fact, the first strategy is inferior to the second. Indeed, using the first, the player can never win on any trial and may lose $1. Its only virtue is that one gets to play many trials before ruin occurs. The second strategy at least affords a positive probability of winning and of increasing an initial fortune. For example, if one began with $1 and quit after the first trial, then the average final fortune would be 0.25, whereas using the smart strategy, it would be 0.99. If, however, we played until we have won or lost $1, then the expected final fortunes would be 0.25 and 0, respectively. It follows that unless the duration of the trials is important, the relative rates of return are insufficient by themselves to determine if a game is one of skill. One must make a more detailed analysis of the performance of the strategies in question.

5. CHARACTERIZING A GAME OF SKILL

In Section 4 we argued that the skill ratio is not a sufficient measure of the degree of skill in a game. Rather, one must make a more detailed analysis of the particular strategies. In this section, we argue that one must be very careful in defining a "no-skill" strategy to use as a basis of comparison with a "skillful" or "smart" strategy. When defining a no-skill strategy, one should restrict attention to strategies which satisfy three conditions:

1. The strategy is well-intentioned (i.e., it does not attempt to deliberately do poorly).
2. The strategy assumes a full knowledge of the rules and scoring system of the game.
3. The strategy can be based on a few simple heuristics that a child could apply after a few minutes of instruction.

The third requirement merits some explanation. As an illustration, we consider the game of video tic-tac-toe. In this game, the player deposits $1 for each play. The player is allowed to choose whether or not to go first and then plays a standard game of tic-tac-toe against a machine that is programmed never to lose and to win if possible. If the player wins, he wins $100,000! Of course, this payoff never occurs. If the game is tied, the $1 is returned, whereas it is lost if the player loses. We now must determine whether this is a game of pure chance or a game of sufficient skill to be legal in Pennsylvania. Using the Kadane approach, we would introduce an ignorance or no-skill strategy and compare it with a smart or optimal strategy. In this case, the optimal strategy is the one that leads to a tie game on every trial. What would be a no-skill strategy? It seems that a dumb (but not suicidal) strategy would be a random-number-generated strategy. The player would pick at random whether to go first or second, and would select boxes at random from the remaining boxes. It is fairly easy to compute that the player can never win, has a $\frac{1}{96}$ probability of a tie game when he is second, and a probability of 0.05 of a tie if he goes first. This game offers a rate of return of 0.03 using the no-skill strategy. The resulting skill ratio is more than 30. Nevertheless, there is near universal agreement that tic-tac-toe is not a game of skill. The reason is that one need only apply one or two simple rules in order to never lose the game (e.g., play first in the center, play in a corner, block the opponent). When one can play optimally or nearly optimally using a few simple rules that can be implemented by a child, one generally considers the game to involve no skill. We argue in Section 6 that this is also the case with electronic draw poker.

6. THE EVALUATION OF SKILL IN VIDEO DRAW POKER

In this section, we deal specifically with the question of whether video draw poker can be considered to be a game of skill. We argue first that the dumb strategy used by Kadane is an unacceptable representative of a no-skill strategy and hence is an unacceptable basis of comparison. In particular, this strategy ignores the possibility of improving hands at no risk. For example, if one were dealt a hand valued at one point (pair of aces), the Kadane dumb strategy would draw no cards. A player with no skill but knowing the scoring rules of poker would keep the two aces and discard the three other cards. This offers the possibility of obtaining a hand of value 25, 10, 2, or 1 point. A more sophisticated strategy is, of course, possible. We assert that playing to enhance the value of a hand is not a skillful strategy; rather it is one that assumes only knowledge of the scoring system and no knowledge of probability theory.

There are 2,598,960 distinct poker hands that can be dealt. Among them, there are 2,317,320 that do not score any points. Of the 281,640 that do score points, 18,696 cannot be enhanced without risk, while 262,944 can be enhanced without risk. We thus define the *enhancement strategy* s_e as follows: Draw no cards if dealt a straight flush, four of a kind, a full house, a flush, or a straight; draw two cards with three of a kind; draw one card with two pairs; draw three cards with a pair of aces; and draw five cards with a hand of no points.

The return from the enhancement strategy can be calculated by theory. It has a rate of return of 0.570, compared with 0.246 from the Kadane dumb strategy. It embodies no skill in or knowledge of probability theory, and hence cannot be thought of as a skillful strategy. Furthermore, this strategy dominates the dumb strategy in all cases. No matter what cards are dealt, the enhancement strategy is guaranteed to do at least as well as the dumb strategy and better in many cases. We regard this enhancement strategy as being a better representative of the no-skill strategies.

We next introduce a series of strategies that use some simple rules for discarding. These strategies are analogous to the optimal tic-tac-toe strategy, which was characterized by a few simple playing rules. The strategies we consider consist of a sequence of discarding rules that are to be applied sequentially, using the first applicable rule. If none applies, then all five cards should be discarded. The rules are

R_0: Keep and enhance any hands worth at least one point.
R_1: Keep a pair; draw three cards.
R_2: Keep a four-card flush; draw one card.
R_3: Keep a four-card straight (consecutive or broken); draw one card.
R_4: Keep an ace; draw four cards.

We create a sequence of strategies s_1, s_2, s_3, and s_4. Strategy s_i is defined by applying rules R_j, $j = 0, \ldots, i$, in that order. This sequence was suggested by Dorian Feldman.

In each case, the expected rate of return can be calculated using combinatorial arguments. The results are given in the following, where s_d represents the Kadane dumb strategy:

S_d: 0.246
S_e: 0.570
S_1: 0.768
S_2: 0.809
S_3: 0.836
S_4: 0.857

We have argued earlier that s_e can be considered a no-skill strategy. Furthermore, we have argued that $S_1, \ldots S_4$ are specified by a few simple heuristics, and hence do not involve skill. That each of these strategies has an expected return of nearly 0.8 (or more) suggests that they must be near the optimal. Indeed, I would estimate that the optimal strategy (the one that maximizes expected return) has a rate of return only slightly above 0.9. Furthermore, the gains of the optimal over the simple heuristic strategies are in large part from events of small probability, for example, discarding an ace from AKQJ and a second ace. While such a strategy serves to increase the overall rate of return, it will not increase by much the expected final fortune when the gambler's ruin formulation is used. Consequently, it would seem that there is very little room for skill in electronic draw poker over and above the simple heuristic rules cited earlier. A final resolution of this issue will, however, have to wait until either Kadane clearly describes his smart strategy so that its performance can be determined exactly or the optimal strategy (in the sense of highest rate of return) and its performance are determined. These performances can then be compared with that of the heuristic strategies defined earlier in this section. Only then can a final determination concerning skills be made.

7. CONCLUSION

Kadane has presented an analysis of video draw poker that shows that it is possible to make a profit playing the game using a smart strategy and that a representative no-skill strategy has a rate of return far below the rate of return of the smart strategy. We have shown that the empirical evidence pointing to video draw poker having an optimal expected payoff larger than the entry fee can be easily explained by sampling variability. Although the dumb strategy is very inferior to the optimal strategy, calculations presented in Section 6 indicate that simple heuristic strategies can perform very well and possibly at a near-optimal level. Indeed, it appears that the difference between the best heuristic strategy listed in Section 6 and the optimum arises mostly from the inclusion of low-probability, high-payoff results. If one focuses on the final fortune of a player, the differences are likely to be further narrowed. This brings us to the conclusion that video draw poker cannot be considered to be a game of skill, even though there are unskillful ways to play it.

3

REJOINDER

Joseph B. Kadane

Lehoczky's comment makes important contributions to the scientific study of electronic draw poker, but neglects, at times, the legal context of my paper. I thank him for the former, and will defend my work from the latter.

1. The randomness assumption: Had I presented calculations that assumed randomness, I would have been vulnerable on cross-examination to questions about how I knew that every hand was equally likely. Since I did not have explicit knowledge of that, it seemed to me legally wise to base my testimony on experimentation rather than calculations. I still think this was wise, Lehoczky's comment notwithstanding.

2. Sample size and accurate approximation of the skill ratio: Lehoczky's rough calculations are quite interesting, and will advance our understanding of electronic draw poker under the randomness assumption. His conclusion, however, that "the number of trials used by Kadane was not adequate to accurately estimate a skill ratio" is unwarranted in the legal context. All three courts found that I had shown that skill is a factor in playing electronic draw poker. None complained that they were worried about sampling error in my "skill ratio." (This is not too strong an argument, since it is not clear how much they understood about sampling error.) Nonetheless, for this specific legal purpose, I believe that my sample size was adequate.

3. The use of the skill ratio: Lehoczky's first point, about the skill ratio being a ratio, is of course correct, but the use of the ratio was not likely to mislead the court. I did report the number of games played and won under each strategy,

From: *Statistics and the Law.* Edited by Morris DeGroot, Stephen Fienberg, and Joseph B. Kadane. Copyright © 1986 by John Wiley & Sons, Inc. Reprinted with permission of John Wiley & Sons, Inc.

but the skill ratio did (and does) seem to be a convenient dimensionless quantity.

His second point has to do with the appropriateness of the expected return as a utility function. Of course other utility functions are possible and interesting (e.g., see Dubins and Savage, 1965), but as a first mode of analysis, the expected return is quite reasonable. In any event, the ratio of expected utilities is one way of describing the relative worth of two strategies, although it is not invariant to adding a constant to the utility function. I do consider it likely, but not certain, that electronic draw poker is subfair (i.e., that the expected payoff of the optimal strategy is less than one). If it is, then the expected return determines the duration of the game. Since the defendant's position is that the game is "for amusement only," one gets more amusement the longer the game, and, hence, the higher the expected return. Thus larger expected returns are arguably in the interest of the player.

4. Skill: This is indeed the most difficult of the questions raised by my paper. I do not know the expected worth of the optimal strategy, nor do I know a simple set of heuristics whose expected worth approaches that of the optimal strategy. But suppose after further work and computer study such a set is found. Suppose, in addition, that it can be taught to a child. Could one say, having applied all our tools of combinational analysis, supermartingales and fast computation, that the result is a skill-less strategy? Perhaps *ex post* one could, but before the work is done, I must say I still think that electronic draw poker requires skill to play well. As I remarked, the *legal* question of how to define skill is still very obscure. But perhaps with further work, solitaire draw poker will become a well-understood game, and hence skill-less, at least for those who read and write technical papers.

ADDITIONAL REFERENCE

Dubins, L., and Savage, L. J. (1965). *How to Gamble If You Must: Inequalities for Stochastic Processes*, New York: McGraw-Hill.

4

CHANCE AND SKILL IN GAMES:
ELECTRONIC DRAW POKER

Marc J. Ware and Joseph B. Kadane

U.S. laws distinguishing gambling (which may be illegal, taxed, or regulated, depending on state law) from gaming (which can be any form of amusement) depend on three elements: consideration, chance, and reward.[1] While consideration (Do you have to pay to play?) and reward (Do you get something valuable if you win?) are relatively clear-cut, the issue of chance, as opposed to skill, is "blurry."[2] Similarly, the laws of the Netherlands and Austria distinguish between games of chance (under a state monopoly) and games of skill (available in private casinos).[3]

Recently, Peter Borm and Ben van der Genugten proposed a general scheme for measuring the extent to which chance predominates over skill in games.[4] This article applies the Borm and van der Genugten idea to electronic draw poker, which is regarded as a "gray area" machine, and has been the subject of extensive litigation.[5]

Part I discusses skill and chance. Part II analyzes electronic draw poker. Part III presents the results of the Borm-van der Genugten calculations as applied to electronic draw poker.

First published in *Jurimetrics Journal*, 2002, by the American Bar Association.

1. Michael William Eisenrauch, Note, *Video Poker and the Lottery Clause: Where Common Law and Common Sense Collide*, 49 S.C. L. Rev. 549, 555 (1998); Ronald J. Rychlak, *Video Gambling Devices*, 37 UCLA L. Rev. 555, 556 (1990).

2. Rychlak, *supra* note 1, at 558.

3. Peter Borm & Ben van der Genugten, *On the Exploitation of Casino Games: How to Distinguish Between Games of Chance and Games of Skill?*, in GAME PRACTICE: CONTRIBUTIONS FROM APPLIED GAME THEORY 19 (Fioravante Patrone et al. eds., 2000).

4. *Id.*

5. *See* Eisenrauch, *supra* note 1, at 560, 573; Rychlak, *supra* note 1, at 567.

1. CHANCE AND SKILL

To what extent is the outcome of a game determined by chance, and to what extent by skill? This seemingly simple question is difficult to answer. Although roulette is determined mainly by chance, chess mainly by skill, and backgammon somewhere in between, the question remains difficult.

Patrick Larkey, Joseph B. Kadane, Robert Austin, and Shmuel Zamir created a simplified form of draw poker and showed that intransitivities among strategies could result. A player using strategy A could regularly defeat a player using strategy B; a player using strategy B could regularly defeat a player using strategy C; yet a player using strategy C could regularly defeat a player using strategy A.[6] In such a game, it is not useful to conceptualize the skill of a player according to the strength of the strategy employed, since the "strength" of a strategy depends on the strategy used against it.

Borm and van der Genugten limit their discussion to games where (1) outcomes can be measured monetarily, thus satisfying consideration and reward, (2) skill can be measured by a long-term average result, and (3) players differ in their long-term results. They also distinguish three kinds of players: a beginner who has just learned the rules, a real average player who represents the majority, and a virtual average player who has been provided with the outcome of whatever random device is involved. Suppose these players have expected results R_1, R_2, and R_3, respectively, with $R_1 \leq R_2 \leq R_3$. Then Borm and van der Genugten define the extent of skill E in a game as:

$$E = \frac{R_2 - R_1}{R_3 - R_1},$$

where $0 \leq E \leq 1$.

Recognizing that the real average player is hard to operationalize, they propose replacing R_2 and R_3 with the expected result R^*_2 of an advanced, almost optimal player, and the expected result R^*_3, of an advanced, almost optimal player who knows the random outcome before deciding on a move. This leads Borm and van der Genugten to their second measure of skill:

$$E^* = \frac{R^*_2 - R_1}{R^*_3 - R_1},$$

which they propose as an approximation to E. Again $0 \leq E^* \leq 1$.

2. ELECTRONIC DRAW POKER

Electronic Draw Poker is played against a machine that displays five cards as if from a deck of ordinary playing cards. Each card may be retained or discarded. Those

---

6. Patrick Larkey et al., *Skill in Games*, 43 MANAGEMENT SCI. 596, 601 (1997).

TABLE I. Skill Points for Various Hands in
Electronic Draw Poker[8]

Type of Hand	Number of Skill Points
Straight Flush	50
Four of a Kind	25
Full House	10
Flush	8
Straight	5
Three of a Kind	3
Two Pairs	2
Pair of Aces	1

discarded are replaced with other cards by the machine. We assume that these cards are chosen by the machine at random, without regard for the consequence to the player of the cards displayed. Table 1 gives the number of skill points awarded as a result of the final hand.[7]

The machine is operated by putting money into it, which causes a point to appear in the "points" counter. Pushing a "play credit" button, one may choose up to ten points to be transferred to the "bet" column. The number of skill points won, according to Table 1, multiplied by the bet, is added to the points counter after the hand is concluded.

In *Commonwealth v. One Electro-Sport Draw Poker Machine*,[9] Kadane was hired to investigate the extent to which skill is required to play these machines. He played 128 times, winning 34 skill points using a simple strategy of never discarding, and winning 159 skill points discarding as best he could. He therefore reported that he performed 159/34 = 4.86 times better with his complex strategy than with his simple one, using this ratio as a measure of skill. The lower courts applied the English rule that any degree of skill suffices and deemed electronic draw poker a game of skill.[10] However, the Pennsylvania Supreme Court found that chance predominated over skill,[11] using the "American rule."[12]

In an academic commentary, John Lehoczky criticized Kadane's analysis by examining five strategies of increasing complexity:[13]

T_0: Keep and enhance any hands worth at least one point.

T_1: Keep a pair; draw three cards.

T_2: Keep a four-card flush; draw one card.

7. Additionally, some machines award 250 skill points for a royal flush.

8. Table 1 is taken from Joseph B. Kadane, *Does Electronic Draw Poker Require Skill to Play?*, in STATISTICS AND THE LAW 333–43 (Morris DeGroot et al. eds., 1986).

9. 465 A.2d 973, 978 (Pa. 1983).

10. 443 A.2d 295 (Pa. Super. Ct. 1981).

11. 465 A.2d at 978.

12. "[T]he American rule...holds that the element of chance is satisfied when chance is the 'dominant' factor in determining the results of the game." Eisenrauch, *supra* note 1, at 563.

13. John Lehoczky, *Comment, in* STATISTICS AND THE LAW, *supra* note 8, at 344–51; *see also* Joseph B. Kadane, *Rejoinder, in id.* at 351–52.

T_3: Keep a four-card straight (consecutive or broken); draw one card.

T_4: Keep an ace; draw four cards.

Assuming each ordering of the cards to be equally likely and using combinatorial methods, Lehoczky found the expected payoffs, S_i, for these five strategies and the expected payoff, S_d, for the Kadane simple strategy to be: $S_d = 0.246$, $S_0 = 0.570$, $S_1 = 0.768$, $S_2 = 0.809$, $S_3 = 0.836$, and $S_4 = 0.857$. Lehoczky argued that because these are simple rules that can be taught to a child, electronic draw poker does not require skill to play.[14] Furthermore, Lehoczky conjectured that the expected payoff of the optimal strategy is slightly above 0.9, so that his strategies achieve nearly the highest attainable payoff.

3. NEW RESULTS

Calculations using commercially available software[15] show that Lehoczky's guess is remarkably accurate. The expected worth of the optimal strategy is 0.9067 (standard deviation 2.18) without an additional payoff for a royal flush and 0.9104 (standard deviation 2.41) with 250 skill points for a royal flush.[16]

To apply the Borm and van der Genugten results, it is necessary to specify what is meant operationally by each of their terms. The Kadane simple strategy and each of Lehoczky's increasingly complex strategies $R_0, \ldots, R_4$ are different ideas about what a beginner's strategy would be. The calculations reported below give the results of all six possibilities for R_1. The expected result, R^*_2, of the "nearly" optimal strategy is taken to be 0.9067.[17]

There are two ways to analyze what it means to inform a player about the random outcome. First, one can think of the random outcome as an ordering of cards from a shuffled deck. The first five are shown face up. If a player decides to discard k cards, the final hand consists of the $5 - k$ cards remaining together with the first k undealt cards. Thus, for example, the tenth card would enter the final hand only if all five of the original cards were rejected. The second way of interpreting

---

14. The calculation of the expected worth of the Kadane simple strategy corresponds well with the result of his calculation: $34/128 = 0.266$.

15. ZamZow Software Solutions, Bob Dancer Presents WinPoker version 6.0, *at* http://www.zamzone.com (last visited Nov. 8, 2002).

16. It should not come as a surprise that the expected worth of the optimal strategy for electronic draw poker is less than one, in view of the Gambler's Ruin theorems. Gambler's Ruin is a set-up in which gambler A starts with i dollars and gambler B with $k - i$ dollars. They play a game in which A wins a dollar with probability p and loses a dollar with probability $q = 1 - p$. They play until one player has all k dollars, and the other is broke. Morris DeGroot shows that the probability that A will ruin B is:

$$\frac{x^i - 1}{x^k - 1}, \text{where } x = \frac{q}{p}$$

If $p > \frac{1}{2}$, which corresponds to positive expected winnings, with moderately high i and k, A is almost sure to ruin B. In the application, no gambling business can allow itself to be in a position where a large number of plays is almost certain to cause bankruptcy. Morris H. DeGroot & Mark J. Schervish, Probability and Statistics 89–92 (3d ed. 2002).

17. *See supra* note 15 and accompanying text.

TABLE 2. Skill Proportion Under Different
Interpretations of Beginner and Randomness

Beginner Strategy	Randomness	Interpretation
	1	2
S_d	0.336	0.327
S_0	0.205	0.198
S_1	0.096	0.093
S_2	0.070	0.067
S_3	0.051	0.049
S_4	0.035	0.034

full information is to think of the ten cards as five pairs, each with one card showing and one not. If the shown card is discarded, the unshown member of the pair
replaces it. By this interpretation, the tenth card in the deck would enter the final
hand whenever the fifth card is discarded, independent of any other discards taking place. Although the player in the real game who does not know the values of
the hidden cards would not distinguish between these two ways of implementing
randomness, the hypothetical player to whom the next five cards are revealed would.
Indeed, simulation shows that optimal play under the first interpretation yields an
expected payoff of 2.212, while the second yields 2.267. Both interpretations are
used below.

Thus we have six interpretations of R_1 and two of R^*_3, yielding the twelve values
for E^* reported in Table 2.

The interpretation of a beginner is more influential on the skill proportion than
the interpretation of randomness. That each of the numbers reported in Table 2
is less than one-half supports the findings of several courts that chance predominates over skill in electronic draw poker.[18] Under the English rule that requires
that skill play no part in the player's resultant success or failure,[19] however, the
fact that each of the numbers in Table 2 is positive also supports a finding that
skill is required.

The availability of the standard deviation of the payoff for the optimal strategy
allows a second line of inquiry. Kadane won 159 skill points in 128 trials of a game
with an expected payoff of 0.9067. Is an outcome this favorable likely if the machine
were dealing cards fairly, or is it possible that he was hoodwinked by being supplied
a machine that awards skill points at an uncommonly high rate?

According to several writers, electronic draw poker machines are often equipped
with a switch allowing the operator to set the "retention ratio" or "house percentage."[20]
To determine whether the machine Kadane studied was set to award skill points at

18. For cases, see Eisenrauch, *supra* note 1, at 563–64. Ben van der Genugten, in his report "Een keuringsmethode
voor behendigheidsatomaten," April, 1997 (in Dutch), EIT, Tilburg University, at pages 22 and 23, suggests that games
with $E^*<0.15$ should be characterized as having chance predominate over skill, and conversely. Using such a rule
would make electronic draw poker a game of skill if the simple strategy T_d or T_0 were taken as the interpretation of
the strategy of a beginner, but not if T_1, T_2, T_3, or T_4 were.

19. Eisenrauch, *supra* note 1, at 563.

20. *Id.* at 565–66; Rychlak, *supra* note 1, at 570.

an uncommonly high rate, suppose Kadane were playing using the optimal strategy. His expected number of skill points would be (128)(0.9067) = 116, with a standard deviation of $(2.18)(\sqrt{128}) = 24.66$. Therefore, his payoff of 159 skill points is

$$\frac{159-116}{24.66} = 1.74$$

standard deviations above the mean. This would rate a verdict of "suspicious but unproved" on the hypothesis that the machine played by Kadane had higher payoffs than a random deal would provide.[21] However, machines that allow the operator to set the expected house percentage in a way that the customer cannot observe would seem to raise an issue of theft by deception or an unfair trade practice.

The calculations reported above certainly strengthen Lehoczky's argument. His prediction that the optimal strategy would have an expected worth slightly over 0.90 is very accurate. The theory of Borm and van der Genugten seems promising as a way to describe the relative importance of skill and chance in games such as electronic draw poker.

21. The sum of many independent plays of the machine, whatever the setting of the switch, is well approximated by a normal distribution using the Central Limit Theorem. The probability that a normal random variable is as high as 1.74 standard deviations or higher is 0.9591. Hence, the probability that Kadane would earn 159 or more skill points if the machine were set to return what a random shuffle would is 0.0409, which is significant at the 0.05 level but not the 0.01 level. *See generally* DeGroot & Schervish, *supra* note 16.

B. CHEATING ON EXAMINATIONS

There are many forums outside of courts in which cases are heard. One of those is the procedures used in colleges and universities to hear cases of misconduct. These vary from institution to institution.

In the case reported here, I was given data on a series of multiple choice examinations, in some of which copying was alleged. The results of the alleged copier, S (for "student"), and the alleged copyees, C, were identified. I did not have seating charts for the examinations. I thought about, but abandoned, the idea of a formal model of the data. Instead, I used five indicators of whether S had copied from various copiees on a large scale. With respect to each indicator, the answer was "no."

One of the frustrations of this analysis is that it does not, and cannot, answer the question of whether small-scale cheating occurred. I can examine the data for over-all patterns, which is what I did. However, whether S copied a single answer from one of the copyees, C, is not a question I can address from the data.

1

AN ALLEGATION OF EXAMINATION COPYING

Joseph B. Kadane

The maintenance of a fair environment for students who are competing in the classroom is an essential function of a modern university. When a breach of academic integrity is alleged, the consequences are very important, to both the academic community and to the persons involved. For the accused, expulsion from the university and inability to enter a profession that requires character checks—law, medicine, the military—can result. For the academic community, a perceived inability to provide an atmosphere free of cheating is debilitating.

In the case reported here, I was asked to assess whether the data support an accusation by classmates of copying multiple-choice examination answers from neighbors on several occasions in a professional program.

DATA AND FORMAL ANALYSIS

Let us call the accused student S. I was retained by S's family and was given the answers by each student on each of 11 examinations taken that semester. S's answers were identified, as were the answers of each student S was alleged to have copied from on each exam on which copying was claimed. I also had the answer sheets giving the correct answers, sometimes more than one for a single question. In addition, I was given erasure data indicating which of S's answers had been erased (sometimes more than one on a single question) and what the final answer had been. Finally, two of the examinations occurred after the accusations had been made, and were taken under special and extra-secure proctoring arrangements, at S's request. I did not have seating charts from the exams, nor did I have, other than for S, cross-examination identification of students.

Methodologically, my first impulse given such a question is to think about a likelihood for the problem. There were about 5 answers per problem and perhaps 60 questions per exam. Hence, there are about 5^{60} (more than 8×10^{41}) possible exam answers by a particular student to a particular exam. Something over 90 students took each exam, so for each exam the number of possible answers by all students is 90 raised to the 5^{60}th power...an enormous number. Finally, there are 11 examinations, so the number of possible datasets is bigger still. Moreover, this calculation does not include any information about erasures or alleged sources of copying. The difficulty of modeling in this space in a way that would be convincing seemed daunting, especially since it would be necessary to develop a model in this big space conditional on copying having taken place and conditional on copying not having taken place.

Even though a formal model seems hopeless, I find it useful to think in Bayesian terms about what I would expect to find in the data if copying took place as alleged and if it did not. Only by finding aspects of the data that are differently probable under these two hypotheses can the data be informative. This less formal analysis is set out in the next section.

PRIORS: WHAT TO LOOK FOR

The use of Bayesian ideas in an informal manner requires that I state the kinds of indications of copying I looked for and the alternative explanations I could imagine. In each case, I intend to make use of the fact that copying is alleged in only some of the examinations.

I first comment on what one might expect to see on S's examinations compared to those of others. If S were copying, one would expect that S's scores on the examinations would be low and in particular would be lower than those of classmates S is alleged to have copied from. Furthermore, one would expect S's performance to be better on those examinations in which copying is alleged as compared to the others. Finally, one would expect that if S had copied as alleged, S would have had lower performance on the last two examinations. On the other hand, because those two examinations were taken after S had been informed of the allegations, it is also possible that S's performance would suffer in the last two examinations if the allegations were false simply because of the psychological pressure of the situation.

All of these statements have to do with how S performs compared to other students in the class. Thus my concern is not with S's score but rather with S's rank in class on an examination. I define this to be 1 plus the number of students who did better than S plus one-half of those who did as well on a given examination. This is the measure of performance I used to compare S to other students. A rank of 1 would indicate that S outperformed every other student on the examination. If 95 students took the examination, a rank of 95 would indicate that every other student did better.

If there were copying, I would expect to see a greater proportion of wrong answers changed to right answers in the examinations in which copying is alleged than in the other examinations. Conversely, if copying did not take place, I would expect the proportion of correction of wrong answers to be about the same in the two sets of examinations.

Finally, I can compute an index of how similar two papers are. For the moment, suppose I have a satisfactory such index. Without any copying, there are many reasons why papers would be similar. First, students may independently know the correct answers. (As teachers, we hope for this.) The students were exposed to the same texts and lecturers. Students study together, or otherwise discuss the course materials. Hence, they would be expected to think similarly about the material. Moreover, some wrong answers are "closer" to the correct ones—that is, are better distractors—than others. Thus, there are several innocent reasons why student papers would be expected to show similarities without copying. One would expect that, if there were copying as alleged, the pairs of examinations of S and the student(s) S is alleged to have copied from would be more similar than those of other pairs of students.

AN INDEX

It remains to specify a measure of similarity between two examinations. To do this, it is important to think about the possible explanations for various patterns of answers that might occur. In conducting this analysis, one feature of the data—the existence of more than one correct answer to certain problems—is important to keep in mind. I distinguish four different configurations:

1. Different answers. I believe this to be evidence against copying, whether both answers are correct, one is correct and another incorrect, or both are incorrect. Moreover, if they are different because one is blank and the other not, this is equally evidence against copying. Score each such case as 0.
2. A shared blank. This I would take to be ambiguous. It is hard for me to imagine why a student would want to copy a blank from another. These are omitted from scoring.
3. A shared correct answer. Shared correct answers might be evidence of cheating but could equally well be evidence that two students independently know the answer. (As teachers, we hope for this sort of thing.) I treat this as an ambiguous circumstance. These are also omitted from scoring.
4. A shared incorrect answer. This I take to be evidence for copying, stronger if the shared incorrect answer is unusual, and weaker conversely. Suppose that the relative frequency of the shared incorrect answer is p_i among other students. I would score each such question by $1 - p_i$.

The index I propose is then the average score among scored questions.

The lower the score, the weaker the evidence of copying, and conversely.

There are several reasons why two students' indexes might be high. One, of course, is that one may have copied from the other. There are other reasons, however. The students are in the same class, studied from the same materials, and heard the same lectures. Some of them may have studied together. All of these reasons can cause them to have high similarities without having copied. The same considerations, however, apply to every pair of students in the class. I would suspect a pair of students of having been involved in copying not on the basis of a high index *per se* but a high index compared to other students in their class. In the case in question, copying was alleged from one to five other students, depending on the examination. If those one to five pairs do not exhibit very high indexes compared to the other pairs of students in the class, I would take this as evidence against copying. Conversely, if one or more of these pairs is extraordinarily high given the distribution of indexes in the class, I would take this as evidence in favor of copying.

To complete the definition, I must say what index to give a pair of students with no scored questions—that is, all their questions are of type 2 and 3. In this case, I would give them the average score for pairs from the class.

Finally, I use the rank of the similarity index for pairs of students to indicate how high the index is compared to the index of other members of the class. Thus, if a particular pair is ranked 1, that means that their examinations' similarity is higher than the similarity of the examinations of every other pair of students in the class. If 95 students take an examination, there are $95 \times 94 \div 2 = 4,465$ pairs of students. Thus, a pair of students whose similarity is ranked 4,465 would have examinations less similar than any other pair of students in the class.

RESULTS

The results concerning scores and similarities are given in Table 1 and Fig. 1. To take the columns in Table 1 from left to right, examinations 8 and 11 are those that took place after allegations of copying were made. S's rank was 7 and 11 in those examinations, out of 90+ students. So there is no evidence that S did poorly in a more highly supervised situation. This is contrary to what one would have expected had copying occurred.

TABLE I. Summary of Examination Results: Scores and Similarities

	Rank by score on the exam of accused copier, S	Rank by score on the exam of alleged copiee	Rank of similarity between alleged copiee and S's answers	Number of students taking exam	Number of questions
1*	6			96	56
2*	13.5			95	39
3	21	48.5	1,090	95	51
		92	925.5		
4	18	38.5	90.5	95	88
		31.5	191		
		71.5	1,741.5		
		38.5	767.5		
		94	2,664.5		
5	20	7	3,596.5	96	63
6*	3			94	79
7	9.5	40.5	3,256	94	65
		29	3,256		
8*	7			93	69
9	5	32	337	95	92
10	2	27	1,605	95	94
11*	11			95	86

* There were no allegations of copying on these exams.

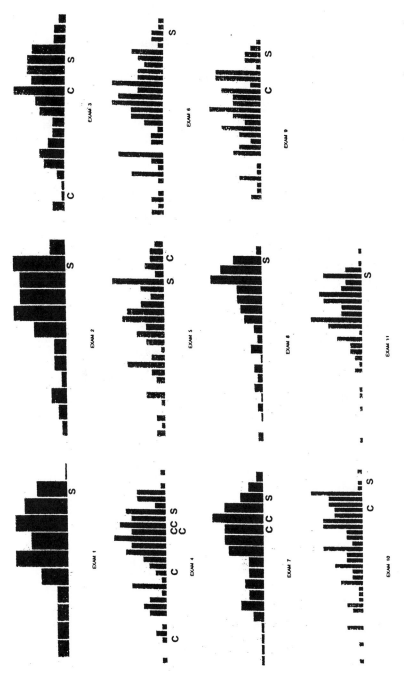

FIGURE 1. Histograms of examination results for each of the 11 examinations. "S" represents S's score, "C" represents the score of the alleged copiees.

In each of the other examinations, as Fig. 1 shows, S scores well into the top quarter of the class and sometimes extremely well. S does not do particularly better on examinations in which examination copying is alleged, again contrary to the hypothesis of examination copying.

Comparing the ranks of alleged copiees to those of S, in each case but one S does better on the examination in question than does the alleged copiee. This again is contrary to what one would expect were there examination copying. Generally one thinks of a weaker student copying from a stronger one. Indeed it requires far-fetched reasoning to imagine why a stronger student would copy from a weaker one.

To appreciate the similarity column, recall that higher similarities reflect greater agreement between the examinations. The similarities are ranked in the same way as were the examinations, so lower ranks indicate higher agreement. As an illustration, the number 1,090 indicates that the pair of S and the first alleged copiee in examination 3 had a similarity with 1,043 pairs higher and 92 pairs the same, for a rank of $1,043 + 1 + 92/2 = 1,090$. These numbers are very high, indicating that the similarity between S's examination answers and those of the alleged copiees are not extraordinarily high. Again this is contrary to what one would have expected had examination copying occurred.

The results on erasures are given in Table 2. In the examinations in which there is not an allegation of examination copying, out of 16 questions with changed answers,

TABLE 2. Erasures of Answers by S

	Examinations in which copying is not alleged				
	Numbers of questions with answers changed				
Exam	wrong → right	wrong → wrong	right → wrong	right → right	total erasures
1	0	0	0	0	0
2	2	1	0	0	3
6	7	0	2	0	9
8	2	0	0	0	2
11	2	0	0	0	2
Total	13	1	2	0	16

	Examinations in which copying is alleged				
	Numbers of questions with answers changed				
Exam	wrong → right	wrong → wrong	right → wrong	right → right	total erasures
3	2	0	0	0	2
4	6	3	0	1	10
5	4	0	0	0	4
7	1	1	0	0	2
9	9 (10 answers)	0	2	0	11 (12 answers)
10	8 (9 answers)	1	1	0	10 (11 answers)
Total	30 (32 answers)	5	3	1	39 (41 answers)

13/16 = 81.25% were changing wrong answers to correct ones. In the examinations in which copying is alleged, there were two questions on each of which S erased two answers. In both cases, the final answer given was correct. It is ambiguous how to count the change in examination 4 from a right answer to another right answer. Depending on how one chooses to treat these two circumstances, the proportion of wrong answers changed to correct ones is 30/39 = 76.92%, 30/38 = 78.95%, 32/40 = 80.0%, or 32/41 = 78.05%. A case might be made for any one of them. For the purposes of this analysis, however, they are numerically similar. In all four cases, the proportion of wrong answers corrected is slightly less among the examinations in which copying is alleged than among examinations in which copying is not alleged. Thus, the erasure evidence fails to support the hypothesis of examination copying and, in fact, supports the contrary hypothesis that no examination copying took place.

CONCLUSION

There are five ways in which the data are examined to see if they are consistent with the hypothesis of examination copying. These are as follows:

- S did well in the two highly proctored examinations, contrary to the hypothesis.
- S is a very strong student and did as well in tests in which copying is not alleged as in those in which it is, again contrary to the hypothesis.
- S did better than each of the alleged copiees except one, again contrary to the hypothesis of copying.
- The similarity between S's examination and the alleged copiees indicates lack of extraordinary similarity compared to other pairs of students in the class. Again this is contrary to the hypothesis of examination copying.
- The proportion of erasures that correct wrong answers is slightly less among the examinations in which copying is alleged than among the examinations in which copying is not alleged. This, too, is contrary to the hypothesis of examination copying.

On the basis of these tests I conclude that the examination data do not support the hypothesis of examination copying. Rather the reverse, they support the conclusion that S was wrongly accused.

What would I have done if some of my indicators had suggested that copying had taken place and others had indicated that it had not? I would have had to think much harder about the degree to which each part of the evidence pointed in each direction, and my conclusions would have had to be much more equivocal.

Finally, a reader might be interested in knowing what actually happened. There was an initial hearing before an honor court, in which most of the evidence of cheating was about body language during the examinations in question. My report was presented to this court, and I was available by telephone for questions had there been any. S was found guilty, and the matter was appealed within the university's structure. At the appeals level, the attorney presenting the case for the prosecution was asked by a member of the appeals court what her view was of the statistical evidence, and she said she thought it was inconsequential. "Oh, I disagree" said the appeals court member. *S* was exonerated. Other than that remark, I have no idea what role my report may have played in that decision.

OTHER PURPOSES, OTHER INDEXES

There are kinds of cheating on multiple-choice examinations other than examination copying—for example, impersonation and theft of the answer key. Each kind of possible cheating and each circumstance requires its own investigation and analysis.

The statistical perspective that underlies many of the articles in this literature is that of significance testing. For example, Angoff (1974) reported that the Education and Testing Service uses one of several indexes and investigates further only if at least one of them departs from the mean by 3.72 standard deviations or more. (Apparently no account is taken of the fact that several tests are being used simultaneously.) If an examinee's test fails this screen, a retest is invited and no further consequence ensues.

Frary, Tideman, and Watts (1977) proposed an index of examination copying that depends on both the number of common wrong answers and the number of common correct answers. Like Angoff, they looked for a significance test computed on the basis of no copying and used high departures from the expected as evidence of copying. For reasons given previously, I am not comfortable with the idea of using common correct answers as evidence of cheating. (Thus, two students who write perfect examinations might be prime suspects to Frary et al. but not to me.)

A general review of the performance of the indexes was conducted in a technical report from the national testing organization ACT, and it found that the theoretical significance levels of the indexes of various authors did not coincide well with the false-positive rates of their benchmark data. The reason for this appears to be that innocent behavior can lead to high indexes under certain circumstances. The adherents of the significance-level approach are thus left in the position of knowing that their analysis exaggerates the weight of evidence on the guilt of the suspected copier without being able to say by how much it does so. By contrast, this article does not use significance testing but instead (implicitly) Bayes's theorem as its fundamental analytic tool. This permits, and in fact requires, consideration of innocent explanations of the phenomena studied. I have not, however, refined the approach quantitatively to the point that I can give a plausible likelihood ratio for examination copying. Much remains to be done, both in the analysis of specific cases and in generalizing from them.

[I thank Dan Cork, Jonathan Forster, and Robert Frary for helpful discussions in the preparation of this article.]

REFERENCES AND FURTHER READING

Angoff, W. H. (1974), "The Development of Statistical Indices for Detecting Cheaters," *Journal of the American Statistical Association*, 69, 44–49.

Frary, R. B., Tideman, T. N., and Watts, T. M. (1977), "Indices of Cheating on Multiple-Choice Tests," *Journal of Educational Statistics*, 2, 235–256.

Hansen, B. A., Harris, D. J., and Brennan, R. L. (1987), "A Comparison of Several Statistical Methods for Examining Allegations of Copying" (Research Rep. Series No. 87–15), Iowa City: American College Testing Program.

C. TAX AUDITS

A government has the right to "audit," i.e., examine the books and financial records of its taxpayers to determine whether enough tax has been paid. But for an auditor to examine each and every record among thousands or millions available is neither efficient nor possible. Hence sampling of some kind is an obviously desirable strategy. The result, when applied to a tax audit, is a probability distribution for the amount owed. However, it is not clear how much the payment should be in such a circumstance.

The chapter in this section addresses the issue of how much to write a check for, proposing a simple family of loss functions for the state. These loss functions are parameterized by a single parameter, the ratio of loss to the state of undercollecting a dollar divided by the loss to the state of overcollecting a dollar. Practices when the paper was written concentrated on ratios of 1, which my coauthors and I think is too low, and 19, which we think is too high. We propose a range of 2 to 4 as more appropriate. I have not seen a discernible change in practice since the paper was published in 1988.

1

STATISTICAL SAMPLING IN TAX AUDITS

Joseph C. Bright, Jr., Joseph B. Kadane, and Daniel S. Nagin

Consider the following situation: A large department store is notified by a state revenue agency that it has been selected for a sales and use tax audit. The auditors conclude that it would be prohibitively time consuming to audit all the store's transactions. Instead they review only a sample of the transactions and estimate the total deficiency based on the sample. Is this a sufficient basis for determining the taxpayer's unpaid liability or must the state review all the transactions to determine the exact amount of tax owed? While the courts with some important qualifications have come down on the side of a complete review, this article suggests a different answer. We argue that audit assessments based on appropriately drawn and analyzed statistical samples do not suffer from the defects that the courts have concluded mar assessments based on nonstatistical samples. We do argue, however, that because of the inherent imprecision of an assessment based on less than a complete review of all records, the calculation of the assessment should include a factor to take into account the risk that the taxpayer has been overassessed. The article suggests an assessment rule that does just this.

The cases and data we rely on involve consumption taxes, primarily sales and use taxes. Such taxes are typically imposed on high-volume, recurring transactions for which sampling is an appropriate tool of analysis. The conclusions we draw, however, are applicable to other taxes, such as the income tax. Indeed, unlike most state and local governments, the Internal Revenue Service has the general authority to use statistical sampling in the conduct of audits of all taxes that it is responsible for administering.[1]

1. It is our understanding, however, that as a matter of practice the IRS typically restricts its use of statistical sampling to audits of large corporate taxpayers.

The article is organized as follows. In section 1 we discuss some key concepts in statistical theory that distinguish statistical estimates from informal, nonstatistical inferences. In section 2 we review and analyze certain judicial decisions that have involved the use of sampling in tax audits. In section 3 we argue that the legal objections to sampling can be addressed if the sampling and analysis are based on well established and routinely applied statistical procedures. In section 4 we present a model that provides an analytic basis for translating sample findings into an overall audit assessment. In section 5 we make some concluding remarks.

1. THE THEORY OF STATISTICAL SAMPLING: AN OVERVIEW

Every day of our lives we make decisions, both trivial and significant, based on inference from a sample. We choose from a luncheon menu based on experience from prior outings to the restaurant, decide what route to take home based on previous experience with travel times at different times of day, pick among potential employees based on experience with individuals with similar backgrounds and personalities, and make important business decisions based on prior experience with comparable products in similar markets. We may not think of this as sampling, but it is. We are drawing inferences based on a limited sample of information we deem relevant to the issue at hand. Occasionally we are wrong, but mostly we are right, which is why we keep doing it.

Statistical sampling is distinguished from casual inference both by method and consequence. The method requires a sampling frame, a random sampling mechanism that is independent of influence by either the taxpayer or the auditor, a sampling plan, and an appropriate analysis. The consequence is that, unlike their counterparts based on casual inferences, conclusions based on statistical samples can be rigorously evaluated in terms of reliability.

A sample frame is a group of items from which a sample is chosen. In the case of a tax audit, the sample frame is the universe of documented transactions available for audit. Generalization from statistical samples extends only to the sampling frame. Thus the application discussed below, in which a tax auditor sampled a taxpayer's inadequate records, is an example of statistical sampling, but the statistical conclusions extend only to the sampling frame—to those records available for examination. The extrapolation of results from records kept to records unkept is made by the courts on legal principles, not statistical ones.

A hallmark of a careful statistical sample is the proper choice of a sample frame. For example, in determining sales tax deficiencies, an auditor might choose to exclude certain untaxed transactions, such as sales of food, if the total food sales are known. However, it is not legitimate to change sampling frames in the midst of an audit.

The second key to adequate statistical sampling is that the items from the frame must be chosen by a random mechanism with known probabilistic properties. The items must be chosen without knowledge of or regard for their tax or other consequences. Failure to observe this principle can invalidate the statistical inference. For example, auditors are often asked to examine records to check certain facts. If they have preconceived ideas about where the problems might be, and exploit either their ideas or their initial findings in deciding where to look further, they are surely entitled to report the discrepancies they find. Nonetheless, they are not entitled to treat the cases they examined as a statistical sample, since they chose them purposely. Proper statistical sampling is a skill distinct from auditing.

The third key to successful statistical sampling is a sampling plan, determined before the process begins, that specifies the frame, the random mechanism to be employed and the probability of selection of each item or set of items, and the rule to be used to decide when to stop sampling. The sampling plan should be so explicit that two auditors with the same plan and the same random number generator for selecting transactions to be included in the sample should report the same results. Deviations from the sampling plan should be avoided.

The simplest sampling plan is called "simple random sampling." In this plan each item has the same chance of appearing in the sample as any other, and items appear or do not appear independently of each other. Another sampling plan is stratified sampling.[2] Here the sampling frame is divided into strata, each of which has its own separate sampling plan. For example, examining all transactions above one million dollars and randomly sampling the others can be thought of as a stratified sampling plan. Preconceived ideas about where problems might exist can be an excellent basis for stratification.

The use of "test periods" as a sampling plan is common in tax audits and requires special comment. In a test period analysis, a few time periods are selected and audit findings are extrapolated to the entire audit period. For this discussion, assume that the test periods are chosen by a random mechanism, even though in practice they are not. As explained above, if they are not randomly chosen, the sample is not a statistical sample. It is typical of a test period design that all the transactions in the selected months are analyzed, while none of those in unselected months are examined. The design is inefficient because a simple random sample among all transactions will provide a more precise estimate of the deficiency, as would a design stratified by month. Furthermore, either alternative design will permit estimation of the reliability of the tax assessment, as discussed later.

Consider the implications of the typical practice of purposely choosing test periods. After long discussions with management, market analysts, and other experts, one might be able to mount a strong argument that the estimated deficiency extrapolated from the test period audit is plausible. Nonetheless, it is important to distinguish inferences based on statistical sampling from the everyday notion of plausibility. Statistical sampling imposes rigorous standards, and a fundamental pillar of those standards is that the sample *not* be purposely chosen.

The final key element of statistical sampling is an analysis appropriate to the chosen design. For even a moderately complex statistical design, the services of a statistical expert or special training for the auditors may be required.

Four elements, then, constitute a statistical sample: (1) the sampling frame, (2) the random sampling mechanism, (3) the sampling plan, and (4) appropriate analysis.

The benefit of conducting a statistical sample is that the samplers can address the following technical questions: (1) What is the probability that the estimate deviates

---

2. See W. G. Cochran, Sampling Techniques (3d ed. John Wiley & Sons, 1977).

from the true value sought by some specific amount? The question concerns the reliability of the estimate. (2) How large a sample must be drawn for the estimate to reach a desired level of precision? This question concerns the sample size necessary to reach a particular level of reliability. We will discuss each of these concepts in turn.

Reliability

Reliability refers to the precision of the estimate. In formal statistical terms, an estimate may meet the standards for making statistical inferences but still be highly unreliable. For example, consider again the problem of estimating a sales tax deficiency based on a random sample of transactions from one month. Suppose that over the course of the month 1,000,000 transactions occurred. Random samples of 10, 100, 1,000, or 10,000 of these transactions will all provide a basis for estimating the deficiency. However, intuition suggests, and statistical theory confirms, that an estimate based on 10 or even 100 transactions may be so imprecise—or in formal statistical terms unreliable—that the sample is useless for assessing taxes.

What statistics adds to intuition is a formal analytical method for quantifying the degree of precision of a specified estimate—such as tax due—computed from a sample of some specific size and having some particular sample characteristics. Concretely, statistical theory provides the ability to make statements about the likely precision of an estimate. Returning to our sales tax auditing sample, suppose that from a sample of 1,000 transactions we estimate a tax deficiency of $500,000. Statistical theory allows us to quantify the imprecision of this estimate with such statements as: There is a 95% chance that the deficiency is as little as $450,000 or as much as $550,000.[3] In short, statistics tells the auditor how likely he is to have erred, and by how much.

Sample Size

Statistical theory not only allows ex post facto quantification of the reliability of an estimate based on a specific sample size, it also provides the basis for determining the sample size required to achieve a desired level of reliability. This capability has practical importance; it provides the tax administrator with an opportunity to control the level of precision of an audit assessment based on a statistical sample. Auditors can draw an initial sample, analyze the sample characteristics, and determine from their analysis the sample size required to meet a specified level of precision. Thus, unlike assessments based on an extrapolation of a test period, audit

---

3. For clarity, it is necessary to discuss some fine points of statistical theory. From the point of view of classical statistical theory, a 95% confidence interval means that if the same procedure is used many times, 95% of the uses will include the true tax deficiency. Confidence intervals do not purport to apply to any specific instance. By contrast, a 95% credible interval means that the probability is 95% that the true tax deficiency is in the interval specified in the given instance. Such intervals require added assumptions and a different, Bayesian statistical framework. Thus a confidence interval tells you what proportion of times you will bracket the right amount if you follow a set procedure and a credible interval tells you what the chances are that you bracketed the right amount this time. See L. J. Savage, The Foundations of Statistical Inference (New York: John Wiley & Sons, 1962), for a general discussion of the distinction. The relation between classical and Bayesian inference is treated in more detail in section 4.

assessments based on statistical sampling may be tested against formal standards of reliability.

2. JUDICIAL DECISIONS

In this section, we review judicial decisions on the use of samples to project audit findings. The principal issues raised are (1) the adequacy of taxpayer records, (2) the statutory authority for auditing, (3) whether the taxpayer consented to the sample, (4) the administrative burden of a complete audit, and (5) the reliability of an assessment based on a statistical sample.

By now, it is incontrovertible that a tax administrator may use sampling to project a tax assessment in the absence of accurate and reliable records kept by the taxpayer. While there was at least one false start when a statute did not require that records be kept,[4] the cases are now legion in which assessments based on samples are enforced against taxpayers who, in one way or another, failed to keep proper records.[5] Even when the lack of records has not been the taxpayer's fault, the courts have permitted sampling by auditors.[6] The cases are bottomed on the common sense premise that a taxpayer cannot escape tax liability through his own failures or even his own misfortunes.

It is equally clear that sampling may be used with a taxpayer's consent, even if implicitly given.[7] Indeed, taxpayer consent is the rule, not the exception. The willingness of most taxpayers to accept sampling suggests not only an interest in keeping the expense and disruption of an audit to a tolerable level but also a reasonable level of confidence in the process itself.

What is not clear, however, is whether, absent statutory authority or taxpayer consent, a tax examiner may assess a liability from a sample when the taxpayer's records are sufficient to determine a liability by a 100% audit. It has been suggested that a 100% audit is not always required when complete records are available,[8] but where the issue has been directly confronted, it has been decided in favor of the taxpayer.

An early and particularly well articulated case was decided by New York's Appellate Division in 1957. In re *Babylon Milk & Cream Co.*,[9] the taxpayer successfully challenged the extrapolation of the results of a fuel tax audit of four "test months" to the remaining eight months of a one-year audit period in which the taxpayer had records. The company, however, failed in challenging an assessment based on a projection over a period for which the taxpayer did not have records.

---

4. State *ex rel.* Foster v. Evatt, 144 Ohio 65, 56 N.E.2d 265 (1944).

5. E.g., Schwegmann Bros. Giant Supermarkets, Inc. v. Mouton, 309 S.2d 686 (La. 1975); Bouchard v. Johnson, 170 A.2d 372 (Me. 1961); Ridolfi v. Director, 1 N.J. Tax 198 (1980); *In re* Babylon Milk & Cream Co., 5 A.D.2d 712, 169 N.Y.S.2d 124 (1957); King Drug of Dayton v. Bowers, 171 Ohio 461, 172 N.E.2d 3 (1961). Other cases are cited in H. Leib, Using Sampling Techniques to Assess State Taxes, 3 J. St. Tax (1985) and L. Fournier & W. Raabe, Statistical Sampling Methods in State Tax Audits, 2 J. St. Tax 115 (1983).

6. *In re* Grecian Square, Inc., 119 A.D.2d 948, 501 N.Y.S.2d 219 (1986); Pato Foods, Inc. v. Lindley, 7 Ohio App. 3d 22, 453 N.E.2d 1274 (1982); Torridge Corp. v. Commissioner of Revenue, 84 N.M. 610, 506 P.2d 354 (1972).

7. E.g., Mitchell Bros. Truck Lines v. Hill, 363 P.2d 49 (Ore. 1961) (fuel use tax case); W. T. Grant Co. v. Joseph, 2 N.Y.2d 196, 140 N.E.2d 244 (1957).

8. E.g., Yonkers Plumbing & Heating Supply Corp. v. Tully, 62 A.D.2d 18, 402 N.Y.S.2d 792 (1978), 674 P.2d 785 (Abs. 1983).

9. 5 A.D.2d 712, 169 N.Y.S.2d 124 (1957).

With respect to the first issue, the taxpayer did not dispute the audit findings for the four months actually examined; rather the question was whether the findings could be projected over the period of an entire year. The court held for the taxpayer, ruling that as a matter of law the tax commissioner had no authority to project the results of a sample if complete records were available. The court stated:

> The records of the petitioner were all available for year 1953 and the exact amount of the understatement of mileage could have been determined for the remaining eight months in the same manner in which it was determined for the four months chosen for the test months. The only reason given for not determining the understatement in this manner was that it would have required additional work.... The use of the average method, at best, produced only an approximation of the amount of the tax owing.[10]

Thus the court took comfort in its conclusion from the observations that there did not seem to be any reason for projecting a sample other than to avoid work and that the projection would result only in an approximation or even, perhaps, a guess.

However, on the second issue the court held for the tax commission, sustaining an assessment based on a projection over 27 months of an examination of two-week records. No doubt because the two-week records were the only ones available to the tax commission, the court did not disparage use of the projection. Rather the court stated: "The result reached by the auditors was consistent with other findings made when the trucks and loads were weighed in 1954 and 1955. The method used by the auditors was a reasonable one in view of the petitioner's failure to keep any records. The petitioner has not demonstrated that the amount of the assessment was to any extent unjustified."[11]

Babylon Milk is a microcosm of many cases that followed it. If a taxpayer fails to keep adequate records, courts usually approve the projection of a sample, often reinforcing the result by pointing out factors supporting the accuracy of the projection. Historically, judges have not often criticized the reliability of a sample when a taxpayer has failed to keep adequate records. A high-water mark of judicial tolerance occurred in New York in *Markowitz v. State Tax Commission.*[12] Three of five judges approved projecting the results of an audit of one day's cash register tapes over a three-year period. With remarkable understatement, the majority observed, "Although it is probably true that cash register tapes of several days would give a better picture of the business of the petitioner and thus his tax liability, exactness is not required where the party's own failure to maintain the proper records prevents it."[13]

A more balanced approach was taken by an Ohio court in *McDonald's of Springfield, Ohio v. Kosydar.*[14] The taxpayer did not have adequate records, and the tax examiner made two assessments based on statutorily authorized spot checks. Pointing to certain statutory language, the court sustained one assessment but struck down the other on the grounds that the second sample was not representative.[15]

---

10. *Id.*
11. *Id.*
12. 54 A.D.2d 1023, 38 N.Y.S.2d 176, 177 (1976), *aff'd*, 44 N.Y.2d 684, 405 N.Y.S.2d 454, 376 N.E.2d 927.
13. *Id.*
14. 43 Ohio 2d 5, 330 N.E.2d 699 (1975).
15. See also Zapitelli v. Lindley, 1981 Westlaw 4376 (Ohio App. N.E.2d 1981).

Recently, New York judges have become more critical than they were in *Markowitz*. In re *Grecian Square, Inc.*,[16] a taxpayer did not keep adequate records and an assessment was made on a sample. However, the court held:

> Here, respondent's auditor found petitioner sales figures to be much lower than other establishments which he audited. Accordingly, he increased petitioner's estimated sales figure by 200%. However, the record does not disclose any specific information concerning the bars which McKenna [the auditor] had audited and found to have been comparable to petitioner's. As best as we can determine, no such information was given petitioner in advance of the hearing. At the hearing, McKenna merely stated that he had estimated sales by calling upon his wide experience in auditing other bars. Considerable latitude is given an auditor's method of estimating sales under such circumstances as exist in this case.... Nevertheless, there was insufficient evidence for the auditor's computation. By the same token, without some information about the size, location, number of employees and nature of the operation, this court is unable to make a determination as to the existence of a rationale [*sic*] basis. Hence, the matter must be remitted to respondent for further testimony of McKenna in accordance with this decision.[17]

Whatever may be the latitude given to auditors when proper records are not kept, as a general rule sampling has not been permitted in the absence of consent if adequate records are kept. Various reasons are given why sampling is not authorized in such cases, but invariably the reasons are accompanied, as in *Babylon*, by judicial comments on the unreliability of the sampling procedure.

For example, in a 1965 Maine case,[18] the records of a wholesaler and retailer of automotive parts and supplies were examined for a four-month period. The examination showed that the tax liability was understated. From their review of the four-month period, the auditors calculated what was described as a margin of error—that is, a ratio of understated liability to reported liability. An assessment was based on a projection of the margin of error over a 23-month period. The court struck down the projected assessment. After observing that the taxpayer had kept adequate records as required by the statute, the court observed: "The legislature required an audit based upon an examination of the taxpayers records and *not the establishment of tax liability by surmise and conjecture*" (emphasis added).[19] The court added: "We understand and appreciate the problem faced by the auditors in such a time-consuming task as they were confronted with but according to our analysis of the sales and use tax statute in its entirety, we find no alternative. This may constitute a serious administrative problem time-wise for the Tax Assessor's Department but it is one for the Legislature to consider and not the Courts."[20]

The court's decision was expressly based on the conclusion that the revenue agency had no statutory authority to base a tax liability on estimates, except in the limited circumstances "where a taxpayer fails to make a report, or where the departure from the State of a taxpayer is imminent." Nonetheless, it is evident from the opinion that the court was driven to its conclusion by anxieties about the reliability

16. 119 A.D.2d 948, 501 N.Y.S.2d 219 (1986) (citation omitted).
17. *Id.*
18. Farrar Brown Co. v. Johnson, 207 A.2d 406 (Me. 1965).
19. *Id.*
20. *Id.*

of sampling procedures and by a suspicion that governmental advocacy of the efficiency of sampling might be an excuse for laziness.

Such reservations were repeated in a 1978 New York case.[21] The taxpayer argued that a sample should not have been used to project liability because his records were adequate to determine any deficiency. Without expressly commenting on the adequacy of the records, the court held for the petitioner and stated: "There is no inflexible rule that an item-by-item audit be made whenever it is possible, but it should be utilized if the records are available and the test check method is insufficient to afford a reasonable calculation of the taxes due."[22]

In 1980, the same court in *Names in the News v. New York State Tax Commission*[23] held:

According to the testimony of an Associate Sales Tax Examiner for the State, the test period approach was utilized by the Tax Bureau because petitioner's records were "too voluminous" and it was the bureau's normal auditing procedure to use a trial period. This same examiner conceded, however, that petitioner's general ledger, purchase invoices, sales invoices and Federal tax returns were available to the bureau and that no request for information or documents was refused or rejected by petitioners, and upon the present record it can only be concluded that the State's auditors had access to petitioner's detailed records for the whole three year period, a vastly different situation from that encountered in *Matter of Meyer v. State Tax Comm.*, [61 A.D.2d 223, 402 N.Y.S.2d 74 (1978),]... where the court approved an analysis of purchases and that in the matter of *Markowitz v. State Tax Comm.*, [54 A.D.2d 1023, 38 N.Y.S.2d 176, 177 (1976), *aff'd*, 44 N.Y.2d 684, 405 N.Y.S.2d 454, 376 N.E.2d 927,]... where we approved a test period or spot check only because of the inadequacies in methods and procedures and failure to maintain proper records.... Under these circumstances, the bureau's use of the test period to compute petitioner's tax liability was plainly *unnecessary, arbitrary and capricious*, and petitioners were entitled to have their tax assessment calculated based upon a detailed audit of their records for the three-year period under consideration. [Mohawk Airlines v. Tully, 429 N.Y.S.2d 759 (App. Div. 1980) [emphasis added]; In re Chartair, Inc. v. State Tax Comm., 65 A.D.2d 44, 411 N.Y.S.2d 41 (App. Div. 1978).][24]

Recently, the law of New York was summarized in In re *Christ Cella, Inc.*:[25]

Section 1138... of the Tax Law governs when external indices tests may be used. The statute provides that: "if a return when filed is incorrect or insufficient, the amount of tax due shall be determined by the tax commission from such information as may be available. If necessary, the tax may be estimated on the basis of external indices, such as stock on hand, purchases, rental paid, number of rooms, location, scale of rents or charges, comparable rents or charges, type of accommodations and service, number of employees or other factors." The department, however, may not use such external indices unless it is "virtually impossible to verify taxable sales receipts and conduct a complete audit" with available records....[26] Petitioner's contention that the

21. Yonkers Plumbing & Heating Supply Corp. v. Tully, 62 A.D.2d 18, 402 N.Y.S.2d 792 (1978).
22. *Id.*
23. 429 N.Y.S.2d 755 (App. Div. 1980).
24. Names in the News v. New York State Tax Comm., 429 N.Y.S.2d 755 (App. Div. 1980) (emphasis added).
25. 102 A.D.2d 352, 477 N.Y.S.2d (1984).
26. *In re* Korba, 84 A.D.2d 655, 444 N.Y.S.2d 312 (1981), *appeal denied* 56 N.Y.2d 502, 435 N.E.2d 1099; *In re* Chartair, Inc., 65 A.D.2d 44, 411 N.Y.S.2d 41 (App. Div. 1978).

department violated this rule by not requesting the records covering the three-year period, but rather proceeding directly to the markup test, has merit. The auditors testified that, "We do markup tests in approximately every restaurant to determine whether the particular restaurant has the proper records or not." This procedure violates *Chartair*, where this court stated: "The honest and conscientious taxpayer who maintains comprehensive records as required has a right to expect that they will be used in any audit to determine his ultimate tax liability." Consequently, the markup test could not be used unless petitioner's records were so insufficient that its sales could not be verified or such records were unavailable.[27]

What authority there is on the subject indicates that the same rules are applied in the case of a claim by a taxpayer for a refund. In an early case, a court determined that where adequate records were available, a petitioner could not claim a refund calculated from a sample, even when the sampling was apparently done on a statistically sound basis.[28] The court therefore directed a complete audit, which yielded a number very close to the result of the statistical sample.

Conversely, a taxpayer without required records was entitled to a refund based on a sample, albeit with assistance from certain statutory language. In *Belgrade Gardens, Inc. v. Koysdar*,[29] a restaurant incorrectly determined the amount of sales tax collected and therefore overpaid the tax. A tax examiner made a "test check" of guests' checks and determined that an overpayment had been made, but the tax commissioner denied the refund claim on the grounds that the taxpayer had not maintained the required records. The court held for the taxpayer, holding that the absence of records does not, in itself, preclude a refund. The court stated:

> The audit period herein exceeded three years. There is no question that a test check conducted to determine the appropriate tax liability for that period could provide an accurate figure for the amount of overpayment. An audit of guest checks and cash register tapes maintained by the taxpayer for the period would provide a more accurate figure than the calculated approximation provided by a test check. However, adequate records were not available in this case, and the Tax Commissioner conducted a test check to fulfill his mandatory duty to investigate the facts in connection with the taxpayer's claim. Neither party has contested the accuracy of the test check....However, the Tax Commissioner proposes that a valid test check conducted by him cannot be used to the advantage of the taxpayer....This court has never held that any specific burden of proof attaches to a taxpayer who claims to have made an erroneous overpayment of sales tax. Indeed, [state law] places the duty of ascertaining the amount of overpayment upon the Tax Commissioner.[30]

Perhaps the closest a court has come to an unconditional authorization of sampling was in *Underwood v. Fairbanks North Star Borough*.[31] After lengthy procedural litigation, the taxpayer agreed to a sales tax audit by the borough, and a judgment was entered in favor of the borough. Shortly thereafter, the borough

---

27. *Id.*
28. Sears, Roebuck & Co. v. City of Inglewood (Los Angeles Super. Ct. 1955), *discussed in* Sprowls, Admissibility of Sample Data into a Court of Law, 4 U.C.L.A. L. Rev. 222, 226 (1957).
29. 38 Ohio 2d 135, 311 N.E.2d 1 (1974).
30. *Id.* at 141–43.
31. 26, 174 P 2d 785 (Alaska 1983)

moved to amend the judgment to correct a clerical mistake in favor of the taxpayer. The taxpayer in response attacked the audit that led to the judgment in the first place. The court stated:

> Underwood argues that the sampling method was unreasonable because a better method of determining tax liability existed, namely, audit all the records. We agree that the figures a full-scale audit would have produced would have been more accurate than the estimate on which the Borough relies, but we see no reason to force the Borough to bear the expense of interpreting all the records Underwood has surrendered. The difference between the deficiency figure a statistically valid estimate produces and the deficiency a full-scale audit suggests will usually be far less than the added expense of conducting a full audit. If courts force a taxing authority to bear this additional cost even though a taxpayer's miscalculations have caused the problem, they rob it of resources with which it could be providing services.... The parties stipulated that Underwood "shall produce for [the Borough] for purpose of audit by [the Borough], all of the documents and records requested to be produced."...After the material is produced, how any tax arrearages are to be proven in court is a matter properly committed to the appropriate rules of evidence. Samples are generally receivable in evidence "to show the quality or condition of the entire lot or mass from which" they are taken. 2 J. Wigmore, Evidence 439, at 522 (Chadbourn rev. ed. 1979). The question becomes whether the sample used was large enough to be statistically reliable and, if so, whether there is something about the sample period that makes it atypical. These questions can go to both the admissibility and to the weight of the evidence. Here the evidence has been stipulated as admissible, and thus the only question is what weight should be given it. Underwood has not contended that the sample is too short a period to be reliable, or that the period was atypical. Therefore, we conclude that the trial court did not err in relying on the evidence presented on the grounds that it was a sample.[32]

The Alaska court perhaps for the first time addressed policy issues that we believe should be at the core of any discussion of the use of statistical sampling. Nonetheless, the unique procedural posture and the effect of the prior court settlement on the decision cloud any claim that the case is reliable precedent for nonconsensual statistical sampling cases.

The judicial responses are not surprising. The courts have not been provided with standards for balancing the interests of tax administrators against the concerns of taxpayers about the reliability of an assessment based on a sample. In this vacuum of standards, the courts have given the benefit of the doubt to the citizen, not to the state. Only where the taxpayer has failed to keep adequate records have the courts generally rejected challenges to the tax administrator's authority to use samples and other analytical methods to make plausible estimates of the taxpayer's liability. The reason has simply been that if the tax is to be enforceable, there is no choice.

When there is a choice—that is, when records are available—the authority has not been found. Probably it could be found if the judges were willing. The no-records cases support the use of sampling on the grounds that the taxpayer failed to comply with a statutory requirement to keep accurate records. However, the tax statutes also require that an accurate return be filed, and in every case in which sampling has not been approved, the sample has determined *definitively* that with

32. *Id.*

respect to the sample the return was not accurate, and by likely inference that the returns for the balance of the period were not accurate either. As a matter of logic, there is no reason why the failure to file an accurate return could not justify the use of sampling. Statistical sampling is a specialized form of circumstantial evidence. As in other cases involving circumstantial evidence, there is no reason per se why it should not be used with appropriate safeguards. Yet no court has so held when complete records are available.

We believe the reason for the failure to approve audits using statistical sampling is that once the facts move beyond a no-choice situation, the courts are quite at sea about how the balance should be struck between administrability and reliability and have been given no help by any legislature or litigant. No statute provides even minimal guidance for sampling—when it may be used by a tax administrator and with what required degree of accuracy. Neither does it appear that any tax administrator has ever argued the case for sampling using well-established statistical theory. Perhaps the only statistically sound presentation was made 30 years ago by the taxpayer in the *Sears Roebuck* refund case noted above—but unsuccessfully. It is no wonder courts have refused to wade in alone.

If statistical sampling is to be used in a nonconsensual situation in which records are available, the guidance gap will have to be filled. The lines must be drawn with an eye to the concerns of both administrators and taxpayers. What are the relevant concerns?

The tax administrator is responsible for collecting the maximum amount of taxes legally due as equitably and efficiently as possible. Sampling techniques properly applied can advance this objective, principally because they allow improved economies in the use of government resources.[33] But the taxpayer's concerns are equally real. Tax assessments based on statistical samples are always estimates and therefore are inherently imprecise. Although we do not believe that the imprecision should be fatal, the taxpayer does have the right to know the magnitude of the uncertainty; he should be protected by appropriate standards requiring a minimum level of precision; and arguably some accommodation should be made in the assessment to compensate for the risk that it exceeds the true liability. For both parties, there are some common objectives. Audits are disruptive, time-consuming, and inherently confrontational. Like the administrator, the taxpayer has an interest in minimizing their duration. Sampling can contribute to achieving this goal.

In the next section, we discuss how properly conducted statistical sampling, combined with well established procedures for extrapolating results, can provide

33. While the courts have placed tight constraints on its use, sampling is routinely used in sales and use tax audits. The reasons are that taxpayers without complete records are prime audit targets and that many taxpayers with complete records consent to the use of sampling. To illustrate sampling's contribution to efficient enforcement, consider the case of Pennsylvania's Department of Revenue. In fiscal year 1984–85, the department conducted about 6,500 sales and use tax audits that identified over $44 million in assessed deficiencies. The total direct personnel commitment charged to these audits was nearly 150 person years. In about 85% of these audits, some sort of test period procedure was used. A restriction of test period auditing to instances where taxpayer records were incomplete would have required a 15-fold increase in auditors at an annual cost of about $75 million—a commitment of 2,500 additional auditors, which is more than the department's entire personnel complement. Alternatively, the no-sampling audit strategy would require a reduction in the number of audits of about 85%, resulting in a direct revenue loss of about $35 million. The direct reduction in audit productivity would have resulted in deficits for the Commonwealth in four of the past eight fiscal years.

the analytical basis for balancing the competing concerns of tax administrators and taxpayers.

3. THE CASE FOR STATISTICAL SAMPLES

In this section we draw upon the concepts developed in section 1 and apply them to the judicial concerns discussed in section 2. The judicial cases have expressed three principal concerns: (1) sampling may be used as a device to rationalize bureaucratic indolence; (2) assessments projected from sample results may be distorted by unobserved or unique factors and, therefore, the sample results are of unknown reliability; and (3) there is no statutory authority for the use of samples where records are complete.

The first objection is that sampling may be used as a rationalization for avoiding a thorough audit or, in the extreme, as a ruse to camouflage laziness. Certainly sampling, like almost all well-established procedures for improving efficiency, can be abused. But statistical theory provides formal quality standards for detecting and insuring against such abuses. Moreover, sampling may be the tax administrator's only effective remedy if a taxpayer files an inaccurate return and simply insists that the tax assessor correct it. Although the practice of filing inaccurate returns may be only a minor nuisance that is adequately dealt with by penalties for understatement, the authority to use statistical sampling in the conduct of an audit can be tactically important when dealing with a taxpayer who conducts a large volume of transactions. Without sampling, it may be literally impossible for a tax examiner with a limited staff to audit an entire period before the statute of limitations runs. With proper safeguards, a tax examiner ought to be entitled to use statistical sampling to avoid such a result. Otherwise, not only will the state's fiscal position be injured, but tax equity will suffer by a shifting of the tax burden from noncomplying to complying taxpayers.

The second judicial concern is whether, in the words of one court, sampling is "plainly unnecessary, arbitrary and capricious."[34] Although perhaps overstated, the criticism goes to the heart of judicial discomfort with sampling. We believe that *statistical* sampling is none of those things. In every taxpayer challenge of sampling we have identified, the sampling procedure did not meet the requirements necessary for drawing formal statistical inferences. In every case, the taxpayer was challenging the extrapolation of findings from a 100% sample of one or more purposely chosen test periods to an entire audit period.

The use of test periods to extrapolate tax assessments is a routine practice in state revenue agencies. Despite the high probability of a successful legal challenge to a test period audit assessment in most jurisdictions, few such challenges are made. The reasons, we believe, are threefold. First, as a general rule, taxpayers have reasonable confidence in the procedure. If the audits are conducted carefully and a sufficient number of test periods are examined, the projected assessments will pass the reasonable person's test of plausibility. Second, the taxpayer has an interest in

---

34. Names in the News v. New York State Tax Comm., 429 N.Y.S.2d 755 (App. Div. 1980).

shortening the duration of the auditor's unpleasant visit. Among other risks, the auditor might turn up some other problem. Third, legal challenges are costly.

If the test period approach to audit sampling has met with general acceptance, why has it failed to pass judicial muster when challenged by the nonconsenting taxpayer? The reason, we believe, is that it is impossible to defend the estimate's reliability in a formal statistical sense. Since test period samples fail to meet the requirements of a random sample, the tax administrator can make no formal quantitative evaluation of the precision of the projected assessment. Thus even a sympathetic court is placed in an awkward position: The taxpayer argues that the estimate may over-state his liability and that records are available for determining the true deficiency, and the tax administrator has no effective rebuttal.

As our discussion in section 1 states, appropriately drawn and analyzed statistical samples provide a basis for clearing the hurdles in the way of quantitative statements about the reliability of an estimate. We believe that if a tax administrator is armed with proper statistical procedures, the courtroom result might be different but for one missing ingredient: guidance on the minimum standards of reliability. As our prior discussion indicates, statistical theory cannot accomplish the impossible. It cannot eliminate the uncertainties inherent in projections. However, it can provide a basis for quantifying the imprecision and for developing strategies to reduce the imprecision to an acceptable level. Even so, what is an "acceptable" level of reliability is a normative issue about which statistical theory is mute. The issue requires the kind of judgment a legislature is particularly well situated to exercise. Among the questions that should be considered are:

> Since sampling inherently results in an imprecise finding, should the assessment be adjusted to reflect the risk of overassessment? That is, should the lowest "reasonable" number be chosen for the assessment?
>
> Where sampling is used, what minimum standard of reliability should be required? Perhaps a 10% chance of an error in excess of 5%?
>
> Should sampling be permitted in all circumstances? Or should it be limited to circumstances where a minimum sales volume or transaction threshold is exceeded?

We address these issues in the next section.

4. STATUTORY STANDARDS FOR THE USE OF STATISTICAL SAMPLING

Statistical sampling cannot provide an exact determination of tax owed. Only an examination of every item in the sample frame will do that—but that is not sampling. In recognition of the inherent imprecision of assessments based on the findings of a statistical sample, three of the revenue agencies we have identified as using statistical sampling—the IRS and the revenue departments in New York and Pennsylvania—adjust the assessment to provide the taxpayer with a considerable degree of protection from the risk of overassessment. The IRS and New York assess the taxpayer an amount designed to reduce the probability of overassessment to .05, and Pennsylvania on at least one occasion reduced the probability to .025. We believe that in concept the practice of these revenue agencies is sound. We will argue below that an audit assessment based on a less than exhaustive sample should reflect the risk of overassessment. We will also develop a formal assessment rule to accommodate the risk and show that the assessment rule of the three revenue agencies is a special case of the more general rule.

Derivation of the rule requires a formal mathematical characterization of both the uncertainty accompanying nonexhaustive sampling and the relative cost of overassessment versus underassessment of the true tax liability. The former is called a probability distribution and the latter a loss function.

In the following paragraphs, we elaborate on what a probability distribution is. We then discuss the central limit theorem—a theorem that gives conditions under which averages tend to have a normal distribution as the sample size increases. In combination, these ideas justify the use of the normal distribution as a representation of the uncertainty about the true tax deficiency, and thus provide the first ingredient for the derivation of an assessment rule. We note, however, that although this brief explanation of statistical ideas should be adequate to understand this article, it cannot substitute for a statistical text, such as DeGroot's *Probability and Statistics*, to which the reader is referred for a more thorough discussion.[35]

Nonexhaustive sampling provides only a probability distribution on the tax owed. The distribution, which is based on the findings of the audited sample, permits us to calculate the probability that the true tax deficiency falls within any specified interval. For example, consider an audit based on a 15% random sample. A probability distribution provides the basis for making such statements as: "There is only a 10% chance that the total deficiency is less than $4,500 or more than $5,500."

The curve in figure 1 is the probability density of the normal distribution; it has the familiar bell shape. This distribution is characterized by two parameters: the mean, μ, and the standard deviation, σ. The mean is at the top of the bell. It is the average of the distribution; and, because of the symmetry of the normal distribution, it is also the median—that is, half the probability is above μ and half is below. The standard deviation σ measures how peaked or spread out the bell is.

The interpretation of the two parameters can be illustrated with a nonauditing example. Consider the distribution of heights in the population. The average height is measured by μ and the variability in heights is captured by σ. The shaded area under the curve measures the proportion of the population whose height is between A and B feet tall. Alternatively, this proportion can be interpreted as the probability that the height of a randomly chosen individual is between A and B feet.

More peaked normal curves correspond to those with smaller standard deviations—that is, to curves depicting populations with less variability. For example, a probability distribution for the height of women only would be more peaked (i.e., have a smaller standard deviation) than for the population of both sexes.

The central limit theorem is a remarkable result of probability theory. It states that under a very general set of conditions, the distribution of averages of observations tends to become more and more like the normal distribution as the number of observations gets larger. The requirements for a random sample discussed in section 2 are sufficient to ensure that the central limit theorem applies. If the observations

35. M. DeGroot, Probability and Statistics (Reading, MA: Addison-Wesley, 1978).

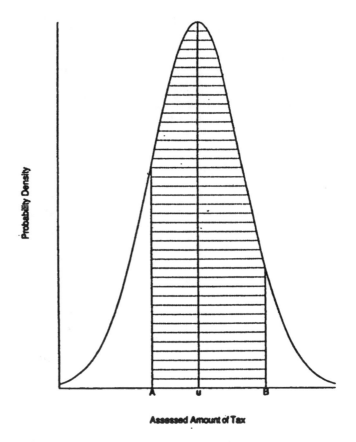

FIGURE I. The normal probability density functions

are transactions, and the datum is the amount of tax deficiency in the transaction, the average of the observed tax deficiencies will tend to have a normal distribution, with some mean and some standard deviation that can be estimated from the audit data. What is remarkable is that this result holds true even if the probability distribution of transaction deficiencies is not even remotely characterized by the normal distribution.

To illustrate the power of the central limit theorem, consider the following stylized example. Assume that a retailer sells only two items, both of which are taxable. One item costs $1, the other $2. Assume further that the tax rate is 5% and that for any given transaction, the retailer either collects all the tax due or fails to collect any tax. Thus an audit of any particular transaction will reveal one of three outcomes: (1) no tax due; (2) 5 cents due; or (3) 10 cents due. Obviously, a plot of the amount due per transaction will not correspond to the bell-shaped normal distribution depicted in figure 1. Instead, all observations will cluster at three points. According to the central limit theorem, however, the *average* deficiency across transactions will tend to be distributed normally as the sample size becomes large. Specifically, if successive samples of equal size are drawn and the average deficiency computed, a plot of these averages will approximate the bell-shaped normal distribution depicted in Figure 1.

The reader may observe, however, that the distribution we have—the distribution of the average of the data—is not the distribution we need—the distribution of the tax deficiency. The central limit theorem provides the basis for calculating the probability that another estimate of the average deficiency calculated from still another independent sample of transactions will fall within a specified range. But this is not what we want. Instead, we want to be able to make some probabilistic pronouncements, based on the audit findings, about the *true* total tax deficiency. After all, that is why the audit was conducted.

The Bayesian school of statistics provides a remedy using a result in probability theory called Bayes's theorem. Bayes's theorem provides the basis for converting probability statements about the size of the deficiency in the sample to probability statements about the true deficiency in the population, given the value of the deficiency actually observed in the sample. A key concept is the prior distribution. In the case at hand, the distribution measures prior opinion about the amount of the tax deficiency. Although the audit may have been prompted by a strong opinion that a substantial deficiency exists, such a surmise will have no standing in substantiating a deficiency to a court. The case for the deficiency must be built from the audit findings. Thus for our purpose, the prior distribution must be characterized by substantial uncertainty. One way to characterize such uncertainty is by a normal distribution with a very large standard deviation.

Bayes's theorem provides a formal mechanism for updating the prior distribution to take account of data—in our case the audit findings from the sampled transactions. The resulting "updated" prior distribution, called the posterior distribution, measures precisely what we desire: the uncertainty about the true deficiency given the results of the audited sample. Further, when both the prior distribution and the distribution of the data are normal, then the posterior distribution is normal. In addition, when the standard deviation of the prior distribution is very large, as has been supposed above, the mean and standard deviation of the posterior normal distribution are, respectively, the mean and standard deviation of the data distribution. Hence, with these assumptions, the posterior distribution is well approximated by a normal distribution whose mean is the average of the sampled observations and whose standard deviation can also be calculated from the sample data.

In conclusion, Bayesian statistics combined with the central limit theorem provides the technical machinery for concluding that the uncertainty about the true deficiency can be reasonably characterized by a normal distribution with a mean equal to the average deficiency per transaction audited times the number of transactions in the universe, and a standard deviation, which can also be estimated from the audited sample.

The standard deviation will be affected by two factors. First, the standard deviation decreases as n increases. This result is intuitively appealing; as more transactions are audited, we expect our uncertainty about the true amount owed to decrease. Indeed, in the extreme case, where all transactions are audited, the standard deviation equals zero—that is, there is *no* uncertainty about the true tax deficiency. A second factor affecting the magnitude of the standard deviation is the inherent variation of the tax owed among individual transactions. For example, consider two different taxpayers. One is a retailer who specializes in selling inexpensive toys, all of which are taxable. The other is a general merchandise retailer who sells items of widely different values, only some of which are taxable. Further assume that the chance that either merchant does not collect the appropriate tax on any given transaction is equal. Statistics confirms the intuition that for any given nonexhaustive sample, the standard deviation on the taxes owed will be less for the first retailer than for the second.

Our first question is how much the taxpayer should be assessed in light of the uncertainty about the exact amount owed. One obvious candidate is the average of the observations multiplied by the number of transactions in the frame. If this rule is used, however, some taxpayers will pay more than they actually owe, and others will pay less. Moreover, it is not possible to tell which is which without doing a complete audit of every taxpayer. The probabilities of overassessment and underassessment are equal, but that gives little solace to the taxpayer who believes he or she has been overassessed.

For several reasons, we submit that the rule for balancing the two possible errors should build from the premise that overassessing is worse than underassessing. First, a cornerstone of tax jurisprudence is that statutes will be strictly interpreted against the state. One of the roots of this principle is that taxpayers need protection from overaggressive tax administrators attempting to collect taxes that the legislature did not intend to be levied. Second, since the burden of proof is on the taxpayer to prove that an assessment, once levied, is invalid, the taxpayer deserves some consideration for taking the risk that an audit assessment based on a statistical sample may overstate his liability. We say this because the taxpayer will have no basis for arguing that he in particular has been overassessed, beyond the observation that there is some chance the assessment is too high. Third, and perhaps most important, we believe that an assessment rule that fails to place greater emphasis on the cost of overassessment compared to underassessment may injure voluntary compliance and taxpayer perceptions of the equity of tax administration. Audited taxpayers will correctly conclude that 50% of their number are being overassessed by an amount equal to the underassessment of the remaining 50%. Such an assessment rule might appear to be more like a lottery than an appropriate system of tax administration.

Our argument that overassessments are worse than underassessments can be mathematically characterized in terms of the "loss function" depicted in figure 2. If the taxpayer is assessed an amount precisely equal to u, then as shown in figure 2, no underassessment or overassessment costs are incurred. This point is the minimum of the loss function. The line to the left of u measures the loss caused by underassessment and that to the right, the loss due to overassessment. Consider a deviation from u of some specified amount x. Consistent with our argument that overassessment is worse than underassessment, the value of the loss function for an assessment $u + x$ exceeds that for an assessment $u - x$. The result is guaranteed by requiring that the line measuring the overassessment cost be steeper (i.e., its slope be greater) than that measuring underassessment cost.

The ratio, k, of the slopes of these two lines measures the cost ratio of an overassessment versus underassessment of a given amount. Thus we suppose that k must necessarily be some number greater than one. Or, stated differently, the loss due to an overassessment of any given amount will always be greater than the loss from an underassessment of that same amount.

In light of the inherent uncertainty in the true amount owed, as illustrated by the probability density function in Figure 1 and the relative losses from overassessment

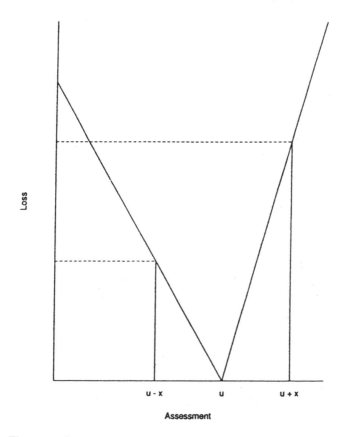

FIGURE 2. The costs of overassessment and underassessment: illustrative loss function

and underassessment as depicted by Figure 2, the problem is to derive an assessment rule that, on average, minimizes the overall cost of under- and overassessment across all audits where statistical sampling is employed. The required assessment rule, which is derived in appendix A, is that the taxpayer should be assessed an amount equal to $\bar{x} - \rho\hat{s}$, where $\bar{x}$ is the average deficiency per sampled transaction times the number of transactions in the universe, ρ is a mathematical function of k, and $\hat{s}$ is the estimate of σ based on the sample audit findings. Table 1 gives ρ for various values of k. It shows that the larger the value of k, the larger the resulting value of ρ. This is reasonable, since as k gets larger the relative consequence of collecting too much is worse, and hence the assessment should decline. To illustrate the application of the rule, consider again a tax deficiency of $100,000 projected from a sample of 5,000 transactions. Assume further that $\hat{s}$—the estimate of σ based on the audited sample—is $10,000. If the cost of overassessment is judged to be twice that of an underassessment of the same amount—that is, $k = 2$—then $\rho = .46$, and in this example, the taxpayer would be assessed $95,400 (= $100,000 − [.46 × 10,000]). Alternatively, if overassessment were responding $\rho = .84$. In this case, the adjustment for uncertainty is larger than for $k = 2$ (because there is an even higher cost attached to overestimating the tax liability), and the taxpayer would be assessed $91,600 (= $100,000 − [.84 × $10,000]).

TABLE 1. The Choice of k and the Resulting Values of ρ and P

Choice of k	ρ	Probability That Too Much Tax Is Collected
1	0	.5
1.5	.25	.4
2	.46	.333
2.5	.55	.29
3	.68	.25
4	.84	.20
5	.96	.17
6	1.08	.14
9	1.28	.10
19	1.67	.05
39	1.96	.025

The last column of Table 1 shows P, the probability that too much tax is collected. Observe that the probability declines as k increases. For example, if $k = 2$, the probability is 1 in 3, but if $k = 4$, it is 1 in 5. Again, the result is reasonable, since as k increases, the adverse consequences of overassessment are greater and the assessment rule should provide the taxpayer with greater protection.

Some further intuition into the mechanics of the rule are revealed by the fact that $k \times P$ equals the probability of underassessment, $1 - P$. Intuitively, this expresses a balance between the probability and relative cost of overassessment (kP) and the cost of underassessment ($1 - P$).

The rule, we submit, is fair to both the state and the taxpayer. It allows the state to collect an amount of tax equal to the mean of taxes owed, with a reduction equal to $\rho\hat{s}$ for the uncertainty that the state imposes on the taxpayer. The magnitude of the reduction can be controlled by the state. Recall from the previous discussion that one of the factors affecting the magnitude of $\hat{s}$ is sample size. This parameter decreases with sample size; in the extreme, where all transactions are sampled, it equals zero. Thus, if the state concludes that the uncertainty adjustment is sacrificing too much revenue, it has the option of drawing larger samples at its own cost.

Similarly, we believe the rule is fair to the taxpayer. Observe that the adjustment for uncertainty, $\rho\hat{s}$, increases in proportion to $\hat{s}$. Thus, as the imprecision of $\bar{x}$, as measured by $\hat{s}$, increases, the adjustment for uncertainty increases. This, we submit, is fair to the taxpayer because the magnitude of the uncertainty adjustment can be controlled by the state in its selection of sample size. If for whatever reason the state chooses to draw a sample that can provide only a highly imprecise estimate of the true deficiency, the taxpayer should not be required to bear all the risk that the assessment may substantially overstate the deficiency. Also observe that for any given value of k there is a *fixed* probability that the taxpayer will be overassessed which is independent of $\hat{s}$. This means that the uncertainty adjustment is always just large enough to maintain a constant probability of overassessment. Thus, as measured by the probability of overassessment, taxpayers are being treated uniformly.

Implementation of the assessment rule requires the choice of a specific value of k, which we believe should be made by the legislature. We will briefly discuss some of the considerations we believe the legislature should take into account and hazard a preliminary suggestion on an appropriate value. One consideration involves balancing the social cost to society, in marginally less funding for government-supported programs, versus the cost to the individual of having to pay marginally more tax than he owes. Another consideration is the extent to which the benefit of the doubt should be given to taxpayers in the exercise of the coercive power of the state. Here it should be kept in mind that what is at stake is money, not the individual's basic civil freedoms. A final consideration is the magnitude of imprecision in the assessment. For reasons that will be discussed below, we believe that minimum standards for precision should be tight.

Some might object to our proposed assessment rule on two grounds: that the choice of k is arbitrary and that on a practical level it would be difficult to implement. We address these two issues in turn.

As we previously noted, when statistical sampling is used, the IRS[36] and the New York State Department of Tax and Finance[37] assess the taxpayer an amount calculated to reduce the risk of overassessment to .05. Inspection of table 1 reveals that such an overassessment probability corresponds to a k of 19. What is the rationale for .05? The use of .05 dates back to R. A. Fisher in his exposition of significance testing.[38] As Raiffa and Schlaiffer put it, "decisions are actually made by treating the numbers .05 and .95 with the same superstitious awe that is usually reserved for the number 13."[39] There is no analysis we know of that supports the use of .05 in the context of tax collection.

Pennsylvania has used varying rules for calculating assessments based on statistical samples. As previously indicated, on at least one occasion, the assessment was calculated to reduce the risk of overassessment to .025.[40] This corresponds to a k of 39. On at least one other occasion Pennsylvania has assessed at the mean, which implies a probability of overassessment of 0.5, and a k of 1.[41] We know of no analysis to support either of these rules.

Thus all three agencies, IRS and the revenue departments in New York and Pennsylvania (with both its rules), have already implicitly adopted our assessment rule but without apparent consideration of the relative costs of over- versus underassessment. While we acknowledge that the choice of k involves a subjective weighing of competing considerations, surely it is better to consider the tradeoffs explicitly. While the assessment rules of the IRS, New York and Pennsylvania appear to be objective, beneath the surface each of them constitutes a probably instinctive choice of k.

We believe that the choice of $k = 19$ is too high and $k = 1$ is too low. In our judgment, the cost of overassessing a taxpayer by a given amount is not, in most cases, 19 times worse than underassessing him by the same amount. Nor, in our judgment, is the cost of overassessing a taxpayer by a given amount equal to the

36. See W. L. Felix & R. Roussey, Statistical Inference and the IRS, 159 *J. Accountancy* 38 (1985).

37. New York State Dep't of Tax. & Finance, EDP Systems Audit Bureau, letter to Pennsylvania Dep't of Revenue, Bureau of Audit, at 4.

38. R. A. Fisher, Statistical Methods for Research Workers (14th ed. New York: Hafner Publishing Co., 1973).

39. 39. H. Raiffa & R. Schlaiffer, Applied Statistical Decision Theory viii (Cambridge, Mass.: MIT Press, 1961).

40. Pa. Dep't of Revenue, Board of Appeals Docket #504817 SUT.

41. Pa. Dep't of Revenue, Board of Appeals Docket #713816 SUT.

cost of underassessing him by the same amount. We think that the appropriate value for k is in the range of 2 to 4.

Would our proposed rule be difficult to implement? Given a policy establishing a value for k, implementation would depend solely on correctly calculating a mean and a standard deviation for the amount of tax owed. The auditors could mechanically apply the rule in calculating the assessment, just as they now do for the IRS and in New York and sometimes in Pennsylvania. The greater difficulty will be in conducting a sampling audit competently, so that the items selected are a random sample chosen from a proper sampling frame.

We now consider whether the legislature should establish any minimum standard of reliability. An argument can be made that no such minimum standard is necessary. On one hand, the taxpayer is protected by an adjustment for uncertainty that increases with the imprecision of the estimate of u. On the other hand, the tax administrator can control the magnitude of this adjustment and thus the resulting loss in assessment revenue by the choice of sample size. Thus the problem reduces to a tradeoff between the cost of sampling and forgone revenue collections.

We reject this argument and urge a legislated floor on precision. First, from a purely technical perspective, our proposed assessment rule builds from the assumption that the probability density function depicted in figure 1 follows the normal distribution. This assumption, which is based on the central limit theorem, may be violated if sample sizes are too small. For any minimum reliability standard we can imagine being enacted, the sample size requirement would be sufficient to provide good insurance against failure of the normality assumption. Second, the loss function depicted in figure 2 assumes that the relative costs of over- and underassessments are characterized by linear functional forms. Aside from simplifying analysis, we can provide no rationale for choosing the linear functional form over, say, a quadratic form. By requiring a minimum level of precision, we may keep within tolerable limits any perceived inequities in the magnitude of the uncertainty adjustment resulting from competing arguments on the appropriate functional form of the loss function. Moving beyond technical observations, there is a fundamental argument. Without a minimum standard for reliability, the raison d'être for sampling—namely, that sampling provides a scientifically proven method for making a valid and reliable estimate of the true amount owed—may be jeopardized. If audit assessments were to build routinely from unreliable estimates of the tax deficiency, the perceived legitimacy of the auditing process would be at risk. Notwithstanding the adjustment for uncertainty, taxpayers might come to perceive the auditing process as a game of chance rather than as a legitimate, albeit unpleasant, exercise of the state's authority to enforce the tax laws. Stated differently, the adjustment for uncertainty can increase the perceived fairness of an audit assessment based on sampling only if the sampling itself is perceived as being thorough. We suggest that one dimension of "thoroughness" is a sample size sufficient to guarantee a minimum level of precision.

The considerations above lead us to believe that the minimum standard of precision should be tight, but not so tight as to be unaffordable. Some calculations should be done with typical cases and costs to determine appropriate sample sizes, from which may emerge an appropriate minimum standard. We intend to pursue this issue in a separate work.

Even with a minimum standard of precision and an adjustment for uncertainty, some taxpayers might still be reluctant about the use of sampling and desire either that a larger sample be drawn or that all transactions be audited. In such circumstances we suggest that the taxpayer have the option of requesting further auditing, with two important provisos: (1) the taxpayer will bear the expense of the additional work, and (2) the statute of limitations on the periods under review would toll. The tax authorities would not be bound to honor the request if the level of effort required for the additional work would be prohibitive or would unduly disrupt the completion of other audit assignments, if the request is only a delaying tactic, or if there is a reasonable basis for concluding that the expense for the additional work would not or could not be paid.

Finally, we ask whether the legislature should limit sampling to taxpayers of some minimum size (assuming records are complete). We believe that the answer is yes. Smaller taxpayers are less likely to have experience with sampling in other business contexts (and therefore to trust the method), to be able to evaluate the quality of the sampling procedures, or to have the resources to challenge an audit assessment if the adequacy of the sampling procedures is in question. Regardless of the soundness of the sampling procedures, such taxpayers might perceive the assessment as arbitrary or oppressive. Put simply, the burden that sampling imposes on the small business is probably too high, since with a little more effort, a 100% audit is possible.

5. CONCLUSIONS

Statistical sampling is a well-established technological tool of the modern world. Tax administration can be made more efficient, fairer, and less intrusive if the tax administrator's technologies for identifying and measuring tax deficiencies are expanded to include controlled use of statistical sampling. We have attempted to identify the major elements of a statutory prescription for "controlled" uses: (1) careful adherence to the requirements of statistical sampling—namely, having well-defined sampling frames, random sampling mechanisms, and sampling plans and using appropriate analyses; (2) an assessment rule that explicitly accounts for the uncertainty inherent in statistical sampling; (3) establishment of minimum standards for acceptable reliability; and (4) a policy that statistical sampling will not be used in audits of small taxpayers.

APPENDIX A

This appendix uses calculus to derive the form of the optimal assessment. Let u be the unknown correct assessment, which by the argument of section 4 is taken to have a normal distribution with mean $\bar{x}$ and standard deviation $\hat{s} = s/n^{1/2}$. If a is the assessment imposed, the loss function is

$$L(a, u) = \begin{bmatrix} u - a \text{ if } u > a & \text{(underassessment)} \\ k(a-u) \text{ if } a > u & \text{(overassessment)} \end{bmatrix}.$$

Then a is to be chosen to minimize the expected value of L, where the expectation is taken over the unknown value of u, as follows.

The expected loss is

$$L^* = \int_{-\infty}^{a} k(a-u)f(u)du + \int_{a}^{\infty}(u-a)f(u)du,$$

where

$$f(u) = \frac{1}{\sqrt{2\pi}\hat{s}}e^{-\frac{1}{2}\left(\frac{x-\bar{x}}{\hat{s}}\right)^2}$$

is the normal density function.

Now taking the derivative of L^* with respect to a, we obtain

$$\frac{dL^*}{da} = k(a-u)f(u)|_{u=a} + \int_{-\infty}^{a} kf(u)du - (u-a)f(u)|_{u=a} + \int_{a}^{\infty}-f(u)du. \tag{1}$$

This equation can be simplified using the fact that $k(a-u)f(u)|_{u=a}$ and $(u-a)f(u)|_{u=a}$ are both zero. Let

$$F_u(x) = \int_{-\infty}^{x} f(u)du.$$

Then

$$\frac{dL^*}{da} = kF_u(a) - (1-F_u(a)).$$

$$F_u(a) = \frac{1}{k+1} \tag{2}$$

is the only solution, and hence the value of a that minimizes the expected loss satisfies

$$a = F_u^{-1}\left(\frac{1}{k+1}\right). \tag{3}^{42}$$

Note that $P = F_u(a)$ is the probability of overassessment. Then (2) can be rewritten as

$$P = \frac{1}{k+1}, \quad \text{or } kP = 1-P,$$

42. See also M.DeGroot, Optimal Statistical Decisions 261 (New York McGraw-Hill Book Co., 1970).

as discussed in section 4.

It is a property of the normal distribution that u can be written

$$u = \bar{x} + \epsilon\hat{s},$$

where ϵ has a standard normal distribution with mean 0 and standard deviation 1. Then

$$F_u(x) = F_\epsilon\left(\frac{x - \bar{x}}{\hat{s}}\right) \quad \text{for all } x.$$

Consequently,

$$\frac{1}{k+1} = F_u(a) = F_\epsilon\left(\frac{a - \bar{x}}{\hat{s}}\right).$$

Thus,

$$a = \bar{x} + \hat{s}F_\epsilon^{-1}\left(\frac{1}{k+1}\right), \quad \text{or,} \quad a = \bar{x} - \rho\hat{s}, \quad \text{where } \rho = -F_\epsilon^{-1}\left(\frac{1}{k+1}\right).$$

The function $F_\epsilon^{-1}(\cdot)$ is called the normal ogive, and is tabulated in many books, among them *Probability and Statistics*.[43]

43. M. H. DeGroot, Probability and Statistics 577 (Reading, MA: Addison-Wesley Publishing Co., 1975).

D. VOTE TAMPERING

In a democracy, the integrity of the voting system is crucial to the maintenance of public confidence in the outcome. While each system of voting has vulnerabilities, paper ballots are particularly vulnerable. In this case, it was alleged that certain ballot boxes were tampered with between the count on election night and a later recount.

The first analyses done on the data were exploratory and informal, but indicated that vote tampering was likely and showed which precincts were likely to have been the subjects of the tampering. The courts, basically using evidence specific to the ballots, found sufficient tampering to reverse the outcome of the election, in the same precincts. Finally, the paper shows an elaborate model of all the races decided in that election, again showing a high probability of vote tampering, again in the suspect precincts. Thus all three methods, although quite different, come essentially to the same conclusion.

1

VOTE TAMPERING IN A DISTRICT JUSTICE ELECTION IN BEAVER COUNTY, PA

Ilaria DiMatteo and Joseph B. Kadane

1. INTRODUCTION

On November 2, 1993, in a general election, Joseph Zupsic apparently defeated Delores Laughlin for the office of District Justice in Beaver County, Pennsylvania, by a vote of 3,783 to 3,747, a 36-vote margin. Laughlin requested a recount, which on January 5, 1994, showed her the victor by 3,793 to 3,747. This article addresses whether the data support the conclusion that sometime in the intervening period, the ballots (which were paper) were tampered with, to Laughlin's benefit.

A District Justice is the only judge that most citizens ever appear before. This court handles civil cases with a value less than $7,000 and screens out inappropriate arrests. Elections for this office are often hotly contested, especially when a vacancy occurs.

The ballot boxes were locked and sealed with special, numbered seals. The boxes were stored in a locked room. However, every ballot box in the county could be unlocked with the same key, copies of which were widely distributed, and there were several keys to the locked storage room. The Board of Elections failed to record the numbers on the seals until after the recount. Finally, a county detective testified that he had obtained seals from the manufacturer by claiming that he needed them for an art project. Thus it is plausible that access to the ballot boxes might have been gained between the election night and January 5, 1994, when the recount was complete. The evidence about whether this occurred then rests on a physical examination of the ballots, on which the courts relied, and on a comparison of the two vote counts, that of November 2 and of January 5, which were conducted by vote-tabulating machines.

A vote is recorded for a candidate if either a box is marked for all candidates of the candidate's party, or if a box is marked specifically for that candidate. For example, if a ballot were marked for the candidates of the Democratic Party, then Zupsic, the Democratic Party nominee, would receive a vote, unless the ballot were marked specifically for Laughlin, in which case she would get the vote. If neither party's box were marked, and neither candidate's box were marked, then an "undervote" would occur, and neither candidate would get a vote. If neither party's box were marked and both candidates' boxes were marked, then an "overvote" would occur, and again neither candidate would get a vote. The tabulation of votes records the number of votes for each candidate, and the numbers of undervotes and overvotes, but not the number of votes gained by each candidate by party designation.

How would an additional ballot mark made by an intruder for a candidate, say candidate A, affect the vote totals? This would depend on how the ballot had been marked by the voter. If the ballot had been marked for candidate A's party, then no change would result. If it had been marked for candidate B's party, then candidate B would lose a vote and candidate A would gain one, a shift of two. If the ballot had been marked for candidate B, then candidate B would lose a vote, and the overvote would increase by one. Finally, if the ballot had been unmarked, then candidate A would gain a vote, and the undervote would decrease by one.

For these reasons, it is reasonable to use as an index of precinct vulnerability the increase in overvote plus the decrease in undervote between the two ballot countings. High values of this index indicate precincts where ballot boxes may have been tampered with. Under the hypothesis that no tampering occurred, we would expect the shift in votes to be centered at 0, not favoring either candidate particularly. A shift favoring one of the candidates only, particularly in highly vulnerable precincts, would, under this argument, favor the hypothesis of vote tampering. Table 1 gives the results for all precincts.

Table 1 suggests that the precincts 56, 114, 115, 138, and 149 are most vulnerable. Examining the last column in Table 1, which indicates the vote shift by precinct between the two counts, shows a very large shift favorable to Laughlin occurring in exactly the suspect precincts. This supports the hypothesis of vote tampering.

The total vote counts were not exactly the same in each precinct in the original count and in the recount. As noted earlier, the total count, as recorded in the court decisions, increased from $3{,}783 + 3{,}747 = 7{,}530$ to $3{,}793 + 3{,}747 = 7{,}540$. This explains why there are precincts with no change in overvote or undervote, yet the shift in vote is not divisible by two. We thank an attentive referee for pointing this out.

Moreover, it is necessary to address the question of whether index numbers and vote shifts of the size reported in Table 1 are unusual. The correlation between the index and the vote shift is .844. We drew 10,000 random permutations of the index; in no case was the absolute correlation with the vote shift greater than .844. Fortunately, other data at hand allow us to examine this issue more precisely. Other races were to be decided in this election in the five suspect precincts, and all of these were counted on November 2 and then recounted on January 5. Concentrating only on the single-office races, 202 other vote counts involving the same precincts were decided at the same time. Among these, the largest index was 3, which occurred once. Thus we conclude that indices of the sizes reported in Table 1 for

TABLE I. Changes in Vote Count by Precinct, November 2, 1993–January 5, 1994

Precincts Voting for District Justice	Increase in Overvote	Decrease in Undervote	Index	Votes Shift Z to L
49	−1	1		2
50		2	2	−3
51		1	1	−2
52				−1
53		−3	−3	2
54				1
55				−1
56	5	8	13	25
82				
84				
93				
114	3	5	8	20
115	3	5	8	24
116		1	1	4
117				−3
118				2
119				−2
138	5	4	9	10
140		3	3	−1
141				
149		7	7	4

NOTE: Blanks indicate zeros.

the suspect precincts are highly unusual. We also calculated the vote shifts between November 2 and January 5 for these 202 precinct races. The largest shifts were two of size 6, followed by two of size 4. Hence, of the vote shifts among the suspect precincts reported in Table 1, only the shift for precinct 149 is within range of the 202 others.

Thus the evidence suggests vote tampering, but the foregoing informal analysis does not indicate the strength of the evidence. To measure this is our task. The article is organized as follows. Section 2 gives the history of the litigation surrounding this matter, Section 3 describes our model, and Section 4 gives the results.

2. LEGAL HISTORY

During the recount process between November 2 and January 5, a total of 87 ballots were challenged, 69 by Zupsic and 18 by Laughlin. On January 10, 1994, Zupsic filed a lawsuit seeking to overturn the Recount Board's awarding of the election to Laughlin.

After three hearings, a three-judge panel of the Court of Common Pleas in Beaver County issued an opinion on April 8, 1994, finding that vote tampering had occurred, that "in all probability, a sufficient number of ballots were altered...so as to change the outcome of the election," and that striking the altered ballots would

TABLE 2. Common Pleas Court Findings of Vote Tampering, by District, from the July 12, 1996, Decision

District number	Voter testimony	Agreed by parties	Court finding	Total
49			1	1
56	1	4	5	10
114			6	6
115	1	4	9	14
116			2	2
138	3	2	1	6
149			6	6
Total	5	10	30	45

unfairly disenfranchise voters. On this basis it set aside the election and ordered a new one.

In a further elaboration of its ruling issued on May 5, 1994, the Court of Common Pleas panel noted that there were five voters who, because of idiosyncratic write-in votes, were able to identify their ballots specifically and testified that their ballots had been altered to include a mark for Laughlin. A total of 45 of the 87 contested ballots had marks for Laughlin inconsistent with the other marks on those ballots.

This decision was appealed by both candidates. The appeal was heard by the Supreme Court of Pennsylvania on September 19, 1994; the opinion was announced on January 22, 1996. The Supreme Court (543 Pa. 216; 670 A.2d 629 (Pa. 1996)) found the Court of Common Pleas's decision was insufficiently precise in that it did not specify which ballots that it found to have been altered and why it thought so. It also ruled that if the Court of Common Pleas found that ballots had been altered, it should award the votes in question to the candidate for whom the voter had intended to vote. With these instructions, the Supreme Court reversed the decision of the Court of Common Pleas to set aside the election, and remanded the case back to the Court of Common Pleas.

On July 12, 1996, the same three-judge panel of the Court of Common Pleas issued an opinion in which it specified 45 of the 87 challenged ballots as having been tampered with: 5 from direct testimony referred to earlier, 10 by agreement between the attorneys for Zupsic and Laughlin, and 30 more by similarity to the previous 15. The distribution of these challenged ballots by district is given in Table 2.

As a result of these changes, Zupsic got 45 more votes and Laughlin 21 fewer votes. From other challenged votes, Laughlin got 5 more votes and Zupsic 1 more vote. The final result was 3,786 for Zupsic and 3,778 for Laughlin. On this basis, the Court declared Zupsic the winner.

Laughlin appealed this decision to the Commonwealth Court of Pennsylvania, an intermediate appellate court. A three-judge panel heard this case and upheld the new decision of the Court of Common Pleas on June 4, 1997 (695 A.2d 476). Finally, the Supreme Court declined to rehear the case on October 14, 1998. On November 17, Zupsic was sworn in as District Justice. He was re-elected, unopposed, in November 1999.

3. MODEL

To assess whether the ballots were altered between the two counts in the race for District Justice, we can examine the conditional classification probabilities of each vote in the second count given its first count classification. If there was no "human intervention" between the two counts, with few exceptions each vote should be classified the same on the second count as on the first. Because the exceptions would occur from machine errors in counting, it is reasonable to believe that when there is a change in a vote's classification, that vote is equally likely to fall in any one of the other categories. On the other hand, if vote tampering occurred, then a clear structure in the transition probabilities, as specified in Section 3.3, would be expected.

To estimate the normal counting error, represented by the transition probabilities, we use the data from the other races. We assume that no vote tampering has occurred in those races and thus they provide information on the likelihood of each vote being classified after the second count in the same category as in the first count, absent vote tampering.

The next three sections present the Bayesian hierarchical model used to analyze the data. Section 3.2 specifies the structure of the data; as stated earlier, the interest is in estimating the transition probabilities for a single ballot, but the observed data represent the aggregate behavior of the ballots (i.e., we observe only the total number of ballots in a voting precinct classified in each category in the two counts). We thus augment the data (Tanner and Wong 1987) by introducing a variable representing single-ballot behavior. Section 3.3 presents the details of the hierarchical model.

3.1 Data Structuring

Consider a generic race r with m candidates and a precinct g. The data are tabulated as follows: for each category i (namely, candidate $1,\ldots$, candidate m, undervote, overvote, and scattered vote), the total number of votes is recorded after the first count, $y_{i.}^{rg}$, and after the second count, $y_{.i}^{rg}$. Table 3 shows the data format for race r and precinct g, where as before we classify as undervote the ballots with no mark for either candidate or candidate's party, as overvote the ballots with more than one

TABLE 3. Data for a Race, r, with m Candidates and Precinct, g

States	First count	Second count
Candidate 1	$y_{1.}^{rg}$	$y_{.1}^{rg}$
$\vdots$	$\vdots$	$\vdots$
Candidate m	$y_{m.}^{rg}$	$y_{.m}^{rg}$
Undervote	$y_{m+1.}^{rg}$	$y_{.m+1}^{rg}$
Overvote	$y_{m+2.}^{rg}$	$y_{.m+2}^{rg}$
Scattered Vote	$y_{m+3.}^{rg}$	$y_{.m+3}^{rg}$

mark, and as scattered the ballots with a preference different from the candidates listed on the ballot.

We can imagine the data in Table 3 in the following way: The totals resulting from first count and the second count are the row and column marginals of a $k \times k$ $(k = m + 3)$ contingency table whose elements z_{ij}^{rg} represent the number of ballots that were classified in category i in the first count and in category j in the second count. Table 4 shows the matrix representation of the data for a race r and precinct g.

The data are therefore augmented by introducing the interior elements Z^{rg} of the contingency table for each race r and precinct g in order to estimate the transition probabilities of a single ballot.

3.2 Model Specification

In this section we present the four stages in the model. First, we introduce some notation used throughout this section. We let $y_1^{rg} = (y_{1 \cdot}^{rg}, ..., y_{k \cdot}^{rg})^T$ and $y_2^{rg} = (y_{\cdot 1}^{rg}, ..., y_{\cdot k}^{rg})^T$ denote the vector of counts in the first count and second count in race r with m candidates and precinct g. (Note that $k = m + 3$ and that in general the number of candidates depends on the race, but for simplicity of notation we write k instead of k^r.)

Also let $Y_1 = \{y_1^{rg}, \forall g, r\}, Y_2 = \{y_2^{rg}, \forall g, r\}$, and $Z = \{Z^{rg}, \forall r, g\}$ be the collection of the results of the first count, the second count, and the collection of the augmented data Z^{rg}.

Likelihood. We specify in this stage the conditional distribution of the results of the second count given the results of the first count and the augmented data. It is obvious that

$$\Pr(y_2^{rg} \mid y_1^{rg}, Z^{rg}) = I_{\left\{JZ^{rg} = y_1^{rg}, \, JZ^{rg^T} = y_2^{rg}\right\}}(y_2^{rg}) \qquad \forall g, r,$$

where $J \in \mathbb{R}^k$ is a vector containing all 1's, and $I_{\{A\}}(x) = 1$ if $x \in A$ and 0 otherwise. The foregoing model simply states that the conditional probability of the column marginal given the row marginal and the interior elements of the table is 1 if the column marginals satisfy the constraints and 0 otherwise.

TABLE 4. Data for a Race, r, with m Candidates and Precinct, g

				First count
				$y_{1 \cdot}^{rg}$
		{not observed}		$\vdots$
		$Z^{rg} = \{Z_{ij}^{rg}\}$	$\cdots$	$y_{i \cdot}^{rg}$
		$\vdots$		$\vdots$
				$y_{k \cdot}^{rg}$
Second count	$y_{\cdot 1}^{rg}$	$\cdots$	$y_{\cdot j}^{rg}$	$\cdots$

Furthermore, we assume that given the latent variable and the results of the first count, the results of the second count are independent across races and precincts,

$$L(Z \mid Y_2, Y_1) = \prod_{rg} I_{\left\{JZ^{rg} = y_1^{rg}, JZ^{rg^T} = y_2^{rg}\right\}}\left(y_2^{rg}\right).$$

In particular, let r' denote the race for District Justice, then write the likelihood as

$$L(Z \mid Y_2, Y_1) = \prod_{g} I_{\left\{JZ^{r'g} = y_1^{r'g}, JZ^{r'g^T} = y_2^{r'g}\right\}}\left(y_2^{r'g}\right) \times \prod_{r \ne r', g} I_{\left\{JZ^{rg} = y_1^{rg}, JZ^{rg^T} = y_2^{rg}\right\}}\left(y_2^{rg}\right). \tag{1}$$

Latent Variables. In this stage we specify a probability model for the latent variables. We assume that for each race r and precinct g, there is a transition probability matrix Q^{rg} that characterizes the shift of each ballot from one category to another between the two counts. Each element q_{ij}^{rg} of this transition probability matrix represents the conditional probability of a ballot classified in category i in the first count shifting to category j in the second count. Obviously it must be that for every $i, \sum_j q_{ij}^{rg} = 1$ for all r and g.

For a race r and a precinct g, given the transition matrix Q^{rg} and the row margins, we assume that the interior elements follow a product of multinomial sampling scheme,

$$\Pr\left(Z^{rg} \mid y_1^{rg}, Q^{rg}\right) = \prod_{i=1}^{k} \frac{y_{i\cdot}^{rg}!}{\prod_{i=j}^{k} z_{ij}^{rg}!} \prod_{j=1}^{k} q_{ij}^{Z_{ij}^{rg}} I_{\left\{JZ^{rg} = y_1^{rg}\right\}}\left(Z^{rg}\right), \tag{2}$$

where q_{ij} is the ijth element of Q^{rg}. Models of this type were discussed by Lee, Judge, and Zellner (1970).

We now describe the structure of the transition probability matrix in the cases of no-cheating, cheating in favor of candidate 1, and cheating in favor of candidate 2. Consider first the case of no vote tampering in race r and precinct g. We assume that $Q^{rg} = P$, where

$$P = \begin{bmatrix} p & \dfrac{1-p}{k-1} & \cdots & \dfrac{1-p}{k-1} \\ \dfrac{1-p}{k-1} & p & \dfrac{1-p}{k-1} & \vdots \\ \vdots & \dfrac{1-p}{k-1} & \ddots & \dfrac{1-p}{k-1} \\ \dfrac{1-p}{k-1} & \cdots & \dfrac{1-p}{k-1} & p \end{bmatrix}. \tag{3}$$

Note that P depends on the race through the number of categories k (its dimensions change with k), but the parameter p characterizing P does not depend on the race or the precinct.

This form of the transition matrix P relies on the assumption that if there is no vote tampering, then the transition probabilities are concentrated on the main diagonal and are uniformly spread on the off-diagonal elements, representing the normal misclassification error. As mentioned before, we assume that no vote tampering has occurred in any of the other races (i.e., in all but the District Justice race). This assumption corresponds to setting $Q^{rg} = P$ for all races $r \neq r'$ and all precincts g. In addition, we assume that p is close to 1.

In the District Justice race (i.e., $r = r'$), we assume that the transition probability matrix $Q^{r'g}$ takes the form

$$Q^{r'g} = P + X_g C_\eta(\gamma_1, \gamma_2, \delta), \tag{4}$$

where X_g is an indicator variable that is 1 if vote tampering occurred in precinct g and 0 otherwise (in which case the transition probability matrix is just P). If $X_g = 1$, then the transition probability matrix changes according to the parameter η which denotes the candidate for whom the vote tampering is in favor. If the shift in votes is in favor of candidate Zupsic, then $\eta = 1$, in which case

$$C_1(\gamma_1,\, \gamma_2,\, \delta) =$$

	Zupsic	Laughlin	Undervote	Overvote	Scattered vote
Zupsic	0	0	0	0	0
Laughlin	$+\gamma_1$	$-(\gamma_1+\gamma_2)$	0	$+\gamma_2$	0
Undervote	$+\delta$	0	$-\delta$	0	0
Overvote	0	0	0	0	0
Scattered vote	0	0	0	0	0

If the shift in votes is in favor of candidate Laughlin, then $\eta = 2$, in which case

$$C_2(\gamma_1,\, \gamma_2,\, \delta) =$$

	Zupsic	Laughlin	Undervote	Overvote	Scattered vote
Zupsic	$-(\gamma_1 + \gamma_2)$	$+\gamma_1$	0	$+\gamma_2$	0
Laughlin	0	0	0	0	0
Undervote	0	$+\delta$	$-\delta$	0	0
Overvote	0	0	0	0	0
Scattered vote	0	0	0	0	0

These two matrices represent the expected increments in the transition probabilities when vote tampering has occurred in a particular precinct. They correspond to the ways in which ballots could be altered to increase the number of votes for one candidate and decrease the number of votes for the other candidate. Suppose, for example, that cheating is in favor of Laughlin (i.e., $\eta = 2$); then a ballot can be altered in the three ways:

- An undervote ballot becomes a vote for Laughlin by simply marking that blank ballot in favor of Laughlin. This leads to an increase of the number of votes for this candidate and therefore an increase of the probability of going from undervote to Laughlin. This increment in the transition probability is represented by the parameter δ.

- A ballot in favor of Zupsic becomes an overvote by adding marks on that ballot. This leads to an increase of the transition probability of a ballot to shift from Zupsic to overvote, which is represented by the parameter γ_2. This also leads to a decrease in the number of votes for Zupsic.
- A party vote for Zupsic becomes a vote for Laughlin by marking that ballot in favor of Laughlin. This shift increases the corresponding transition probability by an amount γ_1. The consequence of this shift is that the number of votes for Laughlin increases and the number of votes for Zupsic decreases.

Thus the parameters γ_1, γ_2, and δ represent the possible increases in transition probabilities from the case of vote tampering. These parameters are constrained as follows to guarantee that the elements q_{ij} are nonnegative and less than 1:

$$\gamma_1, \gamma_2 \in \Gamma = \left\{ (\gamma_1, \gamma_2) \in \mathbb{R}^2 : \gamma_1 \geq 0, \gamma_2 \geq 0, \gamma_1 + \gamma_2 \leq p \right\}$$

and

$$\delta \in [0, p].$$

Modeling the "Cheating" Parameters. In this stage we model the parameters that indicate whether cheating has occurred and, if so, its strength. We start modeling the variables X_g, which indicate for each precinct g whether cheating has occurred in that precinct. Recall that the ballots from a precinct were stored together in a box, and the boxes for the 21 precincts were gathered together in the same room. To alter the result of the election, someone would have had to enter the room, pick a box of ballots, open it, and examine the ballots to alter them. We assume that each box, and thus each precinct, has the same probability θ of being opened and altered,

$$\theta = \Pr(X_g = 1), \quad \forall_g,$$

and that the indicator variables $(X_1, \ldots, X_{21})^T$ are exchangeable,

$$P\left(\sum_g X_g = x \mid \theta \right) = \binom{21}{x} \theta^x (1-\theta)^{21-x} \quad x = 0, \ldots, 21.$$

If vote tampering has occurred, then its direction is given by the parameter η. If $\eta = 1$, then cheating is in favor of Zupsic; if $\eta = 2$, then cheating is in favor of Laughlin. By assuming that η does not depend on the precinct g, we force the precincts to have the same direction of cheating. We assume a priori that η is a Bernoulli random variable with probability τ,

$$\pi_\eta(\eta) = \begin{cases} \tau & \text{if } \eta = 1 \\ 1 - \tau & \text{if } \eta = 2, \end{cases}$$

where τ is known (we have used $\tau = .5$ in estimation).

The parameters γ_1, γ_2, and δ in (4) give a measure of the strength of cheating; they measure how much the transition probabilities in the case of vote tampering differ from the "normal" transition probabilities. They are assumed to be the same for all of the precincts in which cheating has occurred. A priori we assume uniform distributions over the space Γ for γ_1 and γ_2 and a uniform distribution in $[0, p]$ for δ. We also assume a beta(θ_1, θ_2) prior for θ.

Modeling the "Normal" Transition Probability. In the last stage of the hierarchical model we specify a distribution on the transition probability p in (3). We note that p represents the probability that each ballot is classified in the same category in the two counts; if there were no errors in the counting mechanism, we would expect p to be exactly 1. In the real world there are errors, but we still expect this probability to be close to 1 with a very small variance. These considerations help us to elicit a prior for p, which in general is a beta distribution,

$$\pi_p(p) = \text{beta}(\alpha, \beta)$$

with α and β known.

3.3 Estimation with Markov Chain Monte Carlo

The stages of the hierarchical model are summarized as follows:

$$
\begin{cases}
\Pr(Y_2 \mid Z, Y_1) = \prod_{rg} I_{\left\{ JZ^{rg} = y_1^{rg}, JZ^{rg^T} = y_2^{rg} \right\}} \left(y_2^{rg} \right) \\
\Pr(Z \mid Y_1, X_1, ..., X_{21}, \eta, \gamma_1, \gamma_2, \delta, p) \\
\quad = \prod_{r \neq r'} \prod_g \Pr(Z^{rg} \mid y_1^{rg}, p) \\
\quad\quad \times \prod_g \Pr(Z^{r'g} \mid y_1^{r'g}, X_1, ..., X_{21}, \eta, \gamma_1, \gamma_2, \delta, p) \\
\Pr\left(\sum_g X_g = x \mid \theta \right) = \binom{21}{x} \theta^x (1-\theta)^{21-x} \\
\pi(\gamma_1, \gamma_2, \delta, \eta \mid p) = \pi_{\gamma_1, \gamma_2}(\gamma_1, \gamma_2 \mid p) \pi_\delta(\delta \mid p) \pi_\eta(\eta) \\
\pi_p(p) \sim \text{beta}(\alpha, \beta) \\
\pi(\theta) \sim \text{beta}(\theta_1, \theta_2).
\end{cases}
\tag{5}
$$

Simulations from the posterior distributions of the parameters in (5) are obtained using the Metropolis–Hastings algorithm within Gibbs sampling. Although sampling from the conditional posteriors of the parameters p, γ_1, γ_2, δ, and η is straightforward, updating the latent variables Z^{rg} and the indicator variables of precinct subject to tampering, X_g, requires some explanation, as we describe in the Appendix.

After simulating the joint posterior distribution of the parameters, we can estimate the posterior probability of vote tampering. The event "no vote tampering" in the election is formally represented in our model by $\left(x = \sum_{g=1}^{21} X_g = 0 \right)$. In computing the posterior probability of "no vote tampering," we use the marginal posterior distribution of x.

4. RESULTS

In this section we report the MCMC estimate of the parameters in (5), which are based on 50,000 iterations and a burn-in of 10,000 iterations. In each run, plots of the results suggest good mixing. To check convergence we used standard convergence diagnostics from the Bayesian output analysis (BOA) program. These tests include the Geweke convergence test, the Raftery–Lewis test, and the Heidelberg–Welch stationarity and interval halfwidth tests. (More information on BOA is available at http://www.public-health.uiowa.edu/boa/.)

We assumed a beta(50, 3) distribution for the transition probability p, which corresponds to assuming a priori that p has mean .94 and variance .001. Thus we are assuming that around 94% of the votes are counted the second time in the same way that they were the first time, which seems to us a minimal assumption in this context. It turns out (see Fig. 1 below) that the data indicate a higher rate than 94%. Sensitivity analyses to the different prior distributions for p suggest that the estimates of the parameters remain unchanged. Figure 1 depicts the posterior distribution of the parameters in the model.

Table 5 gives the posterior means of the parameters together with the 95% credible intervals. The estimates of the parameters in Table 5 are insensitive also to the choices of the prior distribution of θ; we obtain the same estimates of the parameters when we assume very different priors for θ and thus for the number of cheating precincts, $\sum_g X_g$. Figure 2 displays the prior and posterior distributions on $\sum_g X_g$ corresponding to a $\beta(2, 6)$ for θ in the (a) row, and corresponding to a beta(2, 20) in the (b) row. Note that the posterior distributions, shown in (a2) and (b2), are very similar, although the prior distributions, in (a1) and (b1), are not.

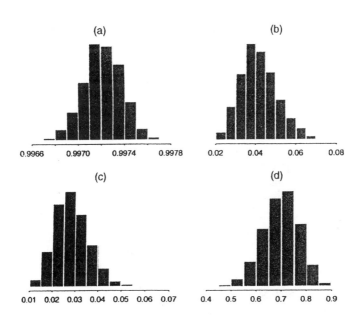

FIGURE 1. Posterior Distribution of (a) p, (b) γ_1, (c) γ_2, and (d) δ.

TABLE 5. MCMC Results for the District Justice Race
Assuming Priors $p \sim$ beta(50,3) and $\theta \sim$ beta(2,6)

	Posterior median	95% credible interval
p	.997	(.997, .998)
γ_1	.041	(.025, .062)
γ_2	.028	(.015, .045)
δ	.697	(.547, .824)
η	2.000	
θ	.243	(.111, .423)

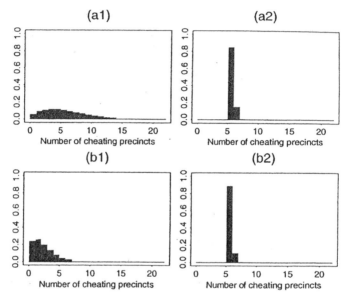

FIGURE 2. Prior (1) and Posterior (2) Distributions on the Number of Precincts in Which Cheating Occurred for (a) a $\theta \sim \beta(2,6)$ Prior and (b) a $\theta \sim \beta(2,20)$ Prior.

We can summarize the findings as follows: Vote tampering has occurred in the District Justice race with probability 1. Furthermore, because $\Pr(\eta = 2 \mid data) = 1$, we can also assess with probability 1 that vote tampering occurred in favor of candidate 2, namely Laughlin, as shown in Table 5.

There is a significant increment of the probability of reclassifying as an overvote a ballot initially classified in favor of Zupsic; the estimate of the parameter γ_2 representing this increment in probability is .029 and has a 95% credible interval (.015, .045). The transition probability γ_1 of a ballot to shift from Zupsic to Laughlin also increases significantly, by .041 (.025, .062). Finally there is a significant increment of the transition probability of a ballot to shift from undervote to Laughlin; the estimate of δ is .697 with 95% credible interval (.547, .824).

These results should be considered jointly. If tampering were associated with very small values of γ_1, γ_2, and δ, then one might find a high probability of tampering, but to such a low degree that it would be in name only. That in the districts

TABLE 6. Comparison of the Court Findings of Vote Tampering and Our
Estimates of the Probability of Vote Tampering by District

District number	Number of altered ballots found by the Court	Posterior probability of cheating	Expected number of altered ballots
49	1	0.00	0
56	10	1.00	22
114	6	1.00	11
115	14	1.00	16
116	2	.07	1
138	6	1.00	9
149	6	1.00	7
Total	45		66

found to have been tampered with, roughly 3% of Zupsic's votes were transformed
into overvotes and 4% of Zupsic's votes were transformed into Laughlin's votes is
suspicious, but that 70% of the undervotes become Laughlin's votes is an extremely
large shift. We consider this strong evidence of real vote tampering.

To determine the precincts in which vote tampering occurred, we look at the
posterior probability $\Pr(X_g = 1)$ for each precinct $g = 1, \ldots, 21$. Table 6 presents
comparisons between the precincts in which the Court found physical evidence of
altered ballots and the precincts in which our model detected alterations in transition probabilities. The Court's findings were based almost entirely on the physical
evidence presented by challenged ballots, whereas our model uses the comparison
between the counts of November 2 and January 5.

Note that our estimates of tampered precincts agree with the Court's decision:
Our model found with probability 1 all of the precincts in which the Court found
at least six altered ballots. Precinct 116 has a very low posterior probability of
vote tampering, and this agrees with the fact that the Court found only two altered
ballots—insufficient evidence considering that the total number of ballots in that
precinct is 305. Similarly in District 49 the posterior probability of vote tampering
is zero, but the Court found one altered ballot out of 331 total ballots.

The posterior probability of vote tampering for all the other precincts is zero
except for districts 82 and 93, which have probability .08 and .09. A possible explanation for this might rely on the small number of ballots in these districts: 69 and
52, compared to more than 300 in most of the other districts.

4.1 Model Validation

In this section we investigate how our model performs on election data in which
no vote tampering is suspected. As the race that might be contested, we choose the
Beaver County vote in the (statewide) Pennsylvania Superior Court, taking the role
of the District Justice election between Zupsic and Laughlin. For this calculation,
we assume that only the Supreme Court and School District Director races had no
tampering, and do not use the data from the other races.

Although the estimates of the model are robust to the prior distributions of p,
γ_1, γ_2, and δ, they are not to the prior on θ. Note that the number of races used here

TABLE 7. Posterior Medians of the Parameters in the Superior Court Race Together
With Their 95% Credible Intervals Obtained Under Different Prior Distributions on θ

	$\theta \sim \text{beta}(2,6)$	$\theta \sim \text{beta}(2,20)$	$\theta \sim \text{beta}(2,50)$
γ_1	.0065 (.0017, .0164)	.0074 (.0010, .0242)	.0083 (.0008, .0723)
γ_2	.0007 (1.6e−05, .0040)	.0010 (2.7e−05, .0157)	.0032 (7.4e−05, .0575)
δ	.0037 (.0001, .0268)	.0062 (.0002, .1946)	.0487 (.0007, .4192)
p	.9967 (.9961, .9972)	.9967 (.9961, .9972)	.9967 (.9962, .9972)
Pr(no cheating)	.01	.32	.81

as comparison data is much smaller than that in the analysis of the District Justice
race. We used three prior distributions on θ, representing very different beliefs on
the number of cheating precincts.

Table 7 shows the effects of the different prior distributions on the estimates of
the parameters. Assuming a $\beta(2, 6)$ distribution for θ, which corresponds to a prior
knowledge on $\Sigma_g X_g$ represented in Figure 2, the posterior probability of no vote
tampering in the Superior Court race is .01. However, the estimates of the cheating parameters are negligible. Thus this prior leads to high probability of a trivial
amount of tampering. On the other hand, assuming a beta(2, 50) for θ distribution
leads to a posterior probability of no vote tampering in the race for Superior Court
of .81 and higher cheating parameters. So in this case, the results suggest a low
probability of substantial tampering.

Figure 3 gives the posterior distributions for the number of altered ballots in
favor of candidate 2 for both Superior Court and District Justice race under different priors for θ. The histograms relative to the Superior Court clearly show that in
all three cases the number of altered ballots is small in magnitude and its posterior
distribution is not affected by the prior choice. In the District Justice race the number of altered ballots in favor of candidate 2 is much higher.

5. DISCUSSION

We find that three different analyses of the District Justice race lead to the same
conclusions. The informal data analysis of Section 1, the Court's findings based on
an examination of particular ballots, and the model of this article all find that vote
tampering occurred in favor of Laughlin and against Zupsic. Furthermore, these
analyses generally agree on the districts involved: 56, 114, 115, 138, and 149. These
results are reasonably insensitive to the priors used.

The corroborating model check used the Supreme Court race. Depending on the
prior used, this analysis found a low probability of a substantial amount of vote
tampering or a high probability of a trivial amount of tampering. Thus we find that
the model strongly distinguishes these datasets.

The model and mode of analysis that we have chosen is strongly influenced
by the particulars of the dataset and the problem we address. There are, however,
important statistical analyses related to other problems of determining a winner in
a close election. Finkelstein (1978) and Gilliland and Meier (1986) have considered

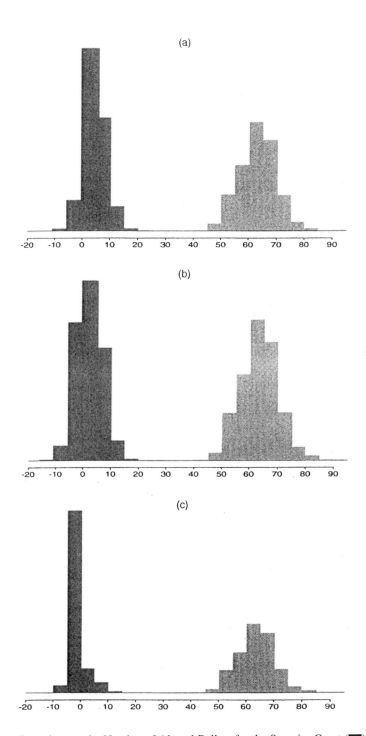

(a)

(b)

(c)

FIGURE 3. Posteriors on the Number of Altered Ballots for the Superior Court (■) and the District Justice (▨) Races Under Different Priors for θ (a) $\theta \sim \beta(2, 6)$; (b) $\theta \sim \beta (2, 20)$; (c) $\theta \sim \beta(2, 50)$.

the probability that an election outcome would be reversed with the elimination of votes from ineligible voters, under the assumption that the ineligible voters are drawn from the same population as the eligible ones.

In a more recent case (Marks v. Stinson, U.S. District Court for the Eastern District of Pennsylvania, 1994 U.S. Dist. Lexis 5273), the Court, after finding evidence of fraud in the administration of absentee ballots, used various methods, including regression (Ashenfelter 1994), to estimate what the result of the election would have been absent the fraud. This led to a reversal of the election outcome, with Marks, not Stinson, seated in the State Senate.

APPENDIX: MARKOV CHAIN MONTE CARLO ESTIMATION

Updating Z^{rg}

Updating the augmented data Z^{rg} for each race r and district g corresponds to updating a matrix of counts with fixed row and column marginals according to a conditional posterior given by

$$\Pr\left(Z^{rg} \mid y_2^{rg}, y_1^{rg}, \ldots\right) \propto \Pr\left(Z^{rg} \mid y_1^{rg}, \ldots\right) I_{\left\{JZ^{rg} = y_1^{rg}, \, JZ^{rg^T} = y_2^{rg}\right\}}\left(y_2^{rg}\right),$$

where $\Pr\left(Z^{rg} \mid y_1^{rg}, \ldots\right)$ is a product of multinomials as specified in (2). Given the current state Z^{rg}, a candidate configuration, $Z^{rg^{(c)}}$ is generated according to the following scheme: choose randomly two rows (i_1, i_2) and two columns (j_1, j_2); given the elements in those rows and columns, choose randomly a sign for the element $Z_{i_1 j_1}^{rg}$. Then the candidate configuration $Z^{rg^{(c)}}$ has the elements in rows (i_1, i_2) and columns (j_1, j_2) altered by 1 depending on the sign. For example, if the sign is positive, then the candidate configuration is given by

$$Z^{rg^{(c)}} = \begin{bmatrix} & \vdots & & \vdots & \\ \cdots & Z_{i_1 j_1}^{rg} + 1 & \cdots & Z_{i_1 j_2}^{rg} - 1 & \cdots \\ & \vdots & & \vdots & \\ \cdots & Z_{i_2 j_1}^{rg} - 1 & \cdots & Z_{i_2 j_2}^{rg} + 1 & \cdots \\ & \vdots & & \vdots & \end{bmatrix}.$$

Clearly the margin constraints are satisfied. The proposal distribution $q(Z^{rg^{(c)}} \mid Z^{rg})$ is therefore defined as

$$q(Z^{rg^{(c)}} \mid Z^{rg}) = \Pr(i_1, i_2, j_1, j_2) \times \Pr(\text{sign} \mid i_1, i_2, j_1, j_2),$$

where $\Pr(i_1,i_2,j_1,j_2) = \Pr(i_1,i_2,)\Pr(j_1,j_2) = \binom{k}{2}^{-2}$ because it corresponds to independent samplings without replacement two elements out of k (with k the dimension of the matrix). To guarantee that the elements of the $Z^{rg^{(c)}}$ are nonnegative, we define

$$\Pr(+\,|\,i_1, i_2, j_1, j_2) = \begin{cases} 1/2 & \text{if } z_{i_1j_1} > 0,\, z_{i_2j_2} > 0,\, z_{i_2j_1} > 0,\, z_{i_1j_2} > 0 \\ 0 & \text{if } z_{i_1j_1} > 0,\, z_{i_2j_2} > 0,\, z_{i_2j_1} = 0,\, \text{or } z_{i_1j_2} = 0 \\ 1 & \text{if } z_{i_1j_1} = 0 \text{ or } z_{i_2j_2} = 0,\, z_{i_2j_1} > 0,\, z_{i_1j_2} > 0. \end{cases}$$

In all of the other cases—namely, when there are two 0's in the same row or column—we do not update Z^{rg}. Once we generate a candidate configuration $Z^{rg^{(c)}}$ the acceptance probability becomes

$$\alpha = \min\left\{1,\ \frac{\Pr(Z^{rg^{(c)}} \,|\, y_2^{rg}, y_1^{rg}, \ldots)}{\Pr(Z^{rg} \,|\, y_2^{rg}, y_1^{rg}, \ldots)} \frac{q(Z^{rg} \,|\, Z^{rg^{(c)}})}{q(Z^{rg^{(c)}} \,|\, Z^{rg})}\right\}.$$

This formula can be further simplified. Suppose, for example, that all elements in rows $(i_1\ i_2)$ and columns $(j_1\ j_2)$ are strictly positive and the sign is +; then it can be easily shown that the acceptance probability of $Z^{rg^{(c)}}$ is given by

$$\alpha = \min\left\{1,\ \frac{z_{i_1j_2}^{rg} z_{i_2j_1}^{rg}}{(z_{i_1j_1}^{rg}+1)(z_{i_2j_2}^{rg}+1)} \frac{q_{i_1j_1} q_{i_2j_2}}{q_{i_1j_2} q_{i_2j_1}}\right\},$$

where the q_{ij}'s are the transition probabilities corresponding to each cell above, as described in (4).

Updating $X_1, \ldots, X_{21}$

In updating the exchangeable indicator variables X_g, we first integrate out analytically the hyperparameter θ. Specifying the prior of $\Sigma_g X_g$ in two stages helps us to better understand the prior assumptions. After integrating out θ, we have the following prior for $\Sigma_g X_g$:

$$\pi\left(\sum_g X_g = x\right) = \binom{21}{x} \frac{B(\theta_1 + x, 21 + \theta_2 - x)}{B(\theta_1, \theta_2)},$$

where $B(\alpha, \beta)$ denotes the beta function. Because updating $\Sigma_g X_g$ obviously corresponds to a change in the single variables X_g, we update $\Sigma_g X_g$ and the X_g's jointly. Let $\Sigma_g X_g = x$ be the current number of cheating precincts, then with equal probability the candidate value, $x^{(c)}$, can take values $x - 1$, x, and $x + 1$. If $x^{(c)} = x + 1$, then one precinct, j, is chosen at random from the set $C_0 = \{g : X_g = 0\}$ and set

$X_j = 1$. If $x^{(c)} = x - 1$, then one precinct, j, is chosen at random from the set $C_1 = \{g : X_g = 1\}$ and set $X_j = 0$. Finally, if $x^{(c)} = x$, then two precincts, (i, j), are chosen at random from the set $C_1 = \{g : X_g = 1\}$ and $C_0 = \{g : X_g = 0\}$, and set $X_i = 0$ $X_j = 1$. In each of these cases the candidate vector $(X_1^{(c)}, ..., X_{21}^{(c)})$ differs from the current vector $(X_1, ..., X_{21})$ only in the element(s) i and/or j, depending on the value of $x^{(c)}$. We report the acceptance probability of a candidate vector $(X_1^{(c)}, ..., X_{21}^{(c)})$ when $x^{(c)} = x + 1$, in which case

$$\alpha = \min\left\{1, \frac{\Pr(y_2^{r'j} \mid X_1^{(c)}, ..., X_{21}^{(c)}, ...)}{\Pr(y_2^{r'j} \mid X_1, ..., X_{21}, ...)} \frac{\pi(x^{(c)})}{\pi(x)} \frac{21-x}{x+1}\right\}.$$

The acceptance probabilities in the other cases are obtained similarly.

To obtain convergence for the Superior Court race data Monte Carlo runs, we had to add an importance sampling component. As is evident from the third column of Table 7, with an informative prior on θ, there is little evidence of cheating in this race. Because the parameters γ_1, γ_2, and δ are all defined as conditional on cheating, if the chain were run using the prior of interest, then it would take a long time to assemble data on these parameters. A solution is to run the chain with a different prior that puts more probability on cheating than the prior of interest, and reweight the observations by the ratio of the prior of interest to the prior used for the runs. This permits the chain to converge, and allows the parameters to be well estimated. We used a beta(2,6) prior for θ for these runs, with 10,000 iterations of burn-in and 50,000 total iterations.

REFERENCES

Ashenfelter, O. (1994), "Report on Expected Absentee Ballots," unpublished manuscript.

Finkelstein, M. O. (1978), *Quantitative Methods in Law,* New York: The Free Press.

Gilliland, D., and Meier, P. (1986), "The Probability of Reversal in Contested Elections," in *Statistics and the Law,* eds. M. H. De Groot, S. E. Fienberg, and J. B. Kadane, New York: Wiley.

Lee, T. C., Judge, G. G., and Zellner, A. (1970), *Estimating the Parameters of the Markov Probability Model from Aggregate Time Series Data,* Amsterdam: North-Holland.

Tanner, M. A., and Wong, W. H. (1987), "The Calculation of Posterior Distribution by Data Augmentation," *Journal of the American Statistical Association,* 82, 528–540.

E. CIVIL INCARCERATION OF VIOLENT SEXUAL OFFENDERS

Current law in several states permits the civil incarceration of violent sexual offenders (rapists and pedophiles) after their criminal sentences are served, if they are found to be likely to reoffend. The issue in this section is the quality of the expert evidence offered in support of such incarceration.

The social stakes in this issue are large. On the one hand, society wishes to protect itself from persons likely to repeat violent sexual acts. On the other hand, a person who has served a sentence for a crime generally ought to be freed. Detaining a person on the grounds that they might, in the future, commit a crime is fraught with danger to a society that values civil liberty.

If it were the case that reoffense could be reliably predicted, the competing interests outlined above might be simultaneously satisfied. However, the paper in this section casts doubt on the current ability of science to make such predictions reliably. It could be that the legal process is demanding more of science than it can currently deliver.

1

EXPERT TESTIMONY SUPPORTING POST-SENTENCE CIVIL INCARCERATION OF VIOLENT SEXUAL OFFENDERS

George G. Woodworth and Joseph B. Kadane

1. CIVIL COMMITMENT OF SEXUAL PREDATORS

There is a clearly discernible public perception that both the criminal justice system and civil commitment of dangerously mentally ill persons has failed to protect the public adequately by failing to incapacitate a class of violent sexual offenders who are likely to re-offend and, lacking the capacity to control their behaviour, are not deterred by the threat of future punishment. In response, as of May 2003, 16 states had enacted some form of civil proceeding under which convicted sexual offenders may be confined following completion of their current offence (Table 1).

These statutes appear to be based on the perception that both the penal code and existing statutes providing for the civil commitment of dangerously mentally ill individuals are inadequate to deal with a uniquely dangerous category of violent offenders. For example, the preamble to the Texas statute asserts,

> The legislature[1] finds that a small but extremely dangerous group of sexually violent predators exists and that those predators have a behavioural abnormality that is not amenable to traditional mental illness treatment modalities and that makes the predators likely to engage in repeated predatory acts of sexual violence. The legislature finds that the existing involuntary commitment provisions of Subtitle C, Title 7, are inadequate to address the risk of repeated predatory behaviour that sexually violent predators pose to society. The legislature further finds that treatment modalities for sexually violent predators are different from the traditional treatment modalities for persons appropriate for involuntary commitment under Subtitle C, Title 7. Thus, the legislature finds that a civil commitment procedure for the long-term supervision and treatment of sexually violent predators is necessary and in the interest of the state.

---

1. Texas Senate Bill 365, Health and Safety Code, Title 11, Section 841.141, www. capitol. state.tx.us/statutes/docs/HS/content/htm/hs.011.00.000841.00.htm.

TABLE 1. Summary of State Sexual Predator Commitment Statutes[1]

State	Standard	Jury Trial[2]	Eligibility[3]
AZ	Reasonable doubt	Yes	Likely to engage in sexual violence
CA	Reasonable doubt	Unanimous	The person is a danger to the health and safety of others in that he or she will engage in sexually violent criminal behaviour.
FL	Clear and convincing	Unanimous	Likely to engage in acts of sexual violence.
IL	Reasonable doubt	Yes	Substantially probable that the person will engage in acts of sexual violence.
IA	Reasonable doubt	Unanimous	Likely to engage in predatory acts constituting sexually violent offences.
KS	Reasonable doubt	Unanimous	Likely to engage in predatory acts of sexual violence.
MA	Reasonable doubt	Unanimous	Likely to engage in sexual offences again.
MN	Clear and convincing	No	Likely to engage in acts of harmful sexual conduct.
MO	Reasonable doubt	Unanimous	More likely than not to engage in predatory acts of sexual violence.
NJ	Clear and convincing	No	Likely to engage in sexual violence.
ND	Clear and convincing	No	Likely to engage in further acts of sexually predatory conduct.
SC	Reasonable doubt	Unanimous	Likely to engage in acts of sexual violence.
TX	Reasonable doubt	Unanimous	Likely to engage in a predatory act of sexual violence.
VA	Reasonable doubt	Unanimous	Likely to commit sexually violent offences.
WA	Reasonable doubt	Unanimous	Likely to engage in predatory acts of sexual violence.
WI	Reasonable doubt	Unanimous	Substantially probable that the person will engage in acts of sexual violence.

[1] Current through May 1, 2003. Source: www.ndaa-apri.org/pdf/sexually_violent_predator_statutes.pdf.
[2] Yes or unanimous means that a jury trial is possible (automatic in MA).
[3] In addition to the presence of a sexual, mental, or personality disorder or abnormality.

Is this perception of a small but extremely dangerous class of offenders accurate? An 11-state Bureau of Justice Statistics study (Beck & Shipley, 1989) indicated that rapists had a 51.5% three-year re-arrest rate for any charge and a 19.0% re-arrest rate for homicide, rape, or assault. The corresponding rates for prisoners convicted of assault were 60.2% for any offence and 22.4% for homicide, rape, or assault. For prisoners convicted of any violent offence the corresponding re-arrest rates were 62.5%, and 21.8%.

The Solicitor General of Canada's pre-publication research summary (Solicitor General of Canada, 1997) of Hanson & Bussière's Meta-Analysis (1998) reported

The research literature on sexual offender recidivism was thoroughly reviewed.... The review included data from a total of 28,972 sexual offenders, although fewer were available for any particular analysis. On average, the sexual offense recidivism rate was low. Given the average 4–5 year follow-up period, 13.4% recidivated with a sexual offense. The recidivism rate for nonsexual violence was 12.2%, and for any recidivism, 36.3%. Rapists were slightly more likely to recidivate sexually (19%) than were child molesters (13%). Among child molesters, however, the rate of sexual offense recidivism was much lower for the incest offenders (4%) than for the boy-victim pedophiles (21%).

For most of the sources in this meta-analysis, recidivation was defined as re-conviction, which may account for the comparatively lower estimates of recidivation rates compared to the BJS study, which used re-arrest as the end point.

A study of 191 child molesters summarized by the Solicitor General of Canada (1996)

> ...found that 42% were reconvicted of a sexual or violent crime during the 15–30 year follow-up period. Ten percent of the total sample of child molesters were first convicted for a sexual/violent crime between 10 and 31 years after release. Not all child molesters recidivated at the same rate. The highest rate of recidivism (77%) was for those with previous sexual offenses, who selected extrafamilial boy victims, and who were never married. In contrast, the long-term recidivism rate for the low risk offenders was less than 20%.

Although the statistical data are mixed, it is clear that there are subclasses of violent sexual offenders who recidivate at comparatively high rates. While this is not unique to sex offenders it is understandable that violent sexual recidivism is viewed as a greater risk than equally probable violent non-sexual recidivism since the risk represented by a hazard is often defined as the product of its probability and the seriousness of its consequences.

For example, according to the Commonwealth of Pennsylvania, Sexual Offenders Assessment Board (2003),

> Sexual assault is a crime that strikes at the very core of the human spirit. In the wake of sexual violence are scarred victims, shattered lives, disrupted families and frightened communities. Media accounts of previously convicted sex offenders committing new violent sexual assaults have become commonplace.

Granting for the sake of argument that there is a dangerous subclass of violent sexual offenders who are unable to control their behaviour and as a consequence likely to commit additional violent offences, what purposes are served by civil incarceration?

Criminal sanctions are generally thought to have three purposes: punishment, deterrence (both of the individual criminal and of others), and incapacitation. The purposes of punishment and deterrence of others (also called general deterrence) are presumably dealt with by the serving of the criminal sentence. Consequently, punishment and general deterrence are not legitimately an issue in a proceeding for civil incarceration. The individuals in question are presumably immune to specific deterrence, because of their inability to control their future behaviour. Hence the purpose of civil incarceration is to incapacitate them, so that they are unable to commit further violent sexual offences.

The idea of civil prior restraint, of incarcerating a person before an offence out of fear that they might offend, is fraught with danger. Every person walking the streets might be so regarded. As a result, the courts generally have narrowly construed the grounds for allowing such incarceration.

2. CONSTITUTIONAL STATUS OF CIVIL COMMITMENT LAWS

The United States Supreme Court has recently clarified the rationale for distinguishing violent sexual offenders from other classes of violent criminals. Statutory language varies somewhat among the 16 states, but generally requires the fact-finder to determine that there is some level of probability (see Table 1) that the offender will commit a violent sexual offence following release from prison and that this is the consequence of a mental abnormality. For example, the Kansas statute permits the civil detention of a person convicted of any of several enumerated sexual offences, if it is proven beyond a reasonable doubt that he suffers from a 'mental abnormality'—a disorder affecting his 'emotional or volitional capacity which predisposes the person to commit sexually violent offences'—or a 'personality disorder,' either of which makes the person likely to engage in repeat acts of sexual violence.

Singling out this particular class of offenders for possible civil confinement has caused concern in some quarters that it circumvents criminal procedural safeguards by using a civil forum to increase a criminal's punishment. However, in Kansas v. Hendricks (1997) and again in Kansas v. Crane (2002), the United States Supreme Court has approved of civil commitment as a constitutionally valid method of enhancing the state's capacity to incapacitate the most dangerous offenders. In the majority opinion in Crane, Justice Breyer wrote,

In *Hendricks,* this Court upheld the Kansas Sexually Violent Predator Act, Kan. Stat. Ann. 59–29a01 et seq. (1994), against constitutional challenge. 521 U.S. at 371. In doing so, the Court characterized the confinement at issue as civil, not criminal, confinement. *Id.,* at 369. And it held that the statutory criterion for confinement embodied in the statute's words 'mental abnormality or personality disorder' satisfied 'substantive' due process requirements. *Id.,* at 356, 360. In reaching its conclusion, the Court's opinion pointed out that 'States have in certain narrow circumstances provided for the forcible civil detainment of people who are unable to control their behavior and who thereby pose a danger to the public [*869] health and safety.' *Id.,* at 357. It said that 'we have consistently upheld such involuntary commitment statutes' when (1) 'the confinement takes place pursuant to proper procedures and evidentiary standards,' (2) there is a finding of 'dangerousness either to one's self or to others,' and (3) proof of dangerousness is 'coupled'…with the proof of some additional factor, such as a 'mental illness' or 'mental abnormality.'

In the Crane opinion, Justice Breyer took some pains to narrow the purpose of civil commitment to the incapacitation of dangerous offenders (as opposed to punishment or deterrence), and to enunciate incapacity to control behaviour as the principle for distinguishing sexually violent offenders from other violent offenders.

Hendricks underscored the constitutional importance of distinguishing a dangerous **sexual offender** subject to civil commitment from other dangerous persons who are perhaps more properly dealt with exclusively through criminal proceedings. That distinction is necessary lest 'civil commitment' become a 'mechanism for retribution or general deterrence'—functions properly those of criminal law, not civil commitment. [As] a critical distinguishing feature…there must be proof of serious difficulty in controlling behaviour. And this…must be sufficient to distinguish the dangerous sexual offender whose serious mental illness, abnormality, or disorder subjects him to civil commitment from the dangerous but typical recidivist convicted in an ordinary criminal case. [534 U.S.–2002].

The Supreme Court has ruled constitutional a civil procedure by which a citizen may be indefinitely deprived of liberty. As a result of its rulings, the decision to commit a sexual offender to civil confinement rests on the determination that the respondent (1) lacks the capacity to resist committing future violent sexual offences, and (2) is likely to commit a violent sexual offence if released. This paper argues that the assessment of the risk of a respondent's re-offending requires a scientifically valid statistical assessment of the conditional probability that the respondent will re-offend given what is known about him and that the specific language of each state's statute fairly narrowly specifies the threshold probability for civil commitment.

3. EVIDENCE OF DANGEROUSNESS

Statutory language concerning civil confinement of sexual predators generally has two components: (1) there is some level of probability that the offender will recidivate, which varies somewhat among jurisdictions (Table 1), and (2) there is some mental abnormality which makes the offender substantially unable to control his behaviour. For example, the Wisconsin statute defines a **sexually violent** person as one 'who has been convicted of a **sexually violent** offence... and who is dangerous because he or she suffers from a mental disorder that makes it substantially probable that the person will engage in acts of sexual violence.'

3.1 Mental Abnormality

The mental abnormality component is based entirely on clinical judgments, which are notoriously unreliable (Dolan & Doyle, 2000), particularly when not constrained by any standards regarding what data are to be used to support the judgment, or any standards regarding the training and disinterestedness of the evaluator. Indeed, the American Psychiatric Association Task Force on Sexually Dangerous Offenders stated in its report (American Psychiatric Association, 1999) that the diagnosis of sexual predator is based on

TABLE 2. Risk Levels, Score Values, and Recidivation Rates for RRASOR, STATIC-99

Risk Level[1]	*RRASOR*		*Static-99*	
	Score	10-year Rate(%)	Score	10-year Rate(%)
Low	0	6.5	0	11
<10%			1	7
Medium	1	11.2	2	13
11–20%			3	14
Moderate	2	21.1		
21–30%				
Medium High	3	36.9	4	31
31–40%			5	38
High	4	48.6	6+	45
>41%	5	73.1		

[1] Static-99 risk level descriptors and recidivism rates are taken from Hanson & Thornton (1999); RRASOR recidivism rates are taken from Hanson & Thornton (1997).

TABLE 3. Risk Levels, Score Values, and Recidivation Rates for MnSOST-R

Risk Level[1]	MnSOST-R Score	Dev[2]	Rep[3]
Low	3 or below	16%	12%
Moderate	4 to 7	45%	25%
8 to 12	8 to 12	63%	50%
High	8 and above	70%	57%
Refer for Commitment	13 and above	88%	72%

[1] MnSOST-R risk level descriptors (with the exception of '8 to 12') and recidivism rates are taken from Table 6 of Epperson *et al.* (2003) and Tables 4 and 5 of Epperson *et al.* (2000).
[2] Recidivation rates based on the development sample.
[3] Recidivation rates based on the aggregate representative sample.

'a vague and circular determination that an offender has a 'mental abnormality' that has led to repeat criminal behavior. Thus, these statutes have the effect of defining mental illness in terms of criminal behavior. This is a misuse of psychiatry, because legislators have used psychiatric commitment to effect non-medical societal ends.'

3.2 Risk of Re-Offence

There are several ambiguities in the probability component. For what event is the probability sought? First, what offences are to count? Presumably only violent sexual ones. The definition of these vary (somewhat) by jurisdiction. Second, is it the violent sex act, the public accusation by a prosecutor, or the conviction in court that triggers recognition of recidivism? The first is impossible to know, while the second and third vary by jurisdiction as well, since they depend on the extent of prosecutorial attention and resources devoted to this kind of crime. Finally, what time period is relevant for this determination? It might range from one year to the rest of a person's life. Each of these ambiguities must be resolved before it makes sense to speak of a probability of re-offence.

By contrast, we think that the level of probability required is (relatively) clear. For example, the Wisconsin Supreme Court held in State v. Curiel, 227 Wis. 2d 389, 597 N.W.2d 697 (1999) that the term 'substantially probable,' as used in Wis. Stat. ch. 980, is unambiguous and must be defined as 'much more likely than not.' Similarly, Justice of the Massachusetts Superior Court Ralph D. Gants, in Commonwealth v. Christopher Reese, April 5, 2001 (13 Mass. L. rep. 195), wrote

This Court defines 'likely to engage in sexual offenses' to mean that there is a substantial likelihood, at least more likely than not, that the respondent will commit a new sexual offense within the immediate future, understood generally to be within the next five years but with a longer time horizon if the anticipated future harm is extremely serious. This definition provides a floor (more likely than not) below which a person may not be civilly committed for fear of future sexual dangerousness...

These opinions reinforce our view that the probabilistic language in the state statutes specifies a risk level substantially above 50%. We agree with Mosteller & Youtz (1990) and Kadane (1990) that the usage of such probabilistic terms can be quantified with reasonable precision—for example, people use the words 'likely' and 'often' to signify something close to 70% probability and the phrase 'more often than not' to signify approximately 60% probability. We will therefore presume that the statutes generally require the fact-finder to determine beyond a reasonable doubt (12 states) or with clear and convincing evidence (4 states) that there is at least a 60 to 70% probability that the respondent will commit an additional violent sexual offence if released. The appropriate probability, in statistical terms, is the conditional probability that a specific offender will commit a violent sexual offence given what is known about him.

However, the California Supreme Court, in People v. Ghilotti (2002), found 'the phrase *'likely'* to engage in acts of sexual violence' (italics added)...connotes much more than the mere *possibility* that the person will re-offend as a result of a predisposing mental disorder that seriously impairs volitional control. On the other hand, the statute does not require a precise determination that the change of re-offence is *better than even*. Instead, an evaluator applying this standard must conclude that the person is 'likely' to re-offend if, because of a current mental disorder which makes it difficult or impossible to restrain violent sexual behaviour, the person presents a *substantial danger*, that is, a *serious and well-founded risk*, that he or she will commit such crimes if free in the community.' However, the dissent of J. Werdegar challenges the majority thinking on precisely this point. See also Monahan & Wexler (1978). Hence, with the exception of California, we believe that the legislatures intend and courts require a probability of roughly 70%.

4. RISK ASSESSMENT

How is the risk to be assessed? There are two possibilities: to use clinical judgment or to use the results of statistical studies. It is generally conceded that expert clinical judgment is inferior to statistical probability assessment (see Dolan & Doyle, 2000 for a recent review); for example, Justice Geraldine Hines of the Superior Court (Suffolk County) of Massachusetts wrote in Commonwealth v. Mujaheed, CA No. 00–2217 (June 26, 2001) (memorandum opinion) (http://www.mass.gov/cpcs/mhp/mhpSDPorders3.html):

> First, [the state's expert] Dr Silverman relied on a discredited clinical methodology in predicting that Mujaheed will commit sex offenses unless he is detained in a secure facility....In rejecting Dr Silverman's opinion, I credit the documentary evidence offered by Mujaheed and the testimony of his expert to the effect that the clinical method has been definitively repudiated as a reliable means of predicting recidivism among sex offenders. See Barefoot v. Estelle, 463 U.S. 880 (1983) ('Psychiatric testimony predicting dangerousness may be countered not only as erroneous in a particular case but also as generally so unreliable that it should be ignored').

If clinical risk assessment is inadmissible the only alternative is statistical risk assessment, which amounts to computing P(Recidivate | Data). Several actuarial (actually statistical) instruments for predicting violent sexual recidivism have been developed in response to the courts' reluctance to accept clinical judgments of the risk of violence. Each of the instruments in common use (RRASOR, STATIC-99, and MnSOST-R) involves a list of risk factors. Each risk factor is either absent or is present to some degree for a given respondent. Each (level of each) risk factor is

1. Number of sex/sex-related convictions (including current conviction):
   One   0
   Two or more   +2
2. Length of sexual offending history
   Less than one year   −1
   One to six years   +3
   More than six years   0
3. Was the offender under any form of supervision when they committed any sex offence for which they were eventually charged or convicted?
   No   0
   Yes   +2
4. Was any sex offence (charged or convicted) committed in a public place?
   No   0
   Yes   +2
5. Was force or the threat of force ever used to achieve compliance in any sex offence (charged or convicted)?
   No force in any offence   −3
   Force present in at least one offence   0
6. Has any sex offence (charged or convicted) involved multiple acts on a single victim within any single contact event?
   No   −1
   Yes   +1
7. Number of different age groups victimized across all sex/sex-related offences (charged or convicted):
   Age group of victims: (check all that apply)
       Age 6 or younger
       Age 7 to 12 years
       Age 13 to 15 years and the offender is more than five years older than the victim
       Age 16 or older
   No age group or only one age group checked   0
   Two or more age groups checked   +3
8. Offended against a 13- to 15-year-old victim and the offender was more than five years older than the victim at the time of the offence (charged or convicted):
   No   0
   Yes   +2
9. Was the victim a stranger in any sex/sex-related offence (charged or convicted)?
   No victims were strangers.   −1
   At least one victim as a stranger   +3
   Uncertain due to missing information   0
10. Is there evidence of adolescent antisocial behaviour in the file?
    No indication   −1
    Some relatively isolated antisocial acts   0
    Persistent, repetitive pattern   +2
11. Pattern of substantial drug or alcohol abuse (12 months prior to arrest for instant offence or revocation):
    No   −1
    Yes   +1
12. Employment history (12 months prior to arrest for instant offence):
    Stable employment for one year or longer   −2
    Homemaker, retired, full-time student, or   −2
    disabled/ unable to work
    Part-time, seasonal, unstable employment   0
    Unemployed or significant history of unem-   +1
    ployment
    File contains no information   0
13. Disciplining history while incarcerated (does not include discipline for failure to follow treatment directive):
    No major discipline reports or infractions   0
    One or more major discipline reports   +1
14. Chemical dependency treatment while incarcerated:
    No treatment recommended/Not enough   0
    time/No opportunity
    Treatment recommended and successfully   −2
    completed or in program at time of release
    Treatment recommended but offender   +1
    refused, quit, or did not pursue
    Treatment recommended but terminated by   +4
    staff
15. Sex offender treatment history while incarcerated:
    No treatment recommended/Not enough   0
    time/No opportunity
    Treatment recommended and successfully   −1
    completed or in programme at time of
    release
    Treatment recommended but offender   0
    refused, quit, or did not pursue
    Treatment recommended but terminated   +3
16. Age of offender at time of release:
    Age 30 or younger   +1
    Age 31 or older   −1

FIGURE I. MnSOST-R risk factors and scoring weights.

weighted and the risk score is the sum of the weights. Figure 1, for example, lists the risk factors and scoring weights for the Revised Minnesota Sexual Offender Screening Tool, or MnSOST-R (Epperson *et al.,* 2000, 2003). Risk factors 1 through 12 are *static,* in the sense that after a civil commitment they cannot change, whereas factors 13 through 16 are *dynamic*—possibly referring to the period after commitment and can change over time. Clearly a risk-assessment instrument that omits dynamic factors will, after civil commitment, always assign the same risk level to an offender, who can therefore never be judged eligible for release.

The developers of RRASOR, STATIC-99, and MnSOST-R have provided tables of recidivation rates for subgroups of offenders (Tables 2, 3, and 4) computed from convenience samples, which we discuss later. Presumably these rates are meant to be estimates of the probability that an individual offender will recidivate given his risk score.

4.1 Construction and Calibration of MnSOST-R, RRASOR, and STATIC-99

RRASOR and STATIC-99 were developed by Karl Hanson and associates using data from five large follow-up studies of sexual offenders in California, Canada, and the UK. Hanson *et al.* selected candidate risk factors via a meta-analysis of 61 follow-up studies (Hanson & Bussière, 1998). Approximately one-third of the variables were significantly related to recidivism ($p < 0.05$) with correlations of 0.10 or greater. The single best predictor was phallometric assessment of deviant sexual preference (plethysmography) (median $r = 0.20$).

RRASOR

Hanson winnowed the pool of candidate variables down to seven by selecting those that had an average correlation of at least 0.10 with sexual offence recidivism, and that could be scored using commonly available information (e.g. offence history, police reports, demographic characteristics)....If several variables were expected to be highly correlated with each other (e.g. never married/currently married) only the variable with the highest correlation was selected (Table 4)....The next step was creating common operational definitions of each of the predictor variables. Stepwise multiple regression was used to reduce the number of risk factors to 4: number of prior convictions for sex offences (2 = 2 or 3, 3 = 4 or more), age less than 25, any male victims, and any extrafamilial victims. Data from 8 follow-up

TABLE 4. RRASOR Risk Factors

Variable	Average r
Prior sex offences	0.19
Any stranger victims	0.15
Any prior offences	0.13
Age (young)	0.13
Never married	0.11
Any non-related victims	0.11
Any male victims	0.11

TABLE 5. RRASOR and STATIC-99 Development and Validation Samples

Development Samples	n	Age	% Rapists	Follow-up (yrs)	Rate	Endpoint
Millbrook, Ontario	191	33.1	0.0	23	0.35	convictions
Canadian Federal 1983/84 releases	316	30.5	n/a	10	0.20	convictions
Institut Philippe Pinel	382	36.2	29.6	4	0.15	convictions
Alberta Hospital Edmonton	363	35.5	23.1	5	0.06	charges
SOTEP (California)	1138	37.6	27.6	5	0.12	charges
Canadian Federal 1991/1994 releases	241	36.8	56.0	2	0.07	charges
Oak Ridge (Penetang)	288	30.4	50.7	10	0.35	charges/ readmissions
Validation sample HM Prison Service (UK)	303	34.3	18.7	16	0.25	convictions

studies (Table 5) were used to estimate recidivation rates (Tables 2 and 3), Hanson
et al. reported using a fairly crude adjustment to accommodate the widely varying
follow-up intervals. There was no attempt to adjust for the use of different endpoints
(convictions or charges).

Static-99

Hanson's Static-99 instrument consists of the four RRASOR risk factors and six
additional risk factors from an instrument called SACJ-Min (never lived with part-
ner for at least two years, any convictions for non-contact sex offences, any stranger
victims, any current non-sexual violence, any prior non-sexual violence, and 4+
sentencing dates). No explanation of how and why these risk factors were selected
is provided. It is worth noting that Hanson stipulated that the instrument is misspecified (Hanson and Thornton, 1999–2002):

> The inclusion of dynamic factors would likely increase the scale's predictive accuracy
> (Hanson & Harris, 1998, Hanson & Harris, in press). Among non-sexual criminals,
> dynamic variables predict recidivism as well or better than static variables (Gendreau
> *et al.*, 1996). The research on dynamic factors related to sexual offending is not well
> developed, but some plausible dynamic risk factors include intimacy deficits (Seidman
> *et al.*, 1994), sexualization of negative affect (Cortoni, 1998), attitudes tolerant of sexual
> assault (Hanson & Harris, 1998), emotional identification with children (Wilson, 1999),
> treatment failure, and non-cooperation with supervision (Hanson & Harris, 1998).

As we pointed out earlier, the omission of dynamic risk factors is particularly
troubling since it makes civil commitment a life sentence.

MnSOST-R

MnSOST-R was developed using data from a complex sample of 256 released sexual
offenders in Minnesota (Epperson *et al.*, 2003). The developmental sample of 256
cases consisted of a random sample of 107 from the cohort of sex offenders released
in 1988 and a random sample of 108 from the 1990 cohort. These prospective

random samples were augmented with 'a sub-sample of (41) offenders readmitted to the Minnesota Department of Corrections during the time when the sample was being put together, regardless of release year.' Re-offence within six years was the endpoint. The selection of candidate risk factors was based on 'an extensive review of the general literature on sex offenders,' from which 14 risk factors were compiled. When the instrument was revised in 1996, there was an additional review of 'the more recent research on the prediction of sexual recidivism,' which resulted in the addition of an unstated number of additional risk factors.

Epperson *et al.* (2000, 2003) used an ad hoc procedure to select and weight risk factors. Each candidate risk factor was cross-tabulated against re-offence. If the re-offence rate associated with an item level was within 5% of the baseline re-offence rate (34.75%) that item level was scored 0. Item levels with re-offence rates 5% greater than the baseline and the adjacent item level were scored +1 (or −1) for each 5% increment above (or below) the baseline rate. An item level with a small n (generally under 10% of the sample) was collapsed with the adjacent item level that had the most similar re-offence rate (see Figure 1).

Subsequently, an independent cross-validation sample was obtained, consisting of 220 sex offenders released in 1992. Estimated recidivation rates in Tables 3 and 5 are based on the aggregate of the three prospective cohort samples (1988, 1990, 1992). Assuming that offenders released in unrepresented years are statistically exchangeable with those released in 1988, 90, and 92, rate estimates based on these data are unbiased. In addition, Table 3 reports Epperson *et al.*'s earlier (2000), biased estimates based on the full developmental sample, which consisted of two prospective cohort samples and a retrospective sample of 41 readmitted offenders (Epperson *et al.*, 2000). These biased estimates have been widely quoted in expert testimony.

MnSOST-R is currently the only prediction instrument for which, to our knowledge, the underlying data have been made available to scholars (Epperson, 2001, private communication). Our reanalysis of that data indicates that, when viewed as a probabilistic predictor, MnSOST-R is misspecified in its treatment of nonviolent molesters who have no history of ever using force to achieve compliance in any sex offence. Table 7 is our computation of crude recidivation rates for the MnSOST-R representative samples (n = 435) employing the broad risk categories recommended by the developer. Table 7 reports smoothed recidivation rates computed via logistic regression.

As shown in Table 7, it appears that nonviolent molesters have a substantially lower probability of recidivation compared to rapists and violent molesters with the same MnSOST-R score. However, the most serious misspecification implicit in Epperson's four-category 'actuarial' model (Table 3) is that it attributes the same recidivation rate to broad categories of offenders, some of whom have substantially lower recidivation rates than the category as a whole. For example, although Table 3 indicates that 57% of sex offenders in the 'high-risk' category (scores of 8 or above) recidivated within 6 years, Table 6 indicates that offenders with scores between 8 and 10 are less likely than not to recidivate.

We argued earlier that the legal standard for civil commitment requires at least a 70% probability of recidivation,[2] which is not attained for MNSOST-R scores

---

2. In states that require a judgment that the offender is *likely to recidivate.*

TABLE 6. MnSOST-R Crude Frequencies[1]

Score	Cases	Recidivists	$\hat{p}$ (%)[2]	RF[3]	$\bar{p}$[4]
−11	1	0	2		
−10	2	0	3		
−9	1	0	3		
−8	5	0	4		
−7	4	0	4		
−6	9	0	5		
−5	9	1	6		
−4	21	2	7		
−3	20	0	8		
−2	33	3	10		
−1	22	3	11		
0	29	9	13		
1	25	4	15		
2	38	6	17		
3	33	2	20	12	12
4	25	8	23		
5	23	6	26		
6	26	7	29		
7	26	4	33	25	28
8	21	12	37		
9	14	6	41		
10	10	5	45		
11	9	5	50		
12	4	1	54	50	43
13	4	1	58		
15	2	1	66		
16	2	1	70		
17	6	5	73		
18	2	2	77		
19	5	5	79		
20	2	1	82		
21	1	1	84		
24	1	1	90	72	73

[1] Source: Raw data for the representative sample, Epperson (2001, private communication).
[2] Logistic regression estimate of recidivation rate.
[3] Relative frequency.
[4] Average logistic estimate.

lower than 15. Thus some offenders in the higher risk category in Table 3 (Refer for Commitment) do not to reach the legal threshold for commitment (Table 6).

4.2 Comments on MnSOST-R, RRASOR, and Static-99

These 'actuarial' risk-assessment instruments, widely used in litigation, were constructed ad hoc with no attention to model specification (non-additivity) and omitted variables (most importantly, type of offenders—whether rapists or child molesters). Candidate variables were selected partly on the basis of cost and availability and as a consequence, one of the most powerful predictors (plethysmography) was omitted.

TABLE 7. Crude Recidivation Rates for the MnSOST-R Representative Sample[1]

MnSOST-R Score	Rapists and Violent Molesters		Nonviolent Molester		All	
	n	Rate(%)	n	Rate(%)	n	Rate(%)
3 and below	207	13.0	45	6.7	252	11.9
4 to 7	94	25.5	6	16.7	100	25.0
8 to 12	54	53.7	4	0.0	58	50.0
13 and above	24	75.0	1	0.0	25	72.0

[1] Source: Underlying data from Epperson, private communication, 2001.

For RRASOR and Static-99 the endpoints in the calibration sample were not consistent and for none of the instruments was the endpoint restricted to violent sexual offences, as specified by all but one of the state statutes (Table 2).

Although Hanson claimed to have created 'common operational definitions' of the risk factors across the eight development and validation samples, it is difficult to see how this could have been done in a completely consistent manner due to the great diversity of jurisdictions and the fact that, for Static-99, if not for RRASOR, the ascertainment must certainly have been based on secondary data and not on the primary records which an assessment would be based on in a commitment hearing.

In every case, estimated recidivation rates are computed for broad score categories and therefore do not in principle provide valid estimates for individual respondents. This is particularly egregious in the case of the MnSOST-R 'high risk' category (scores 8 and up) which, according to Epperson, has 57% crude risk of re-offence. In fact, fewer than 50% of individuals with scores between 8 and 10 recidivate (Table 6).

5. THE USE OF STATISTICAL RISK ASSESSMENT IN EXPERT TESTIMONY

In actual testimony the rates found by these instruments are subjected to ad hoc 'clinical' and 'base-rate' adjustments and are combined in ways that are difficult to understand or to quantify. Testimony of Dr Dennis Doren (see Exhibit 1), who has appeared extensively as a prosecution expert, gives some indication of how they are presented in litigation. Based on a superficially plausible but statistically indefensible medical analogy, Dr Doren appears to base his risk estimate on some function of three risk estimates (RRASOR: 21%, Static-99: 45%, and MnSOST-R: 63%). Dr Doren alluded to but nowhere identified the research that he relied on. Dr Doren dismisses the low RRASOR score in favour of the STATIC-99 score on the grounds of undocumented assertions that the different instruments have different statistical properties for different types of offenders. If this is indeed the case, then the models underlying the instruments are misspecified. His conclusion, that the '... risk is beyond more likely than not to commit a sexually violent act over [the defendant's] lifetime,' is certainly not supported by the 45% recidivation rate associated with the STATIC-99 score; however, he seems to dismiss the MnSOST-R score, which, in his terms, ought to be his best evidence. Dr Doren's cafeteria-style use of the risk assessment instruments obviates the norms in Tables 2 and 3, which, as we argued above, are dubious at best. His risk estimate is some nonlinear function of the three recidivism rates (perhaps their maximum) possibly modulated by a psychopathology instrument. It is, in other words, a new, un-normed and vague instrument.

The testimony of Dr Jack Vognsen in Garcetti v. The Superior Court of Los Angeles County, Respondent; Paul Marentez, Real Party in Interest, B143330, features sheer seat-of-the-pants extrapolation combined with what has elsewhere been styled 'Guided Actuarial Risk Assessment':

> In Dr Vognsen's opinion, Marentez is more likely to reoffend than not, and the percentages are somewhere between 52 to 57 percent in favor of reoffense. The percentages will increase if Marentez resumes his level of drug abuse upon release. Dr Vognsen derived the percentages by using the Static-99, which provided a percentage of 52 percent [the 15-year recidivism rate], plus 5 percentage points based on other 'clinical' factors or 'subjective' risk indicators not included in the Static-99, such as the amount of sexual deviancy demonstrated, the degree of antisocial behavior displayed, and the degree of cooperation with supervision that has occurred.

On the basis of these additional subjective clinical factors, Dr Vognsen adjusted the score upward to reach a level of 57 percent. He characterized the process as utilizing the Static-99 actuarial instrument and adjusting with '[a] dash of clinical judgment.' Dr Vognsen also testified that he looked at Marentez not just as 'numbers and statistics,' but as an individual by considering 'dynamic' and clinical factors. He testified that he used his clinical judgment based on his experience as a psychologist to assess Marentez's current condition. Additionally, Dr Vognsen testified that the risk of 'reoffence' predicted by the Static-99 is not just the risk of Marentez engaging in undetected sexual offence, but the risk of his actually being charged or convicted based on detectable offences. He therefore concluded that, '... the 52 percent chance of reoffence predicted by the Static-99 is very conservative.'

Again, Dr Vognsen's addition of a 'dash of clinical judgment' to the probability estimate creates a new un-normed instrument. If clinical factors distinguish among offenders in the same Static-99 category, then Static-99 is misspecified. Dr Vognsen's speculation that the re-offence rate is conservative is just that—sheer speculation. It is not the subjectivity of Dr Vognsen's opinion that we find problematic, but his failure to make coherent use of evidence to update his subjective probabilities. We believe there are two reasons that states' experts feel the need to 'adjust' statistical risk assessments. First is the perception that the risk assessments are misspecified both by the omission of salient risk factors and by their failure to distinguish between classes of offenders. The second reason is the perception that the calibration samples are not good proxies for the offender populations in other jurisdictions.

Exhibit 1: Extracts from the testimony of Dr Dennis Doren in re The detention of Steven Howell, Fifth Judicial District of Iowa, 29th of October, 2001

Q. What instruments did you apply and with what results?
A. There were three. One is the Rapid Risk Assessment for Sex Offender Recidivism...pronounced RRASOR. The second instrument is called the Static-99...the third is called the Minnesota Sex Offenders Screening tool, Revised.

I'll just call it the Minnesota instrument...[On] RRASOR,...Mr. Howell's score on that was a two....a relatively low risk finding [21.1% recidivation]...From that instrument alone, his risk would not be viewed as more likely than not, in my opinion....[On] Static-99...he scored six...the highest risk category that that instrument measures. [45%]...[On] the Minnesota instrument...Mr. Howell's score was either a plus 8 or plus 10, depending on how one item was scored....[so he falls in] the high risk category but not the highest risk category. [57%]

Q. All right. And what general conclusion did you draw from this phase of your assessment?

A. Putting this information together with other information...I would conclude that in my opinion, his risk is beyond more likely than not to commit a sexually violent act over his lifetime.

Q. Okay. And how do you reach that if in light of, for example, the RRASOR score of two, which is in the low risk, in other words, one of these obviously says he's in the low risk category, how do you explain that conclusion given that score?

A. There are various pieces of research...that indicate to me that there are different pathways...by which someone who is previously convicted of a sex offence becomes a sexual re-offender. The metaphor that I use to describe that is when I go for a checkup for my physical health...the doctor is going to check out risk factors for instance related to my heart...But even if the doctor assesses my risk to my heart to be very low, the doctor needs to check other systems...And if...it's found I have a malignant brain tumor, it doesn't matter there's low risk to my heart. I'm still at very high risk to my health....

The research that I'm aware of would indicate that there are at least two different pathways...for sex offenders. One of those is being driven by sexual interests that are illegal. The classic case...is the...pedophile...driven to have sex with kids. They may be fine, upstanding citizens who never break the law outside of that...And the RRASOR tends to measure that avenue better than the other instruments. The other avenue that is, in my view, demonstrated by research is...the type of individual who is criminal in a variety of ways, including sexual...They do not necessarily have a sexual disorder at all. Their sexual offending is similar to other offending in that...they take what they want when they want it, irrelevant of the effect on anybody else or consequences to themselves....The instruments that are most associated with this dimension are the Static-99, the Minnesota instrument, the psychological test called the Psychopathy Checklist-Revised, abbreviated PCL-R in capitals, and other instruments that are measures of violence potential.

Just pulling that together again, the question was why I could see him as having a more likely than not risk with a RRASOR that was low. The bottom line then is that the RRASOR being low goes along with the fact I did not diagnose a sexual disorder. He does not have any known child victims. Basically, he does not seem to be driven to sexually offend through a sexual disorder. And that's what the RRASOR is telling me. The other instruments, the Static-99 and Minnesota instrument, were in the high to very high area. And that goes along with a diagnosed personality disorder, along with having adult victims, having a relatively high score on the PCL-R.

6. LEGAL AND SCIENTIFIC STANDARDS FOR STATISTICAL RISK ASSESSMENT

The Daubert standard (enunciated by the Supreme Court in Daubert v Merrill-Dow, 509 US 599 (1993)), supersedes the older, general acceptance test of Frye v United States, 54 App D.C. 46 (1923), in federal cases, and has been adopted as a rule of evidence in at least 26 states. Under Daubert, scientific evidence must be reviewed in terms of the following general guidelines: (1) whether the theory or technique can be, and has been, tested (i.e. a determination of its 'falsifiability'); (2) whether the evidence has been subjected to peer review and publication; (3) the 'known or potential error rate' associated with applications of a theory; and (4) the general acceptance of the theory or technique in question. While the primary shift in the U.S. standard of admissibility of scientific evidence has occurred at the federal court level (i.e. from Frye's 'general acceptance' to Daubert's more sophisticated guidelines), changes are also occurring at the level of individual states, which have their own autonomous court systems.

The courts are responsible for conducting a constitutional analysis of the state interest to incapacitate potentially dangerous sexually violent individuals against the due process rights of the individual. Many of the statutes require expert testimony. In such circumstances the rules of evidence admissibility may not apply. However we think the Daubert rules are still important, as they go to the appropriate weight given to such testimony.

Our interest as statistical scientists in the civil commitment of sexual offenders arises from what appears to us to be the statutory requirement that the confinement decision be based on an assessment of the probability that the sexual offender will commit a subsequent violent sexual offence if released after completing his present sentence. As mentioned earlier, most of the 16 state statutes use probabilistic terms to characterize the threshold risk level required for civil commitment, 'likely' being the most common, and require the fact-finder to determine whether the offender exceeds that threshold probability of reoffending. Thus, it appears to us that the statistical expert's job is to provide the fact-finder with the means to determine, as accurately and specifically as possible, the probability that this offender will recidivate given what is known about him.

A probabilistic prediction model is a function of facts and data. 'Facts' are information specifically about the respondent and 'Data' constitute the data base(s) from which the prediction instrument was developed. There is broad agreement among statisticians about standards for study design and conduct, statistical modelling, and generalizability of predictions, a point to which we will return. These standards are intended to produce a valid probabilistic prediction model.

6.1 Proper Specification

A probabilistic prediction model directly or indirectly creates categories of presumably equal-risk individuals: for example, all individuals with MnSOST-R scores of 12, or more generally, all people for whom the model predicts the same probability of recidivation. Let C_p stand for all combinations of facts for which $P(Recidivate|Facts, Data) = p$. The model is correctly specified if there is no way to identify a subset of

C_p that recidivates at a rate higher or lower than p—individuals in C_p are conditionally *exchangeable*. See, for example, Fisher (1956, p. 33). Correct specification in the narrow sense means that the model is correctly specified with respect to a fixed set of facts (such as the four risk factors underlying RRASOR). Correct specification in the broad sense means that no known external variable can be used to distinguish between members of C_p. For example, MnSOST-R, Static-99, and RRASOR are not correctly specified in the broad sense, unless the states' experts quoted earlier have been disingenuous about the need to 'adjust' risk estimates up or down on the basis of characteristics not incorporated in the risk assessment instrument.

Methods for checking model specification are widely available; consequently, developers of a prediction model are obliged to check that the model is correctly specified. Describing risk assessment instruments as 'actuarial' appears to be an attempt to sidestep this point by appeal to an inappropriate model. Actuarial risk estimates consist of rates of occurrence of an adverse event (say an automobile accident) within ad hoc groups of policy holders (for example, families with a teenage male driver). Insurance companies, with few restrictions, can construct groups as they please (for example, teens with good grades); however, there is no implication that all teenage males with good grades have the same risk—policy holders within a category voluntarily agree to pay a premium appropriate to the average risk of that group. This bears no resemblance to sexual offender classification—it is not voluntary, and for that reason treating all members of a heterogeneous risk group as having equal risk is plainly unjust to the lower-risk members of the group. For example, in Table 3 individuals with MnSOST-R scores 8 or higher are said to have 'high risk' of recidivation (57%); however, in Table 6 we estimate that at score 8 the risk is 37% and at scores above 16 the risk is 70% or more. If we were writing insurance, the break-even insurance premium for the high risk group would be $0.57 per dollar of coverage—$0.20 higher than the fair premium for an individual with score 8 who would be well advised to shop for another carrier, an option not open to respondents in a civil commitment hearing. The state has a fundamental obligation to be fair, and treating all members of the 'high-risk category' as equally risky is not only unjust, but economically irrational—confinement for treatment is expensive, in excess of $100,000 per year in Minnesota (Texas Comptroller of Public Accounts), and should be reserved for the highest risk cases.

6.2 Generalizability

A probabilistic prediction is relevant to the fact-finders in a commitment hearing only if it is generalizable from the development data base to the respondent on trial. Generalizability depends on two elements—standardized ascertainment of risk factors and outcomes, and conditional exchangeability of the respondent with his matching subgroup of the development sample. That is, if r is the respondent and M are individuals in the development sample with the same estimated probability of recidivation as the respondent, then the estimate is relevant only if the fact-finders are aware of no meaningful distinction between r and M. This, we think, places a burden on the state's expert to present positive evidence that r and M are exchangeable and, failing that, presents the defence with a means of rebuttal by presenting evidence that r and M are not exchangeable. RRASOR, Static-99 (mostly

Canadian and British), and especially MnSOST-R (a small Minnesota sample) are all vulnerable on this point for several reasons. First, a given state's statutory definition of violent sexual recidivation may not match the definition(s) used in the various data bases. Second, risk factors (see Fig. 1), may have different meanings in different jurisdictions. For example, a 'significant history of unemployment' may not be as significant in a place with chronic high unemployment. The number of prior charges for sexual offences will not have the same meaning in jurisdictions with overstretched police departments and prosecutors compared to jurisdictions which devote more resources to the investigation and prosecution of sexual offences. Thirdly, there are marked cultural differences between, say, south Florida and Minnesota, which might lead a fact-finder in Miami to doubt the relevance of an instrument developed and normed in Minnesota.

6.3 Standardized Ascertainment of Risk Factors and Outcomes

Lack of or failure to adhere to a common protocol for identifying the legally relevant outcome (violent sexual recidivism) and for ascertaining the risk factors may invalidate the risk assessment for a respondent. With respect to the outcome, it seems obvious to us that if the development data base defined recidivation, for example, as arrest for any sexual offence, and the statute governing a commitment hearing defines recidivation differently, for example as the commitment of a violent sexual assault, then the risk estimate is invalid without some sort of data-based statistical adjustment. We have seen no effort on the part of developers to provide guidelines for making this adjustment nor have we heard any mention of this issue from state's experts.

Risk factors with respect to a respondent in a commitment hearing must be ascertained by the same protocol used by the developer of the risk assessment instrument. If not, then the respondent is not exchangeable with his nominally equal-risk counterparts in the development sample and there is no necessity for the fact-finder to conclude that he has the same risk. One problematic area is specifying what sources of information may be relied upon in ascertaining the presence of a risk factor. It is not entirely clear what information sources were consulted to ascertain risk factors in the development samples, particularly Static-99 and RRASOR. For example, whether only official records were used or if, for example, unofficial notes from psychologists' files could be used. Without standards, there is no good reason for a fact-finder to believe that the score assigned to the respondent by a state's expert is the score the respondent would have received as a member of the developmental data base.

7. A MODEST PROPOSAL FOR DEVELOPING AND MAINTAINING A RISK ASSESSMENT DATABASE

It appears to us that reliable, defensible risk assessments require a standardized data base. This idea is widely accepted in health care delivery systems where the effort to develop protocols for collecting a 'minimum data set' of standardized measures of patient characteristics, interventions, and outcomes are common. The

New Jersey Supreme Court undertook a similar project for the purpose of developing a standardized data base to be used in the proportionality review of capital sentences (Baime, 2001).

The goal is to compile a data base to be used for the purpose of assessing an individual respondent's risk of violent sexual recidivation. This data base could also be designed to support systemic studies of the efficacy of community- and prison-based treatment programmes, and of statewide uniformity in the civil commitment of sexual offenders.

Because of the similarity of statutory language in the approximately 15 states with civil commitment laws for sexual offenders, it is possible that a multi-state cooperative programme would enjoy economies of scale. Such benefits will be greatest in the early phases of the project in which a diverse panel of experts and 'stake holders' will develop a list of variables along with protocols for their ascertainment. It is possible that some of the variables will be standardized clinical assessments. However, it is possible that clinical assessments will prove useful only as evidence of inability to control behaviour and not evidence of probability of re-offending. A unique challenge, not faced as acutely by health care minimum data sets, is the ascertainment of long-term outcomes such as arrests, charges, and convictions, possibly many years or decades after release, and possibly in other states or countries.

Presumably the data set will be initialized using retrospective data, as in the development of MnSOST-R. If risk prediction were perfect, then new cases would provide no information since all future recidivists would be committed and all future non-recidivists would be released. However, nothing like perfection is likely ever to be achieved. Consequently, new cases entering the data base will be statistically informative. However, using them to update the prediction system will require sophisticated statistical methods. We presume that prediction models will be developed and updated via logistic regression; however, this is not the only possibility (see Baime, 2001).

By consistently setting a fairly high risk threshold by the use of the word 'likely' that we argued above amounts to about 70% probability of recidivation, legislators have struck a rough balance between the social harm of releasing a future recidivist and the harm of confining a non-violent person who has completed his criminal sentence.

Taking Table 6 at face value, for the sake of argument, the consequences of this legislative decision can be computed. At the 70% threshold, respondents with scores of 16 or greater would be eligible for civil commitment and of those, only those respondents judged substantially incapable of controlling their behaviour would be committed (if these can be identified). Based on the crude rates in Table 6, 16% (16/102) of future recidivists would be confined and 84% released, about 1% of non-recidivists would be confined (3/333), and 99% would be released.

Legislators were clearly setting an evidentiary standard, not calculating utilities. It is difficult to imagine setting the standard much lower considering that the respondent's liberty is at stake. Indeed, a preponderance of evidence standard (scores of 11 and above) would confine 9% of the non-recidivists while still missing 76% of the recidivists. Yet it is unthinkable that a person could be deprived of liberty for less than a preponderance of evidence, except in California. The plain facts are that what appears at face value to be the best available classification instrument is insufficiently sensitive and insufficiently specific to permit the identification of even a majority of recidivists without imposing an unjust burden on non-recidivists.

It goes almost without saying that sexual violence against children is viewed as one of the most despicable crimes imaginable. Thus elected judges and prosecutors seem caught between the proverbial rock and a hard place—the consequences of participating in the release of a sex offender who ultimately re-offends (a Type I error) or of participating in the long-term confinement of a harmless individual (a Type II error) are unbalanced. Long-term confinement of a non-recidivist is unjust to the respondent but doesn't produce the public opprobrium and career-ending potential of being associated with the release of a sexually violent re-offender. Prosecution experts, who are often state employees, may well feel similar pressures. On the defence side, failure to win a release in a civil commitment hearing does not on its face seem to carry the same kind of personal risks. This balancing of sensitivity against specificity is foreseen by Monahan & Wexler (1978). The difficulty we point to here is that the balance struck by the local decision-makers may differ markedly from the balance desired by the legislatures and the appellate courts.

We suggest that these unbalanced consequences combined with the comparatively relaxed rules of evidence in a civil hearing both skew the incentives of the actors in the system and provide an opportunity to skew the presentation of evidence in ways that might not be transparent. In statistical terms, this imposes on the fact-finder the burden of modelling the behaviour of the expert witnesses, particularly experts who are state employees. We argue that requiring strong standards of statistical and scientific evidence will make the weighing of evidence more transparent for the fact-finder.

The basis of civil commitment decisions must be the balancing of public risk and private civil rights. Legislators seem to have struck this balance by the use of fairly precise probability terms. However, there are intense pressures which may lead to skewing decisions in favour of commitment.

It is therefore necessary to set a high standard for expert testimony about the probability of recidivation. We have reviewed existing instruments for estimating the probability of recidivation and find little evidence that they meet such standards.

It is not possible now, and may not be possible in the foreseeable future, to predict violent sexual recidivation with sufficient sensitivity and specificity to protect the public without putting a high burden on non-violent offenders.

If risk assessment is to pass constitutional and Daubert challenges, it must conform to scientific standards. We have suggested how this might be done by the development of a data base of agreed-upon variables and protocols for ascertaining those variables. It will be necessary to impose standards on the use of 'clinical' observations if it is to meet scientific standards of reliability and validity. Perhaps clinical testimony should be restricted to providing proof of inability to control behaviour.

REFERENCES

American Psychiatric Association, ©1999 *Dangerous Sex Offenders: A Task Force Report of the American Psychiatric Association*. Washington, DC: American Psychiatric Association.

Baime, D. 2001 *Report to the Supreme Court Systematic Proportionality Review Project.* www.judiciary.state.nj.us/baime/baimereport.pdf.

Beck, A. & Shipley, B. 1989 *Recidivism of Prisoners Released in 1983.* NCJ116261: Bureau of Justice Statistics.

Commonwealth of Pennsylvania, Sexual Offenders Assessment Board. 2003 *Sex Offender Education Video Transcript.* www.meganslaw.state.pa.us/soab/cwp/view. asp?a=3&q=189962.

Cortoni, F. A. 1998 *The relationship between attachment styles, coping, the use of sex as a coping stategey, and juvenile sexual history in sexual offenders,* unpublished doctoral dissertation, Queen's University, Kingston, Ontario, Canada.

Dolan, M. & Doyle, M. 2000 Violence risk prediction. Clinical and actuarial measures and the role of the Psychopathy Checklist. *British Journal of Psychiatry,* **177,** 303–311.

Epperson, D., Kaul, J., Huot, S., Hesselton, D., Alexander, W. & Goldman, R. 2000 *Minnesota Sex Offender Screening Tool-Revised (MnSOST-R): Development, Performance, and Recommended Risk Level Cut Scores.* www.psychology.iastate.edu/faculty/epperson/ webtotalpackage.pdf.

Epperson, D., Kaul, J., Huot, S., Hesselton, D., Alexander, W. & Goldman, R. 2002 *MnSOST-R Score Recording Sheet.* www.psychology.iastate.edu/faculty/epperson/ onepagescoresheet.doc and www.psychology.iastate.edu/faculty/epperson/MnSOST-RScoring03312000.pdf.

Epperson, D., Kaul, J., Huot, S., Hesselton, D., Alexander, W. & Goldman, R. 2003 *Minnesota Sex Offender Screening Tool-Revised (MnSOST-R): Development, Validation, and Recommended Risk Level Cut Scores.* www.psychology.iastate.edu/faculty/epperson/ TechUpdatePaperl2–03.pdf.

Fisher, R. A. 1956 *Statistical Methods and Scientific Inference.* New York: Hafner.

Gendreau, P., Little, T. & Goggin, C. 1996 A meta-analysis of the predictors of adult offender recidivism: What works. *Criminolgy,* **34,** 575–607.

Hanson, R. & Bussière, M. 1998 Predicting relapse: a meta-analysis of sexual offender recidivism studies. *Journal of Consulting and Clinical Psychology,* **66,** 348–362.

Hanson, R. K. & Harris, A. J. R. 1998 *Dynamic predictors of sexual recidivism.* (User Report 1998–01.) Ottawa, Ontario, Canada: Department of the Solicitor General of Canada.

Hanson, R. & Thornton, D. 1997 *Development of a Brief Actuarial Risk Scale for Sexual Offense Recidivism* (User Report No. 1997–04.) Ottawa, Ontario, Canada: Department of the Solicitor General of Canada. www.psepc-sppcc.gc.ca/publications/ corrections/199704_e.pdf.

Hanson, R. & Thornton, D. 1999 *Static-99: Improving Actuarial Risk Assessments for Sexual Offenders* (User Report No. 1999–02.) Ottawa, Ontario, Canada: Department of the Solicitor General of Canada. www.psepc-sppcc.gc.ca/publications/ corrections/199902_e.pdf.

Kadane, J. B. 1990 Comment: codifying chance. *Statistical Science,* **5,** 18–20.

Kansas v. Hendricks, 521 U.S. 346 (1997).

Kansas v. Crane, 122 S. Ct. 867 (2002).

Monahan, J. & Wexler, D. B. 1978 A definite maybe: proof and probability in civil commitment. *Law and Human Behavior,* **2,** 37–42.

Mosteller, F. & Youtz, C. 1990 Quantifying probabilistic expressions. *Statistical Science,* **5,** 2–12.

People v. Ghilotti, 44 p. 3d 949 (Ca 2002).

Seidman, B. T., Marshall, W. L., Hudson, S. M. & Robertson, P. J. 1994 An examination of intimacy and loneliness in sex offenders. *J. Interpersonal Violence,* **9,** 518–534.

Solicitor General of Canada, 1996 Child molester recidivism. *Research Summary,* **1.** www.psepc-sppcc.gc.ca/publications/corrections/pdf/199670_e.pdf.

Solicitor General of Canada, 1997 Predictors of sex offenses recidivism. *Research Summary,*
    2. www.psepc-sppcc.gc.ca/publications/corrections/pdf/199779_e.pdf.
Texas Comptroller of Public Accounts, Texas Performance Review, Vol. 2, Chapter 7, Section
    PS 8, 'Implement a Civil Committment [*sic*] Procedure for Sexually Violent Predators.
    www.window.state.tx.us/tpr/tpr5/7ps/ps08.html.
Wilson, R. J. 1999 Emotional congruence in sexual offenders against children. *Sexual Abuse:*
    J. Res. Treatment, **11,** 33–47.

F. PATENT MISCONDUCT

This case concerns alleged misconduct by an inventor, adding material to his notebook after it was witnessed in an effort to mislead the Patent Office about when he thought of the added material. The issue boils down to whether the added text lies above or below the witness line on the page.

To support the allegation of inequitable conduct, the defendant measured the concentration of 13 trace elements on the raw paper, the text, the witness line, and the intersection between the text and witness lines. Finding that the intersections were more similar to the text than to the witness line, they proposed that therefore the text lies on top of the witness line, so it was added later, thus supporting the claim of inequitable conduct.

The work on the paper reanalyzes the data, showing that the paper and witness line are very similar, as are the text and the intersections. But this means that the witness line is so thin as to be essentially invisible to the electron beam used to measure the concentrations. Therefore the data would be the same whether the witness line were above or below the text line, so there is no evidence of misconduct in the data.

1

CROSSING LINES IN A PATENT CASE

Joseph B. Kadane

The U.S. Constitution provides authorization for Congress "to promote the progress of science and the useful arts, by securing for limited times to...inventors the exclusive right to their...discoveries." A patent provides for protection for the use of an invention in return for making public enough details that a person skilled in the field could use the invention. To be patentable, an invention must be new and not obvious.

To obtain a patent, an inventor applies to the U.S. Patent and Trademark Office, specifying what the invention is, what it is useful for and what the claimed novelty is. Infrequently, where the question arises of whether the inventor was first to conceive the claimed subject matter, the inventor submits a notebook showing the conception. Each page of such a notebook is signed and dated by a witness, attesting to the fact that the inventor disclosed the ideas on the page to the witness on that date. An inventor then "prosecutes" the patent, responding to issues raised by the patent examiner and amending the claims if necessary. If the patent examiner is satisfied that all the conditions have been met, the Patent Office issues a patent. Because the exchange between the Patent Office and the inventor is not done on an adversarial basis and there is no "opposing party," the inventor is expected to adhere to a high standard of honest dealing with the Patent Office.

Infringement of a patent is the unauthorized use of the patent before its expiration. Typically, the patent holder brings suit against an alleged infringer, asking for an injunction against further infringement and for monetary damages due to the infringement. In defense, the alleged infringer can present several arguments: that the patent should not have been issued in the first place, that inequitable conduct by the patent holder in his dealings with the U.S. Patent Office makes the patent unenforceable, or that the patent does not cover the alleged infringement. The accused infringer can establish invalidity or inequitable conduct defenses only by clear and convincing evidence, because patents have a presumption of validity and charges of inequitable conduct are all too easy to make.

In the present case, the inventor was Jerome H. Lemelson, whose patents are owned after his death by the Lemelson Medical, Education and Research Foundation, Limited Partnership ("Lemelson Foundation Partnership"), which has collected over a billion dollars of license fees on his patents. The defendant was the Intel Corporation; the patent concerned an invention used in the automatic manufacture of computer chips. Among other defenses, Intel argued for inequitable conduct by Lemelson in prosecuting the patent. Intel's charge was that Mr. Lemelson had added certain written material to a page of his notebook after it had been witnessed the first two times (the page was later witnessed by others as well). Thus, Intel argued that Lemelson falsely claimed an earlier date of invention of the added material than is true.

An inspection of the page of the notebook in question (see Figure 1), shows that it was witnessed several times. The specific issue is whether the witness line labeled

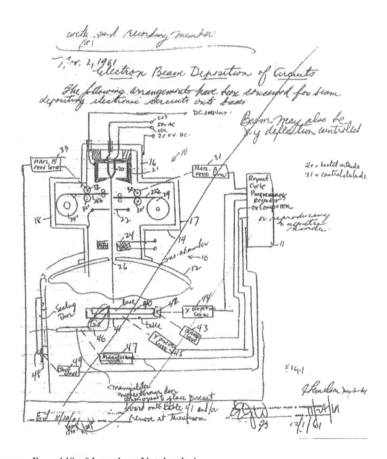

FIGURE 1. Page 140 of Lemelson Notebook A.

"LER," the lower and fainter of the two diagonal lines rising from lower left to upper right, lies above or below the portion of the text on the page reading "manipulator moves through door when open to place circuit board onto table 41."

In support of this allegation, Intel hired a physicist, Dr. Thomas A. Cahill of the University of California at Davis, who trained an x-ray beam at various spots on the page in question, including where the LER line and the text cross. The atoms of the page emit fluorescent x-rays when struck by the x-ray beam. The emissions are detected and analyzed as a mixture distribution to determine the relative amount of each element excited. Dr. Cahill measured the concentrations of 13 trace elements relative to that of argon at points he classified as paper, witness line, text, and the intersection between the witness line and the text (Table 1). The first 17 points are in order taken in a straight line going upward. They are in a group of 26 equally spaced points, nine of which were excluded from analysis because they were unclassified as to what they represented. The last four points are from a series of seven on a line perpendicular to the first line, and exclude three unclassified points.

Dr. Cahill theorized that the elemental concentrations at the intersection would appear most similar to the elemental concentrations of the ink that was on top (applied second). If the intersection matched the ink of the supposedly added text, it would show that the text was on top of the witness line, thus post-dating the early witnessing date.

Intel also hired a statistician, Dr. George R. Fegan of Santa Clara University, to analyze the data. Dr. Fegan did not examine or use the observations classified by Dr. Cahill as paper. Dr. Fegan conducted two t tests for each element, testing the null hypothesis of equality of relative concentration between the intersection and, respectively, the witness line and the text (Table 2). Using the resulting p values as if they were likelihoods, Dr. Fegan then used Bayes' Theorem to compute odds of over a million to one that the text line lies on top of the witness line at the intersection. If true, this would imply that the text had been added after the second witness line, which would substantiate Intel's claim.

It should be noted that Intel's analysis depends on an assumption of shielding, that the ink on the page is sufficiently thick that it would prevent the x-ray beam from reaching and exciting the atoms below the top layer of ink. This is equivalent to the belief that a ball-point pen mark on your hand would obscure your bones in a medical x-ray. A priori this is hard to believe.

I was retained by the Lemelson Foundation Partnership to assess the statistical analysis of Dr. Cahill's data. I found that the data collection suffered from various infirmities, of which the most important was a very small sample size. When a measurement fell below the minimum detection level, Dr. Cahill inserted a value of half that level, and Dr. Fegan treated these values as if they were valid observations, inflating the amount of data apparently available. (In the plots below, I also use half the minimum detection level for such observations.) Additionally, of course, the use of ratios of p values to substitute for likelihood ratios is not justified by Bayes' Theorem and violates the likelihood principle. Nonetheless, Dr. Fegan's analysis did seem to suggest that perhaps the intersections were more like the text than like the witness line.

To examine the data, my first impulse was to plot it. I have 13 elements (but exclude chlorine because I agree with Dr. Cahill that salt (NaCl) can be present

TABLE I. Relative Concentrations

Classification	P	P	P	P	P	I	I	I	I
S	<0.93	<0.83	<0.76	<0.74	<0.813	<1.01	<1.03	<0.79	1.707
Cl	1.488	1.610	1.634	1.901	2.383	1.371	1.458	1.277	1.852
K	3.227	3.485	2.831	2.954	3.845	2.147	2.143	1.746	2.298
Ca	3.252	3.155	2.570	2.732	2.410	2.208	2.165	1.778	2.137
Ti	1.534	1.714	1.362	1.318	2.473	1.594	1.152	0.910	1.398
V	0.060	<0.08	0.079	0.070	0.086	<0.05	0.045	0.054	0.053
Cr	0.094	0.132	0.097	0.112	0.074	<0.06	0.046	0.042	0.057
Mn	0.301	0.309	0.272	0.316	0.372	0.259	0.196	0.154	0.199
Fe	4.183	4.856	5.126	4.305	5.277	4.085	3.715	3.150	4.067
Ni	0.346	0.331	0.246	0.300	0.164	0.094	0.074	<0.057	<0.07
Cu	0.629	0.521	0.526	0.522	0.547	2.104	2.085	1.728	2.087
Zn	0.483	0.504	0.340	0.405	0.363	0.393	0.353	0.285	0.295
Ga	0.282	0.295	0.235	0.388	0.245	0.149	0.086	<0.06	<0.10

Classification	I	W	W	W	P	T	T	T
S	1.242	<0.87	<0.76	<0.81	<1.04	2.233	1.205	1.319
Cl	2.330	1.581	1.955	1.804	1.570	1.332	1.260	1.183
K	2.202	3.012	2.665	2.811	2.429	2.403	2.220	2.914
Ca	2.765	2.455	2.131	2.379	2.155	2.188	1.844	1.969
Ti	1.490	2.415	1.974	2.275	1.196	1.103	1.343	0.935
V	0.034	<0.06	0.062	<0.05	0.048	0.056	0.055	<0.04
Cr	0.078	0.093	0.079	0.073	0.059	0.055	<0.04	0.031
Mn	0.165	0.312	0.285	0.257	0.182	0.181	0.173	0.199
Fe	3.679	4.464	4.430	5.712	3.768	4.837	3.735	4.065
Ni	0.074	0.137	0.184	0.179	0.078	<0.07	0.057	<0.07
Cu	2.467	0.392	0.353	0.348	0.232	2.376	2.533	2.313
Zn	0.319	0.210	0.262	0.188	0.186	0.335	0.374	0.440
Ga	0.081	0.185	0.242	0.216	0.115	0.148	0.084	0.109

Classification	W	W	W	P
S	0.677	<0.92	<0.82	<0.739
Cl	1.794	1.649	1.741	3.159
K	2.578	2.786	2.786	2.047
Ca	2.778	2.567	2.629	2.29
Ti	1.133	1.118	1.389	3.02
V	0.049	0.050	<0.063	0.107
Cr	0.079	0.133	0.106	0.094
Mn	0.291	0.307	0.277	0.279
Fe	4.008	4.420	3.911	3.685
Ni	0.336	0.316	0.246	0.288
Cu	0.737	0.877	0.678	0.819
Zn	0.329	0.439	0.397	0.335
Ga	0.256	0.273	0.224	0.282

Key to Classification: W = Witness line, T = Text line, I = Intersection, P = Paper

TABLE 2. *P* Values of Two-Sample
t-Tests of Witness Line vs. Intersection
and Text vs. Intersection

	Witness vs. Int.	Text vs. Int.
S	0.11697	0.13843
Cl	0.61536	0.17731
K	0.00017	0.08571
Ca	0.14526	0.38340
Ti	0.18395	0.36648
V	0.91863	0.90202
Cr	0.00734	0.29048
Mn	0.00078	0.69592
Fe	0.04791	0.20052
Ni	0.00143	0.32268
Cu	0.00000	0.10315
Zn	0.62588	0.17002
Ga	0.00009	0.29757

from human perspiration). For each of the 12 remaining elements, there are four conditions: witness line, text line, intersection and paper. Hence, 12 groups of four side-by-side scatter plots seemed like a good place to start; these are recorded in Figure 2.

Examining first the paper compared to the witness line, for each element they are remarkably similar. This implies that either, by happenstance, the paper and the ink of the witness line have nearly identical relative concentrations of each of the trace elements, or that the witness line is essentially invisible to the measurement apparatus. The first is hard to believe; the second implies that the intersection between the lines will have readings similar to those of the text, *whichever* line lies on top. Hence the data would have no evidential value whatsoever concerning superimposition.

The Lemelson Foundation Partnership also hired a physicist, Dr. Mark Rivers of the University of Chicago and Argonne National Laboratory. Dr. Rivers performed experiments showing directly that fluorescent x-rays passed through the notebook page readily, so neither the witness ink nor the text ink shielded the x-ray fluorescence from the paper below. Consequently the observations reported by Dr. Cahill and analyzed by Dr. Fegan should include fluorescence from the relevant ink and from the paper in each case. This suggests that the observations should satisfy

observed intersection + observed paper = observed witness line + observed text line,

because each side of this equation has two paper, one witness, and one text reading. To examine this idea further, I bootstrapped a confidence interval for each of the 13 elements. Of the 21 points, there are seven observations of paper, five of the intersection between witness line and text, six of the witness line, and three of the text. A bootstrap sample is computed by drawing seven observations from

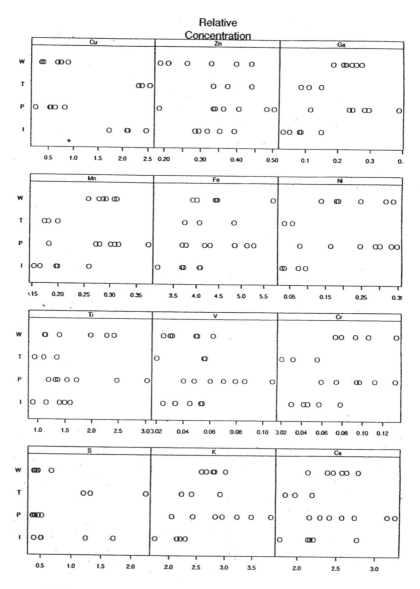

FIGURE 2. Relative concentrations of 12 elements in witness line, text line, paper, and intersection.

the paper (with replacement, so repetition is okay), and finding their median (call it P). Similarly, five observations with replacement are drawn from the intersection observations, and their median, I, is computed. And similarly for W, the witness line median and T for the text median computing the quantity $P + I - W - T$ 5000 times and excluding the highest and lowest 2.5 percent yields the intervals reported for each element. Figure 3 records the results, in the form of a 95 percent confidence

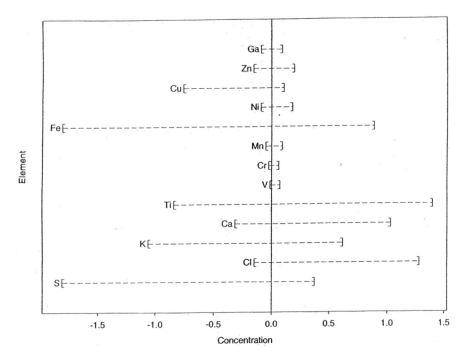

FIGURE 3. 95% confidence intervals around median concentrations.

interval around the median of each of 5000 samples. The results show that each such interval includes zero. I don't usually rely on bootstrapping, but it seemed like a useful technique in this instance.

All of my analyses, and Dr. Rivers's as well, point in the same direction: the witness line is so thin that it does not contribute to the recorded elemental concentrations at any point. The text line does, and a contribution from the text line shows up in the intersections. Neither ink shields the material below it. The witness line could lie above or below the text line at the intersections, and the observed data would be the same. Hence there is no evidence in these data to support Intel's allegation.

After reviewing my declaration and Dr. Rivers's, Intel withdrew Drs. Cahill and Fegan as expert witnesses, and abandoned the contention that these data show inequitable conduct.

ACKNOWLEDGMENTS

I thank Daniel Van Boxel for research assistance, Howard Seltman for help with a figure, and Louis J. Hoffman, Esq., and Victoria Gruver Curtin, Esq., for their helpful comments. I also thank the Lemelson Foundation Partnership and the Lemelson family for permission to reproduce Figure 1.

REFERENCES AND FURTHER READING

DeGroot, M. H. (1973), "Doing What Comes Naturally: Interpreting a Tail Area as a Posterior Probability or as a Likelihood Ratio," *Journal of the American Statistical Association,* Vol. 68, No. 344, 966–969.

Durham, A. L. (1999), *Patent Law Essentials: A Concise Guide,* Westport, Conn: Quorum Books.

Efron, B. and Tibshirani, R. (1993), *An Introduction to the Bootstrap,* London: Chapman and Hall.

Part 5

ENGLISH LAW

The fundamental structure of American law is of course derived from the law of the United Kingdom. Indeed, before the American revolution, the colonists were appealing for the "rights of Englishmen." While a few centuries of developments on both sides of the Atlantic have led to divergences, many of the ideas behind the laws remain common. The first section of this part is a very brief review of the important differences as they relate to the topics of this book.

We then move on to a series of cases that have stirred interest on both sides of the Atlantic. Called variously "cot death" or "sudden infant death syndrome" (SIDS), very young children die in a crib for no apparent reason. Sometimes several children from the same family suffer this fate. Suspicion then falls on the parents that they in fact murdered their children. This suspicion was buttressed in England by Meadow's Law, promulgated by an eminent pediatrician, Professor Sir Roy Meadow. His law states "one cot death is a tragedy, two cot deaths is suspicious, and, until the contrary is proved, three cot deaths is murder." In support of this law, Meadow cites data that the probability of a cot death in an affluent non-smoking family with a mother over 26 years of age is 1 in 8543. Using independence (and this is the key issue), the probability of two or three such deaths becomes minuscule. The first paper in the second section is Ray Hill's recounting of the history of three cot death cases in England. The upshot is that it is likely that there is a genetic predisposition to cot deaths, which would lead to clusters of cot deaths in non-murderous families. The second item in this section is the decision of the Court of Appeal exonerating Sally Clark, one of the mothers convicted of murdering her children.

After this exoneration, the General Medical Council in the UK "struck off" Dr. Meadows, which had the effect of removing his license to practice medicine. It did so on the basis of his giving "misleading and erroneous evidence" in the cot cases. This decision was overturned by the High Court, on the basis that "he acted in good faith." For a review of these actions see the article in the Guardian of

Feb. 17, 2006 at http://society.guardian.co.uk/nhsperformance/story/0,,1512227,00.html. It is interesting to compare the attitudes of the General Medical Council and the High Court to those of the Pennsylvania Courts in the Panitz case reported in section 1E.

Another point of comparison is to the famous Collins case from California. In this case, the alleged probabilities of a number of items of circumstantial evidence were multiplied together (assuming independence without proof) in an effort to show that the probability of another couple answering the description of the accused was small. The Supreme Court of California overturned the conviction; this case completes the section. Thus in both the cot cases and the Collins case, casual use of independence, without justification, led to judicial errors that were later reversed.

Section C of this part records the history of an attempt to teach Bayes Theorem to a jury so that they could properly take into account DNA evidence that was contrary to the other evidence in the trial. The first item is a brief paper by Peter Donnelly, recounting his experiences in leading this effort. It went to trial twice; the first trial was invalidated because the appeals court was dissatisfied with the instructions to the jury. Both decisions are offered here; in the second decision the court comes perilously close to rejecting probabilistic evidence entirely.

A. A VERY BRIEF INTRODUCTION
TO ENGLISH LAW

The final two cases examined in this book are from England. Much of English law is like U.S. law, because until 1776 the original 13 states were English colonies. There are, however, important differences.

Sovereignty in England is undivided. Thus there is no analog in England of the relationship between the U.S. federal government and the states. Additionally, there is no written constitution. Legal rights are therefore principally protected by precedent cases.

Both of the cases discussed below are criminal cases. Less serious criminal cases are heard first in magistrates court. More serious cases like the charges in our two cases (respectively murder and rape) are heard in Crown Court. Both the law and the facts can be appealed to the Court of Appeals (Criminal Division). Cases from the Court of Appeals can, in theory, but rarely in practice, be appealed to the House of Lords (Eddey (1987), pp. 60–63).

The legal profession is divided into two disciplines. Until 1990, solicitors dealt with the business end of the provision of lawyer services: meeting with clients, preparing contracts and wills, preparing cases for court, and appearing in Magistrates Court. When a case is to be heard by Crown Court or higher courts, a solicitor would hire a barrister to argue the case in court. Since that time, there has been a series of moves toward permitting each discipline to perform the functions of the other (Slapper and Kelly (2001), pp. 501–513).

Since 1967, juries in England decide cases by majority vote (Partington (2000), pp. 108, 109).

REFERENCES

Eddy, K. J. (1987). The English Legal System, London: Sweet and Maxwell.
Partington, M. (2000). An Introduction to the English Legal System, Oxford: Oxford University Press.
Slapper, G. and D. Kelly (2001). The English Legal System, 5th edition, London: Cavendish Publishing Limited.

B. CRIMINAL CONVICTIONS AND STATISTICAL INDEPENDENCE

This section considers a series of cases in England concerning cot death, otherwise known as Sudden Infant Death Syndrome (SIDS). In families that had experienced more than one such death, the mothers were charged with murder on that basis alone. The heart of the reasoning was that the probability of such a death is low, and, using independence, the probability of multiple such deaths minuscule. Hence the conclusion of murder.

Later evidence shows a genetic component to SIDS, which means that there is positive correlation between the events of multiple such deaths, which invalidates the assumption of independence.

The introductory essay by Ray Hill gives a history of these cases. Excerpts from the decision in the Sally Clark case give the reasoning of the Court of Appeal, overturning her conviction.

The final item in this section is the decision in the Collins case from California. Here also the issue of independence of events came up in the conviction of a couple based on circumstantial evidence.

1

REFLECTIONS ON THE COT DEATH CASES

Ray Hill

The cases of Clark, Cannings and Patel had many common features. In each case, the mother was acknowledged as being a caring parent, with no history of child abuse. In each case, the husband, family and friends were all fully supportive and convinced of the mother's innocence. In each case, the medical evidence against the mother was decidedly slim. And, in each case, a key prosecution witness was Professor Sir Roy Meadow.

Professor Meadow is an eminent British pediatrician and an advocate of the dictum, known in Britain as "Meadow's law," that "one cot death is a tragedy, two cot deaths is suspicious and, until the contrary is proved, three cot deaths is murder." There is little doubt that such thinking played a part in the charging of Clark, Cannings and Patel. In each case, the first death was recorded as natural and it appears to have been the mere coincidence of a second or third death that served as the trigger for a criminal investigation. Let us consider the three cases in more detail.

SALLY CLARK

Sally Clark, a solicitor from Cheshire, lost her first baby, Christopher, in 1996. There was evidence of a respiratory infection and the death was recorded as natural. In 1998, the Clark's second child, Harry, died at 8 weeks old. Sally was charged with the murder of both babies.

At her trial in November 1999, Professor Meadow claimed that, in a family like Sally Clark's, the chance of two babies both dying a cot death was 1 in 73 million. This figure came from the "Confidential enquiry for stillbirths and deaths in infancy" (CESDI) report[1], an authoritative and detailed study of the deaths of all babies in five regions of England between 1993 and 1996. There it was estimated that the chance of a randomly chosen baby dying a cot death was 1 in 1303. But, if the child was from an affluent non-smoking family with the mother aged over 26, then the chance fell to 1 in 8543. From this last figure it was estimated that the chance of Sally Clark suffering two cot deaths was 1 in 8543 times 8543, i.e. about 1 in 73 million. Meadow went on to say that "in England, Wales and Scotland there are about 700,000 live-births a year, so it is saying by chance that happening will occur about once every hundred years."

Sally was convicted of murdering both babies by a majority of 10 to 2.

The first appeal

At Sally's first appeal, in October 2000, the statistical evidence was challenged on two main grounds. The first was the invalidity of the squaring of 8543, given that the two deaths could not be deemed to be independent events. The second was the likely misinterpretation of what the figures meant anyway, and that jurors may have fallen into the trap of the "prosecutor's fallacy," interpreting "the probability of two deaths, given innocence" as "the probability of innocence, given two deaths." The appeal failed.

I did not become involved in the Clark case until I had read the Appeal Court judgment. I had been shocked, though not surprised, that the jury had returned a guilty verdict. But I was both shocked and surprised that three High Court judges could so fail to see the significance of the flawed statistical evidence and not recognise the impact it must surely have had upon jurors.

The Appeal judges accepted that the 73 million figure was inaccurate but took the view that this was unimportant since the Crown had been using the CESDI data "not for its precise figures, but for a very broad point, namely the rarity of a double cot death." On the issue of the prosecutor's fallacy, they were dismissive, saying that it was "stating the obvious" to say that the two probabilities concerned were not the same and that "you do not need the label 'the prosecutor's fallacy' for that to be clear." Meadow's equating of 73 million with a double cot death once every hundred years was endorsed by the Appeal Court judges as being "a straight mathematical calculation to anyone who knew the birth-rate over England, Scotland and Wales was approximately 700,000 a year." (Far from being a straight calculation, the deduction is in fact complete nonsense, because the figures used by Meadow were not applicable to the whole population, but only to the much smaller sub-population having three special characteristics.)

The second appeal

I was by no means the only person who was outraged by the misuse of statistics in the Clark trial. In October 2001, the Royal Statistical Society (RSS) issued a statement expressing its concern about the use of statistics in the courts, and in the Clark case in particular. In January 2002, the President of the RSS wrote to the Lord Chancellor, who heads the justice system, expressing his extreme concern and calling for the courts to rethink the way they handle statistical evidence.

For my own part, I set out to explain the statistical issues involved in the Clark case in elementary terms. I wrote an article,[2] which I hoped to get published in one of the broadsheet newspapers. I failed to do so but did attract some media attention, which led to my being invited to give a paper at a medical conference on cot deaths. For this second paper,[3] I made a detailed study of the CESDI report, so that I could estimate more accurate probabilities relevant to the Clark case. My two papers were submitted by the defence team at Sally's second appeal.

For the second appeal, there was dramatic new medical evidence. A blood test report, obtained by the original pathologist but not made available to the court, was uncovered from the records at Macclesfield Hospital. This provided evidence that baby Harry had been suffering from an overwhelming bacterial infection. No fewer than nine renowned medical experts were lined up to support the proposition that this was the most likely cause of Harry's death.

The second appeal began on January 28th, 2003. The two grounds of appeal were the new medical evidence and the fresh statistical evidence. The medical evidence was taken first. It soon became clear, on this ground alone, that the prosecution case could not be sustained. Early on the second day, Prosecution Counsel informed the judges that "the prosecution no longer seeks to uphold these convictions." The appeal was won, and before the statistical evidence had even been heard. However, the statistics were considered in the judgment, which read: "It seems likely that if statistics had been argued before us, it would have provided a quite distinct basis on which the appeal had to be allowed." Meadow's evidence was described as "manifestly wrong" and "grossly misleading."

Sally Clark was freed to resume her life with her husband and surviving child. She had spent over 3 years in jail for a crime which had not even happened. A moving account of the whole Sally Clark story is given in John Batt's book.[4]

ANGELA CANNINGS

The trial of Angela Cannings in April 2002 could not have come at a worse time. Sally Clark's first appeal had failed and Angela, a shop assistant from Wiltshire, had lost three babies to sudden unexplained deaths, though she was charged with the murder of only two of them. The medical evidence was even flimsier than in the Clark case, but Angela was convicted of double murder by a unanimous verdict.

I believe that statistics again played a key role—but this time by their very absence. Because of the controversy over the use of statistics in the Clark trial, it was agreed between prosecuting and defence counsels that no references would be made to statistics in Angela's trial. I am sure that this agreement was a major error by the defence team. Although there were many similarities between the cases of Clark and Cannings, one big difference was that the Cannings parents were both smokers. Smoking is a huge risk factor for cot deaths and its presence meant that the prosecution could not have used anything like the 73 million argument in this case. Indeed, any proper use of statistics would have favoured a natural, rather than unnatural, explanation for the deaths.

The danger in avoiding statistics altogether is that jurors may well draw incorrect conclusions for themselves, believing that occurrences of rare events are somehow beyond coincidence. The 73 million figure was still fresh in people's minds and John Batt, who sat through the trials of both Clark and Cannings, is convinced that Meadow's infamous statistic must have entered the privacy of the jury room (see Batt,[4] pages 285–288).

In contrast with her trial, Angela's appeal, in December 2003, could not have come at a better time. Sally Clark's second appeal had been won and Meadow's evidence discredited. The defence now argued statistics—medical statistician Professor Robert Carpenter was a defence witness—and also argued that Meadow had, despite the agreement at trial, "slipped statistics in through the back door" by stating in his evidence that the occurrence of three cot deaths was "very, very rare." The Appeal Court judges agreed and Angela's convictions were quashed.

TRUPTI PATEL

Sandwiched between the successful appeals of Clark and Cannings was the trial of Trupti Patel, a pharmacist from Berkshire, in June 2003. At the time of Trupti's arrest in May 2002, the outlook for her seemed bleak. Both Clark and Cannings were serving life sentences for double murder, and Trupti was accused of suffocating three of her babies. Her solicitor asked me to write a report.

It seemed to me that the best way to counter Meadow's law was to estimate the relative chances of cot death or murder, given that two or more sudden deaths have occurred. Obtaining reliable estimates based on limited data is fraught with difficulty, but my calculations gave the following rough estimates. Single cot deaths outnumber single murders by about 17 to 1, double cot deaths outnumber double murders by about 9 to 1 and triple cot deaths outnumber triple murders by about 2 to 1. So each successive death does give rise to some slightly increased suspicion, but to nothing like the extent that Meadow's law would imply. In particular, when multiple sudden infant deaths have occurred in a family, there is no initial reason to suppose that they are more likely to be homicide than natural.

It turned out that Professor Carpenter had also been solicited for a report and he had independently arrived at a similar conclusion in respect of three deaths. Details of my own calculations may be found in my paper[5] based on my report for the Patel trial.

In the event, this statistical evidence may not have been crucial. By the time of trial, Sally Clark had been cleared and it was now accepted that multiple cot deaths were not as rare as had previously been thought. Also, in the Patel case there was evidence of probable genetic links, with a history of unexplained infant deaths on both sides of the family. A key witness at trial was Trupti's 80-year-old grandmother, flown in from India to testify that five of her 12 children had died within 6 weeks of birth.

However, there was one other issue which threatened to win the day for the prosecution: the question of baby Mia's fractured ribs.

RIB FRACTURES

At the trials of Sally Clark and Trupti Patel, and in other similar cases, an important aspect has been evidence of rib fractures in one of the babies. What happened in the Patel case provides an excellent example of how careful one has to be when drawing conclusions from limited statistical data.

The arguments in such cases typically run as follows. It is accepted that rib fractures are not in themselves a cause of death. The prosecution will claim that they were caused during an act of suffocation, by applying extreme pressure to the chest, or, where the fractures seem unrelated to the cause of death, that they simply provide evidence of previous child abuse. The defence will claim that the fractures most likely occurred during the cardiopulmonary resuscitation (CPR) that is applied in trying to save the baby's life.

Up until the Patel trial, it had generally been accepted that

(1) in cases of known child abuse, it is fairly common to find evidence of rib fractures and

(2) it is very rarely that CPR will cause rib fractures (for example, the judgment at Clark's first appeal had read: "All the experts agreed that it would be very unusual to see a fracture dislocation of ribs from resuscitation").

The prosecution will claim from statements (1) and (2) that, in a case where fractures have been found, the abuse explanation is much more likely than the CPR explanation. But this deduction is totally false: another example of the prosecutor's fallacy. It is easy to devise realistic models in which (1) and (2) are true and yet, in a case of rib fractures where CPR is *known* to have taken place, the CPR explanation is more likely than the abuse.

So, in the Patel trial, there was the potential for a serious misinterpretation of probabilities. But, in a dramatic twist, the prosecution argument fell for a quite different reason, namely that statement (2) was not even correct.

The X-ray evidence of rib fractures in infants can be very hard to see. In the Patel case, it was not initially observed and it took intense scrutiny of the X-rays for the fracture to be identified, and this in a case where fractures were specifically looked for as evidence of abuse. So, might it be the case, where CPR has been applied but where there is no suspicion of foul play, that any rib fractures caused by CPR are not observed simply because there is no good reason to look for them? In short, might rib fractures due to CPR be more common than had previously been thought?

In prosecution evidence at the Patel trial, the eminent paediatric pathologist Professor Tony Risdon had stated that "rib fractures...are extremely unlikely to be caused by resuscitation." He had told police that he had never personally come across a case of a baby's ribs being broken in the hundreds of post-mortems he had carried out. Then, during the trial, he informed the Crown Prosecution Service that, in the previous month alone, he had come across three cases. When he addressed the jury 6 days later, he said that the evidence of the rib fractures "must be removed." It was a decisive moment.

After a trial lasting six and a half weeks, it took the jury less than 90 minutes to find Trupti Patel not guilty.

A NEW BEGINNING

Following Angela Cannings's appeal, in January 2004 the government ordered a review of all similar cases in which Professor Meadow's evidence had helped to secure a conviction. Also, an expert committee was formed under Labour peer Helena Kennedy QC to consider how future sudden infant deaths should be investigated. The Kennedy report was published in September 2004, recommending a new national protocol. One of its aims is to ensure that the tragedy of losing children is never again compounded by innocent parents being accused of their murder.

I conclude with the words of BBC investigative journalist John Sweeney, whose television and radio documentaries played a vital role in the pursuit of justice for the cot death mothers: "One cot death is a tragedy, two cot deaths is a tragedy, and three cot deaths is a tragedy."

REFERENCES

References 2, 3 and 5 are available on request from r.hill@salford.ac.uk.

1. Fleming, P., Bacon, C., Blair, P. and Berry, P. J. (2000) *Sudden Unexpected Deaths in Infancy, the CESDI Studies 1993–1996*. London: Stationery Office.
2. Hill, R. (2002) Why Sally Clark is, probably, innocent. Unpublished.
3. Hill, R. (2004) Cot death or murder—weighing the probabilities. *Developmental Physiology Conference, June 2002*.
4. Batt, J. (2004) *Stolen Innocence; the Story of Sally Clark*. London: Ebury.
5. Hill, R. (2004) Multiple sudden infant deaths—coincidence or beyond coincidence? *Paediatric and Perinatal Epidemiology*, **18**, 320–326.

2

R V. SALLY CLARK

IN THE SUPREME COURT OF JUDICATURE
COURT OF APPEAL (CRIMINAL DIVISION)
REFERENCE BY THE CRIMINAL CASE
REVIEW COMMISSION
UNDER SECTION 9 OF THE CRIMINAL APPEAL ACT 1995

Royal Courts of Justice
Strand, London, WC2A 2LL

Before:
LORD JUSTICE KAY
MR JUSTICE HOLLAND
and
MRS JUSTICE HALLETT
R and SALLY CLARK Appellant
Mr R Spencer QC and Mr M L Chambers
(instructed by CPS, Cheshire)
for the Crown
Miss C P Montgomery QC and Mr J H Gregory
(instructed by Burton Copeland
for the Appellant)

Excerpted: paragraphs 1–7, 110–128 and 195–203.

Lord Justice Kay:

1. On 9 November 1999, Sally Clark was convicted by a majority of 10 to 2 in the Crown Court at Chester of the murder of her baby sons, Christopher and Harry. She appealed against her convictions but her appeal was dismissed on 2 October 2000.

2. There were those, including Mrs Clark's husband, who could not accept that she had killed her children and they continued to strive to demonstrate that the convictions were wrong. In due course, records of the results of microbiological tests performed on samples of Harry's blood, body tissue and cerebrospinal fluid gathered at post mortem were discovered. These had not featured at all in the evidence at

trial because all the lawyers involved on both sides were unaware of their existence. They were submitted to medical experts and this submission gave rise to expert evidence that suggested that Harry may not after all have been murdered but may have died from natural causes. This in turn cast doubt upon the jury's finding that Christopher was murdered.

3. This information was submitted to the Criminal Cases Review Commission ("The CCRC") with an application that the CCRC should refer the case back to the Court of Appeal. The CCRC considered the matter and made such enquiries as seemed appropriate.

4. On 2 July 2002, the CCRC concluded:

> "that there is a real possibility that the Court of Appeal will find that the new evidence renders Mrs Clark's convictions for the murders of Christopher and Harry unsafe."

5. Accordingly it referred the case back to this Court pursuant to its powers under Section 9 of the Criminal Appeal Act 1995.

6. On 28 and 29 January 2003 this court heard the appeal and concluded that the convictions were unsafe and must be set aside. The Crown did not seek a re-trial and accordingly Mrs Clark was released. In view of the public attention given to this case, we stated our reasons for our decision very briefly at the time but indicated that we would give detailed reasons at a later date. This judgment sets out our reasons.

7. The grounds of appeal settled on behalf of the appellant following the referral by the CCRC, as developed before the court, made two essential points. First and principally, the failure to disclose the information contained in the microbiological reports meant that important aspects of the case which should have been before the jury were never considered at trial. They contended that the failure to disclose the evidence and/or the existence of the new evidence rendered the resulting convictions unsafe. Secondly, they contended that statistical information given to the jury about the likelihood of two sudden and unexpected deaths of infants from natural causes misled the jury and painted a picture which is now accepted as overstating very considerably the rarity of two such events happening in the same family.

THE STATISTICAL EVIDENCE

110. Before turning to consider the evidence which it is alleged was not disclosed to the defence and the evidence now available that flows from it, we must record the other feature of the evidence at trial which it is contended may well have had an unfair impact upon the jury's considerations. That evidence was statistical evidence given by Professor Meadow.

111. Professor Meadow was asked about some statistical information as to the happening of two cot deaths within the same family, which at that time was about to be published in a report of a government funded multi-disciplinary research team, the Confidential Enquiry into Sudden Death in Infancy ("CESDI") entitled "Sudden Unexpected Deaths in Infancy" to which the Professor was then writing a Preface.

Professor Meadow said that it was "the most reliable study and easily the largest and in that sense the latest and the best" ever done in this country.

112. It was explained to the jury that there were factors that were suggested as relevant to the chances of a SIDS death within a given family; namely the age of the mother, whether there was a smoker in the household and the absence of a wage earner in the family. None of these factors had relevance to the Clark family and Professor Meadow was asked if a figure of 1 in 8,543 reflected the risk of there being a single SIDS within such a family. He agreed that it was. A table from the CESDI report was placed before the jury. He was then asked if the report calculated the risk of two infants dying of SIDS in that family by chance. His reply was:

"Yes, you have to multiply 1 in 8,543 times 1 in 8,543 and I think it gives that in the penultimate paragraph. It points out that it's approximately a chance of 1 in 73 million."

113. It seems that at this point Professor Meadow's voice was dropping and so the figure was repeated and then Professor Meadow added:

"in England, Wales and Scotland there are about say 700,000 live births a year, so it is saying by chance that happening will occur about once every hundred years."

114. Mr Spencer then pointed to the suspicious features alleged by the Crown in this present case and asked:

"So is this right, not only would the chance be 1 in 73 million but in addition in these two deaths there are features which would be regarded as suspicious in any event?"

He elicited the reply "I believe so."

115. All of this evidence was given without objection from the defence but Mr Bevan QC (who represented the appellant at trial and at the first appeal but not before us) cross-examined the doctor. He put to him figures from other research that suggested that the figure of 1 in 8,543 for a single cot death might be much too high. He then dealt with the chance of two cot deaths and Professor Meadow responded:

"This is why you take what's happened to all the children into account, and that is why you end up saying the chance of the children dying naturally in these circumstances is very, very long odds indeed one in 73 million."

116. He then added:

"it's the chance of backing that long odds outsider at the Grand National, you know; let's say it's a 80 to 1 chance, you back the winner last year, then the next year there's another horse at 80 to 1 and it is still 80 to 1 and you back it again and it wins. Now here we're in a situation that, you know, to get to these odds of 73 million you've got to back that 1 in 80 chance four years running, so yes, you might be very, very lucky because each time it's just been a 1 in 80 chance and you know, you've happened to have won it, but the chance of it happening four years running we all know is extraordinarily unlikely. So it's the same with these deaths. You have to say two unlikely events have happened and together it's very, very, very unlikely."

117. The table that was produced to the jury gave just the figures for probability of a SIDS death in families where one or more factors thought to be relevant were

present together with the figures when there was no such factor. In the CESDI report the table was accompanied by explanatory text but although this was available to the prosecution and the defence, it was not before the jury. It made clear the purpose of the information saying:

"The identification of families at higher risk of SIDS is of importance in allowing the appropriate deployment of scarce health care resources and in attempting to achieve changes in lifestyle or patterns of child care that might reduce this risk."

118. It did not in any way suggest that it provided statistical information that would enable diagnosis of an unnatural death in an individual case.

119. The report also made clear that the figures did not "take account of possible familial incidence of factors other than those included" in the table. It ended with the warning: "When a second SIDS death occurs in the same family, in addition to careful search for inherited disorder, there must always be a very thorough investigation of the circumstances—though it would be inappropriate to assume maltreatment was always the cause."

120. None of these qualifications were referred to by Professor Meadow in his evidence to the jury and thus it was the headline figures of 1 in 73 million that would be uppermost in the jury's minds with the evidence equated to the chances of backing four 80 to 1 winners of the Grand National in successive years.

121. Professor Berry was one of the four editors of the CESDI study. He made the point that simply squaring the figure was an illegitimate over simplification and he drew attention to the qualifications to which we have referred.

122. The trial judge clearly tried to divert the jury away from reliance on this statistical evidence. He said:

"I should, I think, members of the jury just sound a word of caution about the statistics. However compelling you may find them to be, we do not convict people in these courts on statistics. It would be a terrible day if that were so. If there is one SIDS death in a family, it does not mean that there cannot be another one in the same family."

123. This aspect of the case was raised on the first appeal. The areas of attack were threefold. First, evidence was called to show that the statistics were misleading; second, it was said that the evidence was led without regard to the guidance given by this Court in R v Doheny and Adams [1997] Cr App R 369; and third it was contended that the prosecution utilised the statistics in a way that gave rise to the "prosecutor's fallacy" identified in relation to DNA statistical evidence in R v Deen, The Times 10 January 1994.

124. As to the first point, the Court of Appeal (at paragraph 155) concluded:

"The existence of arguments against squaring was known to the jury at trial. Professor Berry made the points to which we have already referred, and the judge reminded the jury about these in his summing-up. But again the precise figures are not important since the Crown was making the broad point that repeated SIDS deaths were very unusual, in which exercise the number of noughts separating the lower risk households from higher risk households did not matter once the overall point was made, as here it was."

125. The court also rejected the second ground which was effectively a complaint that Professor Meadow trespassed beyond his mere expertise. The court said (paragraph 160):

> "No-one would know better than Professor Meadow that this important evidence as to whether these deaths were unnatural lay in the physical finding post–mortem, in the account of the last hours of the infants, and in the evidence and credibility of the parents—it certainly did not lie in statistics. And it is clear from reading his evidence that his conclusions were firmly based on that medical and circumstantial evidence, as we would expect."

126. As to "the prosecutor's fallacy" the court found merit in this argument saying:

> "Therefore we accept that when one is looking post facto at whether two deaths were natural or unnatural, the 1:73 million figure is no help. It is merely a distraction. All that matters for the jury is that when your child is born, you are at a very low risk of a true SIDS death, and at an even lower risk with a second child."

127. The court absolved Professor Meadow of misusing the figure in his evidence but added that "he did not help to explain this limited significance."

128. The court then asked themselves whether the jury might have focussed on that figure to the exclusion of the "real and compelling" evidence in the case. They reminded themselves of the warning given by the judge but concluded that there was some substance in the criticism. Nonetheless the court looked at this matter in the light of all the evidence and concluded that there was an "overwhelming case" against the appellant.

STATISTICAL EVIDENCE

195. Finally we should say a little about the statistical evidence led before the jury. The matter was the subject of only brief argument before us and we certainly heard none of the evidence.

196. It is unfortunate that the trial did not feature any consideration as to whether the statistical evidence should be admitted in evidence and particularly, whether its proper use would be likely to offer the jury any real assistance. Inherent in the evidence were dangers. The jury were required to return separate verdicts on the two counts but the 1 in 73 million figure encouraged consideration of the two counts together as a package. If the jury concluded that one or other death was not a SIDS case (whether from natural causes or from unnatural causes), then the chance that the other child's death was a SIDS case was 1 in 8,543 and the 1 in 73 million figure was wholly irrelevant.

197. In any event, juries know from their own experience that cot deaths are rare. The 1 in 8,543 figure can do nothing to identify whether or not an individual case is one of those rare cases.

198. Generally juries would not need evidence to tell them that two deaths in a family are much rarer still. Putting the evidence of 1 in 73 million before the jury with its related statistic that it was the equivalent of a single occurrence of two such deaths in the same family once in a century was tantamount to saying that without

consideration of the rest of the evidence one could be just about sure that this was a case of murder.

199. If the figure of 1 in 73 million accurately reflected the chance of two cot deaths in the same family, then the whole of the CONI scheme was effectively wasted effort. Seeking to provide guidance and monitoring against the possibility of a second cot death would be taking precautions against a risk that could effectively be discounted.

200. Like the Court of Appeal on the first occasion we are quite sure that the evidence should never have been before the jury in the way that it was when they considered their verdicts. If there had been a challenge to the admissibility of the evidence we would have thought that the wisest course would have been to exclude it altogether.

201. The argument before us would have addressed the question whether the 1 in 73 million figure was misleading in itself quite apart from the use made of it at trial. On the material before us, we think it very likely that it grossly overstates the chance of two sudden deaths within the same family from unexplained but natural causes. There is evidence to suggest that it may happen much more frequently than suggested by that figure although happily the risk remains a relatively unlikely one. The figure of 1 in 73 million was disputed by Professor Berry in his evidence who pointed to the obvious dangers of simply multiplying the risk of one such recurrence by the same figure to obtain the chance of two such deaths. Quite what impact all this evidence will have had on the jury will never be known but we rather suspect that with the graphic reference by Professor Meadow to the chances of backing long odds winners of the Grand National year after year it may have had a major effect on their thinking notwithstanding the efforts of the trial judge to down play it.

202. The Court of Appeal on the last occasion would, it seems clear to us, have felt obliged to allow the appeal but for their assessment of the rest of the evidence as overwhelming. In reaching that conclusion the Court was as misled by the absence of the evidence of the microbiological results as were the jury before it. We are quite satisfied that if the evidence in its entirety, as it is now known, had been known to the Court it would never have concluded that the evidence pointed overwhelmingly to guilt.

203. Thus it seems likely that if this matter had been fully argued before us we would, in all probability, have considered that the statistical evidence provided a quite distinct basis upon which the appeal had to be allowed.

3

THE PEOPLE V. COLLINS

THE PEOPLE, Plaintiff and Respondent, v. MALCOLM RICARDO
COLLINS, Defendant and Appellant

Crim. No. 11176

Supreme Court of California

68 Cal. 2d 319; 438 P.2d 33; 66 Cal. Rptr. 497; 36 A.L.R.3d 1176

March 11, 1968

COUNSEL: Rex K. DeGeorge, under appointment by the Supreme
Court, for Defendant and Appellant.

Thomas C. Lynch, Attorney General, William E. James, Assistant
Attorney General, and Nicholas C. Yost, Deputy Attorney General,
for Plaintiff and Respondent.

JUDGES: In Bank. Sullivan, J. Traynor, C. J., Peters, J., Tobriner, J.,
Mosk, J., and Burke, J., concurred. McComb, J., dissents.

OPINION BY: SULLIVAN

OPINION

We deal here with the novel question whether evidence of mathematical probability has been properly introduced and used by the prosecution in a criminal case.
While we discern no inherent incompatibility between the disciplines of law and
mathematics and intend no general disapproval or disparagement of the latter as an
auxiliary in the fact-finding processes of the former, we cannot uphold the technique
employed in the instant case. **(1a)** As we explain in detail, *infra*, the testimony as to
mathematical probability infected the case with fatal error and distorted the jury's
traditional role of determining guilt or innocence according to long-settled rules.

412

Mathematics, a veritable sorcerer in our computerized society, while assisting the trier of fact in the search for truth, must not cast a spell over him. We conclude that on the record before us defendant should not have had his guilt determined by the odds and that he is entitled to a new trial. We reverse the judgment.

A jury found defendant Malcolm Ricardo Collins and his wife defendant Janet Louise Collins guilty of second degree robbery (Pen. Code, §§ 211, 211a, 1157). Malcolm appeals from the judgment of conviction. Janet has not appealed.[1]

On June 18, 1964, about 11:30 a.m. Mrs. Juanita Brooks, who had been shopping, was walking home along an alley in the San Pedro area of the City of Los Angeles. She was pulling behind her a wicker basket carryall containing groceries and had her purse on top of the packages. She was using a cane. As she stooped down to pick up an empty carton, she was suddenly pushed to the ground by a person whom she neither saw nor heard approach. She was stunned by the fall and felt some pain. She managed to look up and saw a young woman running from the scene. According to Mrs. Brooks the latter appeared to weigh about 145 pounds, was wearing "something dark," and had hair "between a dark blond and a light blond," but lighter than the color of defendant Janet Collins' hair as it appeared at trial. Immediately after the incident, Mrs. Brooks discovered that her purse, containing between $ 35 and $ 40 was missing.

About the same time as the robbery, John Bass, who lived on the street at the end of the alley, was in front of his house watering his lawn. His attention was attracted by "a lot of crying and screaming" coming from the alley. As he looked in that direction, he saw a woman run out of the alley and enter a yellow automobile parked across the street from him. He was unable to give the make of the car. The car started off immediately and pulled wide around another parked vehicle so that in the narrow street it passed within 6 feet of Bass. The latter then saw that it was being driven by a male Negro, wearing a mustache and beard. At the trial Bass identified defendant as the driver of the yellow automobile. However, an attempt was made to impeach his identification by his admission that at the preliminary hearing he testified to an uncertain identification at the police lineup shortly after the attack on Mrs. Brooks, when defendant was beardless.

In his testimony Bass described the woman who ran from the alley as a Caucasian, slightly over 5 feet tall, of ordinary build, with her hair in a dark blonde ponytail, and wearing dark clothing. He further testified that her ponytail was "just like" one which Janet had in a police photograph taken on June 22, 1964.

On the day of the robbery, Janet was employed as a housemaid in San Pedro. Her employer testified that she had arrived for work at 8:50 a.m. and that defendant had picked her up in a light yellow car[2] about 11:30 a.m. On that day, according to the

---

1. Hereafter, the term "defendant" is intended to apply only to Malcolm, but the term "defendants" to Malcolm and Janet.

2. Other witnesses variously described the car as yellow, as yellow with an off-white top, and yellow with an egg-shell white top. The car was also described as being medium to large in size. Defendant drove a car at or near the times in question which was a Lincoln with a yellow body and a white top.

witness, Janet was wearing her hair in a blonde ponytail but lighter in color than it appeared at trial.[3]

There was evidence from which it could be inferred that defendants had ample time to drive from Janet's place of employment and participate in the robbery. Defendants testified, however, that they went directly from her employer's house to the home of friends, where they remained for several hours.

In the morning of June 22, Los Angeles Police Officer Kinsey, who was investigating the robbery, went to defendants' home. He saw a yellow Lincoln automobile with an off-white top in front of the house. He talked with defendants. Janet, whose hair appeared to be a dark blonde, was wearing it in a ponytail. Malcolm did not have a beard. The officer explained to them that he was investigating a robbery specifying the time and place; that the victim had been knocked down and her purse snatched; and that the person responsible was a female Caucasian with blonde hair in a ponytail who had left the scene in a yellow car driven by a male Negro. He requested that defendants accompany him to the police station at San Pedro and they did so. There, in response to police inquiries as to defendants' activities at the time of the robbery, Janet stated, according to Officer Kinsey, that her husband had picked her up at her place of employment at 1 p.m. and that they had then visited at the home of friends in Los Angeles. Malcolm confirmed this. Defendants were detained for an hour or two, were photographed but not booked, and were eventually released and driven home by the police.

Late in the afternoon of the same day, Officer Kinsey, while driving home from work in his own car, saw defendants riding in their yellow Lincoln. Although the transcript fails to disclose what prompted such action, Kinsey proceeded to place them under surveillance and eventually followed them home. He called for assistance and arranged to meet other police officers in the vicinity of defendants' home. Kinsey took a position in the rear of the premises. The other officers, who were in uniform and had arrived in a marked police car, approached defendants' front door. As they did so, Kinsey saw defendant Malcolm Collins run out the back door toward a rear fence and disappear behind a tree. Meanwhile the other officers emerged with Janet Collins whom they had placed under arrest. A search was made for Malcolm who was found in a closet of a neighboring home and also arrested. Defendants were again taken to the police station, were kept in custody for 48 hours, and were again released without any charges being made against them.

(2a) Officer Kinsey interrogated defendants separately on June 23 while they were in custody and testified to their statements over defense counsel's objections based on the decision in *Escobedo* and our first decision in *Dorado*.[4] According to the officer, Malcolm stated that he sometimes wore a beard but that he did not wear a

---

3. There are inferences which may be drawn from the evidence that Janet attempted to alter the appearance of her hair after June 18. Janet denies that she cut, colored or bleached her hair at any time after June 18, and a number of witnesses supported her testimony.

4. *Escobedo v. Illinois* (378 U.S. 478 [12 L.Ed.2d 977, 84 S.Ct. 1758]) was decided on June 22, 1964, four days after the robbery. The investigation was carried on both before and after *Escobedo* but before our first decision

beard on June 18 (the day of the robbery), having shaved it off on June 2, 1964.[5] He also explained two receipts for traffic fines totalling $ 35 paid on June 19, which receipts had been found on his person, by saying that he used funds won in a gambling game at a labor hall. Janet, on the other hand, said that the $ 35 used to pay the fines had come from her earnings.[6]

On July 9, 1964, defendants were again arrested and were booked for the first time. While they were in custody and awaiting the preliminary hearing, Janet requested to talk with Officer Kinsey. There followed a lengthy conversation during the first part of which Malcolm was not present. During this time Janet expressed concern about defendant and inquired as to what the outcome would be *if* it appeared that she committed the crime and Malcolm knew nothing about it. In general she indicated a wish that defendant be released from any charges because of his prior criminal record and that if someone must be held responsible, she alone would bear the guilt. The officer told her that no assurances could be given, that if she wanted to admit responsibility disposition of the matter would be in the hands of the court and that if she committed the crime and defendant knew nothing about it the only way she could help him would be by telling the truth. Defendant was then brought into the room and participated in the rest of the conversation. The officer asked to hear defendant's version of the matter, saying that he believed defendant was at the scene. However, neither Janet nor defendant confessed or expressly made damaging admissions although constantly urged by the investigating officer to make truthful statements. On several occasions defendant denied that he knew what had gone on in the alley. On the other hand, the whole tone of the conversation evidenced a strong consciousness of guilt on the part of both defendants who appeared to be seeking the most advantageous way out. Over defense counsel's same objections based on *Escobedo* and *Dorado*, some parts of the foregoing conversation were testified to by Officer Kinsey and in addition a tape recording of the entire conversation was introduced in evidence and played to the jury.[7]

---

in *People* v. *Dorado* filed on August 31, 1964. Defendants' trial took place in November 1964 after we granted a rehearing in *Dorado* on September 24, 1964, but before our decision on rehearing filed January 29, 1965. (62 Cal.2d 338 [42 Cal.Rptr. 169, 398 P.2d 361].)

5. Evidence as to defendant's beard and mustache is conflicting. Defense witnesses appeared to support defendant's claims that he had shaved his beard on June 2. There was testimony that on June 19 when defendant appeared in court to pay fines on another matter he was bearded. By June 22 the beard had been removed.

6. The source of the $ 35, being essentially the same amount as the $ 35 to $ 40 reported by the victim as having been in her purse when taken from her the day before the fines were paid, was a significant factor in the prosecution's case. Other evidence disclosed that defendant and Janet were married on June 2, 1964, at which time they had only $ 12, a portion of which was spent on a trip to Tijuana. Since the marriage defendant had not worked, and Janet's earnings were not more than $ 12 a week, if that much.

7. Included in the conversation are the following excerpts from Janet's statements:

"If I told you that he didn't know anything about it and I did it, would you cut him loose?"

"I just want him out, that's all, because I ain't never been in no trouble. I won't have to do too much [time], but he will."

"What's the most time I can do?"

"Would it be easier if I went ahead and said, if I was going to say anything, say it now instead of waiting till court time?"

Defendant indicated that he should "go and have trust in [the officer], but maybe I'd be wrong. I mean, this is a little delicate on my behalf."

At another point defendant stated: "I'm leaving it up to her."

At the seven-day trial the prosecution experienced some difficulty in establishing the identities of the perpetrators of the crime. The victim could not identify Janet and had never seen defendant. The identification by the witness Bass, who observed the girl run out of the alley and get into the automobile, was incomplete as to Janet and may have been weakened as to defendant. There was also evidence, introduced by the defense, that Janet had worn light-colored clothing on the day in question, but both the victim and Bass testified that the girl they observed had worn dark clothing.

In an apparent attempt to bolster the identifications, the prosecutor called an instructor of mathematics at a state college. Through this witness he sought to establish that, assuming the robbery was committed by a Caucasian woman with a blond ponytail who left the scene accompanied by a Negro with a beard and mustache, there was an overwhelming probability that the crime was committed by any couple answering such distinctive characteristics. The witness testified, in substance, to the "product rule," which states that the probability of the joint occurrence of a number of *mutually independent* events is equal to the product of the individual probabilities that each of the events will occur.[8] *Without presenting any statistical evidence whatsoever in support of the probabilities for the factors selected*, the prosecutor then proceeded to have the witness *assume*[9] probability factors for the various characteristics which he deemed to be shared by the guilty couple and all other couples answering to such distinctive characteristics.[10]

Applying the product rule to his own factors the prosecutor arrived at a probability that there was but one chance in 12 million that any couple possessed the distinctive

---

Defendant expressed concern during the conversation that any statement by Janet would not necessarily relieve him because he admittedly had been with her all that day since 11:30 a.m. The conversation closed when defendants indicated that they wished more time to think it over.

8. In the example employed for illustrative purposes at the trial, the probability of rolling one die and coming up with a "2" is 1/6, that is, any one of the six faces of a die has one chance in six of landing face up on any particular roll. The probability of rolling two "2's" in succession is 1/6 × 1/6, or 1/36, that is, on only one occasion out of 36 double rolls (or the roll of two dice) will the selected number land face up on each roll or die.

9. His argument to the jury was based on the same gratuitous assumptions or on similar assumptions which he invited the jury to make.

10. Although the prosecutor insisted that the factors he used were only for illustrative purposes—to demonstrate how the probability of the occurrence of mutually independent factors affected the probability that they would occur together—he nevertheless attempted to use factors which he personally related to the distinctive characteristics of defendants. In his argument to the jury he invited the jurors to apply their own factors, and asked defense counsel to suggest what the latter would deem as reasonable. The prosecutor himself proposed the individual probabilities set out in the table below. Although the transcript of the examination of the mathematics instructor and the information volunteered by the prosecutor at that time create some uncertainty as to precisely which of the characteristics the prosecutor assigned to the individual probabilities, he restated in his argument to the jury that they should be as follows:

Characteristic	Individual Probability
A. Partly yellow automobile	1/10
B. Man with mustache	1/4
C. Girl with ponytail	1/10
D. Girl with blond hair	1/3
E. Negro man with beard	1/10
F. Interracial couple in car	1/1000

In his brief on appeal defendant agrees that the foregoing appeared on a table presented in the trial court.

characteristics of the defendants. Accordingly, under this theory, it was to be inferred that there could be but one chance in 12 million that defendants were innocent and that another equally distinctive couple actually committed the robbery. Expanding on what he had thus purported to suggest as a hypothesis, the prosecutor offered the completely unfounded and improper testimonial assertion that, in his opinion, the factors he had assigned were "conservative estimates" and that, in reality, "the chances of anyone else besides these defendants being there,... having every similarity,... is something like one in a billion."

Objections were timely made to the mathematician's testimony on the grounds that it was immaterial, that it invaded the province of the jury, and that it was based on unfounded assumptions. The objections were "temporarily overruled" and the evidence admitted subject to a motion to strike. When that motion was made at the conclusion of the direct examination, the court denied it, stating that the testimony had been received only for the "purpose of illustrating the mathematical probabilities of various matters, the possibilities for them occurring or re-occurring."

Both defendants took the stand in their own behalf. They denied any knowledge of or participation in the crime and stated that after Malcolm called for Janet at her employer's house they went directly to a friend's house in Los Angeles where they remained for some time. According to this testimony defendants were not near the scene of the robbery when it occurred. Defendants' friend testified to a visit by them "in the middle of June" although she could not recall the precise date. Janet further testified that certain inducements were held out to her during the July 9 interrogation on condition that she confess her participation.

Defendant makes two basic contentions before us: First, that the admission in evidence of the statements made by defendants while in custody on June 23 and July 9, 1964, constitutes reversible error under the rules announced in the *Escobedo* and *Dorado* decisions;[11] and second, that the introduction of evidence pertaining to the mathematical theory of probability and the use of the same by the prosecution during the trial was error prejudicial to defendant. We consider the latter claim first.

(3a) As we shall explain, the prosecution's introduction and use of mathematical probability statistics injected two fundamental prejudicial errors into the case: (1) The testimony itself lacked an adequate foundation both in evidence and in statistical theory; and (2) the testimony and the manner in which the prosecution used it distracted the jury from its proper and requisite function of weighing the evidence on the issue of guilt, encouraged the jurors to rely upon an engaging but logically irrelevant expert demonstration, foreclosed the possibility of an effective defense by an attorney apparently unschooled in mathematical refinements, and placed the jurors and defense counsel at a disadvantage in sifting relevant fact from inapplicable theory.

(4) We initially consider the defects in the testimony itself. As we have indicated, the specific technique presented through the mathematician's testimony and advanced

---

11. *Escobedo* v. *Illinois* (1964) 378 U.S. 478 [12 L.Ed.2d 977, 84 S.Ct. 1758]; *People* v. *Dorado* (1965) 62 Cal.2d 338 [42 Cal.Rptr. 169, 398 P.2d 361].

by the prosecutor to measure the probabilities in question suffered from two basic and pervasive defects—an inadequate evidentiary foundation and an inadequate proof of statistical independence.

First, as to the foundational requirement, we find the record devoid of any evidence relating to any of the six individual probability factors used by the prosecutor and ascribed by him to the six characteristics as we have set them out in footnote 10, *ante*. To put it another way, the prosecution produced no evidence whatsoever showing, or from which it could be in any way inferred, that only one out of every ten cars which might have been at the scene of the robbery was partly yellow, that only one out of every four men who might have been there wore a mustache, that only one out of every ten girls who might have been there wore a ponytail, or that any of the other individual probability factors listed were even roughly accurate.[12]

The bare, inescapable fact is that the prosecution made no attempt to offer any such evidence. Instead, through leading questions having perfunctorily elicited from the witness the response that the latter could not assign a probability factor for the characteristics involved,[13] the prosecutor himself suggested what the various probabilities should be and these became the basis of the witness' testimony (see fn. 10, *ante*). It is a curious circumstance of this adventure in proof that the prosecutor not only made his own assertions of these factors in the hope that they were "conservative" but also in later argument to the jury invited the jurors to substitute their "estimates" should they wish to do so. We can hardly conceive of a more fatal gap in the prosecution's scheme of proof. A foundation for the admissibility of the witness' testimony was never even attempted to be laid, let alone established. His testimony was neither made to rest on his own testimonial knowledge nor presented by proper hypothetical questions based upon valid data in the record. (See generally: 2 Wigmore on Evidence (3d ed. 1940) §§ 478, 650–652, 657, 659, 672–684; Witkin, Cal. Evidence (2d ed. 1966) § 771; McCormick on Evidence, pp. 19–20; *Evidence: Admission of Mathematical Probability Statistics Held Erroneous for Want of Demonstration of Validity* (1967) Duke L.J. 665, 675–678, citing *People* v. *Risley* (1915) 214 N.Y. 75, 85 [108 N.E. 200, Ann. Cas. 1916 D 775]; *State* v. *Sneed* (1966) 76 N.M. 349 [414 P.2d 858].) In the *Sneed* case, the court reversed a conviction based on probabilistic evidence, stating: "We hold that mathematical odds are not admissible as evidence to identify a defendant in a criminal proceeding

---

12. We seriously doubt that such evidence could ever be compiled since no statistician could possibly determine after the fact which cars, or which individuals, "might" have been present at the scene of the robbery; certainly there is no reason to suppose that the human and automotive populations of San Pedro, California, include all potential culprits—or, conversely, that all members of these populations are proper candidates for inclusion. Thus the sample from which the relevant probabilities would have to be derived is itself undeterminable. (See generally, Yamane, Statistics, An Introductory Analysis (1964), ch. I.)

13. The prosecutor asked the mathematics instructor: "Now, let me see if you can be of some help to us with some independent factors, and you have some paper you may use. Your specialty does not equip you, I suppose, to give us some probability of such things as a yellow car as contrasted with any other kind of car, does it?...I appreciate the fact that you can't assign a probability for a car being yellow as contrasted to some other car, can you? A. No, I couldn't."

so long as the odds are based on estimates, the validity of which have [sic] not been demonstrated." (Italics added.) (414 P.2d at p. 862.)

But, as we have indicated, there was another glaring defect in the prosecution's technique, namely an inadequate proof of the statistical independence of the six factors. No proof was presented that the characteristics selected were mutually independent, even though the witness himself acknowledged that such condition was essential to the proper application of the "product rule" or "multiplication rule." (See Note, *supra,* Duke L.J. 665, 669–670, fn. 25.)[14] To the extent that the traits or characteristics were not mutually independent (e.g., Negroes with beards and men with mustaches obviously represent overlapping categories[15]), the "product rule" would inevitably yield a wholly erroneous and exaggerated result even if all of the individual components had been determined with precision. (Siegel, Nonparametric Statistics for the Behavioral Sciences (1956) 19; see generally Harmon, Modern Factor Analysis (1960).)

In the instant case, therefore, because of the aforementioned two defects—the inadequate evidentiary foundation and the inadequate proof of statistical independence—the technique employed by the prosecutor could only lead to wild conjecture without demonstrated relevancy to the issues presented. It acquired no redeeming quality from the prosecutor's statement that it was being used only "for illustrative purposes" since, as we shall point out, the prosecutor's subsequent utilization of the mathematical testimony was not confined within such limits.

(3b) We now turn to the second fundamental error caused by the probability testimony. Quite apart from our foregoing objections to the specific technique employed by the prosecution to estimate the probability in question, we think that the entire enterprise upon which the prosecution embarked and which was directed to the objective of measuring the likelihood of a random couple possessing the characteristics allegedly distinguishing the robbers, was gravely misguided. At best, it might yield an estimate as to how infrequently bearded Negroes drive yellow cars in the company of blonde females with ponytails.

The prosecution's approach, however, could furnish the jury with absolutely no guidance on the crucial issue: *Of the admittedly few such couples, which one, if any, was guilty of committing this robbery?* Probability theory necessarily remains silent on that question, since no mathematical equation can prove beyond a reasonable doubt (1) that the guilty couple *in fact* possessed the characteristics described

14. It is there stated that: "A trait is said to be independent of a second trait when the occurrence or nonoccurrence of one does not affect the probability of the occurrence of the other trait. The multiplication rule cannot be used without some degree of error where the traits are not independent." (Citing Huntsberger, Elements of Statistical Inference (1961) 77; Kingston & Kirk, *The Use of Statistics in Criminalistics* (1964) 55 J. Crim. L., C. & P.S. 516.) (Note, *supra,* Duke L.J. fn. 25, p. 670.)

15. Assuming *arguendo* that factors B and E (see fn. 10, *ante*) were correctly estimated, nevertheless it is still arguable that most Negro men with beards *also* have mustaches (exhibit 3 herein, for instance, shows defendant with both a mustache and a beard, indeed in a hirsute continuum); if so, there is no basis for multiplying 1/4 by 1/10 to estimate the proportion of Negroes who wear beards *and* mustaches. Again, the prosecution's technique could *never* be meaningfully applied, since its accurate use would call for information as to the degree of interdependence among the six individual factors. (See Yamane, *op. cit. supra.*) Such information cannot be compiled, however, since the relevant sample necessarily remains unknown. (See fn. 10, *ante.*)

by the People's witnesses, or even (2) that only *one* couple possessing those distinctive characteristics could be found in the entire Los Angeles area.

As to the first inherent failing we observe that the prosecution's theory of probability rested on the assumption that the witnesses called by the People had conclusively established that the guilty couple possessed the precise characteristics relied upon by the prosecution. But no mathematical formula could ever establish beyond a reasonable doubt that the prosecution's witnesses correctly observed and accurately described the distinctive features which were employed to link defendants to the crime. (See 2 Wigmore on Evidence (3d ed. 1940) § 478.) Conceivably, for example, the guilty couple might have included a light-skinned Negress with bleached hair rather than a Caucasian blonde; or the driver of the car might have been wearing a false beard as a disguise; or the prosecution's witnesses might simply have been unreliable.[16]

The foregoing risks of error permeate the prosecution's circumstantial case. Traditionally, the jury weighs such risks in evaluating the credibility and probative value of trial testimony, but the likelihood of human error or of falsification obviously cannot be quantified; that likelihood must therefore be excluded from any effort to assign a *number* to the probability of guilt or innocence. Confronted with an equation which purports to yield a numerical index of probable guilt, few juries could resist the temptation to accord disproportionate weight to that index; only an exceptional juror, and indeed only a defense attorney schooled in mathematics, could successfully keep in mind the fact that the probability computed by the prosecution can represent, *at best*, the likelihood that a random couple would share the characteristics testified to by the People's witnesses—*not necessarily the characteristics of the actually guilty couple.*

As to the second inherent failing in the prosecution's approach, even assuming that the first failing could be discounted, the most a mathematical computation could *ever* yield would be a measure of the probability that a random couple would possess the distinctive features in question. In the present case, for example, the prosecution attempted to compute the probability that a random couple would include a bearded Negro, a blonde girl with a ponytail, and a partly yellow car; the prosecution urged that this probability was no more than one in 12 million. Even accepting this conclusion as arithmetically accurate, however, one still could not conclude that the Collinses were probably *the* guilty couple. On the contrary, as we explain in the Appendix, the prosecution's figures actually imply a likelihood of over 40 percent that the Collinses could be "duplicated" by at least *one other couple who might equally have committed the San Pedro robbery.* Urging that the Collinses be convicted on the basis of evidence which logically establishes no more than this seems as indefensible as arguing for the conviction of X on the ground that a witness saw either X or X's twin commit the crime.

---

16. In the instant case, for instance, the victim could not state whether the girl had a ponytail, although the victim observed the girl as she ran away. The witness Bass, on the other hand, was sure that the girl whom he saw had a ponytail. The demonstration engaged in by the prosecutor also leaves no room for the possibility, although perhaps a small one, that the girl whom the victim and the witness observed was, in fact, the same girl.

Again, few defense attorneys, and certainly few jurors, could be expected to comprehend this basic flaw in the prosecution's analysis. Conceivably even the prosecutor erroneously believed that his equation established a high probability that *no* other bearded Negro in the Los Angeles area drove a yellow car accompanied by a ponytailed blonde. In any event, although his technique could demonstrate no such thing, he solemnly told the jury that he had supplied mathematical proof of guilt.

(1b) Sensing the novelty of that notion, the prosecutor told the jurors that the traditional idea of proof beyond a reasonable doubt represented "the most hackneyed, stereotyped, trite, misunderstood concept in criminal law." He sought to reconcile the jury to the risk that, under his "new math" approach to criminal jurisprudence, "on some rare occasion...an innocent person may be convicted." "Without taking that risk," the prosecution continued, "life would be intolerable...because...there would be immunity for the Collinses, for people who chose not to be employed to go down and push old ladies down and take their money and be immune because how could we ever be sure they are the ones who did it?"

In essence this argument of the prosecutor was calculated to persuade the jury to convict defendants whether or not they were convinced of their guilt to a moral certainty and beyond a reasonable doubt. (Pen. Code, § 1096.) Undoubtedly the jurors were unduly impressed by the mystique of the mathematical demonstration but were unable to assess its relevancy or value. **(3c)** Although we make no appraisal of the proper applications of mathematical techniques in the proof of facts (see *People* v. *Jordan* (1955) 45 Cal.2d 697, 707 [290 P. 2d 484]; *People* v. *Trujillo* (1948) 32 Cal.2d 105, 109 [194 P.2d 681]; in a slightly differing context see *Whitus* v. *Georgia* (1967) 385 U.S. 545, 552, fn. 2 [17 L.Ed.2d 599, 604, 87 S.Ct. 643]; Finkelstein, *The Application of Statistical Decision Theory to the Jury Discrimination Cases* (1966) 80 Harv.L.Rev. 338, 338–340), we have strong feelings that such applications, particularly in a criminal case, must be critically examined in view of the substantial unfairness to a defendant which may result from ill conceived techniques with which the trier of fact is not technically equipped to cope. (See *State* v. *Sneed, supra*, 414 P.2d 858; Note, *supra*, Duke L.J. 665.) We feel that the technique employed in the case before us falls into the latter category.

(1c) We conclude that the court erred in admitting over defendant's objection the evidence pertaining to the mathematical theory of probability and in denying defendant's motion to strike such evidence. The case was apparently a close one. The jury began its deliberations at 2:46 p.m. on November 24, 1964, and retired for the night at 7:46 p.m.; the parties stipulated that a juror could be excused for illness and that a verdict could be reached by the remaining 11 jurors; the jury resumed deliberations the next morning at 8:40 a.m. and returned verdicts at 11:58 a.m. after five ballots had been taken. In the light of the closeness of the case, which as we have said was a circumstantial one, there is a reasonable likelihood that the result would have been more favorable to defendant if the prosecution had not urged the jury to render a probabilistic verdict. In any event, we think that under the circumstances the "trial by mathematics" so distorted the role of the jury and so disadvantaged counsel for the defense, as to constitute in itself a miscarriage of justice. After an examination of the entire cause, including the evidence, we are of the opinion that it is reasonably probable that a result more favorable to defendant would have been reached in the absence of the above error. (*People* v. *Watson* (1956) 46 Cal.2d 818, 836 [299 P.2d 243].) The judgment against defendant must therefore be reversed.

(**2b**) In view of the foregoing conclusion, we deem it unnecessary to consider whether the admission of defendants' extrajudicial statements constitutes error under the rules announced in *Escobedo* and *Dorado*. Upon retrial, the admissibility of these or any other extrajudicial statements sought to be introduced by the prosecution must be determined in the light of the rules set forth in *Miranda* v. *Arizona* (1966) 384 U.S. 436 [16 L.Ed.2d 694, 86 S.Ct. 1602, 10 A.L.R.3d 974]. (*People* v. *Doherty* (1967) 67 Cal.2d 9, 12, 17–21 [59 Cal.Rptr. 857, 429 P.2d 177].) As we have pointed out, the trial herein took place between our first and second *Dorado* decisions (see fn. 4, *ante*). Although defense counsel was commendably alert in basing objections to the admission of the statements upon the decisions in *Escobedo* and *Dorado*, he of course did not have the benefit of our numerous decisions beginning with the second *Dorado* decision expounding various facets of the exclusionary rule. In the event any extrajudicial statements made by defendant are offered in evidence on retrial, the parties will have an opportunity to make a record on pertinent issues subject to prior determination by the court in the light of *Miranda* rules before such statements are received in evidence. It would be fruitless for us to essay such a task at this point when such record does not yet exist.

The judgment is reversed.

APPENDIX

If "Pr" represents the probability that a certain distinctive combination of characteristics, hereinafter designated "C," will occur jointly in a random couple, then the probability that C will *not* occur in a random couple is $(1 - Pr)$. Applying the product rule (see fn. 8, *ante*), the probability that C will occur in *none* of N couples chosen at random is $(1 - Pr) N$, so that the probability of C occurring in *at least one* of N random couples is $[1 - (1 - Pr) N]$.

Given a particular couple selected from a random set of N, the probability of C occurring in that couple (i.e., Pr), multiplied by the probability of C occurring in none of the remaining $N - 1$ couples (i.e., $(1 - Pr) N - 1$), yields the probability that C will occur in the selected couple and in no other. Thus the probability of C occurring in any particular couple, and in that couple alone, is $[(Pr) \times (1 - Pr) N - 1]$. Since this is true for each of the N couples, the probability that C will occur in precisely *one* of the N couples, without regard to which one, is $[(Pr) \times (1 - Pr) N - 1]$ added N times, because the probability of the occurrence of one of several *mutually exclusive* events is equal to the *sum* of the individual probabilities. Thus the probability of C occurring in *exactly one* of N random couples (*any* one, but *only* one) is $[(N) \times (Pr) \times (1 - Pr) N - 1]$.

By subtracting the probability that C will occur in *exactly one* couple from the probability that C will occur in *at least one* couple, one obtains the probability that C will occur in *more than one* couple: $[1 - (1 - \text{Pr}) \text{ N}] - [(\text{N}) \times (\text{Pr}) \times (1 - \text{Pr}) \text{ N} - 1]$. Dividing this difference by the probability that C will occur in at least one couple (i.e., dividing the difference by $[1 - (1 - \text{Pr}) \text{ N}]$) then yields *the probability that C will occur more than once in a group of N couples in which C occurs at least once.*

Turning to the case in which C represents the characteristics which distinguish a bearded Negro accompanied by a ponytailed blonde in a yellow car, the prosecution sought to establish that the probability of C occurring in a random couple was 1/12,000,000—i.e., that Pr = 1/12,000,000. Treating this conclusion as accurate, it follows that, in a population of N random couples, the probability of C occurring *exactly once* is $[(\text{N}) \times (1/12,000,000) \times (1 - 1/12,000,000) \text{ N} - 1]$. Subtracting this product from $[1 - (1 - 1/12,000,000) \text{ N}]$, the probability of C occurring in *at least one* couple, and dividing the resulting difference by $[1 - (1 - 1/12,000,000) \text{ N}]$, the probability that C will occur in at least one couple, yields the probability that C will occur more than once in a group of N random couples of which at least one couple (namely, the one seen by the witnesses) possesses characteristics C. In other words, the probability of *another* such couple in a population of N is the quotient A/B, where A designates the numerator $[1 - (1 - 1/12,000,000) \text{ N}] - [(\text{N}) \times (1/12,000,000) \times (1 - 1/12,000,000) \text{ N} - 1]$, and B designates the denominator $[1 - (1 - 1/12,000,000) \text{ N}]$.

N, which represents the total number of all couples who might conceivably have been at the scene of the San Pedro robbery, is not determinable, a fact which suggests yet another basic difficulty with the use of probability theory in establishing identity. One of the imponderables in determining N may well be the number of N-type couples in which a single person may participate. Such considerations make it evident that N, in the area adjoining the robbery, is in excess of several million; as N assumes values of such magnitude, the quotient A/B computed as above, representing the probability of a second couple as distinctive as the one described by the prosecution's witnesses, soon exceeds 4/10. Indeed, as N approaches 12 million, this probability quotient rises to approximately 41 percent. We note parenthetically that if $1/N = \text{Pr}$, then as N increases indefinitely, the quotient in question approaches a limit of $(e - 2)/(e - 1)$, where "e" represents the transcendental number (approximately 2.71828) familiar in mathematics and physics.

Hence, even if we should accept the prosecution's figures without question, we would derive a probability of over 40 percent that the couple observed by the witnesses could be "duplicated" by at least one other equally distinctive interracial couple in the area, including a Negro with a beard and mustache, driving a partly yellow car in the company of a blonde with a ponytail. Thus the prosecution's computations, far from establishing beyond a reasonable doubt that the Collinses were the couple described by the prosecution's witnesses, imply a very substantial likelihood that the area contained *more than one* such couple, and that a couple *other* than the Collinses was the one observed at the scene of the robbery. (See generally: Hoel, Introduction to Mathematical Statistics (3d ed, 1962); Hodges & Leymann, Basic

Concepts of Probability and Statistics (1964); Lindgren & McElrath, Introduction to Probability and Statistics (1959).)

DISSENT BY: McCOMB

DISSENT:

McCOMB, J. I dissent. I would affirm the judgment in its entirety.

C. PRESENTING BAYESIAN ANALYSIS TO THE JURY

This case, also from England, concerns an experiment, conducted by Peter Donnelly, to introduce Bayes Theorem to a jury in an interactive manner. The first item here is an account by him of his experiences. Also included are two court decisions that have the consequence of invalidating such an interactive approach. But the courts also verge on rejecting probabilistic evidence entirely.

1

APPEALING STATISTICS

Peter Donnelly

THE CRIME

The woman went out to a nightclub with friends, left, and walked home. Part of the walk was on a path along the side of a park, and as she walked through the park someone stopped her to ask the time. She was sure she'd had a good look at the man, and that—as she subsequently told the police—it was a face she would never forget. The man then attacked and raped her, and ran off. She reported the incident to the police, giving a fairly clear description of a man in his early 20s.

The police had no leads at the time, but quite some time later the woman thought she saw the man who had raped her walking along the street. She called the police, but by the time they arrived he had gone.

Some time later again, the police came up with a suspect, Denis Adams, and arranged an identity parade. The victim didn't pick him out. At the committal hearing, he was pointed out to her and she was asked if he matched her description, and she said no. She had described a man in his early 20s, and when asked how old Adams looked to her, she said around 40 (he was in fact 37).

However, there was very strong DNA evidence linking him with the crime and when the case came to trial in 1995, effectively the only incriminating evidence was that his DNA profile matched the DNA evidence found at the scene of the crime. The prosecution forensic scientist had calculated what is called a match probability, that is, the probability that if you pick someone at random, their DNA would match the DNA sample of the assailant. That, according to him, was 1 in 200 million.

It's tempting for people not trained in statistics to get muddled, and to confuse two different probabilities. The first is the probability that a person would match the criminal's DNA profile given that they are innocent. The second is the probability that they are innocent given that they match the DNA profile. The forensic scientist's 1 in 200 million refers to the first probability. But jurors may wrongly think that this is the probability that the defendant is innocent. This misunderstanding is called the prosecutor's fallacy, and can be extremely prejudicial to the defendant. The error was often made in court by lawyers and sometimes even by the forensic scientists, in the early days of DNA evidence. Who knows how often jurors make it in private?

But, even with the correct interpretation, the numbers are saying that the particular DNA profile is extremely rare. If the 1 in 200 million is correct, then it is probable that no one else in Britain has the same profile. Maybe there might be one other person, or just possibly two or even three.

THE TRIAL

This was the only incriminating evidence the jury heard. However, in Adams's defence there was the victim's own identification evidence; also, he had an alibi for the night involved, when his girlfriend said she spent the night with him. So there was a real issue here, for the lawyers in the first instance and ultimately for the jury, somehow to weigh up the prosecution evidence expressed in rather impressive hard numerical terms, against the more traditional defence evidence, expressed in terms juries are more familiar with—reported facts, alibis, and so on. I was asked by the defence barrister how one could compare the two.

I explained that, although difficult, there was a logically correct way: using Bayes's Theorem to reason with uncertainties by converting them into probabilities, and the statistical experts on the prosecution side agreed. The barrister then took the view that this should be explained to the jury, and it fell to me to do so in my evidence, in response to questions from counsel.

This case was quite unusual in having DNA evidence pointing one way and *all* the other evidence—which would often also be incriminating—pointing the other way. It was a really good example of the challenges faced in weighing up evidence—had it been invented as a hypothetical example, I suspect people would have thought it unrealistic! In other DNA cases, once there's a DNA match, there's normally another piece of evidence that incriminates, or at the very least there isn't a whole raft of other evidence that points towards innocence.

Here is an informal way of thinking about the evidence in this case. If the match probability were 1 in 1 million, then that means that, on average, there are 50 people in the UK who match. They may or may not have been in the right area on the night in question, and may or may not have been male and in the correct age range. However, the woman says that her attacker looked nothing like this particular suspect, who also happens to have an alibi, so it might well have been one of the other matching individuals who was the assailant, and a jury may acquit. If the number were 1 in 20 million, there are probably only a handful of people in the UK with that profile. They may be spread all over Britain, some may be women, or very young or very old, so a jury might think it was somewhat unlikely that another one of them was in the same place that night. If 1 in 200 million is right, it's unlikely, although still possible, that there is even one other person in Britain with the same DNA profile. In this case they may well think: if we *knew* there was no one else in Britain with that profile, we'd convict, and then just assume the victim misremembered what her assailant looked like and the girlfriend was lying or mistaken about the alibi.

So, unlike in many DNA cases, it really did matter whether the number was 1 in 2 or 20 or 200 million. The defence argued that for various reasons the number of 1 in 200 million figure was incorrect and, anyway, it is worth remembering that all of the statements about how many other matches there were in Britain exclude close relatives of the person on trial. As it happens, Adams did have a half-brother, whose DNA profile was unknown.

Adams was found guilty. Who knows what the jury thought—in this country you can't ask them. If you believe the 1 in 200 million number and you do all the Bayesian analysis, you might well convict. You might decide that, of the two explanations, the one that says the victim got muddled and the girlfriend was mistaken is the more plausible.

APPEAL, RETRIAL, APPEAL

There was an appeal, however, which was upheld because the Appeal Court felt that the judge hadn't instructed the jury clearly enough. He had told them that they could use Bayes's Theorem or not, as they wished. The Appeal Court said that the judge should have offered the jury more guidance on what to do if they didn't want to use Bayes's Theorem. In passing, they were very negative about the presentation of Bayes's Theorem in court, but, because this was not a matter on which they specifically needed to rule, their comments were not binding for future cases.

The Court of Appeal ordered a retrial, and in the retrial the defence said they wanted to present Bayes's Theorem to the jury. The judge agreed despite the Appeal Court comments.

In the retrial, there was again no dispute among the statistical experts from both sides that in principle Bayes's Theorem was the logically correct way to combine these sorts of evidence. Unlike at the first trial, though, the judge at the retrial asked the experts on both sides to get together and prepare a questionnaire which would help the jury to implement Bayes's Theorem, should they choose to do so. The two prosecution statisticians were from the Forensic Science Service. I knew them well professionally, and we collaborated on writing the questionnaire.

To apply Bayes's Theorem, you need to produce likelihood ratios for various different pieces of evidence. The jury needed to ask themselves a series of questions such as: If he were the attacker, what's the chance that the victim would say her attacker didn't look anything like him? Also: If he *wasn't* her attacker, what's the chance she would say this? As you go through this exercise there are issues about how you agglomerate or separate various pieces of evidence. There's a lot of discussion about this in the legal literature on evidence. Provided one acts rationally (*coherently* is the technical term) the level of agglomeration will not affect the final answer. But that doesn't mean that real people necessarily reason the right way.

We were given instructions by the court about the level of agglomeration, and the jury ended up with about a dozen questions. The questionnaires were produced, and there were boxes where they could enter their numerical assessments, with a formula explaining how to combine them. The jury were told that this was in the experts' view the right way to do the reasoning, but that they were the jury and it was entirely up to them; they didn't have to use the questionnaires if they chose not to.

The episode had some amusing sidelines. It was suggested that it would be helpful to supply the jury (and judge) with basic calculators. Although the total cost was well under £100, this request was so unusual that it seemed to require clearance personally from the Lord Chancellor. Then, during my evidence, we walked the jury through a numerical example—the barrister would suggest token numbers in answer to the questions, and the jury and I entered them in the calculators which were eventually supplied. They seemed to have no difficulty in following this, but at an early stage in the calculation, when I said something to the effect that: "Your calculator should now show the value 31.6," and the jurors all nodded, the judge rather plaintively said: "But mine shows zero."

And, on separate occasions, first the defence barrister and then the learned judge corrected what they saw as a misprint on the questionnaire where we said that "the chance of throwing a six on a fair die" was about 17%. I carefully explained to each that "dice" was plural, and "die" was singular, and that we tended to be very careful about our professional use of language, much to the amusement, or perhaps bemusement, of the jury.

Again Adams was convicted, again there was an appeal—this time unsuccessful—and this time the court did rule on the use of Bayes's Theorem. They came down very heavily against it.

NOT BY MEANS OF A FORMULA

I feel the judgment of the Court of Appeal oversimplified a very complex issue. They said that juries had been weighing up evidence for hundreds of years and that there was no reason to believe they couldn't do it now. They said that the jury should "evaluate evidence and reach a conclusion not by means of a formula, mathematical or otherwise, but by the joint application of their individual common sense and knowledge of the world to the evidence before them."

In a case such as Adams, there is a very live issue. The jury *somehow* have to weigh up the number of 1 in 200 million against ordinary evidence. Maybe, if the number is right, a rational juror would convict; if the number is 1 in 2 million, as the defence argued it might be, a rational person may well acquit.

There are right and wrong ways of reasoning with uncertainty, and there is plenty of documentary evidence that, particularly where small probabilities are concerned, people make mistakes when left to their own devices. So I think it's quite hard to argue in principle that juries should be denied the possibility of having experts explain to them the rules of logic that apply in this sort of situation. After all, juries hear expert witnesses in all sorts of settings: how you interpret financial accounts in fraud cases, how you interpret tire skid marks in a road accident case, and so on. And having an expert explain the logic of reasoning with uncertainty certainly does not reduce the judicial process to a mathematical formula.

Secondly, it was a little disingenuous for the Appeal Court to say that juries have been considering these sorts of problems for hundreds of years. It's only since the advent of DNA profiling that jurors have had to comprehend probabilities like 1 in 200 million. It's not something that is in juries' experience collectively, nor in most people's experience. It is new and it throws up new problems.

So my own view is that it is troubling, as a matter of principle, to exclude evidence explaining how to reason with uncertainties. In the light of extensive evidence about people's routine errors in this setting, it seems inappropriate to deny a defendant the chance to have it explained carefully.

However, the pragmatic question is a very different one. Given the obvious constraints of the courtroom as a forum for teaching (no feedback, no homework, no exams…) one can and should ask whether the quality of judicial decision making would be improved or made worse by subjecting juries to explanations of Bayes's Theorem. Even though juries have little experience of thinking about numbers like 1 in 200 million, trying to explain how they should think may well end up making matters worse.

I should be clear that it was neither my suggestion nor my choice to explain Bayes's Theorem to the jury in the Adams case, and I remain unconvinced that it is a practicable way forward. In Adams, it was the defence barrister's choice, presumably because he felt it would increase the possibility of a favourable verdict for his client.

The Court of Appeal has subsequently provided some guidelines for similar situations. It advocated that judges should summarise cases in the following way. Suppose the match probability is 1 in 20 million. If you believe that number, then on average there will be 2 or 3 people in Britain whose DNA it could be, and probably no more than 6 or 7. That is the effect of DNA evidence: it narrows down the pool of possible suspects from everyone in the UK to 6 or 7 people—plus the accused. Now your job, as a member of the jury, is to decide, on the basis of the other evidence, whether or not you are satisfied that it is the person on trial who was the assailant, rather than one of the few other possible people who match. We don't know anything about the other people who match, although they are probably spread all over the UK, may have been nowhere near the scene of the crime, and some or all may also be ruled out by other factors, for example, gender or age.

I think this way of putting the DNA evidence is extremely helpful. It immediately steers the jury away from the error of the Prosecutor's Fallacy, and it is logically correct. In fact, it amounts to applying Bayes's Theorem to the DNA evidence *before* considering any of the other evidence. In this case a small prior probability of guilt would be appropriate, perhaps the inverse of the UK population. The effect of the evidence that the defendant's DNA matches the criminal's DNA is dramatically to increase the probability of guilt, to 1 in 5 or 6 or 7 in my example above. The juror's job is then to allow for the other evidence, to see whether or not they are convinced beyond reasonable doubt that the defendant is guilty.

But although, ironically given the earlier judgment, this suggestion by the Court of Appeal follows exactly the logic of Bayes's Theorem, it does not require that the

Theorem be explained, and it moves the jury back into realms of reasoning with which they are more familiar. It doesn't solve all the problems, though. It becomes more difficult to put into practice when the match probability is 1 in 1 billion, or in more complex settings where the DNA profile is partially degraded, or involves a mixture from several individuals.

2

R V. ADAMS

COURT OF APPEAL (CRIMINAL DIVISION)

26 APRIL 1996

COUNSEL:

S Tapping for the Crown; R Thwaites QC for the Appellant

PANEL: ROSE LJ, HIDDEN, BUXTON JJ

JUDGEMENT BY-1: ROSE LJ

JUDGMENT-1:

ROSE LJ (reading the judgment of the court): On 24 January 1995, at the Central Criminal Court before His Honour Judge Gordon, this appellant was convicted of rape and was sentenced to seven years' imprisonment. He appeals against that conviction by leave of the single judge.

In summing-up the evidence in relation to the Bayes theorem the judge said this:

"Professor Donnelly said that the right way indeed in his view the only logical way to assess other evidence against statistical and scientific evidence is Bayes theorem. Well, as a statistician I suppose he would say that and he may well be right, it is not for me to say. Whether you decide to use it and do your best to operate Bayes theorem or not is up to you. It is right to say that whatever method you use it involves at the heart of it your judgment of the evidence even if you are using Bayes theorem as you will remember ultimately you have to say how many times is it more likely one question is right than the other, that is at the heart of it."

At page 28F to G the judge said this:

"We will now turn to look at what the defence say about the DNA evidence. What they say really falls into two parts. They say first of all although the random figure,

432

in particular the 200,000,000 one, as a figure looks impressive if you look at the other evidence in this case statistically—that is to say using Bayes theorem—it rapidly becomes less so. They say whatever figures are used in the illustration by Professor Donnelly were revealing in the effect that they had upon the prosecution's figure. They say the use of the Bayes theorem, or something like it, shows that the figure is nothing like as grand as it initially appears."

At 29E he continues:

"...one has really got to look at both those and that means starting with Bayes theorem. Now as I say it is entirely up to you whether you want to use it or not. As a theorem it is agreed in principle between both sides that it is a way of looking at non-statistical matters in statistical terms. Whether it is a practical for you a jury to operate it is as I say something that you will decide for yourselves. You know that Professor Donnelly suggested that you might want to start with the male population of a 10 mile/15 kilometre area and a figure was produced from some local authority of 150,000. This is just giving you illustrations apart from these figures, I am not going to give you anymore you need not worry, it is just to show you how it works to remind you how it works since it was last week that we heard about it. If one says we think there is a 75 percent chance of the person who did it being a local man that would mean that 150,000 would be 75 percent of the relevant population we are talking about. That would mean as a matter of pure mathematics that the relevant population is 200,000 because 150 is 75 percent of 200,000. That is if you took three-quarters. Supposing you said in our view it is a 90 percent chance bearing in mind the hour of the day, where it was etcetera, then as a matter of mathematics— unless I have got them wrong which would not be surprising—you would have a relevant population of 166,666.6 recurring. That is a straight mathematical question depending, like all these things do, on your assessment of matters of fact, here the percentage of the population the percentage chance of it being a local man. No one can tell you that. You are the only ones who can decide that as your start off on the Bayes theorem route.

Then you have to look, do you not, at all the evidence to see the effect that that other evidence has upon the figure that you have now got and you ask yourselves to operate this theorem two questions, do you not? What are the chances of this bit of evidence being given if the defendant is guilty? Secondly, what are the chances of it being given if he is innocent? If the chances are even, is equally likely to be given either way, quite obviously that piece of evidence would be a neutral and would not assist you from one conclusion or another: the statistician and the rest of us would all come to the same obvious conclusion about that. If the answer to one question is more likely than the other you then ask yourselves how many times more likely? These are value judgments for you and you alone. How many times more likely? If guilt is more likely from a piece of evidence you divide, sorry, if guilt is more likely taking the figure that you started out with you divide. If it is three times more likely you divide by three. If innocent you multiply the figure. You see the mathematics you are getting at to produce from your analysis—and I go back to it being your analysis—of the likelihoods there."

At page 45C he said this:

"…while you are fresh and so am I just want to recap very briefly over two aspects of Bayes theorem that I do not think I was very clear about yesterday afternoon. Only those two, it will only take me a moment and then we will move on. Can I pick it up at that stage, and we are dealing with illustrations because the decisions are yours. Professor Donnelly's illustrations where he has taken the 150,000 has made an assumption that you may agree or disagree that there is a 75 percent chance of it being a local man which brings the figure up to fortunately to the round one of 200,000, it being a chance of one in 200,000 at that stage of Bayes theorem that the defendant is the rapist. We have got to that stage, alright. Then you take the other evidence the non-DNA evidence broken down as you find convenient and you ask about it the two questions that I dealt with yesterday: what are the chances of it if guilty? What are the chances if innocent? It is the conclusions you have come to having done that exercise, it is the first matter I want to deal with. If the evidence is more likely to have been given if the defendant is guilty that makes it more likely that he is the rapist. Nothing very complicated about it. Mathematically applying the theorem, this is an illustration. If it is twice as likely to have been given if guilty than if innocent that makes it twice as likely that he is the rapist. And on the figures you have at the moment that would mean that it would bring the figure, alter the figure of one in 200,000 to one in 100,000. You divide it by two and would make him twice as likely as before because of the conclusion you come to on that bit of evidence that it favours guilt. The opposite applies in exactly the same way. Supposing you come to the conclusion in respects of a piece of evidence that it is more likely to have been given if he is not guilty. Obviously that makes it less likely that he is the rapist. Again using the figure twice merely for convenience if you conclude that it is twice as likely to have been given if innocent than guilty then instead of one in 200,000 the chance is doubled, one in 400,000. I hope that is now clear. One in 400,000 because you double the figure which gives you half the chance.

Once you have completed your assessment taking all the different bits, however you choose to breakdown the other evidence—I dealt with all that yesterday—you end up, do you not, with two figures to compare. There is the one in 200,000,000 that it was a random person who was the rapist. That is the DNA figure and I am putting to one side for the purposes of illustration all the criticisms of it that I dealt with yesterday. Let us take that as a figure because it is a neat round one for administrative purposes. You have got one in 20,000,000 that it is a random person. You have got a figure, and I do not know what your figure if you applied Bayes theorem would be, I have not the slightest idea. For a moment or two I am going to call it x as I did yesterday but we will look at two possibilities in just a moment. You have a one in x chance that the defendant was so. So you have got one in 200,000,000 of a random person, one in x Adams. You look to see then which of the two is the more likely and by how much. Let me give you two examples, one way one the other to see how you do. Supposing for the sake of argument that the Bayes theorem figure gave you a result of one in 400,000,000 that it was the defendant. So you would have one in 400,000,000 that it was the defendant but one in 200,000,000 that it was a random man. That would mean that a random man was twice as likely to have committed the offence than the defendant. For the obvious mathematical reason. If taking it the other way round, if the baize calculations and use of the theorem gave you a figure of one in 200,000 and that it was the defendant as against one in 200,000,000 that it was the random man then the defendant would be ten times more likely to have committed the offence than a random man. So there it is.

It all depends does that theorem opinion your being able to give accurate answers in respect of the parts of the evidence to the two questions. What are the chances of it being given if guilty? What are the chances of it being given if not guilty? If you do not feel able to give accurate answers in terms of percentages then the theorem does not work and you could go wrong because when you start multiplying inaccurate figures as Professor Donnelly says you increase the inaccuracy thereby. It is entirely for you, as I stress once again and finally, as to whether that is the best way for you to approach the evidence that you make your judgments about."

Mr Thwaites submits that the judge's summing-up at pages 28 to 31 was, as the judge himself realised, inadequate. In particular he appeared to have forgotten the answer given by Professor Donnelly expressing the probabilities as percentages. In the light of that answer, the passage at 30G to 31D of the summing-up amounted to a misdirection because, the witness having spoken in terms of percentages, the judge was directing the jury as to how many times it was more likely that something had occurred, yet he did not remind the jury of the formula given by Professor Donnelly in relation to the percentages of 10 and 90. In returning to the subject the following morning at 45 to 48 of the summing-up the judge gave the jury no explanation as to the sort of sums which they should do. In a nutshell, Mr Thwaites submits that the judge summed up only part of Professor Donnelly's exposition of the Bayes theorem. And, in speaking of multiplying and dividing at 31C he did not identify what should be multiplied and what should be divided.

For the Crown, Miss Tapping submits that the jury were well aware of the Crown evidence and of the criticisms of it put forward by the defence and they had listened for several days to the expert evidence including that of Professor Donnelly. She accepts that the judge did not put the figures to the jury, but to have done so would necessarily have resulted in suggesting a conclusion either more favourable to the prosecution or to the defence and this would have been to usurp the jury's function. The judge's summing up expressed in words what the defence were trying to say in relation to statistics. If the judge had suggested figures other than those advanced by the defendant this would have strengthened the weight given to the DNA evidence. The judge rightly told the jury that they were not obliged to use Bayes theorem.

It seems to us that the difficulties which arise in the present case stem from the fact that, at trial, the defence were permitted to lead before the jury evidence of the Bayes theorem. No objection was taken by the prosecution. No argument on this point has been addressed to this court. It would therefore be inappropriate for us to express a concluded view on the matter. But we have very grave doubt as to whether that evidence was properly admissible, because it trespasses on an area peculiarly and exclusively within the province of the jury, namely the way in which they evaluate the relationship between one piece of evidence and another. The Bayes theorem may be an appropriate and useful tool for statisticians and other experts seeking to establish a mathematical assessment of probability. Even then, however, as the extracts from Professor Donnelly's evidence cited above demonstrate, the theorem can only operate by giving to each separate piece of evidence a numerical percentage representing the ratio between probability of circumstance A and the probability of circumstance B granted the existence of that evidence. The percentages chosen are matters of judgment: that is inevitable. But the apparently objective numerical figures used in the theorem may conceal the element of judgment on which it entirely depends. More importantly for present purposes, however, whatever the merits or demerits of the Bayes theorem in mathematical or statistical assessments of probability, it seems to us that it is not appropriate for use in jury trials, or as a means to assist the jury in their task. In the first place, the theorem's methodology requires, as we have described, that items of evidence be assessed separately according to their bearing on the accused's guilt, before being combined in the overall formula. That in our view is far too rigid an approach to evidence of the type that a jury characteristically has to assess, where the cogency of (for instance) identification evidence may have to be assessed, at least in part, in the light of the strength of the chain of evidence in which it forms part. More fundamentally, however, the attempt to determine guilt or innocence on the basis of a mathematical formula, applied to each separate piece of evidence, is simply inappropriate to the jury's task. Jurors evaluate evidence and reach a conclusion not by means of a formula, mathematical or otherwise, but by the joint application of their individual common sense and knowledge of the world to the evidence before them. It is common for them to have to evaluate scientific evidence, both as to its quality and as to its relationship with other evidence. Scientific evidence tendered as proof of a particular fact may establish that fact to an extent which, in any particular case, may vary between slight possibility and virtual certainty. For example, different blood spots on an accused's clothing may, on testing, reveal a range of conclusions from 'human blood' via 'possibly the victim's blood' to 'highly likely to be the victim's blood.' Such evidence is susceptible to challenge as to methodology and otherwise, which may weaken or even, in some cases, strengthen the impact of the evidence. But we have never heard it suggested that a jury should consider the relationship between such scientific evidence and other evidence by reference to probability formulas. That such a course would in any event be impossible of sensible achievement by a jury, at least so far as the use of the Bayes theorem is concerned, is demonstrated by the practical application of the stage of that theorem's methodology that involves numerical assessment of the various items of evidence. Individual jurors might differ greatly not only according to how cogent they found a particular piece of evidence (which would be a matter for discussion and debate between the jury as a whole), but also on the question of what percentage figure for probability should be placed on that evidence. Since, as we have pointed out, the translation of an assessment of cogency into a percentage probability of guilt is entirely a matter of judgment and the conferring of a percentage probability of guilt upon one item of evidence taken in isolation is an essentially artificial operation, different jurors might well wish to select different numerical figures even when they were broadly agreed on the weight of the evidence in question. They could, presumably, only resolve any such difference by taking an average, which would truly reflect neither party's view; and this point leaves aside the even greater difficulty of how twelve jurors, applying Bayes as a single jury, are to reconcile, under the mathematics of that formula, differing individual views about the cogency of particular pieces of evidence. Quite apart from these general objections, as the present case graphically demonstrates, to introduce Bayes theorem, or any similar method, into a criminal trial plunges the jury into inappropriate and unnecessary realms of theory and complexity deflecting them from their proper task.

It is these considerations which lead us to the provisional conclusion, uninformed, as we have indicated, by argument, that evidence about the Bayes theorem ought not to have been admitted, without objection. The judge was led into error in that, no doubt, he felt obliged to seek to sum up the evidence to the jury.

That being so, it was, as it seems to us, incumbent upon him to direct the jury both as to the substance of that evidence and as to the way it which it was open to them to use that evidence. It seems to us that, in a summing-up which was otherwise impeccable, he failed in these respects. Because of his conscientious desire to try to ensure that the jury grasped what was, it has to be remembered, the defence argument based on Bayes theorem, he concentrated his directions on that theorem, without indicating to the jury the more commonsense and basic ways in which it would have been open to them to weigh up the relative weight of the DNA evidence. The jury were not properly directed as to the meaning and implications for the prosecution case of an approach based on Bayes. If, as seems entirely possible, the jury abandoned the struggle to understand and apply Bayes, they were left by the summing-up with no other sufficient guidance as to how to evaluate the prosecution case (based as it was entirely on the DNA evidence), in the light of the other non-DNA evidence in the case. This means that their verdict cannot be regarded as safe.

**Denis John Adams, R v. [1997] EWCA
Crim 2474 (16th October, 1997)**

IN THE COURT OF APPEAL CRIMINAL DIVISION
Royal Courts of Justice
The Strand
London WC2

Thursday 16 October 1997

Before:
THE LORD CHIEF JUSTICE OF ENGLAND
(Lord Bingham of Cornhill)
MR JUSTICE POTTS
and
MR JUSTICE BUTTERFIELD

REGINA
– v –
DENIS JOHN ADAMS

Computer Aided Transcription by

Smith Bernal, 180 Fleet Street, London EC4
Telephone 0171–421 4040
(Official Shorthand Writers to the Court)

MR RONALD THWAITES QC and MR MARC BRITTAIN
appeared on behalf of
THE APPELLANT

MR ORLANDO POWNALL and MISS SUSAN TAPPING
appeared on behalf of
THE CROWN

JUDGMENT (AS APPROVED BY THE COURT)

Thursday 16 October 1997

THE LORD CHIEF JUSTICE: On 13 September 1996, in the Central Criminal Court, before His Honour Judge Pownall QC and a jury, the appellant was convicted of rape and sentenced accordingly. He appeals against conviction by leave of the single judge.

The background facts can for present purposes be very briefly summarised. The victim of the offence was a Miss M who was walking home after an evening out on 6 April 1991. Her attacker was a stranger. He approached and asked her the time. She saw his face for a matter of seconds before looking at her watch. He raped her from behind. She reported the attack to the police and a DNA profile was obtained from semen on a high vaginal swab. In October 1993 she attended an identification parade, but did not pick out the appellant or anyone else. At committal proceedings she said that the appellant did not look like the man who had attacked her. The appellant was 37 and the complainant said at one stage in her evidence that he looked at the time of her observation 40 to 42. The description she had given was of a white, clean-shaven man with a local accent, aged 20 to 25. The prosecution case rested entirely upon expert evidence in relation to the DNA sample, which was challenged by the defence. In evidence the appellant gave an alibi for the night of the attack, which he said he had spent with his girlfriend, who also gave evidence before the jury at a trial which resulted in his conviction on 24 January 1995.

The recitation of the facts which we have just given is taken from the judgment of the Court of Appeal when that conviction was the subject of challenge. On 26 April 1996 the Court of Appeal quashed that conviction and ordered a re-trial. The grounds of appeal argued at that time were two-fold: first, that the DNA evidence upon which the Crown had relied was incapable on its own of establishing guilt; and secondly, that the judge had dealt inadequately with evidence of the Bayes Theorem in its application to the facts of the case upon which the defence had relied. In its judgment reported in [1996] 2 Cr App R 467, the court rejected the argument that DNA evidence alone was incapable of establishing guilt, but the court did conclude that the trial judge's directions on the Bayes Theorem evidence had left the jury without adequate guidance as to how to evaluate the DNA evidence in the light of the non-DNA evidence. The court went on to observe per incuriam that it had grave doubt whether the Bayes Theorem evidence was admissible at all. There had been no argument to the effect that it was not, and the court accordingly felt unable to give an authoritative ruling.

The quashing of that first conviction led to a re-trial. In the course of the re-trial the defence once again invited the jury to pay attention to Bayesian evidence directed to calculation of the probabilities of the non-DNA evidence being true or false as the case might be, with a view to calculating the overall probability based on the DNA evidence, and on that occasion provided the jury with a questionnaire to enable them to make the appropriate calculations. We shall come in due course to the grounds upon which this appeal has been mounted.

At the re-trial the Crown's case rested, as it had done at the first trial, solely on DNA evidence based on the vaginal swab which had been taken from the victim. The Crown called a witness, Dr Harris, who gave evidence that the profile from the high vaginal swab matched the profile of the appellant's blood sample. Comparison was made with the database of the white European population, and the chances of another man having the same DNA profile was calculated to be 1 in 2 million. This calculation had been made on the basis of four probes, including a DNA band which had been so weak as to necessitate Dr Harris high-lighting it. If that band were excluded, the probability would be reduced to 1 in 200,000, but Dr Harris's evidence was that there was no reason why that band should be considered unreliable. Later, a fifth probe had been extracted, producing another two bands of DNA. The resulting revised probability was extended to a 1 in 200 million chance that a man unrelated to the applicant had the same profile. The defence had put in evidence the existence of the appellant's brother. The DNA experts acknowledged that he was a complicating factor whose existence reduced the probability of a match to 1 in 220. It was not, however, suggested at the trial that the brother was the assailant of the victim and he did not in any event fit her description.

The Crown called a second prosecution expert, a Mr Lambert, who gave evidence that he had checked Dr Harris's figures with a specialist software package, and it was his opinion that the figures given were reasonable and fair.

That evidence called on behalf of the Crown was the subject of criticism and close questioning on behalf of the defence. In support of those criticisms the defence called a very distinguished statistician, Professor Donnelly. He gave evidence that the figure of 1 in 200,000 was invalid and in his judgment seriously overstated the strength of the DNA evidence. He criticised the small size of the database from which the figure was calculated as too small, with the consequent danger that it would not be truly representative of the population. His opinion was that with a larger database a different figure could have been reached. He was also critical of Dr Harris for having drawn in with a pen one of the bands which had faded when re-examined. He regarded this as a serious flaw and one which would affect any later calculation. This was not a view which Dr Harris accepted, his opinion being that the practice which he had adopted was wholly professional and acceptable on the ground that he had not drawn in something which was not there, but had merely recreated something which had been there so that the computer could recognise it. Professor Donnelly also criticised the Crown's figures as failing to include a proper correction to allow for sampling errors, although he declined, understandably perhaps, to give precise estimates of what correction should have been made. He showed however that a 1% correction would reduce the probability to perhaps 1 in 100 million instead of 1 in 200 million. The view of the prosecution experts was that such a correction would actually have made the DNA evidence stronger, and their opinion was that while some of these criticisms would have had the effect of lowering the figure that they had given, this was more than outweighed by the allowances which had already been made in calculating those figures. This was, as it seems to us, exactly the sort of evidence and the sort of cross-examination which is found in these cases. The findings of the Crown were available for consideration and evaluation by the defence. They had an opportunity to make such criticisms as they thought fit. All those matters were before the jury and a proper subject for their consideration. There was, however, a further and for present purposes a very important dimension to Professor Donnelly's evidence. This relates to his explanation and application of Bayes Theorem. This is a method by which non-DNA evidence could in his opinion be expressed in terms of mathematical probability and so could more readily be applied to the DNA figures so as to reduce the probabilities if the jury judged it appropriate. The transcript shows that there was a long and detailed explanation by Professor Donnelly of how the theorem operated, and he introduced the jury in very considerable detail to the questionnaire which had been prepared. We have had the opportunity to study that document. We observe that it has on it its own instructions as to how it is to be used, instructions which Professor Donnelly explained to the jury in more detail as he gave his evidence. In presenting this appeal Mr Thwaites QC has most helpfully prepared a written summary of his argument and provided the court with the documents upon which that argument relies. We hope the court will not be thought to have neglected the details of the argument if a summary of it is attempted. In essence Mr Thwaites has argued three points. First, he submits that the prosecution in a case of this kind should not be allowed to adduce statistical evidence regarding the random occurrence ratio of a DNA match unless the defence are allowed to call appropriate Bayesian evidence to show how such figures could be reduced in giving effect to the probabilities attached to non-scientific, non-DNA evidence. Secondly, in support of that he submits that the Bayesian approach is logical, sound and approved by expert opinion. Thirdly, he submits that, evidence having been admitted of the Bayesian approach in this case, the judge should have directed the jury fully and not encouraged them to apply their common sense in contradistinction to the Bayesian approach described by Professor Donnelly. It is, we think, convenient to approach those questions in the reverse order. Having summed up the detailed evidence relating to the points concerning the Crown's DNA evidence, at page 24 of the transcript of the summing-up His Honour Judge Pownall said:

"That is an outline of the defence case and if this were any other case that would be virtually all I had to say to you before asking you to retire and consider your verdict, but of course this case is not like any other case and, as I said at the beginning, it is unique, I think; unique because it is the first case in which a jury have been asked to operate Bayes Theorem and that is because it is, I believe, the first case in which the prosecution have relied almost entirely upon the DNA evidence; evidence which they say is positively overwhelming and the Crown have, as they nearly always do in cases involving DNA, put their case into figures and indeed you have heard those figures. The chance of some other man in the white European population of this country having the same matching bands as were found in this case is one in 200 million, and the lower figures if the position is different.

Before you came into this case I was asked by the defence whether I would allow them to put their case into figures and it seemed to me, to cut a long story short, only fair that if I were to allow one side to do so, I should allow the other to also, and the particular way in which they have sought to do that is by the use of Bayes Theorem. That is a result in probability theory, a rule of logic, the point of it being that once certain numerical judgments are made, certain others must necessarily follow. So it is that you have with you a list of questions, lettered almost through, straight through the alphabet from A to X inclusive, in answer to which you may put either a number or a percentage number, as the case may be, and if you were to do it, having done it, you will have reached the last page on which is a formula and, having substituted figures for the letters and, with the help of your calculators, done the multiplying and dividing, you would come to the resulting equation and if you then follow the three steps, as instructed, you will have an end figure and that would be your view of the probability that this defendant was the rapist. In my view it would not be right for me to tell you more about the questionnaire because if I did I would simply be trespassing on what is essentially your task. You have the questions and if you think you can answer them you have the little boxes to put the answers in. I dare say none of you have ever used Bayes Theorem to decide anything in your lives before, nor have you ever given numbers or percentages to your views in logical steps. It is entirely a matter for you whether you use this method to reach your verdict or whether you use the methods which juries in this country have used for many, many years, pretty satisfactorily. It is, perhaps, the difference between what Mr Lambert called the statistical approach and the common sense approach.

I ought to remind you, though, that although Professor Donnelly is the prime advocate for the use of this theorem and, indeed, takes the view that it is the only way of doing it logically, Mr Lambert agrees that Bayes is a logical and consistent way of expressing in figures the non-DNA evidence, but he also thinks that people without statistical training and experience will find it a very difficult and complicated exercise. Although he is not against the exercise, Mr Lambert has very serious misgivings about putting it into practice in a jury trial because, for example, it does not cover all the relevant factors or all the relevant evidence, or all that you might think was relevant. It might be thought too that the questionnaire does not include a box for you to enter your figure for Mr Adams's own evidence and how he gave it, or the difficulties or otherwise that Miss [M] had in being asked to identify her rapist. Those are just examples and there may be others.

If you feel able to use the questionnaire to operate Bayes Theorem and you find it almost as easy as kiss your hand to give the answers, then there you have the opportunity to do it, having not only your own copies but you will have when you go out an extra blank one to fill in your collective view if you want to. If you do not wish to use it that is your privilege and your own private decision and no one will criticise you for not using it. There is absolutely no compulsion on you to use it at all. It is there if you want to use it and follow the instructions given. It was suggested by Mr Thwaites that you might think it only fair to this defendant for at least one of you to do it. I hope he will forgive me if I discourage that and for this reason: Your duty, when it comes first thing tomorrow morning, is to retire, consider your verdict amongst yourselves, all of you together and not with one huddled in a corner with his calculator. All of you should take part in your discussions, each listening to the arguments of the others and in the end reaching your collective verdict, the verdict of you all. If you want to use it then please use it in a collective way so that you are all having an input into it and putting the answers, if you feel able to, in the blank copy that you will have first thing tomorrow morning."

Of that direction Mr Thwaites on behalf of the appellant makes a number of criticisms. First, he criticises the failure of the judge to summarise the effect of the evidence given orally by Professor Donnelly. We have to confess that the task of attempting to summarise the 40 or so pages of evidence given by Professor Donnelly is one that would make any judge quail; the risk of error would be obvious. More fundamentally, however, it seems to us, as it plainly seemed to the judge, that the questionnaire having been explained by its author in very considerable detail, and the jury taken through it without Professor Donnelly trespassing into the jury's domain by suggesting the figures which should go in the various boxes, and the document itself bearing its own instructions, it was unnecessary for him to do more than remind the jury of the fact which must have been prominent in their minds, that the defence were urging this methodological approach to the reaching of their verdict. We cannot on the facts of this case find ground for criticism in the judge's failure to attempt a summary or precis of the evidence given by Professor Donnelly or in his failure to give any more detailed account of the proper approach to the questionnaire.

Mr Thwaites criticises the distinction drawn by the judge between the statistical approach and the commonsense approach, but that does not appear to us to be a fair criticism. It is plain that the Bayesian approach could be described as statistical; it is certainly highly mathematical. The alternative approach is one which would not be based on figures but on a more conventional application of judgment to the various points made by the defence in urging the jury to discount the prosecution evidence. Again, it does not appear to us that the judge erred in his treatment of that matter. Mr Thwaites further criticises the judge's suggestion that if the jury found it "as easy as kiss your hand" to give the answers, then they should use the questionnaire. Mr Thwaites says that that is a somewhat flippant expression to use and that on no showing could the use of the questionnaire be described as so easy an operation. We certainly share the view that the proper use of the questionnaire would be by no means straightforward, and it may be that the judge would have been better to have expressed the matter more solemnly, but the point was clear and the whole passage made it abundantly plain to the jury that if they found the Bayesian approach helpful then they were at complete liberty, having had it explained to them, to use it in their deliberations.

There are other more detailed points that have been made, but having considered the thrust of the summing-up on this aspect of the case we are not persuaded that it is open to criticism. It certainly does not persuade us that the conviction is as a result unsafe.

We turn therefore to the second of the three submissions which we recited, namely that the Bayesian approach is logically sound and approved by expert opinion. We would not for our part wish to take issue with that statement so long as it is applied to appropriate subject matter by persons competent to apply it. We have no reason to doubt, as is stated by a number of highly authoritative experts, that it is a sound and reliable methodological approach in some circumstances. We have, however, the gravest reservations about its use in jury trials in cases such as this.

This brings us to the first of the three submissions which we summarised. It appears to us that there can be no possible ground of objection in principle to the leading of DNA evidence by the Crown, based as it is or should be on empirical statistical data, the data and the deductions drawn from it being available for the defence to criticise and challenge. The more difficult question is whether the fact that the prosecution are permitted to adduce evidence of that kind should lead to the conclusion that the defence should be at liberty to deploy evidence in support of the Bayesian approach to non-scientific, non-DNA evidence, as was done in this case.

We are bound to observe that this is not the first time in which this question has come before the courts. The matter was the subject of consideration, as already mentioned, when the first appeal in this case occurred. The judgment of the court given by Rose LJ turns to this aspect of the matter at page 480G. In a long passage ending on page 482E reasons are given for the court's conclusion that such evidence is inadmissible and ought not to have been admitted. It is unnecessary to read that passage, which speaks for itself and which it would not be easy to improve on as a statement of the difficulties and problems which would arise if reliance on such evidence in cases of this kind became common form.

We have also had our attention directed to a passage in *R v Doheny and Adams* [1997] 1 Cr App R 369, 374G, where Phillips LJ giving the judgment of the court said:

"It has been suggested that it may be appropriate for the statistician to expound to the jury a statistical approach to evaluating the likelihood that the defendant left the crime stain, using a formula which gives a numerical probability weighting to other pieces of evidence which bear on that question. This approach uses what is known as the Bayes Theorem. In the case of *Adams (Denis)* [1996] 2 Cr App R 467 this Court deprecated this exercise in these terms at p482:

'To introduce Bayes Theorem, or any similar method, into a criminal trial plunges the jury into inappropriate and unnecessary realms of theory and complexity deflecting them from their proper task.'

We would strongly endorse that comment."

We note that the judgment given in this case on 26 April 1996 has been the subject of consideration in [1996] Crim LR 898, where the court's observations find favour with the commentator.

In the light of the previous rulings on this matter in this court, and having had the opportunity of considering the evidence in this case, we regard the reliance on evidence of this kind in such cases as a recipe for confusion, misunderstanding and misjudgment, possibly even among counsel, but very probably among judges and, as we conclude, almost certainly among jurors. It would seem to us that this was a case properly approached by the jury along conventional lines. That would involve them perhaps in asking themselves at the outset whether they accepted wholly or in part the DNA evidence called by the Crown. If the answer to that was "no," or uncertainty as to whether the answer was "yes" or "no," then that would be the end of the case. If, however, the jury concluded that they did accept the DNA evidence wholly or in part called by the Crown, then they would have to ask themselves whether they were satisfied that only X white European men in the United Kingdom would have a DNA profile matching that of the rapist who left the crime stain. It would be a matter for the jury, having heard the evidence, to give a value to X. They would then have to ask themselves whether they were satisfied that the defendant in question was one of those men. They would then go on to ask themselves whether they were satisfied that the defendant was the man who left the crime stain, bearing in mind on the facts of this case the obvious discrepancies between the victim's description of her assailant and the appearance of the appellant, the victim's failure to identify the appellant on the identification parade and the evidence of the appellant and the witnesses called by him. Consideration of this last question would of course involve the jury in assessing all the points made concerning the victim's opportunity to see her assailant, the likelihood of her description being accurate or inaccurate in all the circumstances, the significance of her failure to identify the appellant, the strength and weakness of the evidence given by the appellant and his witnesses, and all other matters relied on by the defence. Of course, it is a matter for the jury how they set about their task, and it is no part of this court's function to prescribe the course which their deliberations should take. But consideration of this case along the lines indicated would in our judgment reflect a normal course for a properly instructed jury to adopt. It is the sort of task which juries perform every day, carefully and conscientiously, on the evidence, as they are sworn to do. We do not consider that they will be assisted in their task by reference to a very complex approach which they are unlikely to understand fully and even more unlikely to apply accurately, which we judge to be likely to confuse them and distract them from their consideration of the real questions on which they should seek to reach a unanimous conclusion. We are very clearly of opinion that in cases such as this, lacking special features absent here, expert evidence should not be admitted to induce juries to attach mathematical values to probabilities arising from non-scientific evidence adduced at the trial.

For all these reasons we dismiss this appeal.

INDEX

447